Peter Böger · Ko Wakabayashi · Kenji Hirai (Eds.)

Herbicide Classes in Development

Springer

Berlin
Heidelberg
New York
Barcelona
Hong Kong
London
Milan
Paris
Tokyo

Peter Böger · Ko Wakabayashi · Kenji Hirai (Eds.)

Herbicide Classes in Development

Mode of Action, Targets, Genetic Engineering, Chemistry

With 96 Figures, 2 in Color, and 53 Tables

Springer

Professor Dr. Peter Böger
University of Konstanz
Department of Plant Physiology and Biochemistry
D-78457 Konstanz
Germany

Professor Dr. Ko Wakabayashi
Tamagawa University
Department of Physiology and Biochemistry
Machida-shi, Tokyo 194-8610
Japan

Dr. Kenji Hirai
Sagami Chemical Research Center
Hayakawa 2743-1, Ayase
Kanagawa 252-1193
Japan

ISBN 3-540-43147-0 Springer-Verlag Berlin Heidelberg New York

Library of Congress Cataloging-in-Publication Data

Herbicide classes in development : mode of action, targets, genetic engineering, chemistry / Peter Böger, Ko Wakabayashi, Kenji Hirai (eds.).
p. cm.
Includes bibliographical references.
ISBN 3540431470 (alk. paper)
1. Herbicides. 2. Herbicide-resistant crops. I. Böger, Peter. II. Wakabayashi, K. (Ko), 1938- III. Hirai, Kenji, 1953-

SB951.4 .H425 2002
632'.954 - dc21 2002070471

Springer-Verlag Berlin Heidelberg New York
a member of BertelsmannSpringer Science + Business Media GmbH
http://www.springer.de

Printed in Germany

Cover design: D&P, Heidelberg
Typesetting: SNP Best-set Typesetter Ltd., Hong Kong
SPIN 10774148 31/3130 - 5 4 3 2 1 0 - Printed on acid-free paper

Preface

Chemical pest control is in use in practically every country in the world since agrochemicals play a decisive role in ensuring food supply and protection against damage by pests, insects and pathogenic fungi. Particularly in the half-century since World War II, food production has risen dramatically in most parts of the world. In the last 20 years, the yield of major crops has roughly doubled in Western agriculture and there is still the potential for further achievements, particularly in the developing countries.

The world's cereal and rice production, now more than 2 billion tons/year, has to increase by 2.4% annually to cope with the rising food demand caused mainly by the growing population and improvement of living standards in most of the developing countries. Such a demand for food has to be achieved by higher yields from the restricted arable land already in use. Global farmland resources are about 1.4 billion ha, of which 1.2 billion ha is cultivated with major crops. Experts agree that a future substantial addition of new productive areas is unlikely. Those with a high yield potential are already in use; new fields with a lower output may possibly be obtained by cultivation of arid or cold areas. More recently, new areas of large-scale farmland have been developed in tropical regions of Latin America, primarily in Argentina and Brazil, at the cost of the destruction of tropical rain forest.

The 1980s were an exciting period for the development of modern herbicides, for both industry and academia. Acetolactate synthase (ALS) inhibitors, represented by the sulfonylurea (SU) and imidazolinone (IMI) classes, were introduced into chemical weed control. The start of the widespread use of new acetyl-CoA carboxylase (ACCase) inhibitors such as the phenoxypropionate and cyclohexanedione classes brought about a major turning point in the subsequent evolution of agrochemicals. The discovery of fluoro-modified tetrahydrophthalimides as PPO (= protoporphyrinogen oxidase, Protox) inhibitors, such as flumiclorac-pentyl, is another breakthrough in the explosive development of the next-generation of cyclic imide classes. These new herbicide chemistries, which combine excellent activity with unparalleled lower dosage, crop safety, specific mechanism of action and/or structural high novelty, have been rapidly adopted worldwide and have had an amazing impact on agriculture.

Today, the use rate of modern herbicides is in the range of 100–300 g a.i./ha, with a declining tendency. In particular, the very low use rates of original SU and cyclic imide herbicides have prompted agrochemical researchers to find more highly active compounds, which has led to successive discoveries of as many as 39 kinds of new ALS-inhibiting herbicides, including the triazolopyrimidines and pyrimidyloxybenzoates, and no less than 18 new cyclic imide classes of PPO inhibitors in the 1990s. The chemistry of these ALS and PPO herbicides has been the most dynamic area of research in the past 20 years. The latest phenoxypropionate ACCase inhibitors are applied in the range of 100–150 g a.i./ha and SU and cyclic imide herbicides require an even lower amount, down to 5 g a.i./ha for some commercially active ingredients. Obviously, soil overloading with chemicals or leaching problems is not an issue with such low application doses.

To date, more than 400 herbicides have been registered, or are in the registration process, and these form the active ingredients of thousands of commercial products. Among the registered herbicides whose modes of action are currently understood, 269 kinds of herbicides are used around the world and these are categorized according to their target sites, modes of action, similarity of induced symptoms or chemical classes by the Herbicide Resistance Action Committee (HRAC) in cooperation with the Weed Science Society of America (WSSA).

About ten enzymatic herbicide targets have been characterized in detail, some more may be determined by mode of action studies in the future. Accordingly, the mainstream of herbicide investigation is the search for and synthesis of new structures acting upon these known targets. Therefore, this book should update the state of target-oriented research by dealing with the following topical herbicide classes: (1) ALS inhibitors, (2) carotenogenesis inhibitors (bleaching herbicides), (3) inhibitors of aromatic amino acid biosynthesis (glyphosate), (4) inhibitors of glutamine synthetase (glufosinate), (5) ACCase inhibitors, (6) inhibitors of very long-chain fatty acid biosynthesis, (7) cellulose biosynthesis inhibitors, and (8) PPO (or Protox) inhibitors. This book presents timely physiological and biochemical information on those inhibitors and herbicide classes which are the focus of today's research and development. For example, auxin-type compounds and photosynthesis inhibitors are not dealt with.

Each of the first eight chapters covers, at least in part, the relevant aspects relating to symptoms of herbicidal activity, mode of action to provide a rational approach for weed resistance management, biochemical characteristics of the target enzyme, model assays and cell-free biochemical tests to obtain quantitative phytotoxic inhibition data for larger compound series, and molecular genetics of the herbicide target(s) with special reference to transformed inhibitor-resistant plants.

Development of transgenic herbicide-resistant crops is a strong issue today and will grow in importance. Therefore, Chapter 9, in particular, outlines the methods of how a plant is transformed and a resistant crop is developed, describing the cloning of gene(s), complementation, vector constructs, PCR-mediated gene mutation, selection, and crossings. An extended Chapter 10 provides agrochemical characteristics and major synthetic routes of the typical herbicides cited in Chapters 1–8. Structural evolutions of the inhibitor/herbicide classes belonging to these chapters are chronologically reviewed from the viewpoint of molecular design by illustrating representative compounds for the last decade. However, only some new synthetic pathways for Protox inhibitors will be documented since detailed information up to 1997 has been reviewed in *Peroxidizing Herbicides*, published by Springer in 1999.

Many herbicides and synthetic compounds interfering with plant metabolism exist as optical isomers exhibiting specific phytotoxic or regulatory activities. The *R*-form of phenoxypropionate ACCase inhibitors is far more active than the *S*-isomer. In contrast, the *S*-forms of dimethenamid or metolachlor are active, but not the *R*-form. In addition, glufosinate and bialaphos, produced by fermentation, are optically active compounds and their racemic isomers are less inhibitory. Accordingly, Chapter 11 outlines prominent examples, their enantioselective synthesis and general biological activity. Chapter 12 deals with new considerations on transcuticular penetration based on quantitative analysis of its kinetics by a logistic-kinetic model. The last chapter relates the findings presented in Chapter 6 on chloroacetamides and functionally equivalent novel structures focusing on structure-activity relationships. These are based on herbicidal greenhouse activity and quantitative inhibition of cell-free microsomal fatty-acid elongation.

There is a continuous need for new active ingredients. Changes in agricultural politics, occurrence of herbicide-resistant species and changing toxicological and environmental fate requirements demand the development of more effective, more selective and environmentally benign herbicides. It is believed that members of the modern herbicide classes covered in this volume fulfill these requests. No volume presenting a combination of synthetic chemistry with herbicide physiology, biochemistry and engineered resistance, comparable to the format outlined here, has yet been published. Treatises on the mode of action of herbicide classes included in the present chapters are almost 10 years old. A demand for this book by herbicide researchers can be safely assumed. We also believe that the contributions should be a valuable resource for established colleagues working on plant protection, and for advanced students of organic and agricultural chemistry, as well as plant biochemistry. Finally, it is hoped that readers will be stimulated by the information and messages presented. They may help to further develop integrated weed management practices that deliver a sustained food crop production.

In writing this volume, the editors thank the authors for their outstanding contributions and for making their expertise available. They are also grateful for the help and advice of many colleagues, including graduate students, technicians and many other unnamed colleagues.

P. Böger, K. Hirai, and K. Wakabayashi
Konstanz, Germany, Ayase, Kanagawa and Tokyo/Machida, Japan, April 2002

Contents

9 Genetic Engineering of Herbicide-Resistant Plants
MAMORU HORIKOSHI

10 Major Synthetic Routes for Modern Herbicide Classes and Agrochemical Characteristics
KENJI HIRAI, ATSUSHI UCHIDA, and RYUTA OHNO

Contributors

PETER BÖGER (e-mail: Peter.Boeger@uni-konstanz.de, Tel.: +49-7531-882101, Fax: +49-7531-883042)
Faculty of Biology, University of Konstanz, 78457 Konstanz, Germany

MALCOLM DEVINE (e-mail: malcolm.devine@aventis.com, Tel.: +1-306-4779400)
Biotechnology Research, Aventis CropScience Canada, 203-407 Downey Road, Saskatchewan S7N 4L8, Canada

GÜNTER DONN (e-mail: Guenter.Donn@aventis.com, Tel.: +49-69-3052856, Fax: +49-69-30517207
Aventis CropScience, Biochemistry Research, H 872/N, 65926 Frankfurt/Main, Germany

MARK A. FUCHS
Biology Department, University of Dayton, Dayton, Ohio 45469-2320, USA

DONALD GEIGER (e-mail: donald.geiger@notes.udayton.edu, Tel.: +1-937-2292509, Fax: +1-937-2292225)
Biology Department, University of Dayton, Dayton, Ohio 45469-2320, USA

KENJI HIRAI (e-mail: scrc12gp@alles.or.jp, Tel.: +81-427-424791, Fax: +81-427-427631)
Sagami Chemical Research Center, Hayakawa 2743-1, Ayase, Kanagawa 252-1123, Japan

MAMORU HORIKOSHI (e-mail: horikoshi-mamoru@nichino.co.jp, Tel.: +81-721-56 9000, Fax: +81-721-531414
Biotechnology Laboratory, Research Management Dept. Research Center, Nihon Nohyaku Co., Ltd., 345, Oyamada-cho, Kawachi-Nagano, Osaka 586-0094, Japan

HELMUT KÖCHER
Aventis CropScience, Biochemistry Research, H 872/N, 65926 Frankfurt/Main, Germany

HIROSHI MATSUMOTO (e-mail: hmatsu@biol.tsukuba.ac.jp, Tel.: +81-298-536417, Fax: +81-298-534605)
University of Tsukuba, Institute of Applied Biochemistry, Tsukuba, Ibaraki 305, Japan

BERND MATTHES
Faculty of Biology, University of Konstanz, 78457 Konstanz, Germany

TAKESHIGE MIYAZAWA
Life Science Research Institute, Kumiai Chemical Industry Co., Ltd., Ogasa-gun, Shizuoka-ken 439-0031, Japan

ISHIZU NAKAYAMA
Life Science Research Institute, Kumiai Chemical Industry Co., Ltd., Ogasa-gun, Shizuoka-ken 439-0031, Japan

KOZO NAKAYAMA
Life Science Research Institute, Kumiai Chemical Industry Co., Ltd., Ogasa-gun, Shizuoka-ken 439-0031, Japan

YUKIO NEZU
Life Science Research Institute, Kumiai Chemical Industry Co., Ltd., Ogasa-gun, Shizuoka-ken 439-0031, Japan

RYUTA OHNO
Sagami Chemical Research Center, Hayakawa 2743-1, Ayase, Kanagawa 252-1123, Japan

HIROYOSHI OMOKAWA (e-mail: omokawa@cc.utsunomiya-u.ac.jp, Tel.: +81-28-649 5151, Fax: +81-28-6495155)
Center for Research on Wild Plants, Utsunomiya University, 350 Mine, Utsunomiya 321-8505, Japan

GERHARD SANDMANN (e-mail: Sandmann@em.uni-frankfurt.de, Tel.: +49-69-79824746, Fax: +49-69-798 24822)
Botanical Institute, Johann-Wolfgang-Goethe University, P.O. Box 11932, 60054 Frankfurt/Main, Germany

TSUTOMO SHIMIZU (e-mail: t-shimizu@kumiai-chem.co.jp, Tel.: +81-537-236721, Fax: +81-537-620275)
Life Science Research Institute, Kumiai Chemical Industry Co., Ltd., Ogasa-gun, Shizuoka-ken 439-0031, Japan

ATSUSHI UCHIDA
Sagami Chemical Research Center, Hayakawa 2743-1, Ayase, Kanagawa 252-1123, Japan

Kevin C. Vaughn (e-mail: kvaughn@ars.usda.gov, Tel.: +1-601-6865211, Fax: +1-601-6865422)
USDA Southern Weed Science Laboratory, P.O. Box 225, Stoneville, Mississippi 38776, USA

Ko Wakabayashi (e-mail: kwaka@agr.tamagawa.ac.jp, Tel.: +81-427-398274, Fax: +81-427-398854)
Dept. of Agricultural Chemistry, Tamagawa University, Tamagawa-Gakuen 6-1-1, Machida-shi, Tokyo 194-8610, Japan

Tadakazu Watanabe (e-mail: t.watanabe@agrokanesho.co.jp, Tel.: +81-42-9451315, Fax: +81-42-9451865)
Agro Kanesho Co., Ltd., Research Dept., Shimo-Yasumatsu 852, Tokorozawashi 359-0024, Japan

1 Acetolactate Synthase Inhibitors

Tsutomu Shimizu, Ishizue Nakayama, Kozo Nagayama, Takeshige Miyazawa, and Yukio Nezu

1.1 Introduction

Acetolactate synthase (ALS; EC 4.6.3.8, also referred to acetohydroxy acid synthase; AHAS) is the first common enzyme in the biosynthetic pathway to the branched-chain amino acids; valine, leucine and isoleucine (Fig. 1). The pathway exists in plants and microorganisms such as bacteria, fungi and algae. ALS is the primary target site of action for at least four structurally distinct classes of herbicides including the sulfonylureas (SUs; LaRossa and Schloss 1984; Ray 1984), the imidazolinones (IMs; Shaner et al. 1984), the triazolopyrimidine sulfonamides (TPs; Subramanian and Gerwick 1989) and the pyrimidinylsalicylates (pyrimidinyl carboxy herbicides, PCs; Shimizu et al. 1994b), all of which have been successful in their development as commercial herbicides. The extremely good weed control activity achieved with these herbicides indicates that ALS is a very effective target site for herbicidal action. Deficiency of the pathway of branched-chain amino acids biosynthesis in mammals (Singh and Shaner 1995) shows us that it is the selective target between plants and mammals. ALS is therefore attractive for addressing a large number of goals of modern herbicide research.

Before ALS was elucidated to be the target site of the herbicides, microbial ALSs were actively studied regarding biochemistry and molecular genetics, but, those on plants were limited in some biochemical studies. However, after the elucidation of the target site, study of the molecular genetics of plant ALS was stimulated. Also, through extended studies of microbial ALS, there is a possibility of the above-mentioned chemical compounds being developed as anti-tuberculosis agents (Grandoni et al. 1998).

Many researchers have engaged in studies of the synthesis of chemical compounds, screening and evaluation of active compounds, formulation of compounds, modes of action and effects on mammals and the environment in this field. In addition, plant scientists have examined the application efficacy of these herbicides on the herbicide-resistant weeds from practical aspects. They have enthusiastically investigated the effect of the herbicides on the ALS activities of those weeds as well as the genetics of ALS. Among these studies, the main purpose of this chapter is to review the modes of action of the herbicides, biochemical properties and molecular genetics of the target enzyme

P. Böger, K. Wakabayashi, K. Hirai (Eds.)
Herbicide Classes in Development

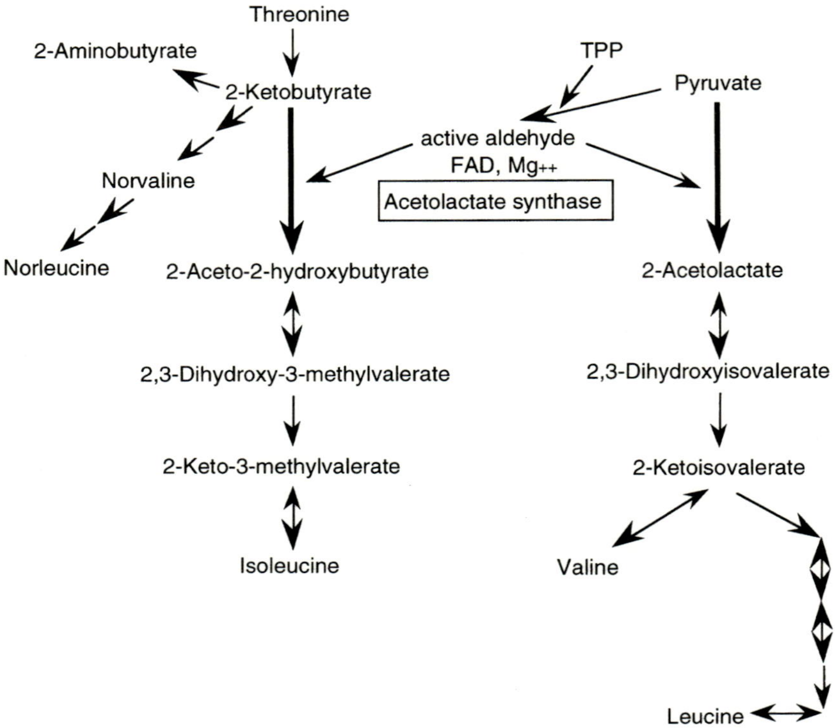

Fig. 1. Biosynthetic pathway of branched-chain amino acids

and herbicide chemistry in recent years. Accordingly, we will focus on the ALS-inhibiting herbicides developed in the late 1990s and will show especially the chemical and biological studies on our PC together with the biochemistry and molecular genetics of the plant ALS. The review of Duggleby and Pang (2000) was most useful for arranging the biochemical and molecular genetic studies of the enzyme.

1.2 ALS-Inhibiting Herbicides Actively Developed in the Late 1990s

The first introduction of an ALS-inhibiting herbicide on the market was a sulfonylurea, chlorsulfuron, in 1982 for use on cereals. From this time until the mid 1990s, 22 SUs, 5 IMs and 2 TPs were developed as commercial products for at least ten crops as well as for vegetation management (Saari et al. 1994). The biological activities of the SUs and TPs are extremely high with field application rates of approximately 10–100 g/ha, while the IMs are approximately tenfold less potent than the SUs and TPs with field application rates of approximately 100–1000 g/ha. This difference in potency is not entirely ascribed to the difference in sensitivity of ALS, because the IMs are 50- to 100-fold less potent inhibitors of ALS than the SUs and TPs (the IMs inhibit plant ALS at concentrations in the micromolar range, whereas the SUs and TPs do so in the nanomolar range; Ray 1984; Shaner et al. 1984). The effectiveness of the IMs in the field is presumed to depend on their facile uptake (Hawkes and Thomas 1990). However, in addition, the inhibition mechanism of ALS by the IM is assumed to be another factor in determining their effectiveness in the field. This will be discussed in a later section.

ALS-inhibiting herbicides commercialized or actively developed in the late 1990s are described in Table 1 [but those shown in the review of Saari et al. (1994) were omitted to avoid duplication]. The SUs have been developed at roughly the same speed compared to the last decade, though Du Pont, the inventor of SUs, reduced their development. Among these SUs, chemical compounds with a sulfonylaminocarbonyl moiety, namely, procarbazone and flucarbazone, are noteworthy based on their new chemical structures (Müller et al. 1992; Amann et al. 2000). On the other hand, development of the IMs has been reduced. This appears to be the reason for their sole development by American Cyanamid. During these periods, the herbicide-resistant weed problems do not appear to have much effect on developing both the SU and the IM. In contrast, the development of the TPs have increased. These herbicides are thought to bind onto the same site on ALS as the SU (Guangfu et al. 1999). Thus, the SU-resistant weed problems may affect TPs development. Their novel chemical structures appear to be the driving force in developing these compounds so far.

Another class of ALS-inhibiting herbicides disclosed in the late 1990s is the PCs developed by Kumiai (Kobayashi et al. 1995; Tamaru et al. 1997; Nezu et al. 1998; Ono et al. 1999). The biological activities of the PCs are as potent as those of the SUs. This high potency is reflected in their inhibitory effect on ALS that requires concentration in the nanomolar range for the inhibition (Shimizu et al. 1997). This will be discussed in later sections. By analogy regarding the chemical structures between the PC and the IM, and partial comprehension of cross-resistance of the SU-resistant weeds against the PC, the PC is categorized in the IM herbicide class. However, this is incorrect. The ALS inhibition mechanism and accurate comprehension of cross-resistance patterns suggest that the PCs are a hybrid of the SU and the IM. Accordingly, we should consider that the PCs have relatively novel properties among the ALS-inhibiting herbicides developed in recent years. In later sections, we will focus on describing our studies on the PCs. If the reader needs information concerning other ALS inhibitors, we recommend the paper by Babczinski and Zelinsli (1991) and the review of Duggleby and Pang (2000).

Table 1. ALS-inhibiting herbicides actively developed in the late 1990s

Common name	Code number	Class	Company	Crop	Entry	Product	Patent[a]	Reference[b]
Iodosulfuron-methyl	AE F115008	SU	AgrEvo	Cereals		1999	1, 2	1
Ethoxysulfuron	HOE 09540	SU	AgrEvo	Rice		1998		2, 3
Procarbazone	MKH 6561	SU	Bayer	Cereals		1999	3	4
Flucarbazone-sodium	BAY MKH 6562	SU	Bayer	Cereals	1997		3	
Flupyrsulfuron	DPX-KE459	SU	DuPont	Cereals	1994			5, 6
Azimsulfuron	DPX-A8947	SU	DuPont	Rice		1997		7
Oxasulfuron	CGA 277476	SU	Novartis	Soybean		1997	4	8, 9
Sulfosulfuron	TKM-19/MON-37500	SU	Takeda/Monsanto	Cereals		1998	5	3, 10
Imazamox	AC 299263	IM	American Cyanamid	Soybean, peanuts		1997		11, 12
Florasulam	DE-570	TP	Dow Agr.	Cereals	1998		6	13
Diclosulam	XDE-564	TP	Dow Agr.	Soybean, peanuts		1998	6	14
Cloransulam-methyl	XDE-565	TP	Dow Agr.	Soybean		1997		15
Pyriminobac-methyl	KIH-6127	PC	Kumiai	Rice		1997	7	16, 17
Bispyribac-sodium	KIH-2023	PC	Kumiai	Rice		1997	8	18, 19
Pyrithiobac-sodium	KIH-2031/DPX-PE350	PC	Kumiai/DuPont	Cotton		1996	9	20, 21
Pyribenzoxime[c]	LGC-40863	PC	LG Chem.	Rice		1998	10	22, 23

[a] 1, Hacker et al. (1996); 2, Ort et al. (1992); 3, Müller et al. (1992); 4, Meyer and Riehen (1992); 5, Ishida et al. (1992); 6, Van Heertum et al. (1992); 7, Saito et al. (1990); 8, Wada et al. (1990); 9, Tamaru et al. (1991); 10, Hur et al. (1995).

[b] 1, Trabold et al. (2000); 2, Hess and Rose (1995); 3, Loubser (1998); 4, Amann et al. (2000); 5, Teaney et al. (1995); 6, Koeppe et al. (1997); 7, Marquez et al. (1995); 8, Brooks et al. (1995); 9, Palmer et al. (1999); 10, Parrish et al. (1995); 11, Brady et al. (1998); 12, Nelson et al. (1998); 13, Lepiece et al. (1999); 14, Shaw et al. (1999); 15, Nelson and Renner (1998); 16, Hanai et al. (1993); 17, Tamaru et al. (1997); 18, Yokoyama et al. (1993); 19, Ono et al. (1999); 20, Takahashi et al. (1991); 21, Nezu et al. (1998); 22, Cho et al. (1997); 23, Koo et al. (1997).

[c] Pyribenzoxime is an oxime ester of bispyribac.

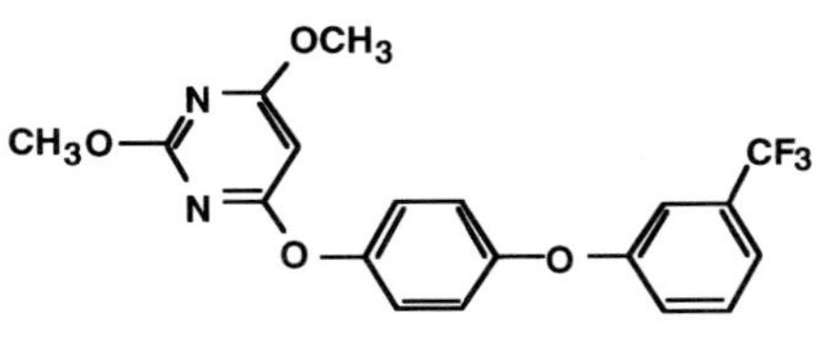

(I)

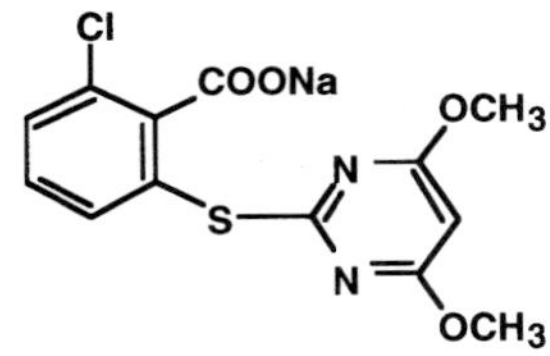

(II) Pyrithiobac-sodium

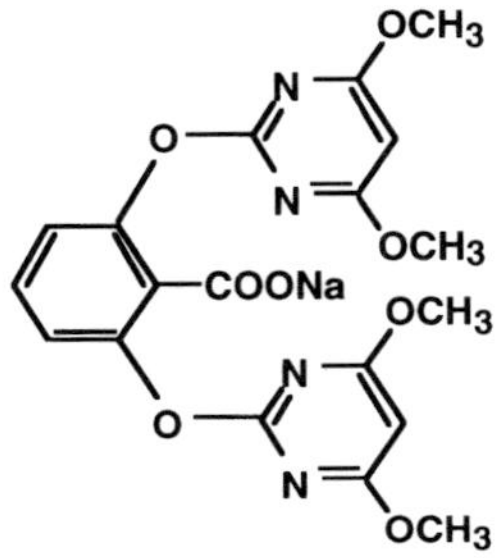

(IV) Bispyribac-sodium

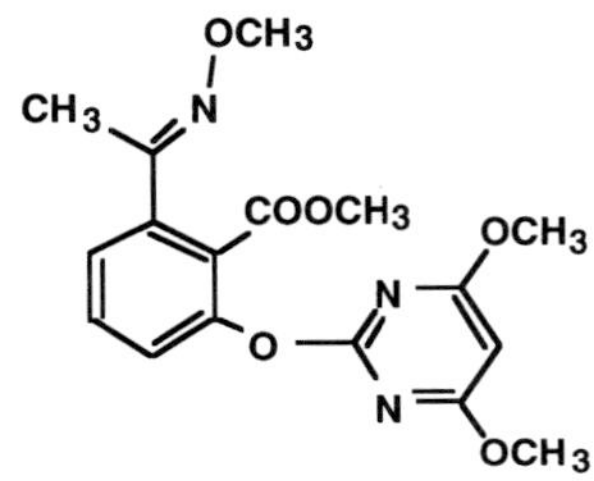

(IV) Pyriminobac-methyl

Fig. 2. Pyrimidinyl carboxy herbicides and their primary lead compound

1.3 Discovery of Pyrimidinyl Carboxy Herbicides (Pyrimidinylsalicylate Class Herbicides)

In the course of our synthetic and bioassay projects on *N*-heteroaromatic compounds, phenoxyphenoxypyrimidine (I) (Fig. 2) was found to show potent herbicidal activity with a post-emergent treatment, but it was phytotoxic against some crops. In elaborating the structure to develop a new herbicide with enhanced crop safety as well as high potency, the syntheses of analogous compounds were performed extensively to disclose the PCs, a novel class of herbicides. They exhibit pre-emergent as well as post-emergent herbicidal activity at very low rates of application. Pyrithiobac-sodium (II), bispyribac-sodium (III) and pyriminobac-methyl (IV) (Fig. 2) are the representatives of the PCs. We will show the course to pyrithiobac-sodium through some lead structures below.

1.3.1 Discovery of the Lead Structures

One of the useful approaches to a new herbicide with a "novel" structure is to make more or less "drastic" modifications in the structures of known herbicides as models. Our first steps in the syntheses started with structural

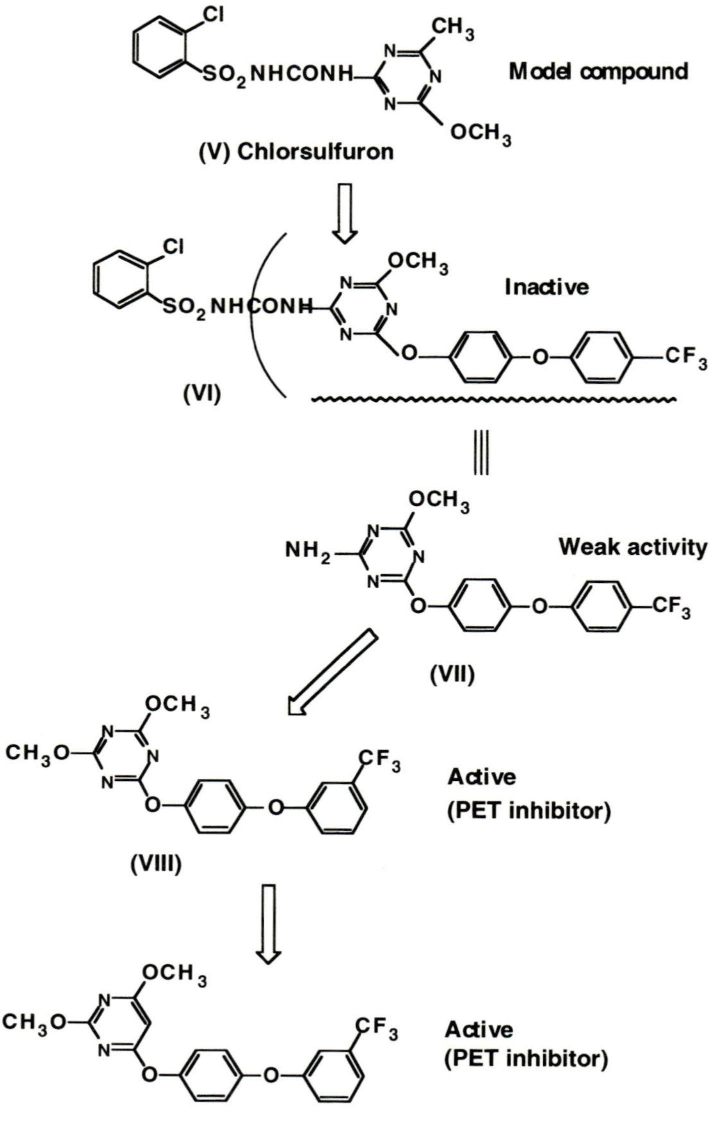

Fig. 3. The steps toward the primary lead of sulfonylureas

modification of the SU such as chlorsulfuron (**V**) (Fig. 3) at their triazine (or nitrogen-heterocyclic) sites, because they are highly active at extraordinarily low rates. We synthesized a variety of SUs possessing substituted triazines (and pyrimidines), but most of the active compounds were found to be already patented.

Among them, the SU (**VI**) (Fig. 3), substituted with a double phenoxy structure, was an important key compound. Although the SU itself (**VI**) showed no herbicidal activity, phenoxyphenoxytriazine (**VII**) (Fig. 3), one of the building blocks of SU (**VI**), showed a herbicidal activity at 4 kg a.i./ha. This was accepted to warrant the initiation of a comprehensive synthetic project around the substituted double phenoxy triazine structures. Although the structure of compound **VII** could be recognized as that belonging to diphenyl ether-type herbicides, the herbicidal symptom observed after its treatment was not related to these caused by diphenyl ethers but was similar to that induced by the SU (**V**). The effects of substituent variations in the substructural ring systems were thus examined systematically for the phenoxyphenoxytriazine (**VII**).

Early stages of the substituent optimization gave dimethoxytriazine (**VIII**) (Fig. 3) and dimethoxypyrimidine (**I**). They were inhibitors of photosynthetic electron transport (PET), but not ALS inhibitors. Dimethoxypyrimidine (**I**) was highly inhibitory, the pI50 values (M) with spinach chloroplast being as high as 7.4. Although it was neither systemic in the plant nor dispersive in the soil, it was highly potent against annual broadleaf weeds. Because of the toxicity against some crops such as corn, however, it was not developed further.

1.3.2 Discovery and Optimizations of the Secondary Lead Structure

As far as substitution (substructural) patterns at the nitrogen-heteroaromatic and the link-end benzene rings were concerned, the optimization was believed to be done in the dimethoxypyrimidine (**I**). The substituent introduction into the central benzene ring was then attempted. Following the examples of diphenyl ether-type herbicides such as acifluorfen-methyl in which the COOMe group is located at the ortho position to the electron-withdrawing NO_2 group in nitrofluorfen, a COOMe group was introduced into the position adjacent to the "electron-withdrawing" pyrimidinyloxy group to give the compound (**X**) (Fig. 4). The introduction of the COOMe group was also expected to endow the systemic property in the plant body to the non-systemic phenoxyphenoxy dimethoxypyrimidine (**I**). Unfortunately, the methyl ester (**X**) was totally inactive, but its "isomer" (**XI**) (Fig. 4) with the 2-pyrimidinyloxy instead of the 4-pyrimidinyloxy substituent in the compound (**X**) was weakly active against broadleaf weeds pre- as well as post-emergently, the activity profiles being similar to those of the SUs (**V**).

At about the same time, our group found a very efficient intermediate (**XII**) (Fig. 4) to synthesize 2-substituted pyrimidines. The methanesulfonyl group in the compound (**XII**) is easily replaced by nucleophilic reagents. One of such compound classes, the thioglycolic acid esters (**XIII**) (Fig. 4), was shown to exhibit ALS inhibitory activity, although they were not very potent (I_{50}: >100 μM). This, along with the ALS inhibitory activity of compound (**XI**),

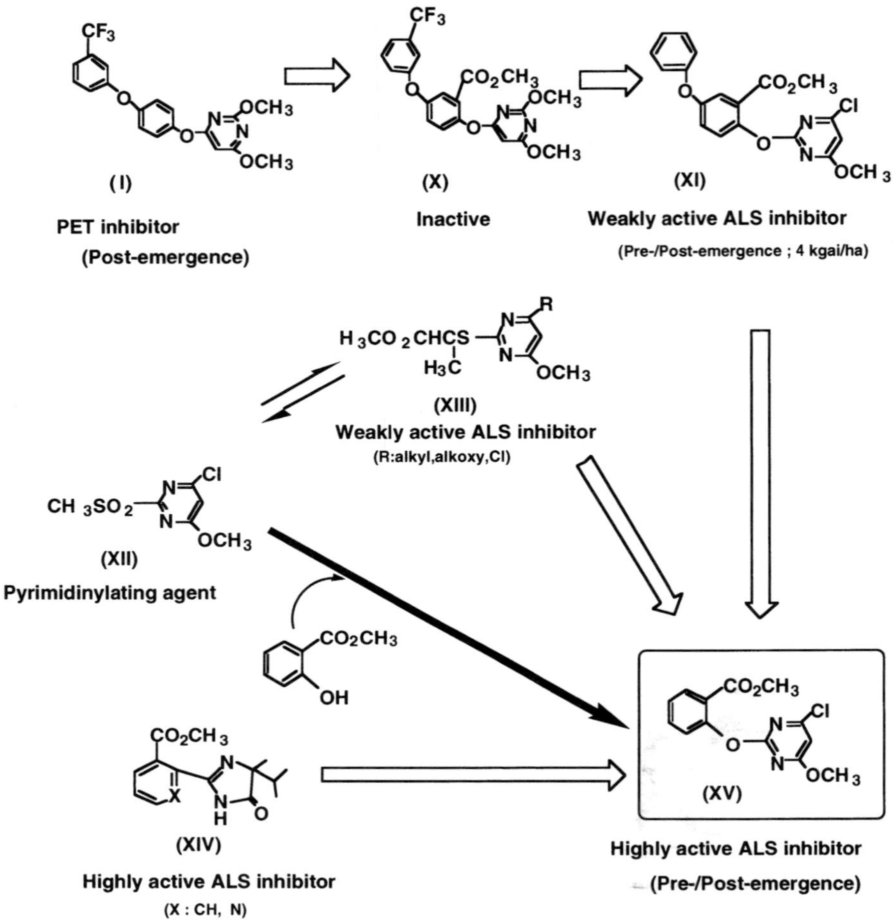

Fig. 4. Structural modification pathways towards the secondary lead compound for inhibition of acetolactate synthase

suggested that an esteric bonding and an appropriately substituted pyrimidine ring are possible requisites for the ALS-inhibiting herbicides.

There is another class of potent ALS-inhibiting herbicides covering a number of oxoimidazolinyl "benzoates" such as imazapyr (**XIV**: X = N) (Fig. 4). In this class of compounds, the electron-withdrawing oxoimidazolinyl group is located at the ortho position of COOR, similar to the pyrimidinyloxy group in compound **XI**. The weak activity of the compound **XI** was then considered to be due to its higher hydrophobicity as well as its larger dimensions than the optimum. Thus, we deleted the phenoxy group from compound **XI** to give compound **XV** (Fig. 4).

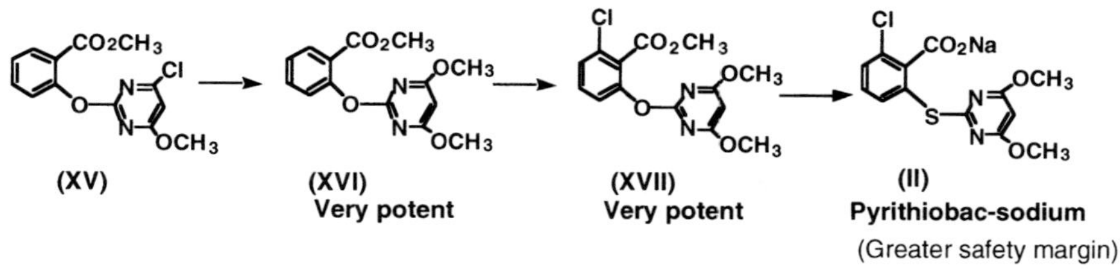

Fig. 5. Safety margin optimization of pyrithiobac-sodium

The pyrimidinyl salicylic acid ester (**XV**) thus obtained was very potent herbicidally as an ALS inhibitor (the free acid of **XV** inhibits ALS potently). At the rate of 1 kg a.i./ha, it kills a variety of grass and broadleaf weed species, pre- as well as post-emergently. We initiated an extensive analogue synthesis project and soon found compound **XVI** (Fig. 5). It was almost as potent as chlorsulfuron (**V**).

1.3.3
Further Optimizations of the PCs

Among conventional substituents, the COOMe group ortho to the pyrimidinyloxy group was best. The pyrimidine ring was better than any other nitrogen heterocycles. The most favorable substitution pattern was the 4,6-dimethoxy-2-pyrimidinyl. The structure of compound **XVI** (Fig. 5) was thought to be optimized as far as the potency was concerned. In fact, it was too potent and the safety margin for crops was not sufficient. Among various trials, the introduction of Cl at the position adjacent to COOMe (**XVII**) (Fig. 5) and the exchange of the O-bridge into the S-bridge, leading to the free acid of compound **II**, were quite successful structural modifications. The introduction of Cl at other positions reduced the potency. The effect of the S-bridge was indeed surprising. The O-bridge was better in terms of the potency against weeds; however, the safety to cotton was extremely high with the S-bridge. The killing effect against grass and broadleaf weeds was reduced by factors of 1/10~1/20 and 1/2~1/4, respectively, with the replacement of the O compound with S, but the toxicity against cotton was 1/250. The compound **II** also showed an excellent systemic property within the plant body which enabled not only the uptake from the roots but also translocation from the roots to the top of the foliar part. Because of this excellent systemic property, it is able to eradicate weeds which were difficult to manage in cotton. Compound **II** was named pyrithiobac-sodium.

As mentioned above, pyrithiobac-sodium was developed through some lead compounds. Among them, compound **I**, having inhibitory activity to PET, was most important. The inhibitory activities towards PET with analogues of this compound have been shown in our paper (Nezu et al. 1996a,b). Chipman et al.

(1998) have proposed that the ALS inhibitors share structural cognates of herbicides that act on the quinone binding site of photosystem II, with the distinguishing feature that the former compounds are generally monoanions and the latter are uncharged. It is very attractive that our course to pyrithiobac-sodium from compound **I** appears to illustrate the proposal of Chipman et al.

1.4 Herbicidal Activity of Pyrimidinyl Carboxy Herbicides

1.4.1 Pyrithiobac-Sodium for Use in Cotton

Pyrithiobac-sodium; sodium 2-chloro-6-(4,6-dimethoxypyrimidin-2-ylthio) benzoate is a herbicide for controlling a wide range of weeds in cotton (Saito et al. 1990; Takahashi et al. 1991; Nezu et al. 1998). This compound provides excellent control of troublesome weeds such as *Ipomoea* spp., *Xanthium strumarium*, *Abutilon theophrasti*, *Sida spinosa*, *Sesbania exaltata*, and *Sorghum halepense*. It can be applied pre- or post-emergence. Soil or foliar treatment with pyrithiobac-sodium at 35–105 g a.i./ha provides excellent control of weeds. Adjuvants such as nonionic surfactants or some petroleum-based adjuvant oils have been shown to play an important role in achieving consistent performance on several weed species when applied post-emergence. A good safety margin for cotton at rates that are effective on weeds has been observed with pre-emergence treatment in both the greenhouse and the field.

1.4.2 Bispyribac-Sodium for Use in Rice

Bispyribac-sodium; sodium 2,6-bis[(4,6-dimethoxypyrimidin-2-yl)oxy]benzoate, is a post-emergence herbicide for the control of a wide range of weeds with excellent selectivity on direct-seeded rice (Wada et al. 1990; Yokoyama et al. 1993; Kobayashi et al. 1995). The low rate of 15–45 g a.i./ha with surfactant has provided outstanding efficacy on *Echinochloa* spp. and can be applied from the one- to seven-leaf stage of the weed. It can also control other troublesome weeds including *Brachiaria* spp., *Cyperus* spp., *Scirpus* spp., *Polygonum* spp., *Sagittaria* spp., *Commelina* spp. and *Sesbania exaltata*. Adjuvants such as nonionic surfactants, silicon-type adjuvants or crop oil concentrate play an important role in enhancing the activity and achieving a consistent performance of this compound. Bispyribac-sodium has high selectivity between rice and *Echinochloa oryzicola* by foliar application under dry-seeded conditions, suggesting that this compound can be used against a wide range of growth stages of *Echinochloa* spp. without rice crop injury.

On the other hand, bispyribac-sodium at even 30 g a.i./ha showed low efficacy against *Leptochloa chinensis* when applied at more than the four-leaf stage, although the herbicidal efficacy of bispyribac-sodium at the two-leaf stage of *L. chinensis* was sufficient. It was considered to be difficult to control *L. chinensis* with bispyribac-sodium at its recommended commercial rate (20–40 g a.i./ha) against *Leptochloa* spp. (Ono et al. 1999). To improve the efficacy of bispyribac-sodium against *Leptochloa* spp., mixture treatments with other post-emergence rice herbicides were evaluated in a greenhouse. Tank-mixture treatment of bispyribac-sodium and fenoxaprop-ethyl showed good efficacy, and the efficacy and rice injury of fenoxaprop-ethyl were not affected significantly by the addition of bispyribac-sodium. However, in the case of tank-mixture treatment of bispyribac-sodium and cyhalofop-butyl, strong antagonism was observed. The efficacy of cyhalofop-butyl decreased markedly by mixing with bispyribac-sodium. These results showed that the tank-mixture treatment of bispyribac-sodium and fenoxaprop-ethyl is considered recommendable, in case infestation of *Leptochloa* spp. is observed in the rice field.

1.4.3 Bispyribac-Sodium for Vegetation Management

Bispyribac-sodium at the rate of 150 g a.i./ha pre-mixed with a nonionic surfactant reduced the vegetative growth of weeds such as *Imperata cylindrica, Digitaria adscendens, Miscanthus sinensis* and *Artemisia princeps* (Tachikawa et al. 1997). The growth reduction persisted for 50 days after the application of this compound when applied 5–10 days after mowing (at 10–20 cm of plant height). Also, bispyribac-sodium controlled a wide range of weed species such as *Solidago altissima, Polygonum lapathifolium, Aeschynomene indica, Paspalum distichum* and *Echinochloa crus-galli* that grew in rice fields or on highway and railroad right-of-ways. The results indicated that bispyribac-sodium can reduce the frequency of mowing in paddy rice levees, and on highway and railroad right-of-ways.

1.4.4 Pyriminobac-Methyl for Use in Rice

Pyriminobac-methyl; methyl 2-[4,6-dimethoxypyrimidin-2-yl]oxy-6-[1-(methoxyimino)ethyl]benzoate, is a selective herbicide with outstanding efficacy on *Echinochloa* spp. in paddy rice (Tamaru et al. 1991,1997; Hanai et al. 1993). This compound has a specific effectiveness against *Echinochloa* spp. during a wide range of growth stages from pre- to late post-emergence with an excellent crop safety in rice. The use rate of pyriminobac-methyl is extremely low in comparison with the recommended rate of molinate and thiobencarb. Pyriminobac-methyl has shown excellent safety on all 11 varieties tested of water-seeded rice and can be applied at any growth stage of rice. There was no observed significant difference in susceptibility to pyriminobac-methyl among rice varieties tested. Pyriminobac-methyl can be used alone or mixed with other rice herbicides such as bensulfuron-methyl. The residual activity of pyriminobac-methyl at 30 g a.i./ha was superior to thiobencarb at 3000 g a.i./ha under flooded conditions in the greenhouse.

1.5 Physiological Plant Response to Pyrimidinyl Carboxy Herbicides

The PCs kill weeds at relatively low application rates as described above. It takes several weeks for complete death of weeds by those herbicides. The meristematic tissues die first causing cessation of growth in sensitive plants, followed by slow necrosis of the mature tissues accompanied by slight chlorosis. These responses of the plants were observed not only in the shoots but also in the roots. The progress of symptoms on *Echinochloa crus-galli* treated by bispyribac-sodium is shown in Fig. 6 (Sadohara 1997). These physiological actions of the PCs are very similar to those of other ALS-inhibiting herbicides.

The growth inhibition of rice seedlings and *Chlorella* by the PCs were alleviated almost completely by simultaneous application of three branched-chain amino acids (Shimizu et al. 1994b). This alleviation was also found in the growth of *Enterobacter agglomerans* isolated from soil as a PC-sensitive bacterium (Yamashita et al. 1994a). These results indicated that the plant was induced to die from starvation of branched-chain amino acids. Hence, we analyzed the amino acid contents in plant cells treated with the PC by high pressure liquid chromatography (HPLC). In the treated tobacco cells, we found changes in free amino acids such as valine, isoleucine, threonine, alanine, and 2-aminobutyrate. Two branched-chain amino acids decreased, whereas threonine, alanine and 2-aminobutyrate notably increased. In addition, an unknown amino acid eluted at the same retention time as the standard norvaline increased in response to these amino acids (Nakayama and Shimizu 1993). 2-Aminobutyrate and norvaline are metabolized through 2-ketobutyrate that accumulates from the biosynthetic pathway of branched-chain amino acids (Fig. 1). Because 2-ketobutyrate accumulates in the SU-treated *Salmonella typhimurium* and is toxic to *S. typhimurium* growth (LaRossa et al. 1987), the accumulation of 2-ketobutyrate in the SU-treated cells has been proposed to be another factor in determining the cytotoxic effect of the SU. However, 2-ketobutyrate has not been shown to accumulate (Shaner and Singh 1993) in the IM-treated corn. There is still no clear explanation of the effect of ALS-inhibiting herbicides on mitosis and photosynthate transport (Singh and Shaner 1995). It might be necessary to examine the effect not only of norvaline but also abnormal amino acids metabolized from norvaline on these physiological processes.

Fig. 6. Progress of symptoms of barnyard grass treated with bispyribac-sodium (40 g a.i./ha). Four days after herbicide application, necrosis is not observed, but strong chlorosis occurs in the plants

1.6 Mode of Action and Selectivity of Pyrimidinyl Carboxy Herbicides

1.6.1 Primary Target

It has been shown that the growth inhibition of pea seedlings and cell suspension cultures of carrot by SUs is alleviated by simultaneous addition of three branched-chain amino acids (Ray 1984; Usui et al. 1991). The addition of three branched-chain amino acids has been shown to reverse the inhibition

Table 2. Inhibition of plant ALS by PC herbicides

	Concentration required for 50% inhibition (nM)			
Enzyme Source	Pyrithiobac	Bispyribac	Pyriminobac	Pyriminobac-Me
Cotton	20	9.1	37	–[a]
Soybean	42	13	38	–
Pea	20	8.3	66	–
Rice (cv. Kinmaze)	15	12	19	59,000
Rice (cv. La-Belle)	17	12	20	–
Wheat	21	12	22	–
Corn	20	12	42	–
Sorghum	15	10	33	–
Morning glory	19	7.1	27	–
Velvetleaf	70	16	41	–
Barnyard grass	–	11	16	169,000

[a] Not tested.

of DNA synthesis of maize cell suspension cultures by the IMs (Shaner and Reider 1986). The growth inhibition of rice seedlings, an algae and a bacterium by the PCs was alleviated by simultaneous application of three branched-chain amino acids as mentioned above (Shimizu et al. 1994b; Yamashita et al. 1994a). These results suggested that the PCs inhibit some steps in the biosynthesis of branched-chain amino acids, especially ALS. Indeed, the PCs including pyrithiobac, bispyribac and pyriminobac strongly inhibited ALS in various plant species at concentrations in the nanomolar range (Table 2; Shimizu 1997). The potency of ALS inhibitions by these PCs is roughly identical to those of the SUs, but not to the IMs.

However, the PCs affected neither ketol-acid reductoisomerase which catalyzes the next reaction step from ALS in the pathway nor the direct acetoinforming enzyme deduced as pyruvate decarboxylase, which has been considered to have the same origin as ALS from pyruvate oxidase. The PCs showed no inhibitory effect on the photosynthetic electron transport system, whereas they inhibited chlorophyll biosynthesis of cotton cotyledons slightly. It might be presumed that PC herbicides did not inhibit chlorophyll biosynthesis directly, but, indirectly through starvation of branched-chain amino acids, which are precursors of some enzymes responsible for chlorophyll biosynthesis (Shimizu et al. 1994b).

1.6.2 Inhibition of Bacterial ALS

The PCs also inhibited the ALS activity of *Pseudomonas aeruginosa* with nearly the same potency as the SUs (Shimizu 1997). This inhibition was approximately tenfold less potent than the plant ALS inhibition. Isozyme II of *Salmonella typhimurium* was also inhibited by the PC at concentrations of 100 μM. At this concentration, the IM had no inhibitory effect on the ALS. These results indicated that the PCs are categorized similar to the SUs with respect to potency of ALS inhibition. It has been reported that the ALS of *Serratia marcescens* is inhibited strongly by the PC but not by the SU and the IM (Yang and Kim 1997); therefore, the sensitivities of bacterial ALS to the ALS-inhibiting herbicides are considered to be different among bacterial sources.

1.6.3 Selectivity

Despite the high selectivity of pyrithiobac for cotton and bispyribac for rice, there were no differences in the sensitivities of ALSs to pyrithiobac between cotton and other plants, and to bispyribac between rice and other plants. The selectivities of pyrithiobac and bispyribac must be determined by other factors. As for pyrithiobac, there is no published paper on its selectivity for cotton. However, oxidative demethylation of the 3,5-dimethoxy moiety has been shown to account for the tolerance of tall morning glory to pyrithiobac (Sunderland et al. 1995). Thus, the same mechanism is assumed to be involved in its selectivity between cotton and other sensitive plants. Regarding bispyribac, translocation of the compound mainly accounts for its selectivity between rice and barnyard grass (unpubl. data). The oxidative detoxification metabolism, similar to that of pyrithiobac (Matsusita et al. 1994), which might be catalyzed by a mixed function oxidase, is presumed to be another factor in the selectivity for rice, because application of P-450 inhibitors such as 1-aminobenzotriazol and piperonylbutoxide reduced its selectivity for Indica-type rice (unpubl. data).

One of the methyl ester compounds of the PC (compound **XVI** in Fig. 5), which has the same herbicidal potency as its free acid, hardly inhibited the activity of ALS separated from esterase. However, this compound inhibited the ALS activity as potently as its free acid, when the esterase was added in the reaction mixture (Nakayama et al. 1993). Thus, the active forms of ester compounds are their free acids. However, pyriminobac-methyl inhibited ALS less potently than its free acid even in the presence of esterase. Pyriminobac-methyl was hardly hydrolyzed by the esterase existing in the soluble fractions of both rice and barnyard grass, whereas it was hydrolyzed by the microsomal fraction of barnyard grass (unpubl. data). Also, the free acid of pyriminobac-methyl was detected in barnyard grass treated with this compound, but not in rice (Mizutani et al. 1998). These results indicate that the selectivity of pyriminobac-methyl between rice and barnyard grass depends on the difference in substrate specificity of the enzyme having esterase activity in the membrane fraction of plants. In all events, further studies are needed for complete confirmation of their selectivity for target plants.

1.7 Biological Characteristics of the Target Enzyme

The target site of action of the SU was first shown to utilize a bacterium (LaRossa and Schloss 1984). This first identification of ALS as the site of action of the SUs prompted studies in plants on the mode of action of the IMs (Shaner et al. 1984), the TPs (Subramanian et al. 1991) and the PCs (Shimizu et al. 1994b) as well as the SUs (Ray 1984). These studies advanced our understanding of the enzyme and the biosynthesis of branched-chain amino acids in plants. ALS has been found not only in bacteria and plants but also in fungi and algae. In this section, we will describe the biochemical properties of plant ALS taking our study as the lead.

1.7.1 Kinetic Studies of Plant ALS

Plant ALS has been shown to exist in the chloroplast of matured leaves (Miflin 1974; Schulze-Siebert et al. 1984; Schulze-Siebert and Schultz 1989; Southan and Copeland 1996). The catalytic properties of plant ALS have been characterized using the ALS of ripening pea seeds (Davies 1964), etiolated pea seedlings (Lee et al. 1991; Shimizu et al. 1994a; Shin et al. 1999), etiolated barley seedlings (Miflin 1971; Durner and Böger 1988), and corn cells (Singh et al. 1988a,b). ALS from etiolated pea seedlings expressed the following enzymological properties (Shimizu et al 1994a). (1) The Km values for pyruvate and thiamine pyrophosphate (TPP) were 1.5 mM and 9.6 μM, respectively, at pH 7.5. These values are lower than those reported by other authors (Duggleby and Pang 2000). However, it is important to consider the Km to change in the assay pH. (2) The rate of acetolactate production did not show a sigmoidal relationship with pyruvate concentration. The Hill coefficient of pyruvate was approximately 1.0 independently of pyruvate concentration and pH value. This result is in contrast to that reported on the ALS of etiolated pea seedlings (Lee et al. 1991) and etiolated barley seedlings (Miflin 1971), where positive cooperativities have been shown. (3) The inhibitions by Leu, Val and isoleucine (Ile) showed negative cooperativities, indicating homotropic allosteric effects in the feedback inhibitions by branched-chain amino acids. (4) In contrast to the Leu/Val combination and the Leu/Ile combination, an antagonistic effect was found in the inhibition by the Val/Ile combination (Fig. 7). (5) The inhibition type of the Leu/Val combination changed depending on the pyruvate concentration. A non-competitive pattern was obtained at low pyruvate concentration. (6) SH inhibitors did not desensitize the inhibitions of ALS by feedback inhibitors. The negative cooperativities of the feedback inhibitions and the sensitivities of the inhibition by leucine to SH reagents are also in contrast to that reported on barley seedlings (Miflin 1971). The results in (3) and (4) indicate that the regulatory centers have two kinds of binding sites for branched-chain amino acids. One of those is presumed to be for Leu and the other either

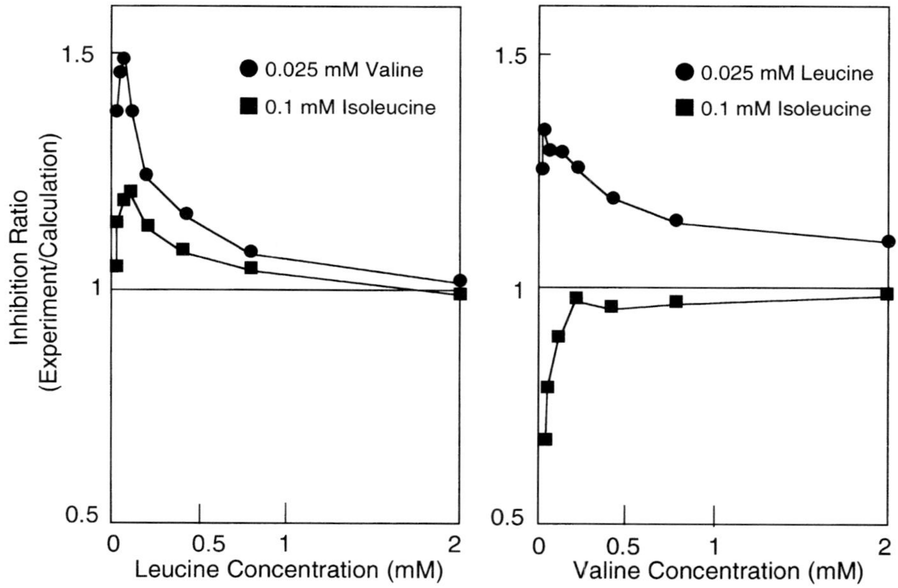

Fig. 7. Variation of synergistic and antagonistic inhibitions of acetolactate synthase from etiolated pea seedlings by branched-chain amino acids. Inhibition ratio was expressed as the ratio of the inhibition percentage which was obtained experimentally by paired amino acids (Experiment) vs. that calculated from Gowing's equation below. Calculation = $IA + IB(1-IA/100)$ where IA is the inhibition percentage given by one of the paired amino acids, and IB is inhibition by another of the paired amino acids. In the case of synergistic inhibition, the inhibition ratios in the figures are over 1, while in the case of antagonistic inhibition they are below 1

Val or Ile. This presumption is reasonable, because the regulatory protomer of plants is over twice as large as that of bacteria and has two domains for acceptance for two feedback inhibitors (Hershey et al. 1999).

1.7.2 Subunit Compositions of Plant ALS

Multiple molecular species of ALS from eukaryotes, which were different forms of a single ALS, were first demonstrated in the ALS of *Neurospora crassa* on gel filtration chromatography (Glatzer et al. 1972). Different forms of ALS have been shown in monocotyledonous plants such as developing maize kernels (Muhitch 1988), sweet corn cultured cells (Singh et al. 1988b), etiolated barley seedlings (Durner and Böger 1988), canola cotyledons (Bekkaoui et al. 1993), etiolated pea seedlings (Shimizu et al. 1986, 1994a; Shin et al. 1999), and wheat shoots (Southan and Copeland 1996). The ALS species of those plants are considered to dissociate to lower molecular weight species from the native enzymes in the absence of flavine adenine dinucleotide (FAD; Singh and

Schmitt 1989; Durner and Böger 1990). This role of FAD has clearly been verified with the ALS isozyme III of *Escherichia coli* (Vyazmensky et al. 1996). We found two molecular species of ALS in etiolated pea seedlings on gel filtration column chromatography. Molecular weights of these two ALS species were approximately 320,000 and 120,000. The higher molecular weight species, designated large ALS, was sensitive to the feedback inhibition by the Leu/Val combination, while the lower molecular species, designated small ALS, was insensitive. The large ALS produced the small ALS during further chromatography on the same column. The Km value for pyruvate and the sensitivity to the feedback inhibition of the large ALS were similar to that of the crude enzyme preparation. Based on these results, the large ALS is considered to be the native enzyme in etiolated pea seedlings and to change to the small one by loss of the regulatory center(s) during purification. It has been reported that the lower molecular species of ALS from developing maize kernels (Muhitch 1988) and etiolated barley seedlings (Durner and Böger 1988) are sensitive to the feedback inhibition, while that of cultured corn cells are insensitive (Singh et al. 1988b). The properties of small ALS in etiolated pea seedlings were similar to that of cultured corn cells.

As described above, various molecular species of plant ALSs have been reported until now, but the molecular weight of the catalytic protomer has been shown to be similar to those of the large protomers of enterobacteria (Singh et al. 1991; Bekkaoui et al. 1993). There is also definite evidence that the regulatory protomer of ALS exists in plants (Hershey et al. 1999) as those of bacteria and microbial eukaryotes (Cullin et al. 1996; Duggleby 1997; Hill et al. 1997). Despite increased knowledge of each protomer, the subunit composition of native ALS of plants is still unclear. Before elucidation of the regulatory protomer, some researchers suggested that the native enzyme was a dimer of the catalytic protomer, from its analogy to those of enterobacteria. However, this cannot explain feedback inhibition by branched-chain amino acids. The native enzyme of plants is considered to be a rather large oligomer with regulatory protomer(s). Regarding the ALS of etiolated pea seedlings that we obtained, it appears that four catalytic protomers with two regulatory ones can explain its native subunit composition. The subunit composition of plant ALS might vary among plant species. At least, ALSs of pea, corn and barley, whose molecular masses of the native enzyme are over 300,000, might resemble that of *Pseudomonas aeruginosa*, because the *Pseudomonas aeruginosa* ALS is composed of larger subunits than the enterobacteria ALS (Arfin and Koziel 1973). This ALS expresses roughly the same sensitivity as those of plant ALSs to inhibitors as described in the preceding section.

1.7.3 Recombinant Systems

There are some advanced studies on biological properties of plant ALS prepared by recombinant systems. Several plant ALS genes encoding for the catalytic protomer have been overexpressed in *Escherichia coli* and their enzymological properties was examined. Findings obtained from these recombinant systems are: (1) the recombinant ALSs of *Arabidopsis* and tobacco are dimeric and insensitive to the feedback inhibitor (Singh et al. 1992; Chang and Duggleby 1997; Chang et al. 1997). (2) The serine to the leucine change at position 214 in the tobacco ALS is responsible for valine resistance (Hervieu and Vaucheret 1996). (3) The *Arabidopsis* recombinant ALS have a low specific activity (Chang and Duggleby 1997). (4) The *Arabidopsis* ALS fused with ketol acid reductoisomerase (KARI) exhibits high activity (Dumas et al. 1997). (5) The *Arabidopsis* recombinant ALS displays negatively co-operative kinetics with respect to pyruvate concentration (the Km value of pyruvate for the first active site is 8 mM and that for the second active site is approximately 100 mM (Chang and Duggleby 1997). (6) The cysteinyl and tryptophanyl residues in the tobacco ALS play key roles in catalytic function of the enzyme (Chong et al. 1998). (7) The tryptophan residue at position 490 in the tobacco ALS is essential for binding of FAD to the enzyme (Chong et al.1999). (8) The disulfide bond is located between two cysteines at position 163 and 309 in tobacco ALS (Chong et al. 2000).

In addition to these studies, the ALS genes responsible for resistance to ALS-inhibiting herbicides have been intensively investigated utilizing the recombinant systems. These will be discussed in a later section. We recommend referring to the reviews of Chipman et al. (1997) and Duggleby and Pang (2000) for detailed information on biochemical properties and molecular genetics of microbial ALS and further important residues in ALS for the catalytic activity of the enzyme.

1.8 Inhibition Mechanism of the Target Enzyme by Pyrimidinyl Carboxy Herbicides

It has been demonstrated that the SU (Durner et al. 1991) and the TP (Subramanian and Gerwick 1989) inhibit plant ALSs activity in the mixed-type with respect to pyruvate in the steady state analysis, while the IM inhibits uncompetitively (Shaner et al. 1984). In extended time-course experiments, these herbicides have been shown to exhibit slow-binding properties to both plant ALS (Muhitch et al. 1987; Hawkes 1989) and bacterial ALS (LaRossa and Schloss 1984). In this section, we will describe our kinetic studies on the inhibition of ALS by the PC together with studies by others.

1.8.1 Inhibition Kinetics with Plant ALS

We have shown the following kinetic results in our studies (Shimizu et al. 1994c, 1995). Pyrithiobac and bispyribac inhibited the ALS of etiolated pea seedlings in the mixed-type (in which both competitive and noncompetitive components are present) with respect to pyruvate by means of a 40-min steady-state analysis. This inhibition pattern was the same as that of a sulfonylurea, chlorsulfuron, but different from that of an imidazolinone, imazapyr. Imazapyr inhibited this enzyme in a uncompetitive manner. The inhibition patterns of these inhibitors are different from those by feedback inhibitors, whose inhibition patterns are partially competitive. The sensitivity of the inhibition by those structurally diverse inhibitors to assay pH differed from that of L-leucine. The small ALS losing its sensitivity to the feedback inhibition was potently inhibited by these inhibitors. These results indicated that the binding sites of these inhibitors on the enzyme are different from those of feedback inhibitors. It has been reported that the inhibition patterns of the SU, the TP and the IM are not competitive with respect to TPP (Nakata 1991; Subramanian et al. 1989; Singh et al. 1989). Pyrithiobac inhibited the ALS of etiolated pea seedlings in the noncompetitive type with respect to TPP (Shimizu et al. 1994c). Imazapyr has been demonstrated to compete with an SU, sulfometuron-methyl for the binding to ALS of *Salmonella typhimurium* (Schloss et al. 1988). In our study, chlorsulfuron competed with bispyribac for the binding to ALS of etiolated pea seedlings. This competition was more potent than that of pyrithiobac (Shimizu et al. 1995). These results suggest two viewpoints. One is that the binding site of the PCs to ALS is located at the allosteric site near the catalytic center. Both the SUs and the TPs might share the binding site with the PCs. Whereas the IMs bind to the site that is somewhat distinct from, but overlaps that of the SUs, the TPs and the PCs. These sites are not at the regulatory subunit, but are considered to be in the vestige of the ubiquinone binding site on the catalytic subunit (Schloss et al. 1988) that lost its role in the enzymatic reaction during the evolutionary process. The other is that the uncompetitive inhibition of the IMs is another factor that might determine their rather potent herbicidal effect on weeds in contrast to that surmised from their inhibitory potency for ALS. The uncompetitive inhibition of an enzyme of a metabolic pathway can have enormously large effects on the concentrations of metabolic intermediates compared to competitive inhibition, under circumstances where their effects on the kinetics of the enzyme are similar (Cornish-Bowden 1986).

On the other hand, it has been shown that despite their reversible nature to the inhibition of ALS (LaRossa and Schloss 1984; Muhitch et al. 1987; Durner et al. 1991), the SU and the IM are slow-binding inhibitors of plant ALS (Muhitch et al. 1987; Hawkes 1989), which inactivate ALS irreversibly after reaching the final steady inhibitions (Durner et al. 1991). The irreversible inactivation of the enzyme has been found either in the presence of pyruvate (Hawkes and Thomas 1990) or in the absence of pyruvate (Ortega et al. 1996). Pyrithiobac and bispyribac inhibited the ALS of etiolated pea seedlings with slow-binding properties (Fig. 8). Pyrithiobac showed the mixed-type pattern with respect to pyruvate in the initial inhibition. The inhibition constants in the initial inhibition by pyrithiobac and bispyribac were 13- to 26-fold larger

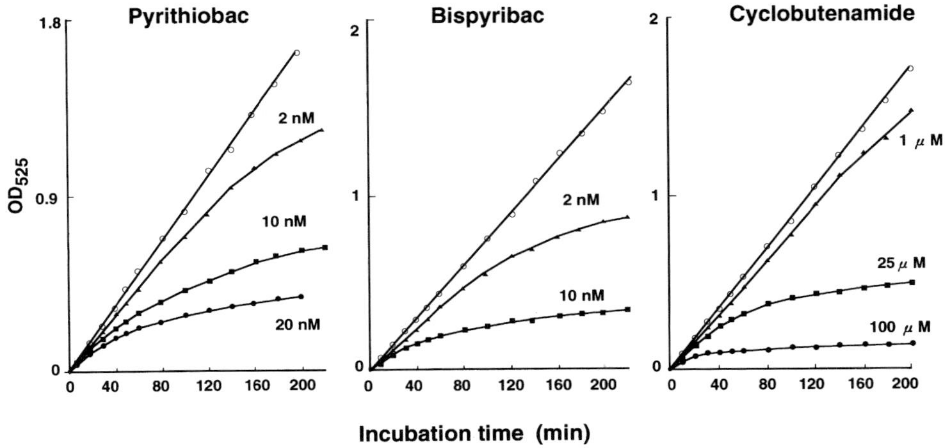

Fig. 8. Assay-time course of ALS of etiolated pea seedlings in the presence of ALS inhibitors

than those in the final steady state. The maximal first-order rate constant (k_1, 0.069 min^{-1}) for transition from the initial to the final steady-state inhibition of pyrithiobac (Shimizu et al. 1994b) was nearly identical to those of the SU and IM (Hawkes 1989). However, the dissociation constant of bispyribac to the ALS of etiolated pea seedlings after reaching the final steady inhibition was nearly identical to the inhibition constant in the initial inhibition (Shimizu et al. 1995). These results might support the hypothesis of Hawkes (1989) that the apparent slow phase of ALS inhibition by ALS-inhibiting herbicides is due to a slow irreversible inactivation of the enzyme rather than isomerization of the enzyme-inhibitor complex to a more tightly bound form as proposed originally (LaRossa and Schloss 1984).

We isolated an ALS inhibitor from *Streptomyces hygroscopicus*, whose chemical structure was deduced to be 2-[*N*-{1-(4-hydroxy-2,3-dioxy-4-cyclobutenyl)ethyl}glycyl]amino-4-ureidobutylamide (we called this compound cyclobutenamide; Yamashita et al. 1994b). This compound inhibited the ALS of etiolated pea seedlings in a competitive manner with respect to pyruvate and competed with bispyribac for the binding to ALS (Shimizu et al. 1995). Notably, this compound exhibited a slow-binding property with nearly the same k_1 value (0.065 min^{-1}) as pyrithiobac (Fig. 8). However, the inhibition constant in the initial inhibition by this compound was 130-fold larger than that in the final steady state. These results advocated the hypothesis of Hawkes, because it seems more reasonable to consider that ALS is inactivated by different kinds of ALS inhibitors with the same k_1 value due to its labile nature, rather than to consider that different kinds of ALS inhibitors have the same k_1 value.

One way to rationalize the slow-binding inhibition followed by irreversible inactivation is to address the dissociation of the regulatory subunit(s) and the catalytic subunits of ALS from its holoenzyme. It is known that ALS requires the regulatory subunit(s) to exhibit high activity (Hershey et al. 1999). The catalytic center is presumed to be formed near the interface between catalytic subunits (Duggleby and Pang 2000), and the herbicidal inhibitors bind close to the catalytic center described above. Consequently, the binding of inhibitors to the enzyme might destabilize the subunit interactions. Taking into consideration that the recombinant plant ALS losing the regulatory subunit(s) was inhibited with nearly the same potency as the native enzyme in the 40-min assay (Kaku et al. 2001) and that it is inhibited by the slow-binding (Chang and Duggleby 1997), the dissociation of catalytic subunits is assumed to be a more important factor for determining the slow-binding property.

1.8.2 Inhibition Kinetics with Bacterial ALS

The PCs also strongly inhibited the ALS activities of *Pseudomonas aeruginosa* and isozyme II of *Salmonella typhimurium* as explained in the preceding section. The inhibition pattern of pyrithiobac for ALS of *Pseudomonas aeruginosa* was noncompetitive with respect to pyruvate. This pattern was the same as chlorsulfuron, but different from that of imazapyr, whose inhibition pattern was uncompetitive (Shimizu et al. 1993). These results were very similar to those found with plant ALS described above, suggesting that both ALS species have a similar sensitivity to ALS-inhibiting herbicides. These results also indicate that the binding site of the PC on the enzyme is located on a similar site of the SU.

1.9 Molecular Genetics of Target Enzyme

ALS is found in bacteria, yeast, fungi, algae and plants. Among these organisms, the enzymes from bacteria and yeast have been extensively studied genetically. The discovery of ALS as a herbicidal target activated molecular genetic studies on the enzyme. ALS genes have been isolated from all of those organisms (Chipman et al. 1997; Duggleby and Pang 2000). In this section, we focus on plant ALS genes including those isolated from rice in our study.

1.9.1 ALS Genes of Plants

The plant ALS genes coding for their catalytic protomers were isolated first from *Arabidopsis thaliana* and tobacco utilizing the yeast ALS gene as a heterologous hybridization probe (Mazur et al. 1987). Since then, a number of plant ALS genes have been cloned and characterized. These plant ALS genes are shown in Table 3. *A. thaliana* has been shown to possess a single copy of

Table 3. Plant species whose sequences of ALS catalytic protomers are elucidated

Plant species	Accession	Herbicide sensitivity	Reference[a]
Nicotiana tabacum	X07644		1
Nicotiana tabacum	X07645		1
Beta vulgaris			2
Arabidopsis thaliana			3
Arabidopsis thaliana	X51514	resistant	4
Brassica napus	X16708		5
Brassica napus	M60068		6
Brassica napus	Z11524		7
Brassica napus	Z11525		7
Brassica napus	Z11526		7
Zea mays	X63553		8
Zea mays	X63554		8
Gossypium hirsutum	Z46959		9
Gossypium hirsutum	Z46960		9
Xanthium sp.	U16280		10
Xanthium sp.	U16279	resistant	10
Amaranthus sp.			11
Amaranthus sp.	U55852	resistant	11
Hordeum vulgare	AF059600 partial sequence		12
Kochia scoparia			13
Kochia scoparia	AF094326	resistant	14
Oryza sativa	AB049822		15
Oryza sativa	AB049823	resistant	15

[a] 1, Lee et al. (1988); 2, Bedbrook et al. (1991); 3, Mazur et al. (1987); 4, Sathasivan et al. (1990); 5, Wiersma et al. (1989); 6, Bekkaoui et al. (1991); 7, Rutledge et al. (1991); 8, Fang et al. (1992); 9, Grula et al. (1995); 10, Bernasconi et al. (1995); 11, Woodworth et al. (1996b); 12, Duggleby et al. (1998, unpubl.,); 13, Fushimi et al. (1997); 14, Foes et al. (1998); 15, Kaku et al. (2001).

ALS, while tobacco and corn have two, canola and cotton have many more. Because tobacco, canola and cotton are allotetraploid species, the presence of multiple ALS genes is partly the result of a combination of genomes derived from their diploid parents. The deduced amino acid sequences are well conserved among plants. They are similar to the catalytic protomer of bacterial and yeast ALS, except for the N-terminal signal peptide sequences, which are required for translocation of the protein to the chloroplast (Duggleby and Pang 2000). On the other hand, the gene coding for the regulatory protomer of ALS has also been cloned and characterized (Hershey et al. 1999). The deduced amino acid of the regulatory protomer is more than twice as large as that of bacteria and has two domains assumed to accept feedback inhibitors.

We have cloned the ALS gene from cultured rice cells by utilizing a partial cDNA (expressed sequence tag; accession number, C72411) obtained from the Ministry of Agriculture, Forestry and Fishery (MAFF) DNA bank of Japan as a homologous hybridization probe (Table 3). This is the second full-length ALS gene isolated from monocotyledonous plants. The deduced amino acid sequence of the rice ALS gene is highly conserved compared with that of corn (Fang et al. 1992) and barley (accession number, AF059600 partial cDNA; Duggleby, unpubl.; Fig. 9). The ALS expressed in *Eschericia coli* from this gene showed similar sensitivity to ALS-inhibiting herbicides compared with that prepared from the natural source (Kaku et al. 2001).

1.9.2
ALS-Inhibiting Herbicide-Resistant Crops (Including *Arabidopsis thaliana*) and Their ALS Genes

A number of plants and cultured plant cells resistant to ALS-inhibiting herbicides have been generated utilizing both conventional mutation breeding method and in vitro cell selection (Newhouse et al. 1991, 1992; Sengnil et al. 1992; Terakawa and Wakasa 1992; Hart et al. 1994; Lavigne et al. 1994; Lange et al 1995; Rajasekaran et al. 1996a; Wright and Penner 1998b). The sulfonylurea-resistant soybeans (Sebastian et al. 1989; Simpson and Stoller 1995), the inbred lines of the imidazolinone-resistant corn (Anderson and Georgeson 1989; Harms et al. 1990; Bright et al. 1992; Currie and Regehr 1995; Currie et al. 1995; Siehl et al. 1996; Krausz et al. 1997; Wright and Penner 1998a) and the imidazolinone-resistant canolas have been developed as commercial products ("STS" soybean, "IMI" corn, Pursuit smart canola) by these methods. Imidazolinone-resistant rice (Croughan 1999; trade name "Clearfield" rice) is being actively developed now. The ALSs from these herbicide-resistant plants have been shown to be insensitive to ALS-inhibiting herbicides. ALS genes encoding for the catalytic protomers have been cloned from some of those plants, and their sequences are shown to differ from their wild types. In most cases, the molecular basis for the herbicide resistance is due to a single amino acid change from the wild-type enzyme. However, in some cases, two changes are identified in two conserved regions. These mutated amino acids deduced from their gene sequences are shown in Table 4. The mutated positions are rearranged into the amino acid sequence of the rice ALS and are shown in Fig. 10. The most commonly encountered mutations involve the residues of alanine at position 96 (A96), proline at position 171 (P171), tryptophan at position 548 (W548) and serine at position 627 (S627). This is the rice ALS numbering; we use this numbering system throughout the discussion. The A96 and S627 mutations were raised by selection with the IM, and the latter was also with the PC, while the P171 and W548 were raised by selection with the SU, and the latter was also with both the IM and the PC.

We obtained PC-resistant cultured cells of rice through bispyribac-sodium selection and cloned a double-mutated ALS gene from the cells (Kaku et al. 2001; Shimizu et al. 2001a). The mutations involve the residues of W548 to leucine (W548L) and S627 to isoleucine (S627I; see Table 4). The mutation of W548L was first isolated together with the mutation of P171A in tobacco (Lee et al. 1988) through the selection by the SU. Since then, this mutation has been found in corn (Bernasconi et al. 1995) and canola (Hattori et al. 1995) and

```
Rice ALS      1 MATTAAAAAAALSAAATAKTGRKNHQRHHVLPARGR-VGAAAVRCSAVSPVTPPSPAPPA
Corn ALS1     1 MATAATAAAA-LTGATTATP--KSRRRAHHLATR-RALAA-PIRCSALSRATPTAP--PA
Corn ALS2     1 MATAAAASTA-LTGATTAAP--KARRRAHLLATR-RALAA-PIRCSAASPAMPMAP--PA
Barley ALS    1 ------------------------------------------------------------
                ... .  .  . .  . ..    .   .  .  .  . .   .    .... .   .  .  ..

Rice ALS     60 TPLRPWGPAEPRKGADILVEALERCGVSDVFAYPGGASMEIHQALTRSPVITNHLFRHEQ
Corn ALS1    54 TPLRPWGPNEPRKGSDILVEALERCGVRDVFAYPGGASMEIHQALTRSPVIANHLFRHEQ
Corn ALS2    54 TPLRPWGPTDPRKGADILVESLERCGVRDVFAYPGGASMEIHQALTRSPVIANHLFRHEQ
Barley ALS    2 ----------------------------------------------TRSPVITNHLFRHEQ
                ........ .... ..... ...... ................****** ********

Rice ALS    120 GEAFAASGYARASGRVGVCVATSGPGATNLVSALADALLDSVPMVAITGQVPRRMIGTDA
Corn ALS1   114 GEAFAASAYARSSGRVGVCIATSGPGATNLVSALADALLDSVPMVAITGQVPRRMIGTDA
Corn ALS2   114 GEAFAASGYARSSGRVGVCIATSGPGATNLVSALADALLDSVPMVAITGQVPRRMIGTDA
Barley ALS   17 GEAFAASGYARASGRVGVCVATSGPGATNLVSALADALLDSIPMVAITGQVPRRMIGTDA
                *******.*** ******* ********************.******************

Rice ALS    180 FQETPIVEVTRSITKHNYLVLDVEDIPRVIQEAFFLASSGRPGPVLVDIPKDIQQQMAVP
Corn ALS1   174 FQETPIVEVTRSITKHNYLVLDVDDIPRVVQEAFFLASSGRPGPVLVDIPKDIQQQMAVP
Corn ALS2   174 FQETPIVEVTRSITKHNYLVLDVDDIPRVVQEAFFLASSGRPGPVLVDIPKDIQQQMAVP
Barley ALS   77 FQETPIVEVTRSITKHNYLVLDVEDIPRVIQEAFFLASSGRPGPVLVDIPKDIQQQMAVP
                *********************** ***** ******************************

Rice ALS    240 VWDTSMNLPGYIARLPKPPATELLEQVLRLVGESRRPILYVGGGCSASGDEL-RWFVELT
Corn ALS1   234 AWDTPMSLPGYIARLPKPPATEFLEQVLRLVGESRRPVLYVGGGCAASGEELC-RFVELT
Corn ALS2   234 VWDKPMSLPGYIARLPKPPATELLEQVLRLVGESRRPVLYVGGGCAASGEE-LRRFVELT
Barley ALS  137 VWDTPMSLPGYIARLPKPPSTESLEQVLRLVGEARRPILYVGGGCAASGEELRR-FVELT
                .**..*.************.** **********.*** *******.***.*.  . *****

Rice ALS    299 GIPVTTTLMGLGNFPSDDPLSLRMLGMHGTVYANYAVDKADLLLAFGVRFDDRVTGKIEA
Corn ALS1   293 GIPVTTTLMGLGNFPSDDPLSLRMLGMHGTVYANYAVDKADLLLAFGVRFDDRVTGKIEA
Corn ALS2   293 GIPVTTTLMGLGNFPSDDPLSLRMLGMHGTVYANYAVDKADLLLALGVRFDDRVTGKIEA
Barley ALS  196 GIPVTTTLMGLGNFPSDDPLSLRMLGMHGTVYANYAVDKADLLLAFGVRFDDRVTGKIEA
                *********************************************.**************

Rice ALS    359 FASRAKIVHIDIDPAEIGKNKQPHVSICADVKLALQGLNALLQQSTTKTSSDFSAWHNEL
Corn ALS1   353 FAGRAKIVHIDIDPAEIGKNKQPHVSICADVKLALQGMNTLLEGSTSKKSFDFGSWHDEL
Corn ALS2   353 FASRAKIVHVDIDPAEIGKNKQPHVSICADVKLALQGMNALLEGSTSKKSFDFGSWNDEL
Barley ALS  256 FASRSKIVHIDIDPAEIGKNKQPHVSICADVKLALQGLNGLLSGSKAQQGLDFGPWHKEL
                **.*.****.*************************** * **  .*.  .  **. *. **

Rice ALS    419 DQQKREFPLGYKTFGEEIPPQYAIQVLDELTKGEAIIATGVGQHQMWAAQYYTYKRPRQW
Corn ALS1   413 DQQKREFPLGYKIFNEEIQPQYAIQVLDELTKGEAIIATGVGQHQMWAAQYYTYKRPRQW
Corn ALS2   413 DQQKREFPLGYKTSNEEIQPQYAIQVLDELTKGEAIIGTGVGQHQMWAAQYYTYKRPRQW
Barley ALS  316 DQQKREFPLGYKTFGEAIPPQYAIQVLDELTKGEAIIATGVGQHQMWAAQYYTYKRPRQW
                ************.. *.* *****************.**********************

Rice ALS    479 LSSAGLGAMGFGLPAAAGASVANPGVTVVDIDGDGSFLMNIQELALIRIENLPVKVMVLN
Corn ALS1   473 LSSAGLGAMGFGLPAAAGAAVANPGVTVVDIDGDGSFLMNIQELAMIRIENLPVKVFVLN
Corn ALS2   473 LSSAGLGAMGFGLPAAAGASVANPGVTVVDIDGDGSFLMNVQELAMIRIENLPVKVFVLN
Barley ALS  376 LSSSGLGAMGFGLPAAAGASVANPGVTVVDIDGDGSFLMNIQELALIRIENLPVKVMILN
                ***.***************.*******************.****.**** ********** .**

Rice ALS    539 NQHLGMVVQWEDRFYKANRAHTYLGNPEC-ESEIYPDFVTIAKGFNIPAVRVTKKSEVRA
Corn ALS1   533 NQHLGMVVQWEDRFYKANRAHTFLGNP-ENESEIYPDFVAIAKGFNIPAVRVTKKSEVHA
Corn ALS2   533 NQHLGMVVQWEDRFYKANRAHTYLGNP-ENESEIYPDFVTIAKGFNIPAVRVTKKNEVRA
Barley ALS  436 NQHLGMVVQWEDRFYKANRAHTYLGNPEN-ESEIYPDFVTIAKGFNVPAVRVTKKSEVSA
                **********************.****    *********.******.********.** *

Rice ALS    598 AIKKMLETPGPYLLDIIVPHQEHVLPMIPSGGAFKDMILDGDGRTVY
Corn ALS1   592 AIKKMLEAPGPYLLDIIVPHQEHVLPMIPSGGAFKDMILDGDGRTVY
Corn ALS2   592 AIKKMLETPGPYLLDIIVPHQEHVLPMIPSGGAFKDMILDGDGRTVY
Barley ALS  495 AIKKMLETPGPYLLDIIVPHQEHVLPMIPSGGAFKDMIMEGDGRTSY
                *******.****************************..*****.*
```

Fig. 9. Alignment and consensus for sequences of rice, corn and barley ALS

Table 4. Mutations in ALS conferring ALS-inhibiting herbicide resistance (1)

Plant species	Mutation[a]	Method[b]	Selection	Rice ALS[c]	Reference[d]
Zea mays	Ala90Thr	CMB	IM	A96T	1, 2
Beta vulgaris	Ala113Thr	SCM	IM	A96T	3
Arabidopsis thaliana	Ala122Val	SDM		A96V	4
Arabidopsis thaliana	Met124Glu	SDM		M98E	5, 6
Arabidopsis thaliana	Met124Ile	SDM		M98I	5, 6
Arabidopsis thaliana	Met124His	SDM		M98H	5
Arabidopsis thaliana	Pro197Ser	CMB	SU	P171S	7
Nicotiana tabacum	Pro196Gln	SCM	SU	P171Q	8
Nicotiana tabacum	Pro196Ala	SCM	SU	P171A	8
Nicotiana tabacum	Pro196Ser	SCM	SU	P171S	9
Beta vulgaris	Pro188Ser	SCM	SU	P171S	3
Brassica napus	Pro173Ser	SDM		P171S	10
Arabidopsis thaliana	Pro197deletion	SDM		P171 deletion	11
Arabidopsis thaliana	Arg199Ala	SDM		R173A	6
Arabidopsis thaliana	Arg199Glu	SDM		R173E	5, 6
Arabidopsis thaliana	Phe206Arg	SDM		F180R	5
Zea mays	Trp552Leu	CMB	IM	W548L	2
Nicotiana tabacum	Trp537Leu	SCM	SU	W548L	8
Brassica napus	Trp557Leu	SCM	SU	W548L	12
Oryza sativa	Trp548Leu	SCM	PC	W548L	13
Gossypium hirsutum	Trp563Ser	SCM	SU	W548S	14
Gossypium hirsutum	Trp563Cys	SCM	SU	W548C	14
Arabidopsis thaliana	Trp574Leu	SDM		W548L	11
Nicotiana tabacum	Trp537Phe	SDM		W548F	15
Arabidopsis thaliana	Trp574Ser	SDM		W548S	4
Arabidopsis thaliana	Trp574deletion	SDM		W548 deletion	11
Zea mays	Ser621Asp	CMB	IM	S627D	1
Arabidopsis thaliana	Ser653Asn	SCM/SDM	IM	S627N	4, 11, 16
Zea mays	Ser621Asn	SCM	IM	S627N	17
Oryza sativa	Ser627Ile	SCM	PC	S627I	13
Arabidopsis thaliana	Ser653Thr	SDM		S627T	18
Arabidopsis thaliana	Ser653Phe	SDM		S627F	18
Arabidopsis thaliana	Ser653deletion	SDM		S627 deletion	11

[a] Amino acids are described by three letters.
[b] The mutated ALS were obtained through conventional mutation breeding (CMB), somatic cell mutation (SCM) or site-directed mutagenesis (SDM).
[c] Amino acids are described by one letter.
[d] 1, Bright et al. (1992); 2, Bernasconi et al. (1995); 3, Wright et al. (1998); 4, Chang and Duggleby (1998); 5, Kakefuda et al. (1996); 6, Ott et al. (1996); 7, Haughn et al. (1988); 8, Lee et al. (1988); 9, Harms et al. (1992); 10, Wiersma et al. (1989); 11, Hand et al. (1992); 12, Hattori et al. (1995); 13, Kaku et al. (2001); 14, Rajasekaran et al. (1996a); 15, Chong et al. (1999); 16, Sathasivan et al. (1990); 17, Dietrich (1998); 18, Y:T. Lee et al. (1999).

```
1       10        20        30        40        50        60        70        80
MATTAAAAAALSAAATAKTGRKNHQRHHVLPARGRVGAAAVRCSAVSPVTPPSPAPPATPLRPWGPAEPRKGADILVEA
**                                                                   ** * *  ***
        90       100       110       120       130       140       150       160
LERCGVSDVFAYPGGASMEIHQALTRSPVITNHLFRHEQGEAFAASGYARASGRVGVCVATSGPGATNLVSALADALLDS
* * **  ******************   * * * *****  ***  ***  *  * * ******* **** ****  **
       170       180       190       200       210       220       230       240
VPMVAITGQVPRRMIGTDAFQETPIVEVTRSITKHNYLVLDVEDIPRVIQEAFFLASSGRPGPVLVDIPKDIQQQMAVPV
 *  ************* ****** ***** ********  * ****   ****** * **** * * *** ***   *
       250       260       270       280       290       300       310       320
WDTSMNLPGYIARLPKPPATELLEQVLRLVGESRRPILYVGGGCSASGDELRWFVELTGIPVTTTLMGLGNFPSDDPLSL
*     ** *    *  *    * *  **  *   * ******   *   *  *********  * ****  *  *   *
       330       340       350       360       370       380       390       400
RMLGMHGTVYANYAVDKADLLLAFGVRFDDRVTGKIEAFASRAKIVHIDIDPAEIGKNKQPHVSICADVKLALQGLNALL
 **************   *****  *********** **** * **** ***  ****** ** * * * *****  * *
       410       420       430       440       450       460       470       480
QQSTTKTSSDFSAWHNELDQQKREFPLGYKTFGEEIPPQYAIQVLDELTKGEAIIATGVGQHQMWAAQYYTYKRPRQWLS
              *  **  *    **  *     * *****  ***** * *** *********** *    *****
       490       500       510       520       530       540       550       560
SAGLGAMGFGLPAAAGASVANPGVTVVDIDGDGSFLMNIQELALIRIENLPVKVMVLNNQHLGMVVQWEDRFYKANRAHT
* ************ **    *   ********** ** ****    ******    ******** **** ** ****
       570       580       590       600       610       620       630       640
YLGNPECESEIYPDFVTIAKGFNIPAVRVTKKSEVRAAIKKMLETPGPYLLDIIVPHQEHVLPMIPSGGAFKDMILDGDG
 ** *       *     *     ** ***       *   **  *** ***   *** **** ***** * * *  ***
 644
RTVY
*
```

Fig. 10. Mutation sites in the ALS gene for resistance to ALS-inhibiting herbicides. *Boldface letters* show the mutation sites. Numbering system is for rice ALS. The *asterisk* indicates a residue conserved among *Arabidopsis*, tobacco, canola, cotton, corn and rice

herbicide-resistant weeds. The latter will be mentioned later. It was first shown that this mutation confers resistance to both the SUs and the IMs using the mutated gene generated by site-directed mutagenesis (Hand et al. 1992). This mutation confers resistance to multiple herbicides (Hattori et al. 1995). At this position, other amino acid substitutions, W548C and W548S, have been found in cotton (Rajasekaran et al. 1996a). On the other hand, the mutation of the S627 position (Sathasivan et al. 1990, 1991) was first found in the IMI-resistant *A. thaliana* (Haughn and Somerville 1990). In contrast to W584L, the mutation of S627N confers resistance to the IM and the PC, but not to the SU and the TP (Mourad and King 1992). This mutation as well as a different amino acid change, S627D, has been reported in IM-resistant corn (Bright et al. 1992; Dietrich 1998). The S627D mutation also confers resistance to the PC, but the resistance level is lower than that of the S627N mutation. The mutations at this position leading to the S627A, S627N, S627T and S627F have been studied in *A. thaliana* by site-directed mutagenesis (Y.T. Lee et al. 1999). Based on the sensitivities of mutated enzymes expressed in *E. coli*, it has been suggested that the size of the amino acid chain at this position determines the resistance. In addition to these substitutions, the deletion of S627 by site-directed mutagenesis has been shown to confer resistance to both the SU and the IM (Hand et al. 1992). There was, however, no report concerning the mutation of S627I found in the PC-resistant rice cells. Regarding the double mutations, two other combinations have been reported in addition to the double mutation of the P171A/W548L pair found in tobacco as described above. Those are the P171S/S627N in *A. thaliana* (Hattori et al. 1992; Mourad et al. 1994) and the A96T/P171S in sugar beet (Wright et al. 1998b). Accordingly, the double mutation (W548L/S627I) found in our study on rice is a new combination of the spontaneous mutations with a novel substitution at the S627 position. The ALS expressed in *E. coli* from this mutated gene showed resistance to multiple herbicides including the PC, the SU and the IM, but it showed stronger resistance to the PC than to the SU and the IM. Bispyribac-sodium had no effect on the enzyme even at 100 μM, which is an approximately 10,000-fold higher concentration than the I_{50} value for the wild-type enzyme (Kaku et al. 2001). The mutation at the S627 position is raised by the IM and the PC, but not by the SU; therefore, it is considered that the PCs share the binding site on ALS with the IMs.

1.9.3 ALS-Inhibiting Herbicide-Resistant Weeds and Their ALS Genes

The SU-resistant weeds were first found in *Kochia scoparia* and *Lactuca serriola* in the field with the repeated use of an SU, chlorsulfuron. Since then, many weed species have developed resistance to the SUs and the IMs (Saari et al. 1994). Several resistant weeds that have been reported in the USA, Australia and Japan during recent years are shown in Table 5. In some cases, the in vivo ALS assay (Gerwick et al. 1993; Simpson et al. 1995) was used to identify

Table 5. Several weeds resistant to ALS-inhibiting herbicides reported in USA, in Australia and in Japan in the late 1990s

Plant species	Common name	Resistance	Reference[a]
Solanum ptycanthum	Nightshade sp.	IM	1
Helianthus annuus	Common sunflower	IM	2
Sorghum bicolor	Shattercane	SU	3, 4
Galium spurium	False cleavers	SU, IM	5
Brassica tournefortii	Mustard sp.	SU	6
Kochia scoparia	Kochia	IM, SU	7
Amaranthus rudis	Common waterhemp	IM, SU, TP	8, 9, 10, 11
Amaranthus palmeri	Palmer amaranth	IM, SU	9
Amaranthus hybridus	Smooth pigweed	IM	12
		IM, SU	13
Amaranthus lividus	Livid amaranth	IM, SU	13
Xanthium strumarium	Common cocklebur	IM, SU, TP	14
Sisymbrium orientale	Indian hedge mustard	SU	6, 15
Limnophila sessiliflora	Kikumo	SU	16
Scirpus juncoides	Inuhotarui	SU	17
Lindenia micrantha	Azetogarashi	SU	18
Lindenia dubia subs. major	America-azena	SU	16
Lindenia dubia	Taketo-azena	SU	16
Lindenia procumbens	Azena	SU	16
Cyperus flaccidus	Hinagayaturi	SU	16
Monochoria korsakowii	Mizuaoi	SU	19

[a] 1, Volenberg et al. (2000); 2, Al-Khatib et al. (1998); 3, Lee CD et al. (1999); 4, Anderson et al. (1998); 5, Hall et al. (1998); 6, Boutsalis et al. (1999); 7, Foes et al. (1999); 8, Foes et al. (1998); 9, Sprague et al. (1997b); 10, Hinz and Owen (1997); 11, Lovell et al. (1996); 12, Manley et al. (1999); 13, Manley et al. (1996); 14, Sprague et al. (1997a); 15, Adkins et al. (1997); 16, Itoh and Wang (1997); 17, Kohara et al. (1999); 18, Itoh et al. (1999); 19, Wang et al. (1997).

herbicide-resistant weeds (Lovell et al. 1996; Uchino et al. 1999). The mutated ALS gene conferring resistance to the SU was first shown in *K. scoparia* (Guttieri et al. 1992). Herbicide-resistant mutations in ALS have now been confirmed in some other herbicide-resistant weeds (Table 6). The weeds possessing the mutated ALS at the P171 position (rice ALS numbering system) were found in the field with the repeated use of the SU, whereas that of the A96 was found using IM. In addition, the mutated ALS at the W548 position was found in the weeds through selection by both the IMs and the SUs. These mutation patterns produced by herbicide applications are very similar to those of herbicide-resistant crops described above. The ALS of *Xanthium strumarium* possessing the A96T mutation has been shown to be resistant to the IM, but generally not to the SU (Bernasconi et al. 1995), as in the cases of the ALS of corn (Bright et al. 1992) and sugar beet (Wright and Penner 1998b) possessing the same mutation. Pyrithiobac and bispyribac inhibited the enzyme of the *X. strumarium* assumed to possess the A96T mutation as potently as the wild-type enzyme (Shimizu et al. 2001b). Pyrithiobac has been shown to inhibit the enzyme of the IM-resistant *Amaranthus hybridus* which is assumed to possess

Table 6. Mutations in ALS conferring ALS-inhibiting herbicide resistance (2)

Plant species	Mutation	Herbicide use	Rice ALS	Reference[e]
Xanthium strumarium	Ala100Thr	IM	A96T	1
Lactuca serriola	Pro197His	SU	P171H	2, 3
Kochia scoparia[a]	Pro189Thr	SU	P171T	4
Kochia scoparia[a]	Pro189Ser	SU	P171S	4
Kochia scoparia[a]	Pro189Arg	SU	P171R	4
Kochia scoparia[a]	Pro189Leu	SU	P171L	4
Kochia scoparia[a]	Pro189Gln	SU	P171Q	4
Kochia scoparia[a]	Pro189Ala	SU	P171A	4
Sisymbrium orientale	Pro/Ile	SU	P171I	5
Brassica tournefortii	Pro/Ala	SU	P171A	5
Scirpus juncoides	Pro/Leu	SU	P171L	6
Lindernia micrantha	Pro179Ala	SU	P171A	7
Lindernia micrantha	Pro179Gln	SU	P171Q	7
Lindernia micrantha	Pro179Ser	SU	P171S	7
Lindernia micrantha	Pro179Lys	SU	P171K	7
Lindernia micrantha[b]	Pro179Gln	SU	P171Q	8
Lindernia procumbens	Pro/Ser	SU	P171S	8
Lindernia dubia subsp. major	Pro/Ser	SU	P171S	8
Lindernia dubia	Pro/Ala	SU	P171A	8
Xanthium strumarium	Ala183Val	(IM)	A179V	9
Kochia scoparia[c]	Asp260Gly		D242G	10
Kochia scoparia[c]	Val276Glu		–[d]	10
Kochia scoparia[c]	Trp487Arg		W465R	10
Kochia scoparia[c]	Asn561Ser		N539S	10
Kochia scoparia	Trp570Leu		W548L	11
Xanthium strumarium	Trp552Leu	IM	W548L	1
Amaranthus sp.	Trp569Leu		W548L	12
Amaranthus rudis	Trp569Leu	(IM)	W548L	13
Sisymbrium orientale	Trp/Leu	(SU)	W548L	5

[a] The ALS catalytic protomer sequence of *Kochia scoparia* has been elucidated by Fushimi et al. (1997) and Foes et al. (1998).
[b] The ALS catalytic protomer sequence of *Lindernia micrantha* has been elucidated by Shibaike (2001).
[c] The ALS possessing all of these four mutations expresses resistance to the SU. It is unknown which mutation is most important for resistance.
[d] Rice ALS does not possess the corresponding amino acid residue.
[e] 1, Bernasconi et al. (1995); 2, Eberlein et al. (1997); 3, Eberlein et al. (1999); 4, Guttieri et al. (1995); 5, Boutsalis et al. (1999); 6, Shibuya et al. (1999); 7, Shibaike (2000); 8, Uchino and Watanabe (1999); 9, Woodworth et al. (1996a); 10, Fushimi et al. (1997); 11, Foes et al. (1999); 12, Woodworth et al. (1996b); 13, Foes et al. (1998).

the same mutation (Manley et al. 1999). Thus, the A96T mutation is considered to confer resistance solely to the IM. On the other hand, the ALS of *Lactuca serriola* possessing the P171H mutation has been shown to be resistant to the SU and to exhibit cross-resistance to the IM and the TP but not to the PC (Eberlein et al. 1997). The P171S mutated ALS of *K. scoparia* was inhibited by bispyribac roughly as potently as the wild-type enzyme but expressed cross-resistance to pyrithiobac (Shimizu et al. 2001b). Therefore, we should consider that the mutations at the P171 position confer resistance to the SU and that the cross-resistance patterns to other ALS-inhibiting herbicides varies depending on the changes in amino acids resulting from mutation and on the tested herbicides. The W548L mutation found in the weeds confers resistance similar to that described in herbicide-resistant crops. There is no report on the S627 mutation in weeds until now.

1.9.4 Genetic Engineering

There are two methods to alter the ALS gene of plants. One is the genetic transformation utilizing recombinant DNA technology. The other is oligonucleotide-mediated gene manipulation. Since the time the herbicide-resistant ALS genes were cloned, the genes have been introduced into various kinds of plants, namely, tobacco (Haughn et al. 1988; Charest et al. 1990; Odell et al. 1990; Brandle et al. 1994), commercial flax (McHughen 1989), canola (Mike et al. 1990), rice (Li et al. 1992), cotton (Rajasekaran et al. 1996b), pea (Polowick et al. 1998), apple (Yao et al. 1999), soybean (Aragao et al. 2000), etc (Mazur and Falco 1989). These plants (except for rice and soybean) were transformed with foreign ALS genes by the *Agrobacterium*-mediated gene transfer that is the frequently used recombinant DNA technology. Rice and soybean were generated by protoplast transformation and particle bombardment, respectively. In contrast to ALS-inhibiting herbicide-resistant plants generated by the conventional breeding method and the in vitro cell selection, there is no commercial product generated by recombinant DNA technology (we have some information that a transgenic cotton and commercial flax are being developed). Instead, the herbicide-resistant ALS genes have been shown to be useful as a selection marker for introducing foreign traits into plants (Li et al. 1992). On the other hand, the oligonucleotide-mediated gene manipulation is a novel method of altering endogenous genes of plants through targeted modification (Beetham et al. 1999; Zhu et al. 1999). It has been shown that the mutation responsible for the IM resistance can be successfully introduced into genes encoding ALS (Zhu et al. 2000). Because this technology does not involve genomic integration of transgenes, the targeted trait is obtained through modifying its normal chromosomal context. When herbicide-resistant plants depend on the mutation of an endogenous gene, this technology as well as homologous recombination appear very important. Another gene technology, namely repression of the ALS activities of plants through antisense inhibition, has been reported (Höfgen et al. 1995).

Acknowledgements. We thank Dr. Peter Porpiglia for critically reading the manuscript and Miss Kazuko Matsumoto for her help with the references. Rice ALS genes were isolated with the help of Dr. Yoshiyuki Tanaka of the National Institute of Agrobiological Science in the course of the MAFF project of Japan.

References

Adkins SW, Wills D, Boersma M, Walker SR, Robinson G, Mcleod RJ, Einam JP (1997) Weeds resistant to chlorsulfuron and atrazine from the north-east grain region of Australia. Weed Res 37:343–349

Al-Khatib K, Baumgartner JR, Peterson DE, Currie RS (1998) Imazethapyr resistance in common sunflower (*Helianthus annuus*). Weed Sci 46:403–407

Amann A, Feucht D, Wellmann A (2000) A new herbicide for grass control in winter wheat, rye and triticale. Z Pflanzenkr Pflanzenschutz 17:545–553

Anderson DD, Nissen SJ, Martin AR, Roeth FW (1998) Mechanism of primisulfuron resistance in a shattercane (*Sorghum bicolor*) biotype. Weed Sci 46:158–162

Anderson PC, Georgeson M (1989) Herbicide-tolerant mutants of corn. Plant Sci Res 31:994–999

Aragao FJL, Sarokin L, Vianna GR, Rech EL (2000) Selection of transgenic meristematic cells utilizing a herbicidal molecule results in the recovery of fertile transgenic soybean [*Glycine max* (L.) Merrill] plants at a high frequency. Theor Appl Genet 101:1–6

Arfin SM, Koziel DA (1973) Acetolactate synthase of *Pseudomonas aeruginosa*: II. Evidence for the presence of two nonidentical subunits. Biochim Biophys Acta 321:356–360

Babczinski P, Zelinski T (1991) Mode of action of herbicidal ALS-inhibitors on acetolactate synthase from green plant cell cultures, yeast, and *Escherichia coli*. Pestic Sci 31:305–323

Bedbrook JR, Chaleff RS, Falco SC, Mazur BJ, Somerville CR, Yadav NS (1991) Nucleic acid fragment encoding herbicide resistant plant acetolactate synthase. US5013659, EI Du Pont de Nemours and Co

Beetham PR, Kipp PB, Sawycky XL, Arntzen CJ, May GD (1999) A tool for functional plant genomics: chimeric RNA/DNA oligonucleotides cause in vivo gene-specific mutations. Proc Natl Acad Sci USA 96:8774–8778

Bekkaoui F, Condie JA, Neustaedter DA, Moloney MM, Crosby WL (1991) Isolation, structure and expression of a cDNA for acetolactate synthase from *Brassica napus*. Plant Mol Biol 16:714–744

Bekkaoui F, Schorr P, Crosby WL (1993) Acetolactate synthase from *Brassica napus*: immunological characterization and quaternary structure of the native enzyme. Physiol Plant 88:475–484

Bernasconi P, Woodworth AR, Rosen BA, Subramanian MV, Siehl DL (1995) A naturally occurring point mutation confers broad range tolerance to herbicides that target acetolactate synthase. J Biol Chem 270:17381–17385; Correction (1996) J Biol Chem 271:13925–13926

Boutsalis P, Karotam J, Powles SB (1999) Molecular basis of resistance to acetolactate synthase-inhibiting herbicides in *Sisymbrium orientale* and *Brassica tournefortii*. Pestic Sci 55:507–516

Brady TM, Cross B, Dohner RF, Finn JM, Ladner DL (1998) The discovery of imazamox, a new broad-spectrum imidazolinone herbicide. ACS Symposium Series 686, Washington, DC, pp 30–37

Brandle AC, Morrison MJ, Hattori J, Miki BL (1994) A comparison of two genes for sulfonylurea herbicide resistance in transgenic tobacco seedlings. Crop Sci 34:226–229

Bright S, William J, Chang MT, Evans IJ, Macdonald MJ (1992) Herbicide resistant plants. WO9208794, Imperial Chemical Industries PLC

Brooks RL, Zoschke A, Porpiglia PJ (1995) CGA-277476: a short residual herbicide for soybean weed control programs. Brighton Crop Protection Conference, Weeds 1, pp 79–85

Chang AK, Duggleby RG (1997) Expression, purification and characterization of *Arabidopsis thaliana* acetohydroxyacid synthase. Biochem J 327:161–169

Chang AK, Duggleby RG (1998) Herbicide-resistant forms of *Arabidopsis thaliana* acetohydroxyacid synthase: characterization of the catalytic properties and sensitivity to inhibitors of four defined mutants. Biochem J 333:765–777

Chang SI, Kang MK, Choi JD, Namgoong SK (1997) Soluble over-expression in *Escherichia coli*, and purification and characterization of wild-type recombinant tobacco acetolactate synthase. Biochem Biophys Res Commun 234:549–553

Charest PJ, Hattori J, Demoor J, Iyer VN, Miki BL (1990) In vitro study of transgenic tobacco expressing *Arabidopsis* wild type and mutant acetohydroxyacid synthase genes. Plant Cell Rep 8:643–646

Chipman D, Barak Z, Schloss JV (1998) Biosynthesis of 2-aceto-2-hydroxy acids: acetolactate synthase and acetohydroxyacid synthase. Biochim Biophys Acta 1385:401–419

Cho JH, Ahn S-C, Koo SJ, Joe KH, Oh HS (1997) LGC-40863: a new broad spectrum post-emergence herbicide. Brighton Crop Protection Conference, Weeds 1, pp 17–20

Chong CK, Chang SI, Choi JD (1998) Functional amino acid residues of recombinant tobacco acetolactate synthase. J Biochem Mol Biol 31:258–263

Chong CK, Shin HJ, Chang SI, Choi JD (1999) Role of tryptophanyl residues in tobacco acetolactate synthase. Biochem Biophys Res Commun 259:136–140

Chong CK, Shin HJ, Chang SI, Choi JD (2000) Determination of the disulfide bond and its possible role in tobacco acetolactate synthase. Arch Biochem Biophys 379:363–366

Cornish-Bowden A (1986) Why is uncompetitive inhibition so rare? A possible explanation, with implications for the design of drugs and pesticides. FEBS Lett 203:3–6

Croughan TP (1999) Herbicide resistant rice. US5952553, Louisiana State University and Agricultural and Mechanical College, Baton Rouge

Cullin C, Baudin-Baillieu A, Gullemete E, Ozier-Kalogeropoulos O (1996) Functional analysis of YCL09C: evidence for a role as the regulatory subunit of acetolactate synthase. Yeast 12:1511–1518

Currie RS, Regehr DL (1995) Methods of measuring the impact of the XA17 gene on imazethapyr injury in corn (*Zea mays*). Weed Technol 9:676–681

Davies ME (1964) Acetolactate and acetoin synthesis in ripening peas. Plant Physiol 39:53–59

Currie RS, Kwon CS, Penner D (1995) Magnitude of imazethapyr resistance of corn (*Zea mays*) hybrids with altered acetolactate synthase. Weed Sci 43:578–582

Dietrich GE (1998) Imidazolinone resistant AHAS mutants. US5731180, American Cyanamid Company

Duggleby RG (1997) Identification of acetolactate synthase small subunit gene in two eukaryotes. Gene 190:245–249

Duggleby RG, Pang SS (2000) Acetohydroxyacid synthase. J Biochem Mol Biol 33:1–36

Dumas R, Biou V, Douce R (1997) Purification and characterization of a fusion protein of plant acetohydroxyacid synthase and acetohydroxyacid isomeroreductase. FEBS Lett 408:156–160

Durner J, Böger P (1988) Acetolactate synthase from barley (*Hordeum vulgare* L.): purification and partial characterization. Z Naturforsch 43c:850–856

Durner J, Böger P (1990) Oligomeric forms of plant acetolactate synthase depend on flavin adenine dinucleotide. Plant Physiol 93:1027–1031

Durner J, Gailus V, Böger P (1991) New aspects on inhibition of plant acetolactate synthase by chlorsulfuron and imazaquin. Plant Physiol 95:1144–1149

Eberlein CV, Guttieri MJ, Thill DC, Mallory-Smith CA, Baerg RJ (1997) Altered acetolactate synthase activity in ALS-inhibitor resistant prickly lettuce (*Lactuca serriola*). Weed Sci 45:212–217

Eberlein CV, Guttieri MJ, Berger PH, Fellman JK, Mallory-Smith CA, Thill DC, Baerg RJ, Belknap WR (1999) Physiological consequences of mutation for ALS-inhibitor resistance. Weed Sci 47:383–392

Fang LY, Gross PR, Chen CH, Lillis M (1992) Sequence of two acetohydroxyacid synthase genes from *Zea mays*. Plant Mol Biol 18:1185–1187

Foes MJ, Liu LX, Tranel PJ, Wax LM, Stoller EW (1998) A biotype of common waterhemp (*Amaranthus rudis*) resistant to triazine and ALS herbicides. Weed Sci 46:514–520

Foes MJ, Liu L, Stoller EW, Wax LM, Tranel PJ (1999) A kochia (*Kochia scoparia*) biotype resistant to triazine and ALS-inhibiting herbicides. Weed Sci 47:20–27

Fushimi T, Nakahira K, Tagawa M, Nawamaki T (1997) Herbicide resistant acetolactate synthase. WO9708327, Nissan Chemical Industries, Ltd

Gerwick BC, Mireles LC, Eilers RJ (1993) Rapid diagnosis of ALS/AHAS-resistant weeds. Weed Technol 7:519–524

Glatzer L, Eakin E, Wagner RP (1972) Acetohydroxyacid synthase with a pH optimum of 7.5 from *Neurospora crassa* mitochondria: characterization and partial purification. J Bacteriol 112: 453–464

Grandoni JA, Marta PT, Schloss JV (1998) Inhibitors of branched-chain amino acid biosynthesis as potential antituberculosis agents. J Antimicrob Chemother 42:475–482

Grula JW, Hudspeth RL, Hobbs SL, Anderson DM (1995) Organization, inheritance and expression of acetohydroxyacid synthase genes in the cotton allotetraploid *Gossypium hirsutum*. Plant Mol Biol 28:837–846

Guangfu Y, Huayin L, Huazheng Y (1999) QSAR and 3D-QSAR analysis of structurally diverse ALS inhibitors: sulfonylureas and triazolopyrimidine-2-sulfonamides. Pestic Sci 55:1143–1150

Guttieri MJ, Eberlein CV, Mallory-Smith CA, Thill DC, Hoffman DL (1992) DNA sequence variation in domain A of the acetolactate synthase genes of herbicide-resistant and -susceptible weed biotypes. Weed Sci 40:670–676

Guttieri MJ, Eberlein CV, Thill DC (1995) Diverse mutations in the acetolactate synthase gene confer chlorsulfuron resistance in kochia (*Kochia scoparia*) biotypes. Weed Sci 43:175–178

Hacker E, Kehne H, Hess M (1996) Herbicides with 4-iodo-2-[3-(4-methoxy-6-methyl-1,3,5-triazin-2-yl)ureidosulfonyl]-benzoic acid esters. WO9641537, Hoechst Schering AgrEvo GmbH

Hall LM, Stromme KM, Horsman GP (1998) Resistance to acetolactate synthase inhibitors and quinclorac in a biotype of false cleavers (*Galium spurium*). Weed Sci 46:390–396

Hanai R, Kawano K, Shigematsu S, Tamaru M (1993) KIH-6127, a new selective herbicide to control barnyardgrass in rice. Brighton Crop Protection Conference, Weeds 1, pp 47–52

Hand JM, Singh BK, Chaleff RS (1992) Herbicide resistant AHAS deletion mutants EP492113, American Cyanamid Company

Harms CT, Montoya AL, Privalle LS, Briggs RW (1990) Genetic and biochemical characterization of corn inbred lines tolerant to the sulfonylurea herbicide primisulfuron. Theor Appl Genet 80:353–358

Harms CT, Armour SL, DiMaio JJ, Middlesteadt LA, Murray D, Negrotto DV, Thompson-Taylor H, Weymann K, Montoya AL, Shillito RD, Jen GC (1992) Herbicide resistance due to amplification of a mutant acetohydroxyacid synthase gene. Mol Gen Genet 233:427–435

Hart SE, Saunders JW, Pemmer D (1994) Herbicide-resistant crops from cell selection. Rev Weed Sci 6:251–263

Hattori J, Rutledge R, Labbe H, Brown D, Sunohara G, Miki B (1992) Multiple resistance to sulfonylureas and imidazolinones conferred by an acetohydroxyacid synthase gene with separate mutations for selective resistance. Mol Gen Genet 232:167–173

Hattori J, Brown D, Mourad G, Labbe H, Ouellet T, Sunohara G, Rutledge R, King J, Miki B (1995) An acetohydroxyacid synthase mutant reveals a single site involved in multiple herbicide resistance. Mol Gen Genet 246:419–425

Haughn GW, Somerville CR (1990) A mutation causing imidazoline resistance maps to the *csr*1 locus *Arabidopsis thaliana*. Plant Physiol 92:1081–1085

Haughn GW, Smith J, Mazur B, Somerville C (1988) Transformation with a mutant *Arabidopsis* acetolactate synthase gene renders tobacco resistant to sulfonylurea herbicides. Mol Gen Genet 211:266–271

Hawkes TR (1989) Studies of herbicides which inhibit branched chain amino acid biosynthesis. British Crop Protection Council Monograph 42, Lavenham Press Limited, Lavenham, UK, pp 131–138

Hawkes TR, Thomas SE (1990) Imidazolinones: factors determining their herbicidal efficacy. In: Barak Z, Chipman DM, Schloss JV (eds) Biosynthesis of branched-chain-amino acids. Balaban, Weinheim, pp 373–389

Hershey HP, Schwartz LJ, Gale JP, Abell LM (1999) Cloning and functional expression of the small subunit of acetolactate synthase from *Nicotiana plumbaginifolia*. Plant Mol Biol 40: 795–806

Hervieu F, Vaucheret H (1996) A single amino acid change in acetolactate synthase confers resistance. Mol Gen Genet 251:220–224

Hess M, Rose E (1995) A new herbicide for broadleaf weed and sedge control in rice. Brighton Crop Protection Conference, Weed 2, pp 763–768

Hill CM, Pang SS, Duggleby RG (1997) Purification of *Escherichia coli* acetohydroxyacid synthase isozyme II and reconstitution of active enzyme from its individual pure subunits. Biochem J 327:891–898

Hinz RR, Owen MDK (1997) Acetolactate synthase resistance in a common waterhemp (*Amaranthus rudis*) population. Weed Technol 11:13–18

Höfgen R, Laber B, Schuttke BL, Klonus AK, Streber W, Pohlenz HD (1995) Repression of acetolactate synthase activity through antisense inhibition. Plant Physiol 107:469–477

Hur CU, Cho JH, Hong SM, Kim HW, Lim YH, Rim JS, Kim JS, Chae SH (1995) Pyrimidine derivatives, process for their preparation and their use as herbicide. EP658549, Lucky Ltd

Ishida Y, Ohta K, Yoshikawa H (1992) Herbicides. EP477808, Takeda Chemical Industries, Ltd

Itoh K, Wang GX (1997) An outbreak of sulfonylurea herbicide resistance in Scrophulariaceae paddy weeds in Japan. 16th Asian-Pacific Weed Science Society Conference 4B, pp 219–221

Itoh K, Wang GX, Ohba S (1999) Sulfonylurea resistance in *Lindernia micrantha*, an annual paddy weed in Japan. Weed Res 39:413–423

Kakefuda G, Ott K, Kwagh J, Stockton GW (1996) Structure-based designed herbicide resistant products. WO9633270, American Cyanamid Company

Kaku K, Shimizu T, Nagayama K, Hukuda A, Tanaka Y (2001) Isolation and expression of cDNA for acetolactate synthase from *Oryza sativa*. Abstract Annual Meeting Pesticide Science Society Japan, Sakai, p 101 (in Japanese)

Kobayashi M, Yokoyama M, Watanabe O, Sadohara H, Wada N (1995) KIH-2023, a new postemergence herbicide in rice (*Oryza sativa*). 15th Asian-Pacific Weed Science Society Conference Proceedings I(A), Kyoto, pp 221–226

Koeppe MK, Barefoot AC, Cotterman CD, Zimmerman WT, Leep DC (1997) Basis of selectivity of the herbicide flupyrsulfuron-methyl in wheat. Pestic Biochem Physiol 59:105–117

Kohara H, Konno K, Takekawa M (1999) Occurrence of sulfonylurea-resistant biotypes of *Scirpus juncoides* Roxb. var. *ohwianus*. T. Koyama in paddy fields of Hokkaido prefecture, Japan. J Weed Sci Technol 44:228–235

Koo SJ, Ahn SC, Lim JS, Chae SH, Kim JS, Lee JH, Cho JH (1997) Biological activity of the new herbicide LGC-40863 (benzophenone O-(2,6-bis((4,6-dimethoxy-2-primidinyl)oxy)benzoyl) oxime). Pestic Sci 51:109–114

Krausz RF, Kapusta G, Matthews JL (1997) Acetolactate synthase-resistant and -susceptible corn (*Zea mays*) response to imazethapyr, imazaquin, chlorimuron, and CGA-152005. Weed Technol 11:810–816

Lange FD, Brown PTH, Loerz H, Kollmorgen JF (1995) In vitro selection for modified amino-acid *Triticum* species. Plant Breed 114:351–354

LaRossa RA, Schloss JV (1984) The sulfonylurea herbicide sulfometuron methyl is an extremely potent and selective inhibitor of acetolactate synthase in *Salmonella typhimurium*. J Biol Chem 25:8753–8757

LaRossa RA, Van Dyk TK, Smulski DR (1987) Toxic accumulation of 2-ketobutyrate caused by inhibition of the branched-chain amino acid biosynthetic enzyme acetolactate synthase in *Salmonella typhimurium*. J Bacteriol 169:1372–1378

Lavigne C, Millecamps L, Manach H, Cordonnier P, Matejicek A, Vasseur J, Gasquez J (1994) Monogenic semidominant sulfonylurea resistance in a line of white chicory. Plant Breed 113: 305–311

Lee CD, Martin AR, Roeth FW, Johnson BE, Lee DJ (1999) Comparison of ALS inhibitor resistance and allelic interactions in shattercane accessions. Weed Sci 47:275–281

Lee EH, Ahn TW, Choi JD (1991) Properties and feedback inhibition of acetohydroxyacid synthase from pea shoots. Korean Biochem J 24:285–291

Lee KY, Tepperman J, Black M, Chui CF, Mazur B, Dunsmuir P, Bedbrook J (1988) The molecular basis of sulfonylurea resistance in tobacco. EMBO J 7:1241–1248

Lee YT, Chang LA, Duggleby RG (1999) Effect of mutagenesis at serine 653 of *Arabidopsis thaliana* acetohydroxyacid synthase on the sensitivity to imidazolinone and sulfonylurea herbicides. FEBS Lett 452:341–345

Lepiece D, Thompson A, Rijckaert G (1999) Florasulam Primus, a new selective herbicide for the control of broad-leaved weeds in young grass. Mededelingen Fac Landbouwkundige Toegepaste Biol Wetenschappen Univ Gent 64:693–712

Li Z, Hayashimoto A, Murai N (1992) A sulfonylurea herbicide resistance gene from *Arabidopsis thaliana* as a new selective marker for production of fertile transgenic rice plants. Plant Physiol 100:662–668

Loubser JW (1998) Activity of chlorsulfuron, ethoxysulfuron and sulfosulfuron towards selected cereal weeds. Appl Plant Sci 12:57–59

Lovell ST, Wax LM, Horak MJ, Peterson DE (1996) Imidazolinone and sulfonylurea resistance in a biotype of common waterhemp (*Amaranthus rudis*). Weed Sci 44:789–794

Manley BS, Wilson HP, Hines TE (1996) Smooth pigweed (*Amaranthus hybridus*) and livid amaranth (*A. lividus*) response to several imidazolinone sulfonylurea herbicides. Weed Technol 10:835–841

Manley BS, Singh BK, Shaner DL, Wilson HP (1999) Imidazolinone resistance in smooth pigweed (*Amaranthus hybridus*) is due to an altered acetolactate synthase. Weed Technol 13:697–705

Marquez T, Joshi MM, Fader TP, Massasso W (1995) Azimsulfuron (DPX-A8947): a new sulfonylurea for post-emergence control of *Echinochloa* species, broadleaf and sedge weeds for southern European rice production. Brighton Crop Protection Conference: Weeds 1:65–72

Matsushita H, Hukai Y, Unai T, Ishikawa K, Yusa Y (1994) The report concerning a novel herbicide, KIH-2023: adsorption, translocation and metabolism of KIH-2023 in rice and wheat seedlings. Abstract of Annual Meeting Pesticide Science Society Japan, Sapporo, p 127 (in Japanese)

Mazur BJ, Falco SC (1989) The development of herbicide resistant crops. Annu Rev Plant Physiol Plant Mol Biol 40:441–447

McHughen A (1989) *Agrobacterium* mediated transfer of chlorsulfuron resistance to commercial flax cultivars. Plant Cell Rep 8:445–449

Meyer W, Riehen S (1992) Sulfonylureas. US5209771, Ciba-Geigy Corporation

Miflin BJ (1971) Cooperative feedback control of barley acetohydroxyacid synthase by leucine, isoleucine, and valine. Arch Biochem Biophys 146:542–550

Miflin BJ (1974) The location of nitrate reductase and other enzymes related to amino acid biosynthesis in the plastids of root and leaves. Plant Physiol 54:550–555

Mike BL, Labbe H, Hattori J, Ouellet T, Gabard J, Sunohara G, Charest PJ, Iyer VN (1990) Transformation of *Brassica napus* canola cultivars with *Arabidopsis thaliana* acetohydroxyacid synthase genes and analysis of herbicide resistance. Theor Appl Genet 80:449–458

Mizutani H, Shinba K, Asano Y, Yusa Y (1998) Metabolism of a herbicide pyriminobac-methyl in rice and barnyard grass seedlings. Abstract Annual Meeting Pesticide Science Society Japan, Matsue, p 106 (in Japanese)

Mourad G, King J (1992) Effect of four classes of herbicides on growth and acetolactate synthase activity in several variants of *Arabidopsis*. Planta 188:491–497

Mourad G, Haughn G, King J (1994) Intragenic recombination in the *csr*1 locus of *Arabidopsis*. Mol Gen Genet 243:178–184

Müller KH, Koning K, Kluth J, Lürssen K, Santel HJ, Schmidt RR (1992) Sulfonylaminocarbonyltriazolinones with oxygen-bound substituents. EP507171, Bayer AG

Muhitch MJ (1988) Acetolactate synthase activity in developing maize (*Zea mays* L.) kernels. Plant Physiol 86:23–27

Muhitch MJ, Shaner DL, Stidham MA (1987) Imidazolinone and acetohydroxyacid synthase from higher plants. Plant Physiol 83:451–456

Nakata M (1991) The mode of action of chlorsulfuron in culture cells of tobacco and hamster. J Pestic Sci 16:583–590

Nakayama I, Shimizu T (1993) Bioactivity of ALS inhibitors (in Japanese). 10th Symposium of Research Committee for the Bioactivity of Pesticides, Nagano, pp 62–70

Nakayama I, Shimizu T, Nakao T, Abe H (1993) The active forms of pyrimidinylsalicylate herbicides for the inhibition of ALS, and the participation of esterase in their activation. Abstract Annual Meeting Pesticide Science Society Japan, Futyu, p 77 (in Japanese)

Nelson KA, Renner KA (1998) Postemergence weed control with CGA-277476 and cloransulam-methyl in soybean (*Glycine max*). Weed Technol 12:293–299

Nelson KA, Renner KA, Penner D (1998) Weed control in soybean (*Glycine max*) with imazamox and imazethapyr. Weed Sci 46:587–594

Newhouse K, Singh B, Shaner D, Stidham M (1991) Mutations in corn (*Zea mays* L.) conferring resistance to imidazolinone herbicides. Theor Appl Genet 83:65–70

Newhouse KE, Smith WA, Starrett MA, Schaefer TJ, Singh BK (1992) Tolerance to imidazolinone herbicides in wheat. Plant Physiol 100:882–886

Nezu Y, Miyazaki M, Sugiyama K, Kajiwara I (1996a) Dimethoxypyrimidines as novel herbicides. Part 1. Synthesis and herbicidal activity of dimethoxyphenoxyphenoxypyrimidines and analogues. Pestic Sci 47:103–113

Nezu Y, Miyazaki M, Sugiyama K, Nobuhide W, Kajiwara I, Miyazawa T (1996b) Dimethoxypyrimidines as novel herbicides. Part 2. Synthesis and herbicidal activity of O-pyrimidinylsalicylates and analogues. Pestic Sci 47:115–124

Nezu Y, Wada N, Yoshida F, Miyazawa T, Shimizu T, Fujita T (1998) Dimethoxypyrimidines as novel herbicides. Part 4. Quantitative structure-activity relationships of dimethoxypyrimidinyl(thio)salicylic acids. Pestic Sci 52:343–353

Odell JT, Caimi PG, Yadav NS, Mauvais CJ (1990) Comparison of increased expression of wild-type and herbicide-resistant acetolactate synthase genes in transgenic plants, and indication of posttranscriptional limitation on enzyme activity. Plant Physiol 94:1647–1654

Ono Y, Yanagisawa K, Kitamura S, Kawamoto A (1999) Herbicide efficacy of bispyribac-sodium against rice weeds. 17th Asian-Pacific Weed Science Society Conference Proceedings I(B), Bangkok, pp 691–694

Ort O, Bauer K, Böringer H (1992) Aryl sulphonylurea compounds, a method of preparing them, and their use as herbicides and growth regulators. WO9213845, Hoechst Aktiengesellschaft

Ortega F, Bastide J, Hawkes TR (1996) Comparison between thifensulfuron methyl-induced inactivation of barley acetohydroxyacid synthase and *Escherichia coli* acetohydroxyacid synthase isozyme II. Pestic Biochem Physiol 56:231–242

Ott KH, Kwagh JG, Stockton GW, Sidorov V, Kakefuda G (1996) Rational molecular design and genetic engineering of herbicide resistant crops by structure modeling and site-directed mutagenesis of acetohydroxyacid synthase. J Mol Biol 263:359–368

Palmer EW, Shaw DR, Holloway C (1999) Influence of CGA-277476 on efficacy of postemergence graminicides. Weed Technol 13:48–53

Parrish SK, Kaufmann JE, Croon KA, Ishida Y, Ohta K, Itoh S (1995) MON 37500: a new selective herbicide to control annual and perennial weeds in wheat. Brighton Crop Protection Conference, Weeds 1, pp 57–63

Polowick PL, Baliski MK, Mahon JD (1998) *Agrobacterium*-mediated genetic transformation of western Canadian pea genotypes. In vitro Cell Dev Biol Anim 34:46A

Rajasekaran K, Grula JW, Anderson DM (1996a) Selection and characterization of mutant cotton (*Gossypium hirsutum* L.) cell lines resistant to sulfonylurea and imidazolinone herbicides. Plant Sci 119:115–124

Rajasekaran K, Grula JW, Hudspeth RL, Pofelis S, Anderson DM (1996b) Herbicide-resistant Acala and Coker cottons transformed with a native gene encoding mutant forms of acetohydroxyacid synthase. Mol Breed 2:307–319

Ray TB (1984) Site of action of chlorsulfuron: inhibition of valine and isoleucine biosynthesis in plants. Plant Physiol 75:827–831

Rutledge RG, Quellet T, Hattori J, Miki BL (1991) Molecular characterization and genetic origin of the *Brassica napus* acetohydroxyacid synthase multigene family. Mol Gen Genet 229:31–40

Saari LL, Cotterman JC, Thill DC (1994) Resistance to acetolactate synthase inhibiting herbicides. In: Powles SB, Holtum JAM (eds) Herbicide resistance in plants. Biology and biochemistry. CRC Press, Boca Raton, pp 83–139

Sadohara (1997) Nominee (bispyribac-sodium): a new post-emergence herbicide in rice. Agrochem Jpn 71:18–19

Saito Y, Wada N, Kusano S, Miyazawa T, Takahashi S, Toyokawa Y, Kajiwara Y (1990) Pyrimidine compounds, and herbicidal method and compositions. US4932999, Kumiai Chemical Industry Co, Ltd and Ihara Chemical Industry Co, Ltd

Sathasivan K, Haughn GW, Murai N (1990) Nucleotide sequence of an acetolactate synthase gene from an imidazolinone-resistant *Arabidopsis thaliana* var. Columbia. Nucleic Acids Res 18:2118

Sathasivan K, Haughn GW, Murai N (1991) Molecular basis of imidazolinone herbicide resistance in *Arabidopsis thaliana* var. Columbia. Plant Physiol 97:1044–1050

Schloss JV, Ciskanik LM, Van Dyk DE (1988) Origin of the herbicide binding site of acetolactate synthase. Nature 331:360–362

Schulze-Siebert D, Schultz G (1989) Formation of aromatic amino acids and valine from $^{14}CO_2$ or 3-[U-14C]phosphoglycerate by isolated intact spinach chloroplasts. Plant Sci 59:167–174

Schulze-Siebert D, Heineke D, Scharf H, Schultz G (1984) Pyruvate-derived amino acids in spinach chloroplasts. Plant Physiol 76:465–471

Sebastian SA, Fader GM, Ulrich FJ, Forney UD, Chaleff RS (1989) Semidominant soybean mutation for resistance to sulfonylurea herbicides. Crop Sci 29:1403–1408

Sengnil K, Usui K, Ishizuka K (1992) Selection of bensulfuron methyl-tolerant rice cells and their acetolactate synthase response. Weed Res Jpn 37:232–238

Shaner DL, Reider ML (1986) Physiological responses of corn (*Zea mays*) to AC 243, 997 in combination with valine, leucine, and isoleucine. Pestic Biochem Physiol 25:248–257

Shaner DL, Singh BK (1993) Phytotoxicity of acetohydroxyacid synthase inhibitors is not due to accumulation of 2-ketobutyrate and/or 2-aminobutyrate. Plant Physiol 103:1221–1226

Shaner DL, Anderson PC, Stidham MA (1984) Potent inhibitors of acetohydroxyacid synthase. Plant Physiol 76:545–546

Shaw DR, Bennett AC, Grant DL (1999) Weed control in soybean (*Glycine max*) with flumetsulam, cloransulam, and diclosulam. Weed Technol 13:791–798

Shibaike H (2000) Abstract of 2nd meeting of herbicide resistant weeds in annual meeting weed science society of Japan (in Japanese)

Shibuya K, Yoshioka T, Yoshio A, Sotoh S, Yoshida S, Hashiba T (1999) Analysis of acetolactate synthase genes of sulfonylurea herbicides-resistant and -susceptible biotypes in *Scirpus juncoides* subsp. *juncoides*. J Weed Sci Technol 44:S72–73

Shimizu T (1997) Action mechanism of pyrimidinyl carboxy herbicides. J Pestic Sci 22:245–256

Shimizu T, Nakayama I, Nakao T, Abe H (1986) Acetolactate synthase of etiolated pea seedlings. Abstract Annual Meeting Japan Society of Bioscience, Biotechnology and Agrochemistry, Kyoto, p 229 (in Japanese)

Shimizu T, Nakayama I, Nakao T, Yamashita K, Nagayama K, Abe H (1993) Kinetics studies on the inhibition of bacterial ALS by pyrimidinylsalicylic acids. Abstract Annual Meeting Pesticide Science Society Japan, Futyu, p 76 (in Japanese)

Shimizu T, Nakayama I, Nakao T, Abe H (1994a) Partial purification and properties of acetolactate synthase of etiolated pea seedlings. J Pestic Sci 19:187–196

Shimizu T, Nakayama I, Nakao T, Nezu Y, Abe H (1994b) Inhibition of plant acetolactate synthase by herbicides, pyrimidinylsalicylic acids. J Pestic Sci 19:59–67

Shimizu T, Nakayama I, Wada N, Nakao T, Abe H (1994c) Kinetic studies on the inhibition of acetolactate synthase by pyrimidinylsalicylic acids. J. Pestic Sci 19:257–266

Shimizu T, Yamashita K, Kato H, Hashimoto N, Abe H, Nakayama I (1995) Interaction of acetolactate synthase and its inhibitors. Abstract Annual Meeting Pesticide Science Society Japan, Tokyo, p 136 (in Japanese)

Shimizu T, Kaku K, Nagayama K, Tanaka Y (2001a) Selection of PC herbicide resistant rice cells and their ALS sensitivities to herbicides. Abstract Annual Meeting Pesticide Science Society Japan, Sakai, p 96 (in Japanese)

Shimizu T, Kaku K, Takahashi S, Nagayama K (2001b) Sensitivities of ALS prepared from SU- and IMI-resistant weeds against PC herbicides. J Weed Sci Technol 46:S32–33 (in Japanese)

Shin YS, Chong CK, Choi JD (1999) Separation and characterization of two forms of acetolactate synthase from etiolated pea seedings. J Biochem Mol Biol 32:393–398

Siehl DL, Bengtson AS, Brockman JP, Butler JH, Kraatz GW, Lamoreaux RJ, Subramanian MV (1996) Patterns of cross-tolerance to herbicides inhibiting acetohydroxyacid synthase in commercial corn hybrids designed for tolerance to imidazolinones. Crop Sci 36:274–278

Simpson DM, Stoller EW (1995) Response of sulfonylurea-tolerant soybean (*Glycine max*) and selected weed species to imazethapyr and thifensulfuron combinations. Weed Technol 9:582–586

Simpson DM, Stoller EW, Wax LM (1995) An in vivo acetolactate synthase assay. Weed Technol 9:17–22

Singh BK, Schmitt GK (1989) Flavin adenine dinucleotide causes oligomerization of acetohydroxyacid synthase from Black Mexican sweet corn cells. FEBS Lett 258:113–115

Singh BK, Shaner DL (1995) Biosynthesis of branched chain amino acids: from test tube to field. Plant Cell 7:935–944

Singh BK, Stidham MA, Shaner DL (1988a) Assay of acetohydroxyacid synthase. Anal Biochem 171:173–179

Singh BK, Stidham MA, Shaner DL (1988b) Separation and characterization of two forms acetohydroxyacid synthase from Black Mexican sweet corn cells. J Chromatogr 444:251–261

Singh BK, Newhouse KE, Stidham MA, Shaner DL (1989) Acetolactate synthase-imidazolinone interaction. Br Crop Protection Counc Monogr 42:87–95

Singh BK, Lumanglas A, Wang BS (1991) Production of a monocot-specific monoclonal antibody against acetohydroxyacid synthase and its use in the purification and characterization of the enzyme. Proc Natl Acad Sci USA 88:4572–4576

Singh BK, Szamosi I, Hand JM, Misra R (1992) *Arabidopsis* acetohydroxyacid synthase expressed in *Escherichia coli* is insensitive to the feedback inhibitors. Plant Physiol 99:812–816

Southan MD, Copeland L (1996) Physical and kinetic properties of acetohydroxyacid synthase from wheat leaves. Physiol Plant 98:824–832

Sprague CL, Stoller EW, Wax LM (1997a) Common cocklebur (*Xanthium strumarium*) resistance to selected ALS-inhibiting herbicides. Weed Technol 11:241–247

Sprague CL, Stoller EW, Wax LM, Horak MJ (1997b) Palmer amaranth (*Amaranthus palmeri*) and common waterhemp (*Amaranthus rudis*) resistance to selected ALS-inhibiting herbicides. Weed Sci 45:192–197

Subramanian MV, Gerwick BC (1989) Inhibition of acetolactate synthase by triazolopyrimidines. ACS Symp Ser 389, Washington, DC, pp 277–288

Subramanian MV, Loney V, Pao L (1989) Mechanism of action of 1,2,4-triazolo[1,5-a]pyrimidine sulfonamide herbicides. Br Crop Protection Counc Monogr 42:97–100

Subramanian MV, Loney-Gallant V, Dias JM, Mireles LC (1991) Acetolactate synthase inhibiting herbicides bind to the regulatory site. Plant Physiol 96:310–313

Sunderland S, Burton JD, Coble HD, Maness EP (1995) Physiological mechanism for tall morning glory (*Ipomoea purpurea*) resistance to DPX-PE350. Weed Sci 43:21–27

Tachikawa S, Miyazawa T, Sadohara H (1997) Vegetation management by KIH-2023 in rice levees, and highway and railroad right-ways. 16th Asian-Pacific Weed Science Society Conference Proceedings 2A, Kuala Lumpur, pp 114–117

Takahashi S, Shigenatsu S, Mirita A, Nezu Y, Claus JS, Williams CS (1991) KIH-2031, a new herbicide for cotton. Brighton Crop Protection Conference, Weeds 1, pp 57–62

Tamaru M, Kawamura N, Sato M, Tachikawa S, Yoshida R, Takabe F (1991) Pyrimidine and triazine derivatives and herbicidal composition containing the same. EP435170, Kumiai Chemical Industry Co, Ltd and Ihara Chemical Industry Co, Ltd

Tamaru M, Inoue J, Hanai R, Tachikawa S (1997) Studies of the new herbicide KIH-6127. 4. Crystal structure of KIH-6127 and quantitative structure-activity relationship of the iminoxy moiety of KIH-6127 derivatives. J Agric Food Chem 45:2777–2783

Teaney SR, Armstrong L, Bentley K, Cotterman D, Leep D, Liang PH, Powley C, Summers J, Cranwell S, Lichtner F, Stichbury R (1995) DPX-KE459: a new sulfonylurea for postemergence grass and broadleaf weed control in cereals. Brighton Crop Protection Conference 1, pp 49–56

Terakawa T, Wakasa K (1992) Rice mutant resistant to the herbicide bensulfuron methyl (BSM) by in vitro selection. Jpn J Breed 42:267–275

Trabold K, Hacker E, Hess M, Huff HP (2000) A new sulfonylurea for weed control in cereals. Z Pflanzenkr Pflanzenschutz 17:701–707

Uchino A, Watanabe H (1999) Mutation in the acetolactate synthase genes of the biotypes of *Lindernia* spp. resistant to sulfonylurea herbicide. J Weed Sci Technol 44:S80–81

Uchino A, Watanabe H, Wang G, Itoh K (1999) Light requirement in rapid diagnosis of sulfonylurea-resistant weeds of *Lindernia* spp. (Scrophulariaceae). Weed Technol 13:680–684

Usui K, Suwangwong S, Watanabe H, Ishizuka K (1991) Effect of bensulfuron methyl, glyphosate and glufosinate on amino acid and ammonia levels in carrot cells. Weed Res Jpn 36:126–134

Van Heertum JC, Gerwick BC, Kleschick WA, Jhonson TC (1992) Herbicidal alkoxy-1,2,4-triazolo[1,5-c]pyrimidine-2-sulfonamides. US5163995, DowElanco

Volenberg DS, Stoltenberg DE, Boerboom CM (2000) *Solanum ptycanthum* resistance to acetolactate synthase inhibitors. Weed Sci 48:399–401

Vyazmensky M, Sella C, Barak Z, Chipmand M (1996) Isolation and characterization of subunits of acetohydroxy acid synthase isozyme III and reconstitution of the holoenzyme. Biochemistry 35:10339–10346

Wada N, Kusano S, Toyokawa Y (1990) Pyrimidine derivatives and herbicidal method and compounds. US4906285, Kumiai Chemical Industry Co, Ltd and Ihara Chemical Industry Co, Ltd

Wang GX, Kohara H, Itoh K (1997) Sulfonylurea resistance in a biotype of *Monochoria korsakowii* an annual paddy weed in Japan. Brighton Crop Protection Conference, Weeds 1, pp 311–318

Wiersma PA, Schmiemann MG, Condie JA, Crosby WL, Moloney MM (1989) Isolation, expression and phylogenetic inheritance of an acetolactate synthase gene from *Brassica napus*. Mol Gen Genet 219:413–420

Woodworth AR, Bernasconi P, Subramanian M, Rosen B (1996a) A second naturally occurring point mutation confers broad-based tolerance to acetolactate synthase inhibitors. Plant Physiol 111:S105

Woodworth AR, Rosen BA, Bernasconi P (1996b) Broad range resistance to herbicides targeting acetolactate synthase (ALS) in a field isolate of *Amaranthus* sp. is conferred by a Trp to Leu mutation in the ALS gene (accession No. U55852). Plant Physiol 111:1353

Wright T, Penner D (1998a) Corn (*Zea mays*) acetolactate synthase sensitivity to four classes of ALS-inhibiting herbicides. Weed Sci 46:8–12

Wright T, Penner D (1998b) In vitro and whole-plant magnitude and cross-resistance characterization of two imidazolinone-resistant sugarbeet (*Beta vulgaris*) somatic cell selections. Weed Sci 46:24–29

Wright TR, Bascomb NF, Penner D, Sturner SF (1998) Biochemical mechanism and molecular basis for ALS-inhibiting herbicide resistance in sugarbeet (*Beta vulgaris*) somatic cell selections. Weed Sci 46:13–23

Yamashita K, Nagayama K, Shimizu T, Toyo-oka K, Abe H (1994a) Biological activity of a novel ALS inhibitor, cyclobutenamide that is produced by *Streptomyces hygroscopicus*. Abstract Annual Meeting Pesticide Science Society Japan, Sapporo, p 39 (in Japanese)

Yamashita K, Nagayama K, Wada N, Abe H (1994b) A novel ALS inhibitor produced by *Streptomyces hygroscopicus*. Nippon Nogeikagaku Kaishi 68:658 (in Japanese)

Yang J, Kim S (1997) Effect of pyrimidylsalicylate on the valine sensitive acetolactate synthase purified from *Serratia marcescens*. J Biochem Mol Biol 30:13–17

Yao JL, Cohen D, van den Brink R, Morris B (1999) Assessment of expression and inheritance patterns of three transgenes with the aid of techniques for promoting rapid flowering of transgenic apple trees. Plant Cell Rep 18:727–732

Yokoyama M, Watanabe O, Kawano K, Shigematsu S, Wada N (1993) KIH-2023, a new post-emergence herbicide in rice. Brighton Crop Protection Conference, Weeds 1, pp 61–66

Zhu T, Peterson DJ, Taglia L, Clair GS, Baszczynski CL, Bowen B (1999) Targeted manipulation of maize genes in vivo using chimeric RNA/DNA oligonucleotides. Proc Natl Acad Sci USA 96:8768–8773

Zhu T, Mettenburg K, Peterson DJ, Tagkiani L, Baszczynski CL (2000) Engineering herbicide-resistant maize using chimeric RNA/DNA oligonucleotides. Nat Biotechnol 18:555–558

2 Bleaching Herbicides: Action Mechanism in Carotenoid Biosynthesis, Structural Requirements and Engineering of Resistance

GERHARD SANDMANN

2.1 Herbicidal Effect and Mode of Action

Carotenoids are essential components for the assembly of the photosynthetic apparatus of green plants. They are constituents of the photosynthetic reaction centers and the antenna (Siefermann-Harms 1987). There, they serve as accessory pigments in light harvesting and play an active role in electron transfer processes of photosystem II (Tracewell et al. 2001). An essential function of carotenoids is protection of the photosystems against photooxidation caused by excited triplet state chlorophyll (Demmig-Adams et al. 1996). Especially under high light stress, protection by carotenoids is essential. A turnover of the carotenoids in light was demonstrated recently (Simkin et al. 2000). Compounds which even partially inhibit carotenoid synthesis may prevent the formation of enough carotenoids to ensure efficient photoprotection. As a result, the degradation of chlorophyll depending on the intensity of illumination leads to the typical bleaching symptoms in plants. Thus, the herbicidal effect caused by all compounds which inhibit the formation of cyclic carotenoids is the decline of photosynthetic activity.

Any enzyme involved in the reaction sequence not only to lycopene with the full length double-bond system, but also to bicyclic α- and β-carotene is a potential herbicide target (Fig. 1). It has been shown that inhibition of lycopene cyclase resulted in the same bleaching effect, as does the inhibition of phytoene desaturase (Windhövel et al. 1997). A single compound is known to inhibit phytoene synthesis. In vitro studies demonstrated that squalestatin, a well-known inhibitor of squalene synthase (Sidebottom et al. 1992), is a non-competitive inhibitor of phytoene synthase with an I_{50} value of 15 μM (Neudert et al. 1998). Commercially important bleaching herbicides are found among the phytoene desaturase inhibitors. Prominent compounds are norflurazon (4-chloro-5-methylamino-2-(3-trifluoromethylphenyl)-pyridazin-3(2H)one) used in cotton, fluridone (1-methyl-3-phenyl-5-(3-trifluoromethylphenyl)-4(1H)-pyridinone) for aquatic weed control and diflufenican (*N*-(2,4-difluorophenyl)-2-(3-trifluoromethylphenoxy)-nicotinamide) which is a

For bleaching herbicides affecting the 4-hydroxyphenylpyruvate dioxygenase (HPPD-inhibitors) see Chapter 10, pp. 221–229.

P. Böger, K. Wakabayashi, K. Hirai (Eds.)
Herbicide Classes in Development

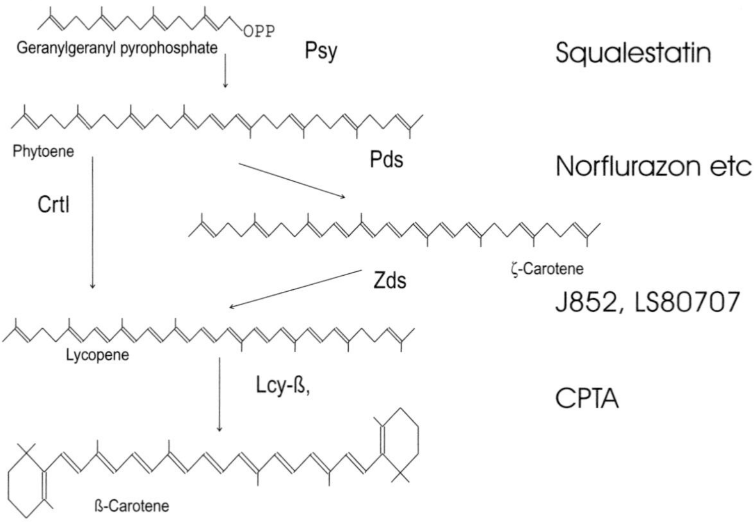

Fig. 1. Reaction sequence of β-carotene formation. The desaturation steps are carried out by a single enzyme CrtI in bacteria (except cyanobacteria) and by two subsequent enzymes *Pds* and *Zds* in plants. Inhibitors of the individual enzymes are indicated

successful commercial product for the control of *Galium* and other broad-leaved weeds in cereals. The in vitro I_{50} values of highly active phytoene desaturase inhibitors are in the range of 0.01–0.1 μM (Sandmann and Böger 1992). Two different ζ-carotene desaturase inhibitor structures are available, LS 80707 (ethyl-*cis*-5-methyl-6-ethyl-2-phenyl-5,6-dihydropyran-4-one-carboxylate) and pyrimidine derivatives like J852 (4-(3-methyl-propoxy)-2-isopropylamino-6-methylpyrimidine). CPTA (2-(4-chlorophenylthio)-triethyl amine HCl) and analogues are the only potent lycopene cyclase inhibitors. For the 4-methylphenylthio derivative, it was shown that substituted amines are non-competitive inhibitors of the enzyme with respect to the substrate lycopene (Schnurr et al. 1998).

2.2 Interaction of Inhibitors with Carotene Desaturation

In plants, two highly homologous desaturases are involved in the desaturation sequence from phytoene to lycopene, a phytoene desaturase, PDS and a ζ-carotene desaturase, ZDS (Sandmann 2001a). They are phylogenetically related to the corresponding desaturases of cyanobacteria. All other carotenogenic bacteria including fungi possess a single desaturase, CRTI, which carries out all four desaturation steps from phytoene to lycopene (Sandmann 1994). Different functional types of inhibitors exist which cause the accumulation of phytoene in plants. Among them are the classical herbicides which directly interact with phytoene desaturase and also compounds like benzoylcyclohexanediones. They are inhibitors of *p*-hydroxyphenylpyruvate dioxygenase, an enzyme in the biosynthetic pathway to quinones (Schulz et al. 1993). The relevance of decreased quinone formation on the phytoene desaturation reaction will be discussed later. Discrimination between both functional types of phytoene-accumulating compounds is possible only when the inhibition of the enzymatic reactions is analyzed. Carotenogenic enzymes including phytoene desaturase are integral membrane proteins and therefore difficult to isolate from plants and microorganisms and to determine in vitro. However, a cell-free assay was developed by functional expression of a plant-type phytoene desaturase in *Escherichia coli* (Sandmann et al. 1996). Another transformant was used to synthesize phytoene, the substrate of the reaction. This in vitro system is useful for enzyme kinetic studies in the presence of inhibitors and for structure-activity investigations. A similar in vitro assay system with overexpressed enzymes has been developed for ζ-carotene desaturase (Breitenbach et al. 1999), and lycopene cyclase (Schnurr et al. 1998). For inhibitor studies with *p*-hydroxyphenylpyruvate dioxygenase, the assay was adapted to a plant enzyme from maize which was more than 100-fold purified (Barta and Böger 1996). It has been shown for several derivatives that benzoylcyclohexanediones are competitive inhibitors of *p*-hydroxyphenylpyruvate dioxygenase with K_i values in the nM range.

A third type of bleaching with simultaneous phytoene accumulation is known from the *Arabidopsis* "immutans variegation" mutant (Wetzel et al. 1994). The enzyme target of this mutation has been identified recently. It is a plastidic alternative oxidase (Carol et al. 1999; Wu et al. 1999). Its participation in carotenoid desaturation in early developmental stages has been shown. Therefore, this alternative may be a novel target for the design of a different functional type of bleaching herbicides.

In cyanobacteria, algae and higher plants, desaturation of phytoene and ζ-carotene proceeds via hydrogen abstraction forming two double bonds at each side of the symmetrical molecule (Fig. 1). It has been shown that nicotinamide adenine dinucleotide (NAD) and nicotinamide adenine dinucleotide phosphate (NADP) are acceptors for hydrogen in the in vitro reaction (Schneider et al. 1997). Recently, it has been found that plastoquinone with a more than 20-fold higher affinity as NADP is an even better cofactor for phytoene desaturase as indicated by the K_m values (Breitenbach et al. 2001). Participation of plastoquinone in ζ-carotene desaturation has already been shown before (Breitenbach et al. 1999). In vitro enzyme kinetic studies with phytoene desaturase revealed that the type of inhibition by different bleaching herbicides is noncompetitive with respect to the substrate phytoene (Sandmann et al. 1989; Kowalczyk-Schröder and Sandmann 1992). However, competition with the inhibitors was observed for the cofactors. This was the case either for NADP (Ogawa et al. 2001) or for plastoquinone (Breitenbach et al. 2001). Figure 2 shows a double-reciprocal Lineweaver-Burke plot of phytoene desaturase activity versus the concentration of decyl plastoquinone. One set of experi-

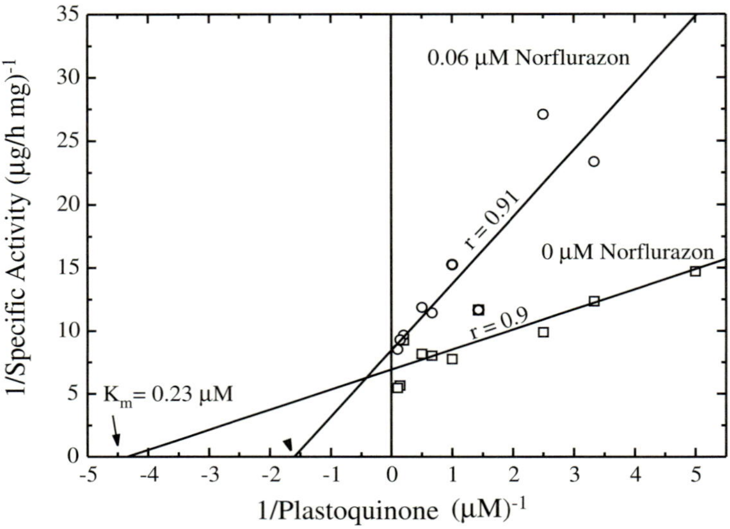

Fig. 2. Enzyme kinetic analysis of interaction of norflurazon with the cofactor decyl plastoquinone at the *Synechococcus* phytoene desaturase

ments was carried out in the absence, the other in the presence of norflurazon. Both graphs intercept close to the *y*-axis, which indicates competition between plastoquinone and the inhibitor.

The prominent role of plastoquinone in phytoene desaturation explains why treatment with *p*-hydroxyphenylpyruvate dioxygenase inhibitors or mutation of a plastidic alternative oxidase impairs phytoene desaturation and accumulates phytoene. Figure 3 presents a model of the desaturation reaction including regeneration of plastoquinone in the chloroplast. For desaturation of phytoene and ζ-carotene the presence of plastoquinone is important as cofactor. Thus, inhibition of plastoquinone synthesis should have a negative effect on desaturation of phytoene. Reduced plastoquinone must be regenerated. It can be oxidized by the photosynthetic electron transport (PET) or by an alternative oxidase (AO; Josse et al. 2000). The latter enzyme must operate as a quinol:oxygen oxidoreductase in early developmental stages of the leaf when carotenoid biosynthesis is important to produce cyclic carotenoids as an essential component for the construction of the photosynthetic apparatus. Therefore, the inactivation of this plastoquinol oxidase prevents phytoene desaturation in developing leaves.

The reaction mechanism of ζ-carotene desaturation is very similar to phytoene desaturation. Furthermore, it could be demonstrated that phytoene desaturase inhibitors are also able to inhibit ζ-carotene desaturase and vice versa (Simkin et al. 2000). The pyrimidine compound J852 inhibited both enzymes with similar K_i values, whereas norflurazon was more than 1000-fold less effective as an inhibitor of ζ-carotene desaturase.

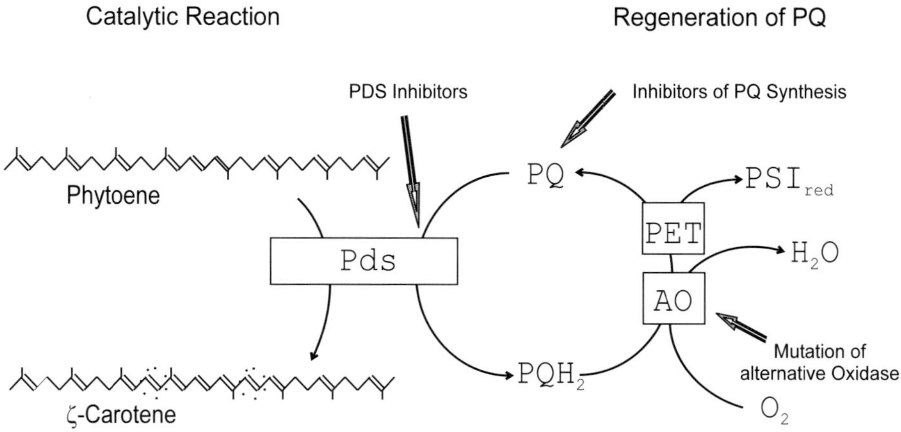

Fig. 3. Inhibition targets in the catalytic reaction of phytoene desaturase (*PDS*) and the regeneration of plastoquinone (*PQ*). *PET* photosynthetic electron transport; *AO* alternative oxidase; *PSI* photosystem I

2.3 Structural Requirements for an Inhibitor of Phytoene Desaturase

An extensive review on the relationship of chemical structure and herbicidal activity of phytoene desaturase inhibitors has been presented several years ago (Sandmann and Böger 1992). Since then, novel compounds have been identified as phytoene desaturase inhibitors. Among them are 2,6-diphenylpyridines (Kawamura et al. 1991; Babczinski et al. 1992), diarylpyrimidines (Kawamura et al. 1993), tetrahydropyrimidinones (Babczinski et al. 1995a), 1,1′-biphenyl derivatives (Laber et al. 1999), phenylpyrimidinones (Burdge 2000) and diphenylpyrrolidinones (Ogawa et al. 2001).

Typical and efficient bleaching compounds possess a central five- or six-membered heterocycle carrying one or two substituted phenyl rings. Several resistant mutants which were selected against bleaching herbicides exhibit different cross-resistance to other bleachers (Chamovitz et al. 1993; Sandmann and Fraser 1993). This finding is a very strong indication for a common binding site of phytoene desaturase for most chemical classes of bleaching herbicides. Attempts have been made to construct hypothetical models for this binding site by overlay of selected structurally dissimilar inhibitors (Mitchell 1995; Laber et al. 1999). In three-dimensional models, steric requirements of inhibitors have been suggested, in order to fit into the binding niche or lipophilic regions of the binding site for interaction with peripheral substituents of bleaching herbicides. Recently, the influence of modifications of the six-membered central heterocycle on the inhibition of phytoene desaturase activity has been investigated by comparison of inhibitory activities of substituted diphenyl pyrimidines, pyridines, 4(3H)-pyrimidinones and triazines (Sandmann 2001b). It has been proposed that inhibitory activity is influenced by the negative charge of the atoms in the heterocycle to which both phenyl rings are connected. By comparison of fluridone derivatives it has already been concluded that the electron density of the substituent at position 4 determines the inhibitory activity (Sandmann and Böger 1992). An overlay of the structures reveals that the keto group in compounds like fluridone can be replaced by the imine groups of various heterocycles (Babczinski et al. 1992). Other influences of the conjugated system of the heterocycle seem to be minor since the nonaromatic tetrahydropyrimidinones (Babczinski et al. 1995b) and ketomorpholines (Sandmann and Mitchell 2001) are also very potent inhibitors of phytoene desaturase. In many bleachers the substituent at the six-membered central ring opposite the keto group or an electronically equivalent N imine (Babczinski et al. 1992) is sterically well defined. It was suggested that either a methyl group at the para position as in phenylpyrimidinones and phenyltetrahydropyrimidines or a meta methylamino group as in phenylpyrimidazinones which cover a similar space results in an optimum inhibition of phytoene desaturase (Sandmann and Böger 1992). The comparison of compounds with a modified central ring showed that the optimum inhibition of phytoene desaturase is reflected by a similar diagonal length from the negatively charged region across the central heterocycle. Another indication for this spatial requirement is the stereospecific inhibitory property of certain substituents of the heterocycle in ketomorpholines (Sandmann and Mitchell 2001) and diphenylpyrrolidinones (Ogawa et al. 2001).

In Fig. 4 all known characteristics of an optimized inhibitor of phytoene desaturase are summarized. Most known bleaching compounds fit quite well as for example (Fig. 4, right) fluridone, flurochloridone, 2-phenyl-4-(3-chlorophenyl)-5-ethylaminopyrimidine (PCEP; Sandmann 2001b) and the 2-(3-trifluoromethylphenyl)-4-methylthio-6-(3-chlorophenyl)-pyridine (TMCP; Babczinski et al. 1992). Other compounds like diflufenican are more difficult to put into this general model. In the case of diflufenican (Fig. 4, right), the carbonyl group of the central heterocycle may be represented by the carboxyimine structure. The diagonal length across the cycle B including this carboxy group would be similar in the other selected molecules (Fig. 4, right). A region of a total of six C- and hetero-atoms (shaded in grey) spans from one end of the substituted central heterocycle with CO or N=C group to the opposite side. In the nonplanar heterocycle of ketomorpholines the optimum length is only five atoms (Sandmann and Mitchell 2001) with the 5-methyl group in the (5S)-form. In the diflufenican molecule, the 3-trifluoromethylphenoxy group would represent ring A and the 2,4-difluorophenyl group ring C of Fig. 4. Finally, the question may be asked whether inhibitors of phytoene and ζ-carotene desaturases can be modeled as analogues of plastoquinone since they are competitive inhibitors of the latter.

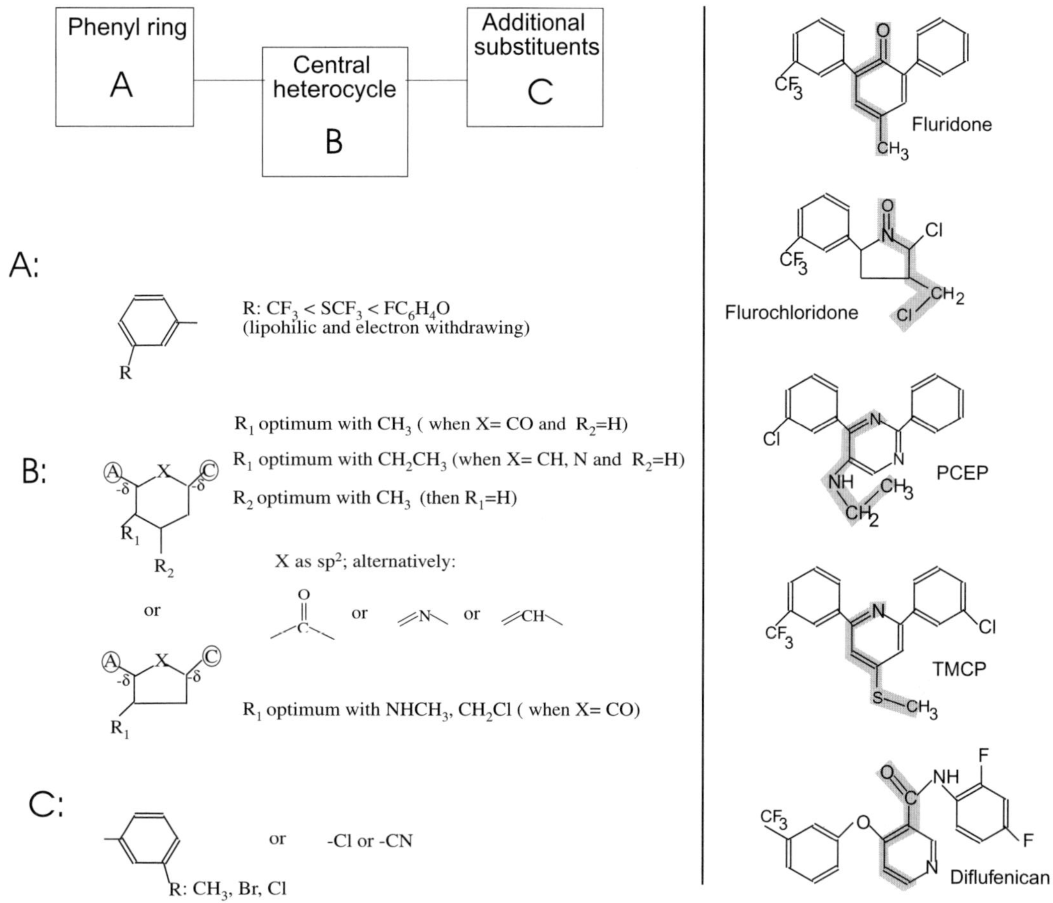

Fig. 4. Common structural elements of inhibitors of phytoene desaturase (*left*) and structures of selected inhibitors (*right*). The diagonal spanning region across the central heterocycle is *shaded*

Structural similarities can be found between some photosystem Inhibitors and phytoene desaturase inhibitors which all interfere at a plastoquinone binding site. The best example is fluometuron which has in vitro I_{50} values of 2 μM as an inhibitor of photosystem II and of 10 μM as a phytoene desaturase inhibitor (Sandmann and Böger 1992). With a series of substituted phenoxyphenyl derivatives the herbicidal functions including inhibition of photosynthetic electron transport and phytoene desaturase activity could be correlated to essential substituents and their positions in the molecule.

2.4
Strategies for Genetic Engineering of Herbicide Resistance by Modification of the Carotenogenic Pathway

Screening new inhibitors of metabolic reactions which exert herbicidal effects on plants is not a big problem. However, one of the crucial steps in the development of new herbicides is the selectivity; finding crop plants which are less affected than their weeds. This tolerance of a plant may be due to different mechanisms. One reason may be a decreased herbicide concentration at the target site. This can be caused by restricted uptake and translocation or by degradation of the herbicide or modification including conjugation to an inactive compound. A second possibility is a modification at the target enzyme. The target enzyme may be mutated to a less susceptible protein or may be overproduced. Higher concentrations of the target enzyme results in an increased binding of the inhibitor. This leads to a decreased concentration of unbound herbicide at the target enzyme. Due to our increasing understanding of plant carotenogenesis at the molecular level, it is now possible to engineer the target enzymes for bleaching herbicides in different ways to obtain tolerant plants. The different strategies which were followed successfully will be described in the next sections.

2.4.1
Overexpression of a Susceptible Lycopene Cyclase in *Synechococcus*

Overexpression of a susceptible enzyme has been achieved for lycopene cyclase in *Synechococcus* (Windhövel et al. 1997). The gene of a bacterial lycopene cyclase which is inhibited by triethyl amine derivatives like the plant lycopene cyclases was placed under a strong promoter and integrated into the genome. The wild type and the resulting transformants were treated with 2-(4-chlorophenylthio)-methyldiethylamine hydrochloride. I_{50} values for inhibition of cyclic carotenoid formation in the wild type was around 0.83 μM. For the transformant, a value of 31 μM was obtained. In this case, a 37-fold resistance was achieved simply by increasing the synthesis of the target enzyme.

2.4.2
Selection of Mutants with Resistant Phytoene Desaturase and Gene Transfer into Tobacco

Plant species resistant against bleaching herbicides are unknown to date. Therefore, unicellular cyanobacteria are very convenient for generation of mutants with a herbicide-resistant phytoene desaturase. Various lines of resistant mutants of *Synechococcus* (Linden et al. 1990; Chamovitz et al. 1993; Sandmann and Fraser 1993) and *Synechocystis* (Martinez-Férez et al. 1994) have been selected. The nature of the mutation was characterized by combined in vivo and in vitro analysis of carotenogenesis in the presence of bleaching herbicides. In most of the mutants the affinities for norflurazon binding to phytoene desaturase are decreased. I_{50} values for inhibition of carotenoid formation in whole cells were compared with K_i values, the dissociation constant of the enzyme-inhibitor complex measured in vitro. Similar K_i and I_{50} values indicate that the affinity for norflurazon binding to the enzyme phytoene desaturase is decreased. If protein overproduction is the reason for resistance in the mutants, the K_i values of the mutants were expected to be in the range of the value for the wild-type enzyme because K_i in the in vitro assays is independent of the enzyme concentration.

Table 1 compiles the characteristics of some of the *Synechococcus* mutants (Linden et al. 1990; Chamovitz et al. 1993). The mutants selected against norflurazon (NFZ) all exhibit a different degree of resistance with a factor of resistance of up to around 70. In addition to the strong resistance to the selection herbicide norflurazon, each of the mutant strains exhibited a unique cross-resistance profile, either to fluridone, diflufenican, difunone and flurochloridone. Mutant NFZ49 exhibited no cross-resistance to any other bleaching herbicide. The different patterns of herbicide cross-resistance found for the mutants indicate that the inhibitors of phytoene desaturase may be attached

Table 1. Characterization of *Synechococcus* mutants resistant against phytoene desaturase herbicides

Mutant	I_{50}	(FR)	K_i	(FR)	Cross-resistance (FR > 3)	Sequence modification
NFZ[a]4	4.56	(41.5)	3.28	(36.4)	FCD	Val 403→Gly
NFZ20	1.12	(10.2)	0.72	(8.0)	DFF, DFN, FCD	Leu 320→Pro
NFZ42	2.42	(22.0)	2.13	(23.7)	FCD	Arg 195→Pro
NFZ49	7.74	(70.4)	6.88	(76.4)	None	Leu 436→Arg
FRD[b]5	0.16	(6.7)	0.03	(1.0)	DFF, DFN, FCD, FTM	promoter structure

WT: I_{50}/K_i values (μM) for norflurazon (NFZ) 0.11/0.09, fluridone (FRD) 0.02/0.02. FCD, flurochloridone; DFF, diflufenican; DFN, difunone

[a] All NFZ strains were selected against norflurazon.

[b] The FRD strain against fluridone.

to the same binding region, but interact with different amino acids. Resistance could be attributed to sequence modifications in the gene for phytoene desaturase of the NFZ mutants. Single amino acid exchanges were found across the coding region. Mutants resistant against norflurazon were also selected from another cyanobacterium, *Synechocystis* (Martinez-Férez et al. 1994). In these mutants, several single point mutations all modified the same amino acid at position 195 of the phytoene desaturase polypeptide to other amino acids. For the *Synechococcus* NFZ mutants it was shown that with an increasing degree of resistance the activity of the modified phytoene desaturases declined (Chamovitz et al. 1993). The mutated strains synthesized lower amounts of colored carotenoids which had a negative impact on photosynthetic oxygen evolution (Sandmann et al. 1993).

In mutant FD5, resistance was only evident in vivo. The K_i value determined enzymatically was quite similar to that of the wild type. For FD5, genetic analysis revealed a deletion in the promoter region that contains the putative −35 and −10 transcription-regulating elements. Resistance in this strain could be a result of an overexpression of phytoene desaturase, the target protein. The amount of phytoene desaturase in cells of FD5, as determined with an antiserum, was at least 20-fold higher than in cells of the wild-type strain.

The gene of the norflurazon and fluridone resistant phytoene desaturase was cloned from mutant NFZ4. It was extended with a transit sequence for plastid import and put under the control of a constitutive promoter before it was used for transformation of tobacco (S. Römer, Universität Konstanz, unpubl. results). The resulting transformants were resistant against both norflurazon and fluridone. The I_{50} for norflurazon of the homozygote seedlings was 2 μM compared to 0.06 μM for wild-type tobacco which indicates a 12-fold higher resistance. I_{50} values for fluridone were 15 μM in the transformant and 6.5 in the control (=2.3-fold increased resistance). These values correspond more or less with the factors of resistance for the *Synechococcus* NFZ4 mutant which are 24 for norflurazon and 4 for fluridone (Table 1).

2.4.3 Naturally Resistant Phytoene Desaturase from Bacteria and Genetic Engineering of a Resistant Tobacco

During the evolution of carotenoid biosynthesis, two completely different phytoene desaturases with unrelated amino acid sequences have been acquired. In contrast to the plant-type phytoene desaturase PDS which also exists in algae and cyanobacteria, the bacterial CRTI type is also a ζ-carotene desaturase, catalyzing a four-step desaturation of phytoene to lycopene (Fig. 1). Another major difference is the cofactor requirement which is flavin adenine dinucleotide (FAD) for the bacterial enzyme instead of quinone for the plant phytoene desaturase. Since bleaching enzymes compete for plastoquinone at the plant enzyme, the bacterial enzyme should be unaffected by these herbicides. Inhibition experiments are shown in Fig. 5 for plant-type PDS (left) and CRTI

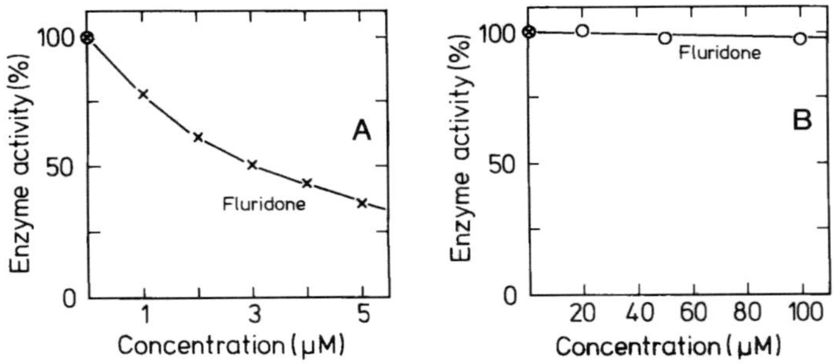

Fig. 5. In vitro inhibition of expressed plant-type phytoene desaturase PDS (**A**) and bacterial CrtI (**B**) by fluridone

from bacteria (right). As expected from in vivo experiments, isolated PDS is susceptible to fluridone in μM concentrations. In contrast, concentrations of up to 100 μM fluridone have no significant inhibitory effect on the bacterial phytoene desaturase.

Engineering of resistance against bleaching herbicides by transformation of a plant with the naturally resistant bacterial phytoene desaturase gene was realized for tobacco (Misawa et al. 1993). Since the endogenous phytoene desaturase of tobacco is located and operates in the thylakoid membranes, the 5′-region of the bacterial phytoene desaturase gene was fused to the sequence for a transit region to ensure plastid import and placed under the control of a constitutive plant-specific promoter. Several regenerated tobacco transformants were isolated and analyzed. The chimeric gene product was expressed and processed in the transgenic plants; the production and processing of the corresponding protein upon chloroplast import could be demonstrated by Western blotting. Immunogold localization showed the targeting in the chloroplasts and the location of the gene product CRTI in the thylakoids (Misawa et al. 1993). The foreign desaturase was enzymatically active in tobacco. Its expression prevented bleaching by norflurazon even at concentrations of 50 times the I_{50} concentration for wild-type tobacco. Thus, when the endogenous phytoene desaturase is inhibited by a bleaching herbicide, the foreign bacterial enzyme effectively takes over the catalysis of the desaturation reactions. For the CRTI transformants, no impairment of growth nor any indication of reduced fitness was observed.

An interesting feature of the CRTI-tobacco is its cross-resistance against many individual herbicidal phytoene desaturase inhibitors (Misawa et al. 1994). The effect of norflurazon, fluridone, flurtamone, flurochloridone and diflufenican on the formation of colored carotenoids in wild-type tobacco and the CRTI transformant is shown in Table 2. Due to the proposed mechanism of desaturation (Fig. 3), no resistance against benzoylcyclohexanediones was obtained. Since the foreign bacterial phytoene desaturase comprises ζ-

Table 2. I_{50} values (M) for inhibition of colored carotenoid formation in tobacco

Herbicide	Control	CrtI
Norflurazon	3.3×10^{-7}	$>10^{-4}$
Fluridone	2.3×10^{-8}	1.2×10^{-5}
Flurtamone	4.7×10^{-8}	$>10^{-4}$
Flurochloridone	1.8×10^{-7}	$>10^{-4}$
Diflufenican	3.8×10^{-8}	$>10^{-4}$
CH309	4.2×10^{-8}	4.8×10^{-8}
J852	1.3×10^{-6}	$>10^{-4}$
LS80707	7.4×10^{-7}	$>10^{-4}$

carotene desaturase activity (Fig. 1), the CRTI-tobacco plants were also resistant against the ζ-carotene desaturase inhibitors J852 and LS80707.

The universal application of the bacterial phytoene desaturase gene to engineer resistance against bleaching herbicides was demonstrated also for cyanobacteria. CRTI was integrated into the genome of *Synechococcus* (Windhövel et al. 1994). Analysis of herbicide resistance with transgenic cultures showed that their content of colored carotenoids was unaffected by herbicide concentrations in the μM range.

2.5
Conclusion and Perspectives

The tremendous progress in molecular genetics of the carotenoid biosynthetic pathway in the recent years had a deep impact on our understanding of the mode of action of bleaching herbicides.

The cloning of carotenogenic genes opened the possibility to modify the carotenoid biosynthetic pathways in plants. Different genetic engineering strategies resulted in transgenic plants which were highly resistant against several bleaching herbicides which target phytoene desaturase and ζ-carotene desaturase.

The membrane-bound phytoene and ζ-carotene desaturases which are very difficult to isolate as functional enzymes from plant tissue were heterologously expressed in an active form and characterized for their reaction mechanism and inhibitor susceptibility. Our understanding of the catalytic action especially the competition of the inhibitors with the cofactor plastoquinone gives a first orientation for rational design of new bleaching compounds.

The highly active heterologous PDS is the basis for much simpler in vitro assays compared to the use of plant tissue as an enzyme source. This in vitro system can now be applied to structure-activity investigations in order to identify common structural elements of PDS inhibitors which are essential for interaction at the enzyme target site.

Overexpression of PDS in a heterologous host and purification of the recombinant protein will supply sufficient enzyme material to attempt protein crystallization with infiltrated herbicides. This approach should lead to an exact three-dimensional model of the herbicide binding region. Once achieved, the resulting knowledge will provide a solid basis for successful rational design of new bleaching herbicides. Furthermore, it will be possible to tailor resistance against individual bleaching herbicides into phytoene desaturases by site-directed mutagenesis.

References

Babczinski P, Heinemann U, Sandmann G, Kawamura S, Hamada T, Sato R, Sanemitsu Y (1992) Inhibition on carotenoid biosynthesis by interaction of 2,6-diphenyl-pyridine derivatives with phytoene desaturation. J Agric Food Chem 40:2497–2499

Babczinski P, Sandmann G, Schmidt RR, Shiokawa K, Yasui K (1995a) Substituted tetrahydropyrimidinones: a new class of compounds inducing chlorosis by inhibition of phytoene desaturase. 1. Biochemical and biological results. Pestic Biochem Physiol 52:33–44

Babczinski P, Blunk M, Sandmann G, Shiokawa K, Yasui K (1995b) Substituted tetrahydropyrimidinones: a new herbicidal class of compounds inducing chlorosis by inhibition of phytoene desaturation. 2. Structure-activity relationships. Pestic Biochem Physiol 52:45–59

Barta IC, Böger P (1996) Purification and characterization of 4-hydroxyphenylpyruvate dioxygenase from maize. Pestic Sci 48:109–116

Breitenbach J, Kuntz M, Takaichi S, Sandmann G (1999) Catalytic properties of an expressed and purified higher plant type ζ-carotene desaturase from *Capsicum annuum*. Eur J Biochem 265: 376–383

Breitenbach J, Zhu C, Sandmann G (2001) The bleaching herbicide norflurazon inhibits phytoene desaturase by competition with the cofactors. J Agric Food Chem 49:5270–5272

Burdge EL (2000) The mode of action of RH-1965: a new phenylpyrimidinone bleaching herbicide. Pestic Manage Sci 56:245–248

Carol P, Stevenson D, Bisanza C, Breitenbach J, Sandmann G, Mache R, Coupland G, Kuntz M (1999) Mutations in the *Arabidopsis* gene *immutans* cause a variegated phenotype by inactivating a chloroplast terminal oxidase associated with phytoene desaturation. Plant Cell 11: 57–68

Chamovitz D, Sandmann G, Hirschberg J (1993) Molecular and biochemical characterization of herbicide-resistant mutants of cyanobacteria reveals that phytoene desaturation is a rate-limiting step in carotenoid biosynthesis. J Biol Chem 268:17348–17353

Demmig-Adams B, Gilmore AM, Adams WW (1996) In vivo function of carotenoids in higher plants. FASEB J 10:403–412

Josse E-M, Simkin AJ, Gaffé J, Labouré A-M, Kuntz M, Carol P (2000) A plastid terminal oxidase associated with carotenoid desaturation during chromoplast differentiation. Plant Physiol 123:1427–1436

Kawamura S, Hamada T, Sato R, Sanemitsu Y (1991) 2,6-Diphenylpyridines: a new class of herbicides. J Agric Food Chem 39:2279–2281

Kawamura S, Sato J, Hamada T, Sakaki M, Sanemitsu Y (1993) Synthesis and herbicidal activity of some 2,4-diarylpyrimidines. J Agric Food Chem 41:288–291

Kowalczyk-Schröder S, Sandmann G (1992) Interference of fluridone with the desaturation of phytoene in membranes of the cyanobacterium *Aphanocapsa*. Pestic Biochem Physiol 42:7–12

Laber B, Usunow G, Wiecko E, Franke W, Franke H, Köhn A (1999) Inhibition of *Narcissus pseudonarcissus* phytoene desaturase by herbicidal 3-trifluoromethyl-1,1′-biphenyl derivatives. Pestic Biochem Physiol 63:173–184

Linden H, Sandmann G, Chamovitz D, Hirschberg J, Böger P (1990) Biochemical characterization of *Synechococcus* mutants selected against the bleaching herbicide norflurazon. Pestic Biochem Physiol 36:46–51

Martinez-Ferez I, Vioque A, Sandmann G (1994) Mutagenesis of an amino acid responsible in phytoene desaturase from *Synechocystis* for binding of the bleaching herbicide norflurazon. Pestic Biochem Physiol 48:185–190

Misawa N, Yamano S, Linden H, de Felipe MR, Lucas M, Ikenaga H, Sandmann G (1993) Functional expression of the *Erwinia uredovora* carotenoid biosynthesis gene *crtI* in transgenic plants showing an increase of β-carotene biosynthesis activity and resistance to the bleaching herbicide norflurazon. Plant J 4:833–840

Misawa N, Masamoto K, Hori T, Ohtani T, Böger P, Sandmann G (1994) Expression of an *Erwinia* phytoene desaturase gene not only confers multiple resistance to herbicides interfering with carotenoid biosynthesis, but also alters xanthophyll metabolism in transgenic plants. Plant J 6:481–489

Mitchell G (1995) Phytoene desaturase. A model for the optimization of inhibitors. In: Baker DA, Fenyes JG, MobergWK, Cross B (eds) Synthesis and chemistry of agrochemicals IV. ACS Symposium Series, vol 584. Am Chem Soc, Washington, DC, pp 161–170

Neudert U, Martinez-Ferez I, Fraser PD, Sandmann G (1998) Expression of an active phytoene synthase from *Erwinia uredovora* and biochemical properties of the enzyme. Biochim Biophys Acta 1392:51–58

Ogawa H, Yamada I, Arai K, Hirase K, Moriyasu K, Schneider C, Sandmann G, Böger P, Wakabayashi K (2001) Mode of bleaching phototoxicity of herbicidal diphenylpyrrolidinones. Pestic Manage Sci 57:33–40

Sandmann G (1994) Carotenoid biosynthesis in microorganisms and plants. Eur J Biochem 223:7–24

Sandmann G (2001a) Carotenoid biosynthesis and biotechnological application. Arch Biochem Biophys 385:4–12

Sandmann G (2001b) Bleaching activities of substituted pyrimidines and structure-activity comparison to related heterocyclic derivatives. Pestic Biochem Physiol (in press)

Sandmann G, Böger P (1992) Chemical structure and activity of herbicidal inhibitors of phytoene desaturase. In: Draber W, Fujita T (eds) QSAR in the development of agricultural chemicals. CRC Press, Boca Raton, pp 357–371

Sandmann G, Fraser PD (1993) Differential inhibition of phytoene desaturase from diverse origins and analysis of resistant cyanobacterial mutants. Z Naturforsch 48c:307–311

Sandmann G, Mitchell G (2001) In vitro inhibition studies of phytoene desaturase by bleaching ketomorpholine derivatives. J Agric Food Chem 49:138–141

Sandmann G, Linden H, Böger P (1989) Enzyme-kinetic studies on the interaction of norflurazon with phytoene desaturase. Z Naturforsch 44c:787–790

Sandmann G, Kuhn M, Böger P (1993) Carotenoids in photosynthesis: Protection of D1 degradation in the light. Photosyn Res 35:185–190

Sandmann G, Schneider C, Böger P (1996) A new nonradioactive assay of phytoene desaturase to evaluate bleaching herbicides. Z Naturforsch 51c:534–538

Schneider C, Böger P, Sandmann G (1997) Phytoene desaturase: heterologous expression in an active state, purification, and biochemical properties. Prot Exp Purif 10:175–179

Schnurr G, Böger P, Sandmann G (1998) Interaction of 2-(4-methylphenoxy)-triethylamine and related compounds with its herbicide target in the carotenoid biosynthetic pathway. J Pest Sci 23:113–117

Schulz A, Ort O, Beyer P, Kleinig H (1993) SC-0051, a 2-benzoylcyclohexane-1,3-dione bleaching herbicide, is a potent inhibitor of the enzyme p-hydroxyphenylpyruvate dioxygenase. FEBS Lett 318:162–166

Sidebottom PJ, Highcock RM, Lane SJ, Procopiou PA, Whatson NS (1992) The squalestatins, novel inhibitors of squalene synthase produced by a species of *Phoma*. II. Structure elucidation. J Antibiot 45:648–658

Siefermann-Harms D (1987) The light-harvesting and protective functions of carotenoids in photosynthetic membranes. Physiol Plant 69:561–568

Simkin AJ, Breitenbach J, Kuntz M, Sandmann G (2000) In vitro and in situ inhibition of carotenoid biosynthesis in *Capsicum annuum* by bleaching herbicides. J Agric Food Chem 48:4676–4680

Tracewell CA, Vrettos JS, Bautista JA, Frank HA, Brudvig GW (2001) Carotenoid photooxidation in photosystem II. Arch Biochem Biophys 385:61–69

Wetzel CM, Jiang C-Z, Meehan, LJ, Voytas DF, Rodermel SR (1994) Nuclear-organelle interactions: the *immutans* variegation mutant of *Arabidopsis* is plastid autonomous and impaired in carotenoid biosynthesis. Plant J 6:161–175

Windhövel U, Geiges B, Sandmann G, Böger P (1994) Expression of *Erwinia uredovora* phytoene desaturase in *Synechococcus* PCC7942 leading to resistance against a bleaching herbicide. Plant Physiol 104:119–125

Windhövel U, Sandmann G, Böger P (1997) Genetic engineering of resistance to bleaching herbicides affecting phytoene desaturase and lycopene cyclase in cyanobacterial carotenogenesis. Pestic Biochem Physiol 57:68–78

Wu D, Wright DA, Wetzel C, Voytas DF, Rodermel S (1999) The *immutans* variegation locus of *Arabidopsis* defines a mitochondrial alternative oxidase homolog that functions during early chloroplast biogenesis. Plant Cell 11:43–55

3 Inhibitors of Aromatic Amino Acid Biosynthesis (Glyphosate)

DONALD R. GEIGER and MARK A. FUCHS

3.1 Introduction

Glyphosate (Glp), (*N*-phosphonomethyl)glycine, is a nonselective, broad spectrum herbicide discovered in 1971 (Baird et al. 1971) and introduced in 1974 (Franz et al. 1997). Effectiveness, along with outstanding environmental and safety qualities, has made Glp one of the most successful commercial herbiides. Essentially nontoxic to mammals, birds, fish, insects, and most bacteria, the herbicide is readily broken down in soil to ammonia, carbon dioxide and inorganic phosphate (Franz et al. 1997; Giesy et al. 2000; Williams et al. 2000). Glp was one of the first commercially important herbicides whose site of action was characterized as a single, defined target enzyme in plants, and is the only herbicide known to inhibit 5-enolpyruvylshikimate 3-phosphate synthase (EPSPS) (phosphoenolpyruvate: 3 phosphoshikimate 1-carboxyvinyl transferase; E.C. 2.5.1.19) which catalyzes the penultimate reaction of the shikimate (Shk) pathway (Amrhein et al. 1980, 1982; Steinrücken and Amrhein 1980; Gruys and Sikorski 1999) in certain bacteria and plants. Because of these unique favorable characteristics, Glp has become and is likely to continue to be one of the most widely used and studied herbicides in the world (Woodburn 2000).

Glp is a systemic herbicide, which moves throughout the plant in the phloem (Gougler and Geiger 1981; Devine 1989) and accumulates in translocation sinks (Schulz et al. 1990) including regions of growth, carbon accumulation and high metabolic activity (Sprankle et al. 1975). Uptake and phloem mobility appear to be well accounted for by the intermediate permeability model, which was proposed by Tyree et al. (1979) and developed by Kleier (1988) and Hsu et al. (1988). The mode of action of Glp, the sequence of events leading to injury and destruction of a plant following exposure to Glp, starts within minutes but develops gradually over days to weeks. Stunted growth, tissue chlorosis, leaf abscission, and tissue necrosis are characteristic indications of Glp toxicity. Early studies of physiological effects of Glp focused on the disruption of the Shk pathway with particular emphasis on the disruption of aromatic amino acid synthesis (Jaworski 1972; Holländer and Amrhein 1980). Subsequent studies in sugar beet revealed that, in addition to the arrest of protein synthesis (Cole et al. 1980), Glp also causes inhibition of photosynthetic carbon metabolism (Servaites et al. 1987). The relative amount of damage attributable

P. Böger, K. Wakabayashi, K. Hirai (Eds.)
Herbicide Classes in Development

to each of these separate, yet complementary, processes remains to be resolved in various plants and biotypes (Geiger and Bestman 1990; Siehl 1997).

As our knowledge of enzyme reaction mechanisms has improved, de novo design of potent enzyme inhibitors has increased dramatically (Sikorsky and Gruys 1997), but translation of this knowledge to herbicide development is much more difficult. In spite of a great deal of research that has led to new insights into Glp's mode of action, the mechanism of inhibition at the primary site of action has not been fully characterized nor has the herbicide molecule been improved upon (Duke 1990; Sikorski and Gruys 1997; Maier 2000). In particular, research is needed to understand the metabolic diversity and flexibility of higher plants as these affect herbicidal action from inhibition of the target enzyme to final plant death.

This review focuses on a range of topics including the general timing and nature of Glp-induced symptom development in plants, molecular characteristics of EPSPS and the unique interaction of the enzyme with Glp, which brings about its herbicidal action. The latter two topics provide background for rational design of herbicides based on inhibition of EPSPS. Examination of the diverse metabolic responses to Glp among plants provides a basis for identifying key metabolic disruptions that give rise to the herbicidal effects of Glp. The review concludes by addressing mechanisms of resistance, including procedures used for the development of Glp-resistant plants, and by examining the issue of acquired tolerance.

3.2 Symptoms of Herbicidal Activity

Typical Glp-induced effects on plant organs and processes and their approximate times of appearance are given in Table 1. Glp-induced injury symptoms follow a general pattern starting with gradual development of visible symptoms such as stoppage of growth and yellowing of tissues. Subsequently, symptoms worsen until eventual browning and deterioration of plant shoot and root tissue occurs several days to weeks after Glp application. Timing and relative intensity of specific symptom development depend on a number of factors including the plant species, age and stage of development (Jordan et al. 1997), environmental conditions (Caseley and Coupland 1985), membrane penetration (Hess 1985), dose and concentration (Cranmer and Linscott 1990, 1991), and adjuvants applied (Gaskin and Holloway 1992; Riechers et al. 1994; Laerke and Streibig 1995; Leaper and Holloway 2000). The range and variety of symptoms among species can be seen in reports that describe responses of cocklebur (Nafziger and Slife 1983), tomato (Mollenhauer et al. 1987; Schulz et al. 1990), soybeans, corn, and barley (Kitchen et al. 1981), yellow nutsedge (Cañal Villanueva et al. 1985), Canada thistle (Carlson and Donald 1988), and bean (Brecke and Duke 1980).

Even though noticeable symptom development is usually subtle and relatively slow, the composite of injury symptoms can provide insights into the

Table 1. Injury symptoms typically found in plants in response to a lethal dose of glyphosate

Symptom or Process	
Effects and Targets	**Time of Appearance**
Plant growth	
Stoppage of stem elongation, leaf initiation, and leaf growth	1 to 2 days
Photosynthetic pigment synthesis	
Absence of pigment in new leaves and shoot tips	1 to 2 days
Growth form	
Epinasty of present and new stems and petioles	several days; progressive
Chlorophyll	
Bleaching in both mature and immature leaves	2 to 3 days; progressive
Necrosis	
Tissue death in leaf blades, veins, stem vascular traces, roots	1 to 3 days
Water uptake	
Decrease in uptake rate from soil	1 to 3 days; progressive
Loss of turgor	
Wilting of leaf blades, petioles, stems	several days
Abscission	
Detachment of vegetative and floral buds, leaves	5 to 7 days; progressive
Organ death and dessication	
Dessication of shoot and roots	one to several weeks

lethal consequences of the herbicide on particular biochemical and physiological disturbances occurring within the plant (Hoagland and Duke 1982; Cole et al. 1983). Analysis of distinct symptom development is founded on the belief that the effects of Glp on key biochemical processes differ among plant species. Such contrasts in injury symptoms have been observed between sugar beet (*Beta vulgaris*) and velvetleaf (*Abutilon theophrasti*) as described by Fuchs et al. (unpubl.). Sugar beet is considered to be highly susceptible to the herbicidal action of Glp (Shieh et al. 1991; Madsen et al. 1995; Geiger et al. 1999) while the problem weed species velvetleaf (Mitich 1991) is more tolerant to GlP. Although the primary mode of Glp action, inhibition of EPSPS, is likely to be the same in both species as evidenced by the accumulation of Shk, contrasting physiological responses result in significant dissimilarities in the development of damage symptoms and lethal toxicity between the two species (Fuchs et al., unpubl.).

3.3 Mode of Action of Glyphosate

3.3.1 Overview of the Mode of Action

Mode of action of a herbicide refers to the sequence of events leading to control or to death of a plant following treatment (Franz et al. 1997). Included in the concept are the absorption, translocation, biochemical target(s) and metabolism of the herbicide and the accompanying changes in growth and structure that result from herbicidal action. Unlike contact herbicides which function only locally, Glp acts not only at the site of application but also at sites to which it is translocated throughout the plant (Baird et al. 1971). Once through the cuticle, the permeability of plasma membranes to Glp enables it to enter the phloem (Tyree et al. 1979; Kleier 1988), be translocated throughout the plant and accumulate in translocation sinks along with sucrose (Gougler and Geiger 1981). The herbicide is taken up into cells by passive and possibly by active transport mechanisms (Sterling 1994) and accumulates in cytosol, vacuoles and plastid stroma (Schulz et al. 1990).

Upon entering plastids, Glp disrupts the Shk pathway (Amrhein et al. 1982) by inhibiting EPSPS (Steinrücken and Amrhein 1980), an enzyme that transfers the enolpyruvyl moity from phosphoenolpyruvate (PEP) to shikimate-3-phosphate (S3P) to form EPSP and inorganic phosphate (Fig. 1). The understanding of the biochemical mode of action of Glp is based largely on studies involving its interactions with the Shk pathway, in which EPSPS plays a central role. The nature, biochemistry and products of this pathway have been reviewed in detail by Haslam (1974), Boudet et al. (1985), Bentley (1990), and Singh et al. (1991) and revisited recently by a number of authors (Herrmann and Weaver 1999; Knaggs 1999; Schmid and Amrhein 1999; Siehl 1999). Located in the chloroplasts of higher plants, the Shk pathway consists of a series of seven biochemical reactions (Fig. 1) that convert PEP and erythrose 4-phosphate (E4P), to chorismate. The latter compound contributes to plant metabolism by way of a number of branch pathways, including those for the production of lignin, flavanoids, alkaloids and three crucial aromatic amino acids, phenylalanine, tyrosine, and tryptophan.

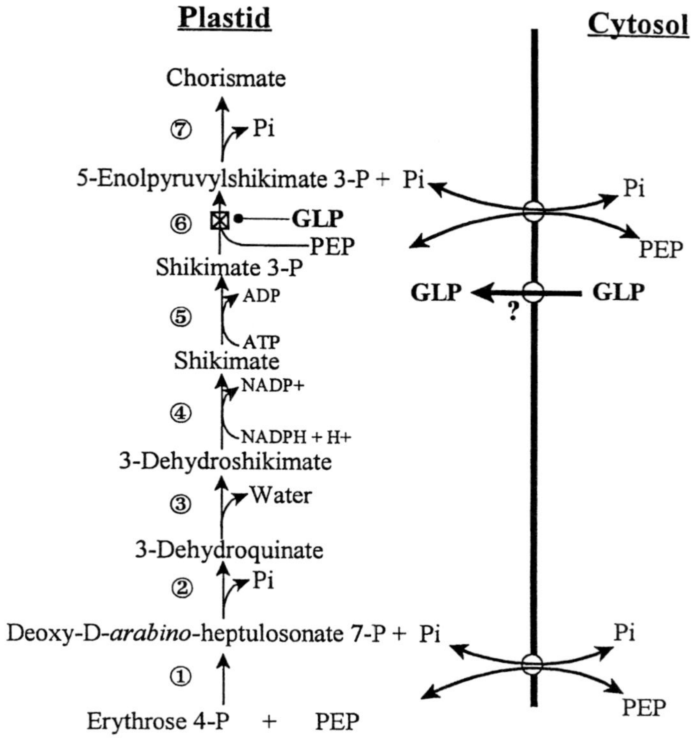

Fig. 1. The seven reaction steps of the main trunk of the shikimate pathway. Enzymes are **1**: 3-deoxy-D-arabino-heptulosinate 7-phosphate, **2**: 3-dehydroquinate synthase, **3**: 3-dehydroquinate dehydratase, **4**: shikimate dehydrogenase, **5**: shikimate kinase, **6**: 5-enolpyruvylshikimate 3-phosphate synthase (EPSPS), **7**: chorismate synthase. In higher plants, steps 3 and 4 are catalyzed by a bifunctional enzyme, 3-dehydroquinate dehydratase-shikimate dehydrogenase. *Circles* represent possible carriers located in the inner membrane of the chloroplast envelope

Two sources of disruptive secondary effects stemming from inhibition of EPSPS are the main causes of injury to the plant (Fig. 2). One source of herbicidal effects is the blockage of chorismate synthesis, which affects various aspects of plant metabolism associated with products of the Shk pathway (Jensen 1986). A second source is the disruption of the photosynthetic carbon reduction (PCR) cycle, which affects photosynthesis and regulation of carbohydrate metabolism (Servaites et al. 1987). Interactions among the many aspects of plant metabolism that depend on the Shk pathway account for the considerable variety of lethal effects of Glp.

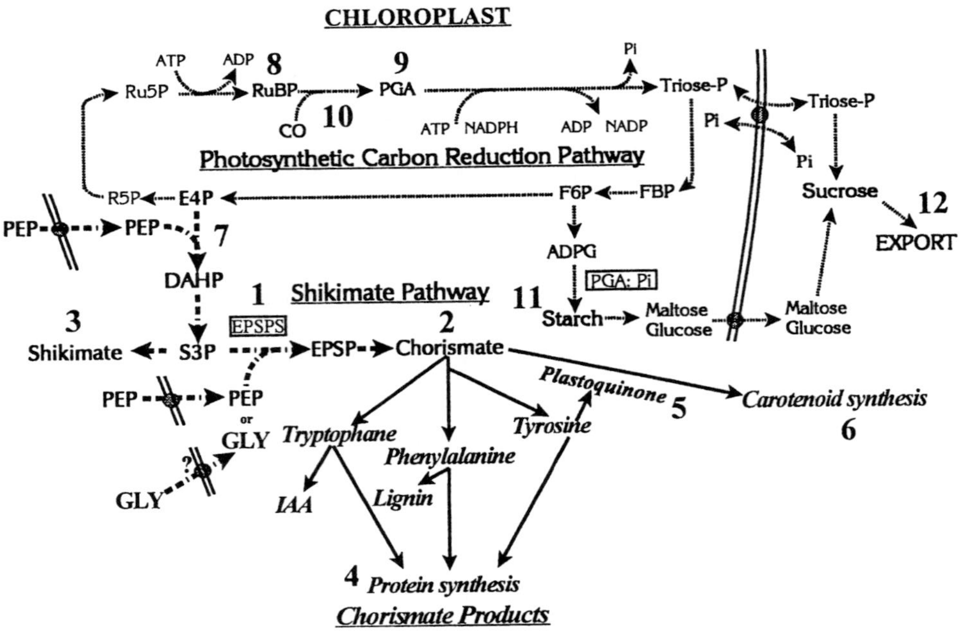

Fig. 2. Biochemical pathways involved in primary mode of herbicidal Gly action and in the disruption of metabolism initiated by inhibition of EPSP synthase (EPSPS). Secondary effects of the inhibition of EPSPS disrupt both the PCR pathway and the Shk pathway. *Numbers* refer to reactants and steps in pathways involved in symptom development and lethal herbicide effects. *Shaded circles* represent possible carriers located in the inner membrane of the chloroplast envelope.

3.3.2
Primary Mode of Action

3.3.2.1
Biochemical Characteristics of the Target Enzyme

EPSPS, which is essential for the synthesis of aromatic amino acid and many other aromatic compounds, is found in a variety of autotrophic organisms including bacteria, fungi, algae and higher plants. The discovery that Glp exclusively inhibits this enzyme opened up a series of questions regarding the molecular interaction needed for herbicidal effectiveness (Franz et al. 1997).

In bacteria and plants, EPSPS consists of a monomeric protein of 46–51 kDa molecular mass (Franz et al. 1997). Studies of the enzyme were greatly facilitated by the availability of large amounts of enzyme protein made possible by cloning the aroA gene from *E. coli* and over-expressing it in *E. coli* strains. The enzyme catalyzes a chemically unusual reversible reaction, which transfers a carboxyvinyl group from PEP to the 5-alcohol group of S3P to yield 5-enolpyruvylshikimate 3-phosphate and Pi. It is of interest that in *E coli* and the other organisms for which EPSPS enzymes have been characterized, the reaction is catalyzed without the aid of metals or other known cofactors. The only other enzyme known to catalyze a similar transfer with PEP is UDP-*N*-acetylglucosamine enolpyruvyl transferase (MurA) (E.C. 2.5.1.7), which transfers the carboxyvinyl portion of PEP to the C-3 alcohol group of UDP-*N*-acetylglucosamine during bacterial cell wall peptidoglycan biosynthesis. These two transfer reactions are chemically unusual because they proceed by C–O bond cleavage rather than by P–O bond cleavage as in most PEP-utilizing enzymes (Franz et al. 1997). Further, the *E. coli* enzyme was shown to operate by a random kinetic mechanism through a single, kinetically competent intermediate (Sikorski and Gruys 1997). While formation of the products is favored, the reaction is slowly but completely reversible (Franz et al. 1997) and can be assayed in both forward and reverse directions (Sikorski and Gruys 1997). Franz et al. (1997) concluded that "all of the available experimental evidence unequivocally demonstrates that EPSPS utilizes a simple addition-elimination reaction and proceeds through a single, kinetically competent tetrahedral intermediate as originally proposed by Sprinson et al."

3.3.2.2
Structural Characteristics of the Target Enzyme

The structural basis for the reaction mechanism has been investigated by three-dimensional, X-ray structure studies performed by Stallings et al. (1991). The authors used crystallography to show that EPSPS from *E. coli* folds into two distinctive hemispherical globular domains, each with a 25 Å radius. Each domain is composed of three copies of a β-α-β-α-β-β-folding unit, and joined by a double-stranded hinge. The six parallel α-helices, with their macrodipoles, are oriented together in such a way as to create a significant electropositive attraction for anionic ligands (Franz et al. 1997). Because of this feature, it is likely that all EPSPS substrates and inhibitors studied to date, including Glp, are multicharged anions. The positive charge gradient formed upon closure may then help guide the ligands to the active site near the hinge region. It is thought that during turnover the two domains exclude water and thus avoid ready hydrolysis of PEP or EPSP to pyruvate. In spite of the detailed description reported by Stallings et al. (1991), the specific active site of the enzyme was not identified.

Franz et al. (1997) have reviewed the participation of the four substrates S3P, PEP, EPSP and phosphate in the EPSPS reaction, their recognition requirements and their molecular characteristics. The 3-phosphate moiety of S3P is an extremely critical functional group for binding and catalysis. The overall size and ionic character of any 3-phosphate replacement group plays a critical role in its catalytic effectiveness. In fact, methods used in S3P and EPSP binding studies have failed to show significant interaction between PEP and free EPSPS. Additional studies with microcalorimetry and equilibrium dialysis, however, have demonstrated the ability of the free enzyme to form a weak EPSPS·PEP binary complex. Binding of PEP is increased approximately 20-fold in the presence of S3P. Charged anionic centers are very important for recognition at the PEP site and amino acid residues Lys-22, Arg-100 and Arg-124 may be involved in PEP recognition. Equilibrium fluorescence change suggests that EPSP forms a tight binary complex with EPSPS to induce a macromolecular conformational change in the enzyme that is not detected with S3P or PEP under comparable conditions. Phosphate, the fourth substrate, has been found to have the weakest affinity of the four EPSPS substrates tested.

Although EPSPS has been the object of study for over three decades, details of the enzyme mechanism remain unresolved. Up to the present, the three-dimensional structure of EPSPS was known only in its unliganded form, which does not reveal the active site of the enzyme. A recent study by Schöenbrunn et al. (2001) of EPSPS co-crystallized with S3P alone and EPSPS co-crystallized with Glp clearly determined the structure of these complexes and helped clarify some unresolved points. From these, the structure of the quaternary EPSPS·S3P·Pi·formate complex was determined at 1.6 Å and the ternary EPSPS·S3P·Glp complex was determined at 1.5 Å. Comparing this new liganded EPSPS structures with the earlier unliganded structures, the authors found that the two domains of EPSPS approach each other having a screw-like movement with the active site emerging in the interdomain cleft. The authors reason that S3P, not negatively charged ions or Glp, triggers the domain closure that leads to an accumulation of positive charges in the cleft that attracts negatively charged molecules, including Glp or PEP to the active site cavity. Schönbrunn et al. (2001) base this conclusion on the observation that both the EPSPS bound with S3P without Glp and the EPSPS·S3P·Glp complex are in the closed configuration. Based on the analogy with the conserved binding residues in the mechanistically related MurA, Schöenbrunn et al. (2001) further proposed that the strictly conserved residues Lys-22, Arg-24, Asp-313, Arg-344, Arg-386 and Lys-411 in EPSPS are involved in binding PEP while the remaining residues of Arg-100, Asp-242 and Asp-384 appear to be involved in easing domain closure.

3.3.2.3
Interaction Between 5-Enolpyruvylshikimate 3-Phosphate Synthase and Glyphosate

Inhibition of EPSPS by Glp was first reported by Amrhein in 1980 (Amrhein et al. 1980; Steinrücken and Amrhein 1980). Glp inhibits EPSPS in a slowly reversible reaction. Prolonged dialysis of the ternary complex restores enzyme activity, confirming that the binding of Glp is reversible. A large negative entropy term for the formation of the ternary complex indicates that Glp fits very well into its binding site.

The reaction shows a number of unique properties based on the nature of the Glp binding site. Glp shows competitive kinetic behavior with respect to PEP, but is uncompetitive with regard to S3P (Boocock and Coggins 1983; Steinrücken and Amrhein 1984). Consistent with this observation, Glp has been shown to bind preferentially to the EPSPS·S3P complex to form a tight ternary EPSPS·S3P·Glp complex (Steinrücken and Amrhein 1984), a trait since confirmed by a variety of biophysical methods (Sikorski and Gruys 1997). Franz et al. (1997) present evidence that the Glp dianion is the actual species present in the active ternary complex with Glp and is the herbicidally active species in the plant under physiological conditions.

Both S3P and EPSPS build up in the presence of Glp but due to the uncompetitive behavior of S3P they combine to form the EPSPS·S3P complex to which Glp binds preferentially (Franz et al. 1997). Binding of Glp to the EPSPS·S3P complex induces a substantial conformational change (Anderson et al. 1988, Della-Cioppa and Kishore 1988). In order to hold the bound enzyme in a completely closed state, potent inhibitors of EPSPS must also function like molecular staples, holding the two-hinged globular domains together through electrostatic and H-bonding interactions (Franz et al. 1997). A series of dipole-dipole, hydrogen-bonding and electrostatic interactions severely restrict the rotational freedom of Glp and the several amino acid side chains involved in the complexes. The ternary complex is fairly rigid compared to the native enzyme. ^{31}P NMR confirms that signals from the Glp phosphonate and the S3P 3-phosphate are shifted dramatically. Synergistic binding to form the ternary complex with Glp is about 4000-fold greater than observed in the ternary complex with PEP. The increased binding accounts for the uncompetitive kinetic behavior for Glp versus S3P that is observed in *E. coli* EPSPS ternary complex with Glp (Gruys et al. 1992).

An important issue related to the mechanism of inhibition is whether or not Glp functions as a transition state inhibitor as proposed by Steinrücken and Amrhein (1984). The latter view, which has become established in the literature and which historically defined the primary mode of Glp action, has been challenged by a recent report by Sikorsky and Gruys (1997). These authors conclude that Glp cannot simply function as a ground-state mimic of PEP because Glp binds more tightly to EPSPS than does PEP and does not inhibit any other PEP-dependent enzymes. Instead, they proposed the view that Glp and PEP each has a unique binding interaction with EPSPS and that the affinity of Glp with S3P for EPSPS is not related to any transition or intermediate state involved in catalysis. Sikorski and Gruys (1997) cite biochemical evidence indicating that Glp is able to act as a potent competitive inhibitor against PEP because of the fortuitous mode of its binding to EPSPS. They conclude that, while there is substantial overlap between the PEP and the Glp binding domains, a substantial portion of the Glp binding site is separated from those amino acid residues intimately involved in PEP binding and catalysis. The binding of Glp is proposed to be an "adventitious allosteric interaction". In support of the difference in binding, the authors point out that while some modification of the PEP structure is tolerated, even minor structural changes in the Glp skeleton lead to a significant loss of inhibitor action and herbicidal activity.

A recent report by Schönbrunn et al. (2001) arrives at some conclusions that are at odds with the above findings of Sikorsky and Gruys (1997). This very high-resolution crystallographic study of EPSPS structure, described in Section 3.3.2.1, allows visualization of the interactions of S3P and Glp ligands with the target EPSPS enzyme. Because the active site is formed by the closure of the enzyme, the Glp binding site is dominated by charged residues from both domains of the EPSPS molecule. Schönbrunn et al. (2001) report that Glp binds close to S3P without perturbing the structure of the active-site cavity that is seen in the Glp-free EPSPS·S3P complex. The overall structure of the EPSPS·S3P complex is virtually identical to that of the ternary complex with Glp, both configurations showing the closed state of the enzyme. The authors state that the binding of the phosphonate and carboxylate moieties of Glp corresponds to the phosphate and carboxylate groups involved in the binding of PEP. The conclusion that Glp occupies the PEP binding site (Fig. 3 of Schönbrunn et al. 2001) is based on a comparison of the PEP binding site on mechanistically related MurA with the Glp binding site on EPSPS.

A commentary on the article of Schönbrunn et al. (2001) by Alibhai and Stallings (2001) reviews past data concerning the two areas of disagreement. The issues are what triggers the closure of the two domains that form the active site and whether Glp and PEP occupy the identical binding site. Concerning the latter, Alibhai and Stallings (2001) cite evidence to the contrary, including the ability of Glp to bind to EPSPS·EPSP to form the EPSPS·EPSP·Glp complex (Sammons et al. 1995). Alibhai and Stallings (2001) also observed that Glp is an uncompetitive inhibitor versus EPSP in the reverse reaction. The claim that Glp and PEP both occupy the identical binding site has already been addressed by Sikorsky and Gruys (1997) in their argument against Glp functioning as a transition state inhibitor. Further, research cited by Sikorski and Gruys (1997) and Franz et al. (1997) has shown that there is no direct correlation between potency of Glp inhibition and the catalytic efficiency at the PEP site for various forms of EPSPS.

In commenting on the issue of what triggers domain closure, Alibhai and Stallings (2001) cite previous works which suggest that the binding of Glp with the EPSPS EPSPS·S3P complex induces a substantial change in the conformation of the enzyme (Anderson et al. 1988). No fluorescence change occurs upon the binding of S3P or Glp alone. It seems that if S3P alone were to trigger closure, access of Glp or PEP to the solvent-inaccessible active site would be blocked. Della-Cioppa and Kishore (1988) showed that, in contrast to the free pre-enzyme, the pre-EPSPS·S3P·Glp complex was not readily transported into the chloroplast stroma. Presumably the enzyme, when in the closed form, is hindered from making the transient changes that are needed for it to be imported across the membrane. Taken together, these results suggest that EPSPS undergoes a macroconformational change upon binding of S3P along with Glp. Alibhai and Stallings (2001) suggest that the crystallized enzyme with S3P but without Glp may be closed because of the phosphate and formate bound at the PEP binding site. Clearly, further structural studies with various substrate complexes will be needed to address these two points of contention.

3.3.2.4 Molecular Requirements for Herbicidal Activity of Glyphosate

Important aspects of the structure of Glp required for binding to EPSPS have been reviewed by Franz et al. (1997) and were described above in relation to the mechanism of Glp interaction with the enzyme-substrate complexes. The X-ray crystallographic studies of Stallings et al. (1991) and Schönbrunn et al. (2001), described above, also provide a particularly important guide to the structural features required for binding of Glp for herbicidal activity. The conclusion of Franz et al. (1997) that the Glp dianion is the herbicidally active component in all Glp-based herbicides also helps specify structural requirements for herbicidal activity. It is thought that only those Glp derivatives or analogs that readily break down to this dianion under physiological conditions on or within a plant will be effective as herbicides.

Considerable research has been done to discover other inhibitors of EPSPS that have equally desirable qualities but different agronomic uses (Siehl 1997). Enzyme-directed synthesis of new EPSPS inhibitors has yielded an elegant series of inhibitors but no desirable herbicides (Sikorski and Gruys 1997). They observed that even small changes in the molecular structure of Glp either lessened or eliminated the molecule's herbicidal effectiveness. Maier (2000) reached a similar conclusion from his systematic study of the effects of changes to the Glp molecule on biological activity. Because of its unique binding mechanism with EPSPS and the negative results of extensive searches for an improved herbicide based on inhibition of EPSPS, Sikorski and Gruys (1997) found it difficult to imagine how Glp could have been designed solely on the basis of knowledge of the role of PEP and the EPSPS mechanism. While some modification of PEP structure is tolerated, even minor changes in Glp structure lead to a significant loss in inhibitor potency and reduced herbicidal activity (Sikorski and Gruys 1997). Only two closely related analogs, *N*-hydroxyglyphosate and *N*-amino-glyphosate show properties nearly comparable to Glp. To date, no analogue or derivative of Glp has been identified that is more potent at inhibiting a single enzyme at such a crucial site than Glp itself (Franz et al. 1997).

In contrast to the pessimistic view concerning possible development of a better molecule to bind the active site, recent progress in identifying details of the mechanism of action of Glp has opened possibilities for progress in new approaches to inhibitors of EPSPS that may be of value as herbicides (Schönbrunn et al. 2001). Alibhai and Stallings (2001) conclude their commentary on X-ray crystallographic studies of the mechanism of Glp binding to EPSPS by stating that the new structures of the enzyme-ligand complexes are significant new tools for the rational design of novel inhibitors. Studies on the binding of the tetrahedral reaction intermediates to EPSPS have demonstrated that tapping into the structural determinants involved in S3P and Glp recognition could lead to inhibitors of picomolar affinity (Anderson and Johnson 1990). Alibhai and Stallings (2001) also note that, with some wisdom, Schönbrunn et al. (2001) propose an alternative strategy for structure-based inhibitor design that makes use of the induced-fit mechanism triggered by the binding of ligands to EPSPS. By the use of studies that spatially identify residues responsible for the conformational changes in the MurA structure and mapping them on the EPSPS structure (Schönbrunn et al. 2000a,b) Eschenburg and Schönbrunn (2000) have identified residues that might be important for conformational changes and provide new patterns for herbicides that block domain closure and the formation of the EPSPS catalytic site.

3.3.3 Secondary Physiological Consequences of Inhibition of 5-Enolpyruvylshikimate 3-Phosphate Synthase

Given the many secondary consequences of Glp-induced inhibition of EPSPS and the physiological diversity of plants, it is not surprising to find considerable variation in responses to Glp both among and within species. Differences in response among species reflect both physiological (Ferreira and Reddy 2000; Fuchs et al., unpubl.) and cellular differences (Westwood and Weller 1997; Feng et al. 1999; see Fig. 2; key steps are identified in the text by number). Glp-induced inhibition of EPSPS and the Shk pathway has two distinct but related effects, inhibition of chorismate synthesis and deregulation of carbon entry into the Shk pathway. The former aspect, inhibition of the synthesis of aromatic amino acids and other compounds made from chorismate (Cole et al. 1980), generally has received considerably more attention with respect to the herbicidal action of Glp. Deregulation of the PCR pathway has received less attention (Siehl 1997). Both, however, have the potential to play an important role in relation to plant death.

3.3.3.1 Inhibition of Chorismate Synthesis

In higher plants EPSPS is synthesized in the cytosol and is transported into plastids with the aid of a chloroplast transit peptide (Mousdale and Coggins 1985; Della-Cioppa et al. 1986; Holländer-Czytko and Amrhein 1987). For disruptive effects to occur enough Glp must enter plastids and reach a concentration that will inhibit the Shk pathway in the sources and sinks of the plant. Under ordinary growing conditions a major portion of the carbon fixed by plants flows through the Shk pathway (Singh et al. 1991) and the derived aromatic amino acids produce a major portion of plant dry mass (Boudet 1985;

Jensen 1986). In this way the pre-chorismate portion of the Shk pathway is directly linked to other key areas of plant metabolism (Jensen 1986; Singh 1991). The pattern of compounds synthesized from the Shk pathway differs among plant species and conditions, so the ensemble of metabolites derived from Shk form a rather unique signature of the species (Herrmann 1995).

Disruption of the Shk pathway by Glp results directly from inhibition of EPSPS (Fig. 2, **1**) (Steinrücken and Amrhein 1980). Inhibition of EPSPS blocks the conversion (**7**) of E4P from the PCR cycle and PEP from the cytosol (Fischer et al. 1997) to chorismate (**2**). The latter is a key source of carbon for aromatic amino acid synthesis (**4**) (Holländer and Amrhein 1980). As a result, secondary products derived from chorismate (**4, 5, 6**) are depleted (Amrhein et al. 1982) and Shk (**3**) accumulates (Amrhein et al. 1980). Aromatic amino acid synthesis pathways also contribute precursors for a number of metabolites such as auxin growth regulators, lignin, plant defense compounds, UV protectants and plastoquinone (**5**) which is essential for the synthesis of carotenoids (**6**).

3.3.3.2
Depletion of Photosynthetic Carbon Reduction Cycle Intermediate Metabolites

Besides affecting the products of the shikimate pathway, Glp-induced reduction of chorismate synthesis initiates a signal to 3-deoxy-D-arabino-heptulosonate 7-phosphate (DAHP) synthase to enhance the flow of reactants into the shikimate pathway (Jensen 1986). Flow of metabolites into the pre-chorismate portion of the Shk pathway begins with the condensation (**7**) of PEP and E4P. The latter comes from the PCR cycle while most plastids must rely on the cytosol for their supply of PEP, which is moved across the plastid envelope by a special PEP/Pi translocator (Fischer et al. 1997). Dependence on the PCR cycle as the source of E4P connects the Shk pathway to carbohydrate metabolism (Geiger and Servaites 1994) and provides another avenue for production of damage symptoms. Unrestricted carbon flow into the Shk pathway causes "back" accumulation of S3P and Shk (**3**) (Holländer and Amrhein 1980) as carbon entry into chorismate is disrupted by Glp activity.

This enhanced flow of E4P has been shown to lower the concentration of phosphorylated intermediates of the PCR cycle (**8, 9**) in sugar beet (Servaites et al. 1987) and velvetleaf (Fuchs et al., unpubl.). This seriously disrupts photosynthesis (**10**) (Geiger et al. 1987) and the allocation of carbon from the PCR cycle to key processes in plant metabolism, particularly starch synthesis (**11**) in sugar beet (Geiger et al. 1986). A lesser degree of inhibition of these processes was observed in velvetleaf where the depletion of phosphorylated intermediates of the PCR cycle was less (Fuchs et al., unpubl.). Further studies are needed to determine the relative importance of the deregulation of the entry of carbon from the PCR cycle into the Shk pathway in the development of damage symptoms in various species and biotypes.

Regulation of entry of compounds from the PCR cycle into each of the pathways that depend on the cycle is critical to avoid a detrimental drain of carbon and phosphate from the cycle (Rao and Terry 1995) and disruption of the integration of photosynthetic carbon metabolism with overall plant function (Badger et al. 1984; Leegood and von Caemmerer 1988; Servaites et al. 1991; Sharkey 1998). For example, the decrease of phosphoglyceric acid (PGA), measured in sugar beet, increases the Pi:PGA ratio, inhibits ADPG pyrophosphorylase and thereby slows starch synthesis (Preiss 1988). Inhibition of photosynthesis slows sucrose synthesis and translocation (**12**) in sugar beet (Geiger and Servaites 1994). When inhibition of photosynthesis is severe, as regularly happens in sugar beet, photoinhibition occurs (Shieh et al. 1991; Madsen et al. 1995; Geiger et al. 1999). Photoinhibition is particularly noticeable on the second day when the photosynthesis rate begins to decrease as light intensity nears the midday level.

3.3.3.3 Development of Secondary Damage Symptoms

Effects of the herbicidal action of Glp on a number of important plant processes are listed in Table 2, along with causal mechanisms thought to be associated with them. The rapidity and severity with which a plant process is affected differ among species and biotypes, suggesting that lethal mechanisms show related differences. For example, if in a given biotype or species Glp causes photosynthesis to decrease rapidly along with PGA cycle intermediates, it is likely that the deregulation of carbon entry into the Shk cycle is important in the development of damage symptoms. If this is not the case then other mechanisms likely are a cause of serious damage symptoms. The order of appearance and time course of damage symptom development also can be used to distinguish whether a given disruption results directly from Glp action or is a consequence of plant deterioration.

3.3.3.4 Bases of Development of Lethal Symptoms Among Species

Although Glp is lethal to most plants, the pattern and intensity of symptom development commonly differ among plant species. Death results from Glp-induced disruption of essential plant processes, but it is not clear which processes are critically affected in each species. Experience indicates that disruptions which affect root sink processes can limit uptake of water and minerals (Foley et al. 1983; Nafziger and Slife 1983; Fernandez et al. 1994) while those that affect source leaf processes can interfere with the steady supply of assimilated carbon (Geiger et al. 1999). Carbon allocation between shoot and root in support of integrated growth and metabolism is highly regulated (Ho 1988; Farrar and Williams 1991; Schulze et al. 1991; Geiger et al. 1996). In a healthy plant, the regulated contribution of photosynthetic carbon assimila-

Table 2. Survey of principal secondary effects of Gly on plant processes. The use of or designates alternatives observed in plants with contrasting responses

Physiological Processes	
Time of Inhibition	**Proposed Mechanisms**
Shikimate and shikimate-3-phosphate accumulation	
rapid: starts within minutes and accumulates	direct consequence of EPSPS inhibition
Inhibition of protein synthesis (inhibited aromatic amino acid synthesis)	
rapid: starts soon but develops gradually	direct consequence of EPSPS inhibition
Decreased photosynthesis rate in leaves exposed to Gly	
rapid: within hours	lowered level of RuBP in PCR cycle
or	
gradual: over several days to weeks	decrease in stomatal conductance and water uptake appears to be a significant factor
Decreased starch synthesis in leaves exposed to Gly	
rapid: initiated within several hours, with a loss of cycling after 1 day	diversion of carbon from PCR cycle lowers PGA and PGA:Pi ratio
or	
gradual: detectable by the end of day 1 with a loss of cycling after several days	decrease of photosynthetic carbon assimilation rate
Decreased carbon translocation	
rapid: export decreases with photosynthesis and with starch availability	decreased synthesis of sucrose from photosynthesis and from starch
or	
gradual: import decreases step wise and continues for several days	inhibition of growth and protein synthesis in growing sinks
Decreased Gly translocation	
rapid: after a day	inhibition of carbon export, self-limitation and photoinhibition
or	
gradual: after several days	progressive loss of photosynthesis and carbon export
Decrease in photosynthetic pigments (chlorophyll, carotenoids)	
begins within days and can last for several days to weeks	inhibition of porphyrin compounds needed for synthesis and reduced amount of protective carotenoids in tissues

tion by source leaves and water and mineral acquisition by roots maintain the nutritional balance essential to the life of the plant. Disruption of this metabolic balance by Glp can be lethal (Geiger and Servaites 1994).

The effects of Glp on sugar beet and velvetleaf plants are compared in Table 3 with a view to identifying the chief factors responsible for the lethal herbi-

Table 3. Comparison of nature and timing of various damage symptoms and causal factors in sugar beet and velvet leaf plants. Glyphosate was applied to source leaves

SUGAR BEET	VELVETLEAF
SOURCE LEAF	
Process or condition affected by glyphosate	
Photosynthetic carbon assimilation	
• inhibited rapidly	• inhibited gradually over several days
• decreases with light intensity; likely due to photoinhibition	• photoinhibition appears not to be a significant factor
• decrease in RuBP is a significant factor	• increased stomatal resistance is a significant factor
• internal leaf CO_2 increases as photosynthesis decreases	• internal leaf CO_2 decreases over several days
Carbon availability	
• reduced rapidly on the first day	• reduced gradually over several days
Leaf cellular integrity	
• decreased steadily and lost by day five	• remains relatively intact until sudden loss on day seven
Factors significant in lethal action	
• decreased availability of source leaf carbon leads to loss of leaf cellular integrity	*• extended availability of source leaf carbon prolongs ability of the plant to respond to stress*
• decreased carbon export lessens export of glyphosate	*• extended carbon export increases export of glyphosate*
SINK ORGANS	
Process or condition affected by glyphosate	
Availability of carbon for sink processes in shoots and roots	
• decreases rapidly to near zero by the end of the first day	• decreases gradually over several days
Physiological processes in shoot and root sinks	
• moderate level of sink leaf chlorosis	• severe sink leaf chlorosis
• gradual decrease in root water uptake over a 4-day period	• gradual decrease in root water uptake over a week period
Factors significant in lethal action	
• early inhibition of export to sinks helps limit translocation of glyphosate to sinks (self limitation)	*• prolonged export to sinks increases import of glyphosate to sink and increases sink damage*
• rapid decrease in carbon available to roots plus lethal effect on sink processes gradually disrupts root integrity	*• glyphosate accumulation in roots eventually disrupts root integrity*

cidal action of Glp. Disruption of the PCR cycle in sugar beet source leaves is a major factor in the lethal action of Glp, leading to a loss of carbon availability that compounds the toxic effects of Glp on the sinks. By contrast, in velvetleaf disruption of the PCR cycle occurs to a lesser extent and the major lethal action is disruption of metabolism in sink tissues. In both species a major cause of death appears to be the severe limitation of water uptake similar to that reported by Nafziger and Slife (1983) and Fernandez et al. (1994). The cause of the limitation of water uptake seems to differ between these two species (Fuchs et al., unpubl.). Availability of carbon severely limits the cellular integrity of both source and sink tissue in sugar beet while continued export and accumulation of Glp destroys root cellular integrity in velvetleaf.

The rapid and complete inhibition of photosynthesis in sugar beet appears to result in a higher degree of susceptibility to Glp whereas the delayed inhibition in velvetleaf contributes to increased to tolerance.

3.4 Mechanisms for Resistance and Tolerance to Glyphosate

Warwick (1991) distinguishes between resistance and tolerance in regard to the degree of susceptibility of plants to herbicides. Resistance is defined as the ability to withstand a normal dose of the herbicide used in the field. This most often occurs as a result of a cellular alteration at the herbicides site of action. Tolerance is then a reduced susceptibility to the lethal effects of a herbicide that results from other causes such as lower uptake and translocation of the herbicide in the plant, herbicide detoxification or differences in plant metabolism. Tolerance often occurs in biotypes that have evolved through selective pressure in response to herbicide use (Powles et al. 1998; Lorraine-Colwill et al. 2001). In addition to reduced uptake or to destruction, plants conceivably could attain tolerance either by avoiding the herbicide's primary mode of action at the beginning of the chain of physiological events or by escaping the damaging secondary consequences of initial inhibition by the herbicide. In the case of Glp, only plants in which inhibition of EPSPS does not occur fail to show symptoms (Padgette et al. 1996).

3.4.1 Development of Commercially Valuable Glyphosate-Resistant Plants

Resistance to Glp has been conferred on certain crop species by genetic transformation of plants with a variant form of EPSPS that circumvents the primary mode of action of Glp (Padgette et al. 1995, 1996). The attempts to develop Glp-resistant crops, starting in the early 1980s, have been reviewed by Padgette et al. (1996) and Bradshaw et al. (1997). Three genetic transformation strategies were used – overproduction of EPSPS, introduction of a Glp-degradation gene and introduction of an EPSPS with decreased affinity for Glp. The first two approaches, however, never produced a level of resistance that was high enough for commercial use (Shah et al. 1986; Barry et al. 1992; Padgette et al. 1996). Thus, preventing Glp from binding to EPSPS seemed to be the only sure approach for producing resistance and an extensive search for Glp tolerant EPSPS mutants began.

Initial attempts to produce resistance to Glp by mutagenesis of *Arabidopsis* seeds followed by selection on the herbicide proved to be unsuccessful for Glp (Padgette et al. 1996). Previously, the method was used successfully to develop resistance to sulfonylurea and imidazolinone herbicides. It appears that single-nucleotide changes that confer resistance on EPSPS are extremely rare. Similar selection attempts with *E. coli* produced a variant form of EPSPS highly tolerant to inhibition by Glp. The bacterial mutant that had an amino acid substitution of alanine for glycine at position 96 of the *E. coli* EPSPS enzyme showed high resistance to Glp (Padgette et al. 1991). Modification of *E. coli* or plant genes for EPSPS in this manner inhibited binding of Glp and conferred a level of resistance that required a 500-fold increase in Glp concentration to produce a 50% inhibition of EPSPS (Padgette et al. 1991, 1996). We now know that glycine 96 (Gly-96) is in the molecular binding domain for Glp (Schönbrunn et al. 2001). In this case, however, the alteration that lessened binding of Glp to the enzyme also reduced catalytic efficiency. Unfortunately, in petunia the altered enzyme was about 72 times less catalytically efficient than the wild type. Even though transgenic plants expressing these EPSPS genes exhibited high levels of tolerance, field tests showed that the level of Glp resistance was only twice the field application rate of Glp, again not sufficient for commercial application (Padgette et al. 1991).

The approach that was eventually successful in developing a commercial Glp tolerance in EPSPS depended on the insights provided by extensive research into the molecular mechanism of the enzyme. The new direction in the ongoing search was oriented to generating mutants with EPSPS that were both Glp-tolerant and had lowered appK_m(PEP). Random mutagenesis of petunia EPSPS genes was conducted in actively growing *E. coli* cultures (Padgette et al. 1996). Out of the millions of initial transformants produced by the mutagenesis schemes, not a single Glp-resistant enzyme was obtained that had a near-normal appK_m(PEP), that is one with $<20\,\mu$M. Thus, the level of tolerance developed with plant genes was unacceptable. This research, including work with sets of *B. subtilis* and petunia mutants, provided evidence that indicated there is no direct correlation between Glp inhibition potencies and the catalytic efficiency at the PEP site for EPSPS (Sikorski and Gruys 1997).

As a consequence, yet another approach was pursued, namely, research aimed at finding a variant EPSPS that had low appK_m (PEP) and high appK_i (Glp). The breakthrough finally came after Schulz et al. (1985) reported that a number of bacteria exhibited a reduced sensitivity to Glp. It was reasoned that populations of these bacteria, which are able to grow actively in the presence of Glp, are likely to contain a naturally modified, highly efficient Glp-resistant EPSPS enzyme. Some of these bacteria were likely to have an EPSPS with the desired combination of low appK_m (PEP) and high appK_i (Glp). A screen with target parameters of appK_m (PEP)~$15\,\mu$M and appK_i (Glp)/appK_m (PEP) ratio of >100 successfully located genes for EPSPS with naturally occurring tolerance. The EPSPS with the highest tolerance was found *in Agrobacterium* sp, strain CP4, and had an appK_m (PEP) of $12\,\mu$M and an appK_i (Glp) of 2.7 mM (Barry et al. 1992; Padgette et al. 1996). These parameters compared with values of $5\,\mu$M and $0.4\,\mu$M, respectively, for wild-type petunia (Padgette et al. 1987).

In practice, Roundup Ready plants that are resistant to Glp were produced by transforming plants with a naturally occurring, Glp-resistant EPSPS gene derived from the soil bacterium *Agrobacterium* sp. strain CP4 (Padgette et al. 1996). The ratio of k_{cat}/K_m for PEP is reduced only by a factor of 10 in relation to the native petunia enzyme but Glp affinity decreases nearly 7000-fold (Sikorski and Gruys 1997). In transformed plants, the transgenic CP4 enzyme catalyzes a reaction that bypasses the Glp-induced block of the plant's own EPSPS. To develop the lead Glp-resistant soybean progenitor line Monsanto 40-3-2, the plasmid for the transformation of the plant genome was prepared by fusing the 5′ end of the CP4 EPSPS gene to a chloroplast transit peptide sequence derived from petunia (Padgette et al. 1996). The transit peptide is cleaved in the chloroplast to produce the mature EPSPS enzyme. The plasmid used to transform the parental soybean line contained two copies of the CP4 gene and a gene coding a β-glucuronidase from *E. coli*, driven by a plant promoter. The DNA was introduced by a particle accelerator and the presence of β-glucuronidase marker was used as evidence of transformation. Shoots of the R_0 transformants were grown to maturity and screened for Glp resistance and the presence of the CP4 gene. The R_0 progenitor received two DNA inserts located at different locations in the genome. R_1 progeny were evaluated for tolerance to Glp and used to supply seed for the R_2 progeny. Subsequent analysis showed that one insert, which had the gene for β-glucuronidase, was lost. The remaining insert derived from the plasmid contained a portion of the E35S promoter, the chloroplast transit peptide from petunia EPSPS and the CP4 EPSPS. F_2 progenies of crosses between 40-3-2 and other soybean lines indicate that the insertion behaves as a single dominant gene inherited in Mendelian fashion (Padgette et al. 1996).

Similar molecular biology techniques are now used to produce a number of Glp resistant forms, including sugar beet (Mannerlöf et al. 1997). The constructs described by these workers, pMON17204 and pMON17209, both contain the CP4 EPSPS gene and the gene for glyphosate oxidase. Sugar beet plants with the Roundup Ready construct not only fail to develop the usual injury symptoms of Glp action but do not show the usual physiological effects on the Shk pathway or the PCR cycle in response to Glp at the usual effective dose (Madsen and Jensen 1995; Geiger et al. 1999).

3.4.2
Tolerance to Field Doses of Glyphosate in Field-Grown Plants

Heap (1997) describes herbicide tolerance (resistance by his terminology) in a weed population as the naturally occurring, inheritable ability of some weed biotypes within a population to survive a normally effective herbicide dose as a result of selection pressure. Bradshaw et al. (1997) presented reasons why the probability of development of naturally occurring plants with a high degree of Glp tolerance can be considered more remote for Glp than for other herbicides. In fact, Glp was used during the previous two decades without a single report of evolution of highly tolerant weed biotypes (Franz et al. 1997).

Naturally occurring biotypes of field bindweed, *Convulvulus arvensis*, with a high degree of Glp tolerance were reported by DeGennaro and Weller (1984). Westwood and Weller (1997) studied the causes for differences in heritable, stable tolerance to Glp in five biotypes of field bindweed that showed a fourfold range of increased tolerance. No differences in gross uptake or translocation patterns were found and the level of tolerance was similar between plants and cell suspension cultures developed from them. The authors concluded that tolerance is based on multiple, unidentified mechanisms at the cellular/metabolic level. Populations of plants show a considerable range of Glp-tolerance levels (DeGennaro and Weller 1984; Pratley et al. 1999), thus providing a basis for selection of biotypes that have a high degree of tolerance with increased use of the herbicide. For example, tolerance differences in biotypes presumably were sufficient for development of a high degree of Glp tolerance in perennial ryegrass, *Lolium perenne*, as a consequence of recurrent selection over an 11-year period (Johnston and Faulkner 1991). The mechanism for tolerance was not identified.

For reasons still unknown, in recent years a growing number of weeds are showing signs of a high degree of Glp tolerance (Preston et al. 1996; Powles et al. 1998; Lee and Ngim 2000). The Weed Science website on 15 June 2001 reported Glp resistance in three species of weeds, *Lolium rigidum*, *Eleusine indica*, and *Conyza canadensis*. A high degree of Glp tolerance was first reported for *L. rigidum*, a valued pasture grass, which is also a weed in cereal crops in southern Australia. The population of highly tolerant biotypes was found in a field in Echuca, northern Victoria, Australia where Glp had been applied for pre-sowing control of weeds at least ten times in the previous 15 years (Pratley et al. 1996, 1999). Plants proved to be tenfold more tolerant to Glp than susceptible biotypes. A substantial degree of tolerance to Glp already exists in more than 30% of the fields in southern Australia (Pratley et al. 1999) as a result of intense selection pressure applied to a species that appears to be well suited for developing tolerance (Powles et al. 1998). Preston et al. (1996) reported that, because of the extensive and repetitive use of herbicides, biotypes of *L. rigidum* have been found to be highly tolerant to the majority of herbicides currently used. This trait is sometimes ascribed to induction of several herbicide-degrading enzymes (Preston et al. 1996). Tardif et al. (1997) cited plant characteristics, including wide distribution, diverse genetic background, a requirement for cross-pollination that facilitates outcrossing and high selection pressure from the use of the same herbicides as factors for promoting development of plant populations with a high degree of herbicide tolerance.

Subsequently, a second population of highly tolerant rigid ryegrass with a 7- to 11-fold increase in tolerance was discovered in an orchard in Orange, NSW, Australia by Powles et al. (1998). Glp had been successfully applied about 40 times over a 15-year period before the highly tolerant biotype was observed (Lorraine-Colwill et al. 1999). The tolerant population showed a 2.5-fold cross-tolerance to diclofop-methyl (Powles et al. 1998). Recently, a highly tolerant biotype of goosegrass, *Eleusine indica* (L.) Gaertn was reported in Teluk Intan, Malaysia after 3 years of repeated usage of Glp, often at high dosage (Lee and Ngim 2000). An on-site field trial confirmed the existence of a highly tolerant biotype population with 8- to 12-fold tolerance in relation to wild type.

A number of factors, including the mechanism of herbicide action, appear to cause different species or biotypes of plants to evolve a high degree of herbicide tolerance at different rates (Pratley et al. 1999). While the dissociation constant of the ternary complex with EPSPS may vary by as much as three orders of magnitude between plant and bacterial enzymes (Herrmann and Weaver 1999), plant enzymes generally show similar responses to Glp. Lorraine-Colwill et al. (1999) examined the basis for the biotype *of L. rigidum* with a high degree of tolerance as reported by Powles et al. (1998). Activities, gene expression and sensitivity to Glp of EPSPS were indistinguishable in susceptible and tolerant plants. A 20% increase in extractable EPSPS activity was observed 16 h after Glp was applied but the values for the activity were indistinguishable between tolerant and susceptible biotypes. Likewise, the activity and expression of DAHP synthase, the first enzyme of the Shk pathway, were essentially the same in both susceptible and tolerant biotypes. While Shk levels increased in both biotypes, levels in the tolerant plants were lower and returned to the pretreatment level by 7 days after treatment (Lorraine-Colwill et al. 1999). The difference could not be attributed to differences in the characteristics of the EPSPS enzyme but are thought possibly to reflect a difference in the access of Glp to the target enzyme. Studies by Feng et al. (1999) concluded that neither uptake nor translocation nor metabolism of Glp plays a major role in the high level of tolerance to Glp in *L. rigidum*.

These plants seem to somehow circumvent the effects of the secondary disruption of plant processes and the development of the damage syndrome (Lorraine-Colwill et al. 1999). These authors suggested reduced movement of Glp to its site of action in the plastid as a possible mode of tolerance. It may be significant that the tolerant plants showed a fourfold cross tolerance to 2-hydroxy-3-(1,2,4-triazol-1-yl)propyl phosphonate, a structural analog of Glp that is a herbicide with a mode of action unrelated to Glp. The chloroplast inner envelope membrane contains a number of metabolite transporters that mediate exchange between cytosol and chloroplast. These include the chloroplast triose phosphate/phosphate translocator and the recently discovered phosphoenolpyruvate/phosphate translocator (Fischer et al. 1997). Conceivably, both Glp and the Glp analog may be subject to the same transport restriction in the chloroplast membrane. Such a mechanism is consistent with the conclusion that tolerance for Glp in rigid ryegrass involves differences at the cellular/metabolic level (Westwood and Weller 1997; Powles et al. 1998; Feng et al. 1999).

3.5 Summary

More than 30 years of research has revealed many things but also left many questions unanswered about the mechanisms of Glp herbicidal action. The discovery of Glp has been instrumental not only in providing an effective herbicide but also given a means of peering into the inner workings of plant biochemistry and physiology. EPSPS has been shown to be the single target site of the herbicide and basis of the mode of action of Glp, even though the specific molecular and enzymological interactions are still being studied. The inhibition of EPSPS causes damage symptoms and herbicidal action through at least two distinct secondary effects on plant metabolism – disruption of the PCR cycle and the inhibition of chorismate and products dependent on it. The corresponding symptoms reflect different pathways for development of damage symptoms, which demonstrates metabolic flexibility in plants. Plants resistant to Glp are characterized by their having an altered EPSPS that has markedly reduced affinity for Glp binding but is able to bind PEP adequately. Advances in studying the configuration of EPSPS with bound ligands has not only given insights into the mechanism of action but also opened a new avenue for rational design of herbicides based on inhibition of EPSPS (Alabhai and Stallings 2001). An alternative strategy for structure-based inhibitor design of molecules that inhibit EPSPS makes use of the induced-fit mechanism triggered by the binding of ligands to EPSPS. Although plants are not likely to develop EPSPS that is resistant to Glp (Bradshaw et al. 1997), reports of increasing numbers of Glp-tolerant populations in agriculture are a cause for concern about the intensive use of a single herbicide.

Acknowledgments. We thank Tracey Reynolds, Kenneth Gruys, June Bourque and Jerome Servaites for valuable critiques and discussions of topic areas in this review and the Monsanto Company for generous research support.

References

Alibhai MF, Stallings WC (2001) Closing down on glyphosate inhibition – with a new structure for drug discovery. Proc Natl Acad Sci USA 98:2944–2946

Amrhein N, Deus B, Steinrücken HC (1980) The site of inhibition of the shikimate pathway by glyphosate. Plant Physiol 66:830–834

Amrhein N, Holländer-Czytko H, Leifeld J, Schulz A, Steinrücken HC, Topp H (1982) Inhibition of the shikimate pathway by glyphosate. In: Boudet AM, Ranjeva R (eds) Journées internationales d'etudes du Groupe Polyphenols. Bulletin de Liason, vol 2, Toulouse, France, pp 21–30

Anderson KS, Johnson KA (1990) Kinetic competence of the 5-enolpyruvylshikimate-3-phosphate synthase tetrahedral intermediate. J Biol Chem 265:5567–5572

Anderson KS, Sikorski JA, Johnson KA (1988) Evaluation of 5-enolpyruvoylshikimate-3-phosphate synthase substrate and inhibitor binding by stopped-flow and equilibrium fluorescence measurements. Biochemistry 27:1604–1610

Badger MR, Sharkey TD, von Caemmerer S (1984) The relationship between steady-state gas exchange of bean leaves and the levels of carbon-reduction-cycle intermediates. Planta 160:305–313

Baird DD, Upchurch RP, Homesley WB, Franz JE (1971) Introduction of a new broad-spectrum post-emergence herbicide class with utility for herbaceous perennial weed control. Proc North Cent Weed Control Conf 26:64–68

Barry G, Kishore G, Padgette S, Taylor M, Kolacz K, Weldon M, Re D, Eichholtz D, Fincher K, Hallas L (1992) Inhibitors of amino acid biosynthesis: strategies for imparting glyphosate tolerance to crop plants. In: Singh BK, Flores HE, Shannon JC (eds) Biosynthesis and molecular regulation of amino acids in plants. Am Soc Plant Physiol, Rockville, MD, pp 139–145

Bentley R (1990) The shikimate pathway – a metabolic tree with many branches. Crit Rev Biochem Mol Biol 25:307–384

Boocock MR, Coggins JR (1983) Kinetics of 5-enolpyruvylshikimate-3-phosphate synthase inhibition by glyphosate. FEBS Lett 154:127–133

Boudet AM, Graziana A, Ranjeva R (1985) Recent advances in the regulation of the prearomatic pathway. In: Van Sumere CF, Lea PJ (eds) Annual proceedings of the phytochemical society of Europe: the biochemistry of plant phenolics. Clarendon Press, Oxford, pp 135–159

Bradshaw LD, Padgette SR, Kimball SL, Wells BH (1997) Perspectives on glyphosate resistance. Weed Technol 11:189–198

Brecke BJ, Duke WB (1980) Effect of glyphosate on intact bean plants (*Phaseolus vulgaris* L.) and isolated cells. Plant Physiol 66:656–659

Cañal Villanueva MJ, Fernandez Muñiz B, Sanchez Tames R (1985) Effects of glyphosate on growth and the chlorophyll and carotenoid levels of yellow nutsedge (*Cyperus esculentus*). Weed Sci 33:751–754

Carlson SJ, Donald WW (1988) Glyphosate effects on Canada thistle (*Cirsium arvense*) roots, root buds, and shoots. Weed Res 28:37–45

Caseley JC, Coupland D (1985) Environmental and plant factors affecting glyphosate uptake, movement and activity. In: Grossbard E, Atkinson D (eds) The herbicide glyphosate. Butterworths, London, pp 92–123

Cole DJ, Dodge AD, Caseley JC (1980) Some biochemical effects of glyphosate on plant meristems. J Exp Bot 31:1665–1674

Cole DJ, Caseley JC, Dodge AD (1983) Influence of glyphosate on selected plant processes. Weed Res 23:173–183

Cranmer JR, Linscott DL (1990) Droplet makeup and the effect of phytotoxicity of glyphosate in velvetleaf (*Abutilon theophrasti*). Weed Sci 38:406–410

Cranmer JR, Linscott DL (1991) Effects of droplet composition on glyphosate absorption and translocation in velvetleaf (*Abutilon theophrasti*). Weed Sci 39:251–254

DeGennaro FP, Weller SC (1984) Differential susceptibility of field bindweed (*Convolvulus arvensis*) biotypes to glyphosate. Weed Sci 32:472–476

Della-Cioppa G, Kishore GM (1988) Import of a precursor protein into chloroplasts is inhibited by the herbicide glyphosate. EMBO J 7:1299–1305

Della-Cioppa G, Bauer SC, Klein BK, Shah DM, Fraley RT, Kishore GM (1986) Translocation of the precursor of 5-enolpyruvylshikimate-3-phosphate synthase into chloroplasts of higher plants in vitro. Proc Natl Acad Sci USA 83:6873–6877

Devine MD (1989) Phloem translocation of herbicides. Rev Weed Sci 4:191–213

Duke SO (1990) Overview of herbicide mechanisms of action. Environ Health Perspect 87: 263–271

Eschenburg S, Schönbrunn E (2000) Comparative X-ray analysis of the un-liganded fosfomycin-target MurA proteins. Struct Funct Genet 40:290–298

Farrar JF, Williams JHH (1991) The effects of increased atmospheric carbon dioxide and temperature on carbon partitioning, source-sink relations and respiration. Plant Cell Environ 14: 819–830

Feng PCC, Pratley JE, Bohn JA (1999) Resistance to glyphosate in *Lolium rigidum*. II. Uptake, translocation, and metabolism. Weed Sci 47:412–417

Fernandez CJ, McInnes KJ, Cothren JT (1994) Carbon balance, transpiration, and biomass partitioning of glyphosate-treated wheat (*Triticum aestivum*) plants. Weed Sci 42:333–339

Ferreira JFS, Reddy KN (2000) Absorption and translocation of glyphosate in *Erythroxylum coca* and *E. novogranatense*. Weed Sci 48:193–199

Fischer K, Kammerer B, Gutensohn M, Arbinger B, Weber A, Häusler RE, Flügge U-I (1997) A new class of plastidic phosphate translocators: a putative link between primary and secondary metabolism by the phosphoenolpyruvate/phosphate antiporter. Plant Cell 9:453–462

Foley ME, Nafziger ED, Slife FW, Wax LM (1983) Effect of glyphosate on protein and nucleic acid synthesis and ATP levels in common cocklebur (*Xanthium pensylvanicum*) root tissue. Weed Sci 31:76–80

Franz JE, Mao MK, Sikorski JA (1997) Glyphosate: a unique global herbicide. ACS Monograph 189, American Chemical Society, Washington, DC

Gaskin RE, Holloway PJ (1992) Some physicochemical factors influencing foliar uptake enhancement of glyphosate mono(isopropylammonium) by polyoxyethylene surfactants. Pestic Sci 34:195–206

Geiger DR, Bestman H (1990) Self-limitation of herbicide mobility by phytotoxic action. Weed Sci 38:324–329

Geiger DR, Servaites JC (1994) Diurnal regulation of photosynthetic carbon metabolism in C3 plants. Annu Rev Plant Physiol Plant Mol Biol 45:235–256

Geiger DR, Kapitan SW, Tucci MA (1986) Glyphosate inhibits photosynthesis and allocation of carbon to starch in sugar beet leaves. Plant Physiol 82:468–472

Geiger DR, Tucci MA, Servaites JC (1987) Glyphosate effects on carbon assimilation and gas exchange in sugar beet leaves. Plant Physiol 85:365–369

Geiger DR, Koch KE, Shieh W-J (1996) Effect of environmental factors on whole plant assimilate partitioning and associated gene expression. J Exp Bot 47:1229–1238

Geiger DR, Shieh W-J, Fuchs M (1999) Causes of self-limited translocation of glyphosate in Beta vulgaris plants. Pestic Biochem Physiol 64:124–133

Giesy JP, Dobson S, Solomon KR (2000) Ecotoxicological risk assessment for Roundup herbicide. Rev Environ Contam Toxicol 167:35–120

Gougler JA, Geiger DR (1981) Uptake and distribution of N-phosphonomethylglycine in sugar beet plants. Plant Physiol 68:668–672

Gruys KJ, Sikorski JA (1999) Inhibitors of tryptophan, phenylalanine, and tyrosine biosynthesis as herbicides. In: Singh BK (ed) Plant amino acids: biochemistry and biotechnology. Marcel Dekker, New York, pp 357–384

Gruys KJ, Walker MC, Sikorski JA (1992) Substrate synergism and the steady-state kinetic reaction mechanism for EPSP synthase from *Escherichia coli*. Biochemistry 31:5534–5544

Haslam E (1974) The shikimate pathway. Butterworth, London

Heap IM (1997) The occurrence of herbicide-resistant weeds worldwide. Pestic Sci 51:235–243

Hermann K (1995) The shikimate pathway as an entry to aromatic secondary metabolism. Plant Physiol 107:7–12

Herrmann KM, Weaver LM (1999) The shikimate pathway. Annu Rev Plant Physiol Plant Mol Biol 50:473–503

Hess FD (1985) Herbicide absorption and translocation and their relationship to plant tolerances and susceptibility. In: Duke SO (ed) Weed physiology, vol 2. Herbicide physiology. CRC Press Inc, Boca Raton, pp 191–214

Ho LC (1988) Metabolism and compartmentation of imported sugars in sink organs in relation to sink strength. Annu Rev Plant Physiol Plant Mol Biol 39:355–378

Hoagland RE, Duke SO (1982) Biochemical effects of glyphosate [N-(phosphonomethyl) glycine]. In: Moreland DE, St John JB, Hess FD (eds) Biochemical responses induced by herbicides. American Chemical Society, Washington, DC, Symposium Series, no 181, pp 175–205

Holländer H, Amrhein N (1980) The site of the inhibition of the shikimate pathway by glyphosate. I. Inhibition by glyphosate of phenylpropanoid synthesis in buckwheat (*Fagopyrumm esculentum* Moench). Plant Physiol 66:823–829

Holländer-Czytko H, Amrhein N (1987) 5-enolpyruvylshikimate 3-phosphate synthase, the target enzyme of the herbicide glyphosate, is synthesized as a precursor in a higher plant. Plant Physiol 83:229–231

Hsu FC, Kleier DA, Melander WR (1988) Phloem mobility of xenobiotics II. Bioassay testing of the unified model. Plant Physiol 86:811–816

Jaworski EG (1972) Mode of action of *N*-phosphonomethyl-glycine: inhibition of aromatic amino acid biosynthesis. J Agric Food Chem 20:1195–1198

Jensen R (1986) The shikimate/arogenate pathway: link between carbohydrate metabolism and secondary metabolism. Physiol Plant 66:164–168

Johnston DT, Faulkner JS (1991) Herbicide resistance in the Graminaceae – a plant breeder's view. In: Caseley JC, Cussans GW, Atkins RT (eds) Herbicide resistance in weeds and crops. Butterworth-Heinemann, Oxford, pp 319–330

Jordan DL, York AC, Griffin JL, Clay PA, Vidrine PR, Reynolds DB (1997) Influence of application variables on efficacy of glyphosate. Weed Technol 11:354–362.

Kitchen LM, Witt WW, Rieck EE (1981) Inhibition of chlorophyll accumulation by glyphosate. Weed Sci 29:513–516

Kleier DA (1988) Phloem mobility of xenobiotics. I. Mathematical model unifying the weak acid and intermediate permeability theories. Plant Physiol 86:803–810

Knaggs AR (1999) The biosynthesis of shikimate metabolites. Nat Prod Rep 16:525–560

Laerke PE, Streibig JC (1995) Foliar absorption of some glyphosate formulations and their efficacy on plants. Pestic Sci 44:107–116

Leaper C, Holloway PJ (2000) Adjuvants and glyphosate activity. Pestic Manage Sci 56:313–319

Lee LJ, Ngim J (2000) A first report of glyphosate-resistant goosegrass (*Eleusine indica* (L.) Gaertn) in Malaysia. Pestic Manage Sci 56:336–339

Leegood RC, von Caemmerer S (1988) The relationship between contents of photosynthetic metabolites and the rate of photosynthetic carbon assimilation in leaves of *Amaranthus edulis* L. Planta 174:253–262

Lorraine-Colwill DF, Hawkes TR, Williams PH, Warner SAJ, Suton SB, Preston C (1999) Resistance to glyphosate in *Lolium rigidum*. Pestic Sci 55:486–503

Lorraine-Colwill DF, Powles SB, Hawkes TR, Preston C (2001) Inheritance of evolved glyphosate resistance in *Lolium rigidum* (Gaud). Theor Appl Genet 102:545–550

Madsen KH, Jensen JE (1995) Weed control in glyphosate tolerant sugarbeet (*Beta vulgaris* L.). Weed Res 35:105–111

Madsen KH, Heitholt JJ, Duke SO, Smeda RJ, Streibig JC (1995) Photosynthetic parameters in glyphosate-treated sugarbeet (*Beta vulgaris* L.). Weed Res 35:81–88

Maier L (2000) What are the requirements in the glyphosate molecule in order for it to be herbicidally active? Heteroatom Chem 11:454–469

Mannerlöf M, Tuvesson S, Steen P, Tenning P (1997) Transgenic sugar beet tolerant to glyphosate. Euphytica 94:83–91

Mitich LW (1991) Velvetleaf. Weed Technol 5:253–255

Mollenhauer C, Smart CC, Amrhein N (1987) Glyphosate toxicity in the shoot apical region of the tomato plant. I. Plastid swelling is the initial ultrastructural feature following in vivo inhibition of 5-enolpyruvylshikimic acid 3-phosphate synthase. Pestic Biochem Physiol 29:55–65

Mousdale DM, Coggins JR (1985) Subcellular localization of the common shikimate-pathway enzymes in *Pisum sativum* L. Planta 163:241–249

Nafziger ED, Slife FW (1983) Physiological response of common cocklebur (*Xanthium pensylvanicum*) to glyphosate. Weed Sci 31:874–878.

Padgette SR, Huynh QK, Borgmeyer J, Shah DM, Brand LA, Re DB, Bishop BF, Rogers SG, Fraley RT, Kishore GM (1987) Bacterial expression and isolation of *Petunia hybrida* 5-enolpyruvylshikimate-3-phosphate synthase. Arch Biochem Biophys 258:564–573

Padgette SR, Re DB, Gasser CS, Eichholtz DA, Frazier RB, Hironaka CM, Levine EB, Shah DM, Fraley RT, Kishore GM (1991) Site-directed mutagenesis of a conserved region of the 5-enolpyruvylshikimate-3-phosphate synthase active site. J Biol Chem 266:22364–22369

Padgette SR, Kolacz KH, Delanneay X, Re DB, LaVallee BJ, Tinius CN, Rhodes WK, Otero YI, Barry GF, Eichholtz DA, Peschke VM, Nida DL, Taylor NB, Kishore GM (1995) Development, identification and characterization of a glyphosate-tolerant soybean line. Crop Sci 35:1451–1461

Padgette SR, Re DB, Barry GF, Eichholtz DE, Delannay X, Fuchs RL, Kishore GM, Fraley RT (1996) New weed control opportunities: development of soybeans with a Roundup Ready gene. In: Duke SO (ed) Herbicide-resistant crops: agricultural, economic, environmental, regulatory, and technological aspects. CRC Press, Boca Raton, pp 53–84

Powles SB, Lorraine-Colwill DF, Dellow JJ, Preston C (1998) Evolved resistance to glyphosate in rigid ryegrass (*Lolium rigidum*) in Australia. Weed Sci 46:604–607

Pratley J, Baines P, Eberbach P, Incerti M, Broster J (1996) Glyphosate resistance in annual ryegrass. In: Virgona J, Michalk D (ed) Proc of the 11th Annu Conf Grasslands Soc New South Wales. The Grasslands Society of NSW, Wagga Wagga, Australia, p 126

Pratley J, Urwin N, Stanton R, Baines P, Broster J, Cullis K, Schafer D, Bohn J, Krueger R (1999) Resistance to glyphosate in *Lolium rigidum*. I. Bioevaluation. Weed Sci 47:405–411

Preiss J (1988) Biosynthesis of starch and its regulation. In: Stumpf PK, Conn EE (eds) The biochemistry of plants, vol 14. Academic Press, New York, pp 181–254

Preston C, Tardif JT, Christopher T, Powles SB (1996) Multiple resistance to dissimilar herbicide chemistries in a biotype of *Lolium rigida* due to enhanced activity of several herbicide degrading enzymes. Pestic Biochem Physiol 54:123–134

Rao IM, Terry N (1995) Leaf phosphate status, photosynthesis, and carbon partitioning in sugar beet. Plant Physiol 107:1313–1321

Riechers DE, Wax LM, Liebl RA, Bush DA (1994) Surfactant-increased glyphosate uptake into plasma membrane vesicles isolated from common lambsquarters leaves. Plant Physiol 105: 1419–1425

Sammons RD, Gruys KJ, Anderson KS, Johnson KA, Sikorski JA (1995) Reevaluating glyphosate as a transition-state inhibitor of EPSP synthase: identification of an EPSP synthase·EPSP·glyphosate ternary complex. Biochemistry 34:6433–6439

Schmid J, Amrhein N (1999) The shikimate pathway. In: Singh BK (ed) Plant amino acids: biochemistry and biotechnology. Marcel Dekker, New York, pp 147–169

Schönbrunn E, Eschenburg S, Krekel F, Luger K, Amrhein N (2000a) Role of the loop containing residue 115 in the induced-fit mechanism of the bacterial cell wall biosynthetic enzyme MurA. Biochemistry 39:2164–2173

Schönbrunn E, Eschenburg S, Luger K, Kabsch W, Amrhein N (2000b) Structural basis for the interaction of the fluorescence probe 8-anilo-1-naphthaline sulfonate (ANS) with the antibiotic target MurA. Proc Natl Acad Sci USA 97:6345–6349

Schönbrunn E, Eschenburg S, Schuttleworth WA, Schloss JV, Amrhein N, Evans JNS, Kabsch W (2001) Interaction of the herbicide glyphosate with its target enzyme 5-enolpyruvylshikimate 3-phosphate synthase in atomic detail. Proc Natl Acad Sci USA 98:1376–1380

Schulz A, Kruper A, Amrhein N (1985) Differential sensitivity of bacterial 5-enolpyruvylshikimate-3-phosphate synthases to the herbicide glyphosate. FEMS Microbiol Lett 28:297–301

Schulz A, Munder T, Holländer-Czytko H, Amrhein N (1990) Glyphosate transport and early effects on shikimate metabolism and its compartmentation in sink leaves of tomato and spinach plants. Z Naturforsch 45:529–534

Schulze W, Stitt M, Schulze E-D, Neuhaus HE, Fichter K (1991) A quantification of the significance of assimilatory starch for the growth of *Arabidopsis thaliana* (L) Heyn. Plant Physiol 95:890–895

Servaites JC, Tucci MA, Geiger DR (1987) Glyphosate effects on carbon assimilation, ribulose bisphosphate carboxylase activity, and metabolite levels in sugar beet leaves. Plant Physiol 85:370–374

Servaites JC, Shieh W-J, Geiger DR (1991) Regulation of photosynthetic carbon reduction cycle by ribulose bisphosphate and phosphoglyceric acid. Plant Physiol 97:1115–1121

Shah DM, Horsch RB, Klee HJ, Kishore GM, Winter JA, Tumer NE, Hironaka CM, Sanders PR, Gasser CS, Aykent SA, Siegel NR, Rogers SG, Fraley RT (1986) Engineering herbicide tolerance in transgenic plants. Science 233:478–481

Sharkey TD (1998) Photosynthetic carbon reduction. In: Raghavendra AS (ed) Photosynthesis: a comprehensive treatise. Cambridge University Press, Cambridge, pp 111–122

Shieh WJ, Geiger DR, Servaites JC (1991) Effect of glyphosate on carbon assimilation and metabolism during a simulated natural day. Plant Physiol 97:1109–1114

Siehl DL (1997) Inhibitors of EPSP synthase, glutamine synthetase and histidine synthesis. In: Roe RM, Burton JD, Kuhr RJ (eds) Herbicide activity: toxicity, biochemistry and molecular biology. IOS Press, Amsterdam

Siehl DS (1999) The biosynthesis of tryptophan, tyrosine, and phenylalanine from chorsimate. In: Singh BK (ed) Plant amino acids: biochemistry and biotechnology. Marcel Dekker, New York, pp 171–204

Sikorski JA, Gruys KJ (1997) Understanding glyphosate's molecular mode of action with EPSP synthase: evidence favoring an allosteric inhibitor model. Acc Chem Res 30:2–8

Singh BK, Siehl DL, Connelly JA (1991) Shikimate pathway: why does it mean so much to so many? In: Miflin BJ, Miflin HF (eds) Oxford surveys of plant molecular and cell biology. Oxford University Press, New York, pp 143–185

Sprankle P, Meggitt WF, Penner D (1975) Absorption, action, and translocation of glyphosate. Weed Sci 23:235–240

Stallings WC, Abdel-Meguid SS, Lim LW, Shieh HS, Dayringer HE, Leimgruber NK, Stegemann RA, Anderson KS, Sikorski JA, Padgette SR, Kishore GM (1991) Structure and topological symmetry of the glyphosate target 5-enolpyruvylshikimate-3-phosphate synthase – a distinctive protein fold. Proc Natl Acad Sci USA 88:5046–5050

Steinrücken HC, Amrhein N (1980) The herbicide glyphosate is a potent inhibitor of 5-enolpyruvyl-shikimic acid-3-phosphate synthase. Biochem Biophys Res Comm 94:1207–1212

Steinrücken HC, Amrhein N (1984) 5-Enolpyruvylshikimate-3-phosphate synthase of *Klebsiella pneumoniae*. 2. Inhibition by glyphosate [*N*-(phosphonomethyl)glycine]. Eur J Biochem 143: 351–357

Sterling TM (1994) Mechanisms of herbicide absorption across plant membranes and accumulation in plant cells. Weed Sci 42:263–276

Tardif FJ, Preston C, Powles SB (1997) Mechanisms of herbicide multiple resistance in *Lolium rigidum*. In: Prado RD, Jorin J, Garcia-Torres L (eds) Weed and crop resistance to herbicides. Kluwer, Dordrecht, pp 117–124

Tyree MT, Peterson CA, Edgington LV (1979) A simple theory regarding ambimobility of xenobiotics with special reference to the nematicide, oxamyl. Plant Physiol 63:367–374

Warwick SI (1991) Herbicide resistance in weedy plants: physiology and population biology. Annu Rev Ecol Syst 22:95–114

Westwood JH, Weller SC (1997) Cellular mechanisms influence differential glyphosate sensitivity in field bindweed (*Convolvulus arvensis*) biotypes. Weed Sci 45:2–11

Williams GM, Kroes R, Munro IC (2000) Safety evaluation and risk assessment of the herbicide Roundup and its active ingredient, glyphosate, for humans. Reg Toxicol Pharmacol 31:117–165

Woodburn AT (2000) Glyphosate: production, pricing and use worldwide. Pestic Manage Sci 56: 309–312

4 Inhibitors of Glutamine Synthetase

GUENTER DONN and HELMUT KÖCHER

4.1 Introduction

Due to the chemical inertness of the N_2 molecule, the availability of metabolically accessible inorganic nitrogen sources was a key limiting factor for plant life on earth. The large scale conversion of N_2 into NH_3 and NO_3^- and their use as nitrogen fertilizer has been an agricultural standard for less than a century. The highly efficient recycling and detoxification of NH_3 released by metabolic processes, as glycine-serine conversion in photorespiration, catabolic metabolism of amino acids and nucleic acids and nitrite reduction, is essential for plants, due to the limiting availability of nitrogen, and the toxicity and volatility of ammonia.

During the past 25 years, it has become evident that for ammonia assimilation and detoxification in plants glutamine synthetase (GS) is the key enzyme. The discovery of potent GS inhibitors and the generation of GS-deficient barley mutants revealed the central role of plant glutamine synthetase in nitrogen assimilation. By this means, it became evident that GS is a potential target for novel herbicides (Lea 1991). One potent glutamine synthetase inhibitor, phosphinothricin (glufosinate), which was discovered as the biologically active amino acid of tripeptides produced by at least three different *Streptomyces* species was introduced as a nonselective herbicide for broad spectrum post-emergent weed control in 1984. After elucidation of phosphinothricin-inactivating enzymes and their respective phosphinothricin-resistance genes in two of the producer strains, attempts were started to use these genes to generate selectivity in crop species for this naturally occurring amino acid with herbicidal activity. For more than 5 years glufosinate-tolerant canola and maize varieties have been grown on farmers' fields in North America and glufosinate is used as a selective herbicide for post-emergent weed control in these varieties.

4.2 Plant Glutamine Synthetase Isoforms and Their Function

Plant glutamine synthetases (GS; E.C. 6.3.1.2) consist, like all known eukaryotic glutamine synthetase enzymes, of eight subunits (McNally and Hirel 1983). The molecular weight of the subunits varies in the range of 38–45 kDa, depend-

P. Böger, K. Wakabayashi, K. Hirai (Eds.)
Herbicide Classes in Development

ing on the species and the subcellular localization of the respective isoforms. At least one cytosolic isoform (GS_1) and one located in the plastids (GS_2) can be distinguished in most higher plants (Peterman and Goodman 1991), whereas their relative abundance varies considerably between species. In some species a root-specific isoform (GS_R) can be distinguished and in legumes at least one nodule-specific isoform (GS_{N1}) was discovered. The polypeptides of the isoforms are encoded by a small multigene family localized in the nucleus. The chloroplast-specific polypeptides are synthesized in the cytoplasm as precursor molecules. The leader peptide sequence is cleaved, after the transfer of the polypeptide chain into the chloroplast has occurred (Forde and Cullimore 1989). The expression of the GS_2 gene is enhanced by light and high sucrose levels (Oliveira and Coruzzi 1999).

Each subunit of the enzyme has an active center with binding capacity for the substrates. The enzyme converts glutamic acid and ammonia into glutamine by the formation of a high energy intermediate (glutamyl phosphate). The phosphorylation requires ATP and Mg^{2+} ions (Fig. 1).

By this reaction, ammonia is withdrawn and glutamate is converted into its amide glutamine. Because membranes are permeable for ammonia (Kleiner 1981), the high affinity of the enzyme for its substrates (K_m 3–5 μM) and the high abundance of the enzyme at the sites of ammonia generation (chloroplasts and root nodules of legume species), the leakage of the valuable nitrogen as gaseous ammonia into the environment is prevented. On the other hand, glutamine is a transport compound for nitrogen in the cells as well as an important intermediate for the synthesis of amino acids and nucleotides. Most importantly, glutamine is the substrate for glutamate synthase (Fig. 2), an enzyme which synthesizes two equivalents of glutamate from glutamine and 2-oxoglutaric acid (Miflin and Lea 1980).

Glutamate is a general amino donor for transaminases and, in addition, it is an essential precursor for several amino acid synthesis pathways as well as for the biosynthesis of chlorophyll.

Due to the high affinity of glutamine synthetase for its substrates, glutamate is the predominant acceptor for ammonia released in a plant cell. The main sources of ammonia in plants are the light-dependent photorespiratory glycine–serine conversion and nitrite reduction and to a lower degree catabolic reactions. According to Keys et al. (1978), 90% of the ammonia recycled in plants originates from photorespiratory glycine-serine conversion. Therefore, it is evident that blocking of the glutamine synthetase will affect photosynthetic-active plant tissues stronger, by far, than nonactive tissues or plants in the dark.

$$\text{L-glutamate} + NH_3 + \text{ATP} \xrightarrow[\text{GS}]{Mg^{2+}} \text{L-glutamine, ADP} + P_i$$

Fig. 1. Synthesis of glutamine; *GS* glutamine synthetase

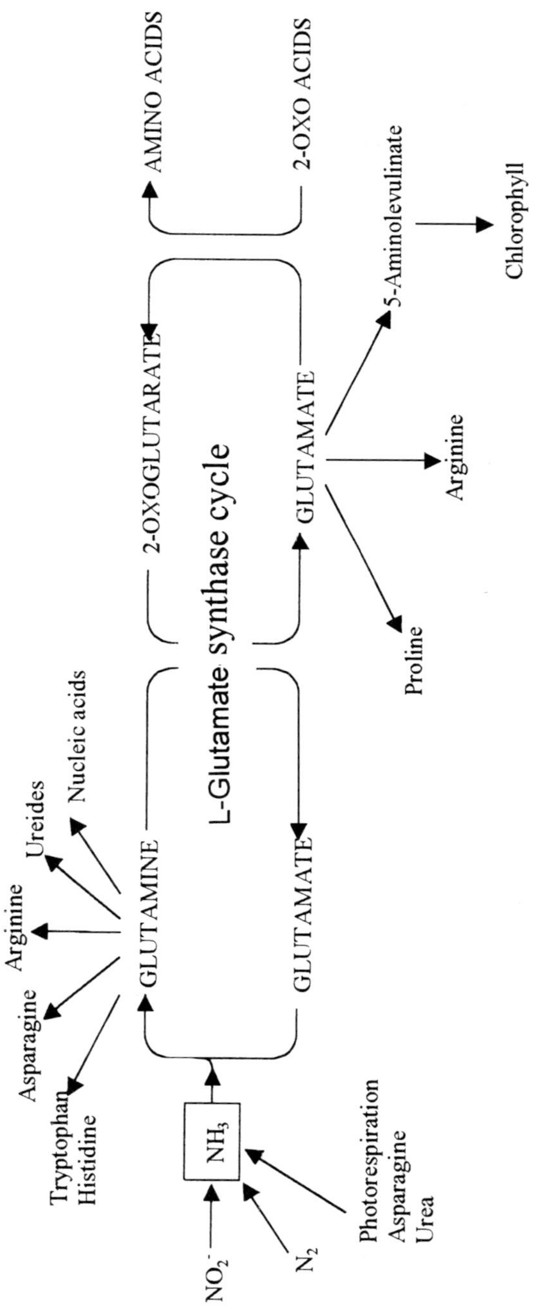

Fig. 2. Central role of the glutamine synthetase/glutamate synthase cycle in plant N metabolism. (Lea 1993)

GS_2-deficient barley mutants which were isolated under conditions which suppress photorespiration grow normally under nonphotorespiratory conditions (2% O_2, 0.7% CO_2), but mutants with less than 40% of the wild-type GS_2 activity show severe phytotoxic symptoms when grown under normal atmospheric conditions in full light (Wallsgrove et al. 1987). The mutants show a significant increase in the level of detectable ammonia in their leaves, depending on the light intensity. Interestingly, under photorespiratory conditions this increase in ammonia level is correlated with the development of phytotoxic symptoms, whereas under nonphotorespiratory conditions the development of phytotoxic symptoms is suppressed even though the ammonia level is elevated (Lea and Ridley 1989).

4.3 Glutamine Synthetase Inhibitors

Structural analogues of glutamic acid which inhibit glutamine synthetase activity in vitro were discovered almost 50 years ago. Methionine sulfoximine has been known as a potent inhibitor of prokaryotic GS since 1952 (Pace and McDermott 1952; Leason et al. 1982). Phosphonic acid derivatives with structural analogy to glutamate were synthesized and their inhibitory effects on GS activity were demonstrated by Mastalerz (1959; see Fig. 3).

In the late 1960s the team of Prof. Zaehner at the University of Tübingen discovered a tripeptide produced by *Streptomyces viridochromogenes* due to

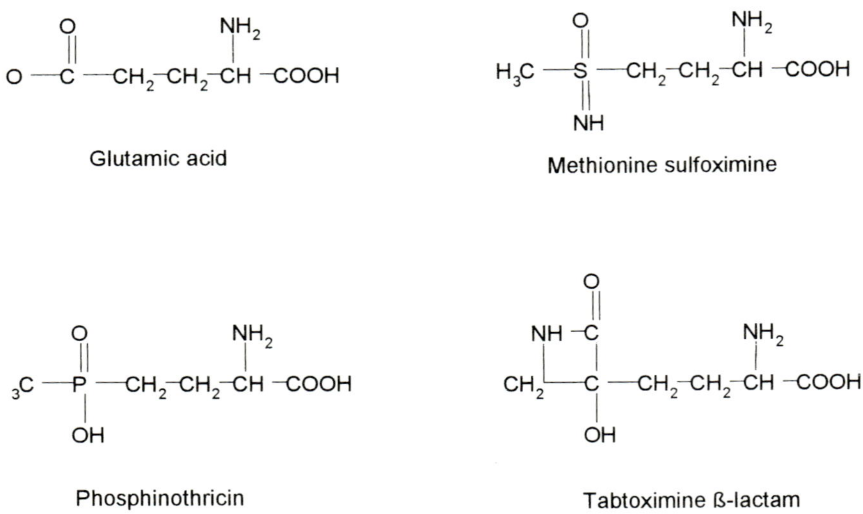

Fig. 3. Glutamate and some analogues with reported inhibitory activity on plant glutamine synthetase

its inhibitory activity against bacteria. The peptide consists of two alanine residues linked to a unique amino acid which was named phosphinothricin (Bayer et al. 1972), whereas the tripeptide phosphinothricyl-alanyl-alanine was named later bialaphos. The hypothesis that phosphinothricin may be a potential GS inhibitor due to its structural analogy to glutamate was tested by Bayer et al. and the high inhibitory activity for bacterial GS was demonstrated.

Bayer et al. concluded, therefore, that phosphinothricin is the biologically active amino acid of the tripeptide, despite the fact that its inhibitory effect on bacterial growth was 1000 to 10,000 times weaker than the bactericidal effect of the tripeptide. It was concluded that this striking difference is a consequence of the active uptake of the tripeptide via bacterial peptide carriers, whereas in bacteria no active transport system for glutamic acid and its analogue exists.

Independently from the research activities dedicated to *Streptomyces viridochromogenes*, a Japanese research team at Meiji Seika Kaisha Company discovered a *Streptomyces* strain producing an antibiotic which showed biological activity comparable to phosphinothricyl-alanyl-alanine (Niida et al. 1973). The strain was named *Streptomyces hygroscopicus* and the biologically active compound was identical to the tripeptide from *S. viridochromogenes* and was named bialaphos (Ogawa 1973a,b).

A novel phosphinothicin producing *Streptomyces* strain which produces a different tripeptide (phosalacine), in which one alanine molecule is replaced by leucine, was described by Omura et al. (1984a,b).

An interesting GS inhibitor was identified as a causative agent of phytotoxic symptoms of a phytopathogenic strain of *Pseudomonas syringae* pv. *tabaci* (Langston-Unkefer et al. 1984). These bacteria cause a halo of senescing tissue in the vicinity of the infection site which is then colonized by the bacteria. The inhibitor tabtoximine β-lactam (Fig. 3) shows structural analogy to glutamate.

4.4 Discovery of the Herbicidal Activity of Phosphinothricin and Bialaphos

In the mid-1970s, phosphinothricin was synthesized in the central research laboratories at Hoechst. The resulting racemic DL-phosphinothricin was tested for its herbicidal activity profile in the Biological Research Department of the Agricultural Division.

Its activity as a soil herbicide for pre-emergent weed control was weak even at the high dosages used in the primary screening. In contrast, its activity as a foliar herbicide for post-emergent weed control was striking. Secondary screening on a broad range of weed species and the selectivity test in the greenhouse revealed the strong herbicidal activity against almost all weed species tested and showed that the compound had no selectivity in field crops. After field trials had confirmed the broad spectrum weed control potential of DL-phosphinothricin, the further development of the compound as a nonselective post-emergent herbicide was initiated (Schwerdtle et al. 1981).

The product was introduced to the market under the common name glufosinate ammonium in 1984 as a nonselective post-emergent herbicide for directed spray application in vineyards and its use was later extended to orchards and plantation crops; subsequently, other uses were developed.

In Japan, Meiji Seika investigated the herbicidal activity of bialaphos (Takematsu et al. 1979a,b). As a consequence of the good performance of the natural product for weed control after foliar application, the tripeptide was developed as a foliar herbicide. It was introduced to the market in 1984 under the trade name Herbiace (Mase 1984).

The discovery of the herbicidal activity of phosphinothricin (glufosinate) and its commercial exploitation triggered an intensive search for further GS inhibitors. To date, the natural compound and the synthetic racemic analogue of L-phosphinothricin (glufosinate) are still the most efficient molecules, whereas all discovered derivatives showed weaker herbicidal activity or no activity at all.

4.5 Mode of Glutamine Synthetase Inhibition

For methionine sulfoximine (MSO) and later for phosphinothricin, GS inhibition was studied in detail. Ronzio and Meister (1968) developed a model of GS inhibition for MSO in which they postulated that MSO mimics the activated glutamate. They concluded that MSO inhibits GS in two steps. The first step is reversible, where the inhibitor competes with glutamate at the binding site. The second step is irreversible. In binding studies using radiolabelled MSO and ^{32}P-labelled ATP, they isolated an MSO derivative from heat or acid-denatured GS protein, which was then identified as MSO phosphate. Similar experiments were made by Manderscheid and Wild (1986) using phosphinothricin as inhibitor. They confirmed the two-step reaction. Again, the initial binding to the GS enzyme was competitive, whereas the phosphorylated phosphinothricin was irreversibly bound to the enzyme. Manderscheid and Wild concluded that each of the eight subunits of the GS enzyme is able to bind one phosphinothricin molecule (Fig. 4).

Acaster and Weitzman (1985) determined K_i values for GS_1 and GS_2 inhibition by glufosinate and found only small differences: in maize (42% GS_1, 58% GS_2 activity) the K_i value was 2.0 μM for GS_1 and 4.0 μM for GS_2. In barley (9% GS_1, 91% GS_2 activity), the K_i value was 3.5 μM for GS_1 and 6.0 μM for GS_2.

Ridley and McNally (1985) selected plant species with different in vivo susceptibility towards glufosinate and determined the ratios of GS_1 and GS_2 and the K_i values of the isolated isoenzymes for glufosinate. In most species GS_2 was the predominating isoenzyme. The K_i values were all in a similar

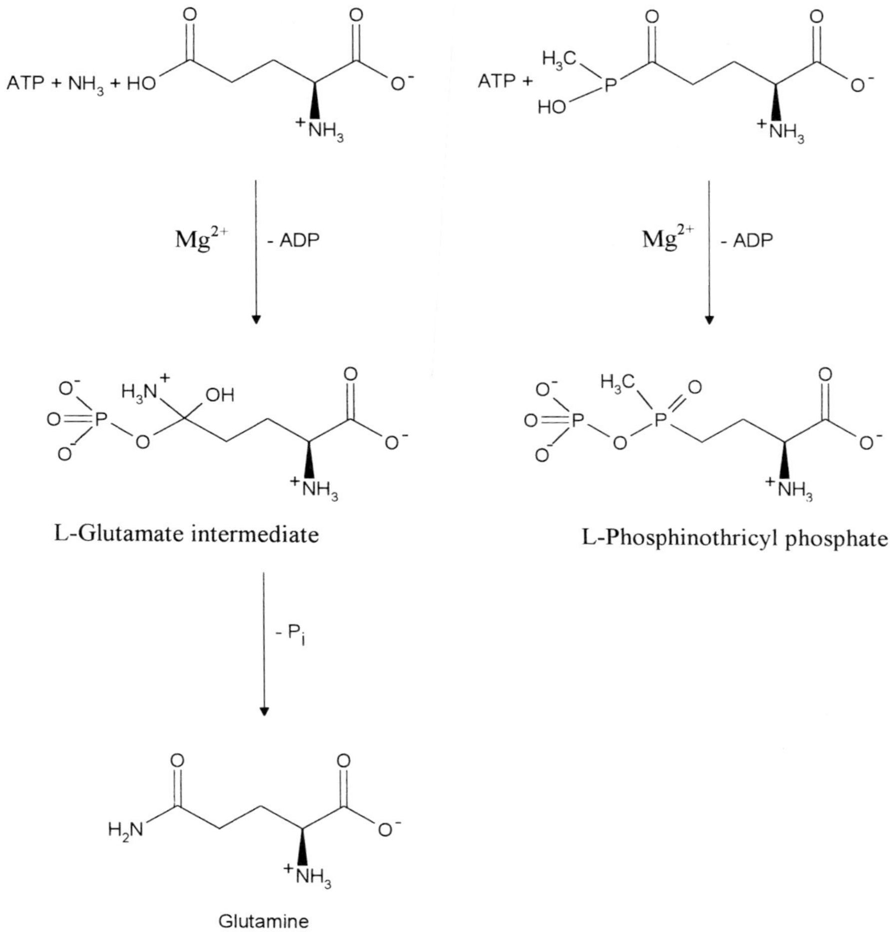

Fig. 4. Structure of glutamylphosphate and of phosphinothricyl phosphate. (Hörlein 1994)

range and not correlated to the different in vivo susceptibility of these plant species.

Only the L-enantiomer of the racemic DL-homoalanin-4-yl(methyl)phosphonic acid (glufosinate), which is identical to the naturally occurring amino acid phosphinothricin, acts as an inhibitor of GS. The tripeptide bialaphos does not inhibit GS itself. After foliar uptake, the peptide is cleaved and the GS inhibitor L-phosphinothricin is released.

Therefore, both commercially available herbicides reveal their activity is due to the presence of the same active ingredient.

4.6 Effects of Glutamine Synthetase Inhibitors in Plants

4.6.1 Visible Symptoms of Herbicidal Action

The time course and the pattern of symptom development following glufosinate treatment depends on weed species and environmental conditions. Within 2 days of glufosinate application or earlier, faint pale green or yellowish discolorations appear on the leaves, often beginning in the interveinal zones. These initial symptoms subsequently develop into leaf chlorosis and desiccation (necrosis). The appearance of foliar desiccation symptoms indicates a perturbation of plant membrane functions soon after application of the herbicide. Depending on the weed species, chlorotic and desiccated leaf zones can appear simultaneously, whereas in other species, typically in grass weeds, extensive chlorosis develops initially and is followed later by desiccation which starts from the leaf tips. Complete death of the weeds usually occurs between 1 and 2 weeks after herbicide treatment.

4.6.2 Physiological Effects of GS Inhibition in Plants by Phosphinothricin

Following treatment with glufosinate, plants kept in the light show a marked increase in ammonia levels and a decrease in the levels of glutamine, glutamate, asparagine, aspartate, alanine, glycine and serine in the leaf tissue within a few hours. Levels of branched-chain and aromatic amino acids, also of lysine and arginine, increase at the same time. These changes are paralleled by a rapid drop of photosynthetic CO_2 fixation. These effects have been shown to occur both in C_3 and C_4 plants, but ammonia accumulation and photosynthesis inhibition are less rapid in C_4 than in C_3 plant species (Köcher and Lötzsch 1985; Wendler et al. 1990; Shelp et al. 1992).

In glufosinate-treated plants maintained in light, ammonia levels were up to ten times higher 4h after treatment, and 1 day after treatment up to two orders of magnitude higher than in control plants. Accumulation of ammonia and the development of visible symptoms of phytotoxicity were much slower in plants which were transferred to darkness immediately after treatment. However, ammonia levels and phytotoxicity symptoms increased rapidly when these treated plants were again exposed to light after 1 day (Köcher 1989). These findings are in accordance with the fact that nitrite reduction to ammonia and ammonia generation during the photorespiratory glycine-serine conversion are dependent on light coupled with the knowledge that inhibition of GS prevents ammonia assimilation and reassimilation into organic N compounds. Ammonia in high concentration is regarded as toxic to plants (Jungk 1984). As proposed by Roberts and Pang (1992), ammonia in high concentration may cause a collapse of the pH gradient between cytoplasm and vacuole. As a consequence, a perturbation of membrane transport processes across the tonoplast membrane will occur, resulting in cytotoxic effects.

Based upon field reports, it appears that glufosinate activity is higher when sprays are made on days with strong sunlight. These observations may be explained by the linkage of glufosinate action to light-dependent physiological processes, namely photorespiration (Keys et al. 1978).

The above-mentioned rapid decrease in tissue levels of glutamine, glutamate, asparagine, aspartate, alanine, serine and glycine after glufosinate treatment was regarded as a direct consequence of the inhibition of the glutamine synthetase/glutamate synthase cycle and of rapid catabolism of these amino acids to ammonia. The aromatic and branched-chain amino acids as well as lysine and arginine, on the other hand, are amino acids with a slow catabolism, and their accumulation in glufosinate-treated plants appeared to be a consequence of protein hydrolysis. Lacuesta et al. (1989) found a 40% reduction in protein content 48 h after treatment of plants with glufosinate. Possibly, protein catabolism was activated as a consequence of the amino-N shortage after glufosinate application.

The rapid inhibitory effect of glufosinate on photosynthetic CO_2-fixation has found considerable interest. In contrast to phenylureas, triazines or other PS II inhibitors, glufosinate does not directly interfere with photosynthetic electron transport. Whereas CO_2-fixation was inhibited within a few hours after spraying plants with glufosinate, the photosynthetic electron transport of chloroplasts prepared from these plants did not decrease earlier than 48 h after spraying the herbicide, suggesting that changes in the photosynthetic apparatus are a secondary effect. Since it is well known from in vitro studies that ammonia can uncouple photophosphorylation, due to a decrease in ΔpH across the thylakoid membrane, it was originally believed that the inhibition of photosynthesis by glufosinate might be caused by the strongly elevated ammonia levels in the leaf tissue. However, the data from chlorophyll fluorescence emission measurements in vivo did not indicate an uncoupling of photophosphorylation (Lacuesta et al. 1992). Accordingly, in trials with mustard plants it was found that photosynthesis inhibition by glufosinate could be prevented, when the plants were kept under nonphotorespiratory conditions (atmosphere with 1000 ppm CO_2, 2% O_2). This was despite the fact that ammonia concentrations had accumulated, which under photorespiratory conditions would have been accompanied by a strong inhibition of photosynthesis (Wild et al. 1987). Furthermore, in petiole feeding trials with excised mustard leaves under normal atmospheric (photorespiratory) conditions it was shown that inhibition of photosynthesis by glufosinate could be reduced markedly if the herbicide was fed in combination with glutamate or glutamine. When exogenous glutamate or glutamine were given, even more ammonia was produced in glufosinate-treated plants than without the feeding of these amino acids. From this, it was concluded that high ammonia levels were not the cause for the rapid decrease of photosynthetic CO_2 fixation after glufosinate application (Sauer et al. 1987; Wild and Wendler 1990).

Based on these findings, it was suggested that photosynthesis inhibition under normal atmospheric, hence photorespiratory, conditions was mainly due to a block or restriction of carbon flow through the photorespiratory pathway, most likely due to depletion of NH_2 donors necessary for the conversion of glyoxylic acid to glycine. This was further corroborated by research with mutants of C_3 plants which lacked chloroplastic glutamine synthetase. These mutants were unable to assimilate the ammonia released during the conversion of glycine to serine in photorespiration. The mutants showed severe symptoms of stress when exposed to air under photorespiratory conditions, but grew normally when photorespiration was suppressed by an increase of CO_2 in the air (Lea 1991).

It was proposed that the inhibition of photosynthetic CO_2 fixation as a consequence of blocked or restricted photorespiratory glyoxylate-glycine conversion is due to insufficient recycling of carbon from the photorespiratory pathway back to the Calvin cycle and/or due to an inhibition of ribulose-bisphosphate carboxylase activation by glyoxylate, which cannot be further metabolized to glycine (Wendler et al. 1992; Wild and Wendler 1993). At present, the evidence is still insufficient to give a final answer to this question.

From the data available, it can be concluded that glufosinate, as a consequence of GS inhibition, leads to plant death by multiple interference with plant metabolism:

1. impairment of membrane functions by ammonia accumulation
2. decreased peptide, protein and nucleotide biosynthesis by a lack of organic N donors for transamination and transamidation reactions
3. increase in proteolysis
4. rapid inhibition of photosynthetic CO_2 fixation as a consequence of an impairment of the photorespiratory pathway, followed by permanent damage of the photosynthetic apparatus.

4.7 Attempts to Generate Selectivity for Glufosinate

Due to its biological properties, namely its efficiency in broad spectrum weed control, the rapid and complete biodegradability in the biosphere and its low toxicity for nontarget organisms (Dorn et al. 1992), attempts were initiated to find a means which would allow the use of glufosinate as a selective herbicide for post-emergent weed control in annual field crops. The use of this herbicide with a unique mode of action would give the farming community new options in weed control and the compound could be used as a building block for agricultural production conditions where weed control is optimized to reliably fulfill economical as well as ecological criteria. From the very beginning when this approach was pursued, it was attractive to search for a resistance gene, which, if present in the chosen crop, would protect the crop safely and, in addition, would allow the option of weed control irrespective of the developmental stage of the weeds and the crop. Compared to the pre-existing selective herbicides which were found empirically and which often under suboptimal application conditions have limitations either in crop selectivity or in weed control efficacy, the generation of herbicide-tolerant crops appeared to be a superior alternative.

4.7.1 Attempts to Select Glufosinate Tolerant Mutants

Whereas in vitro mutant selection in order to generate herbicide-tolerant crops is a reliable approach for several classes of herbicides with a different mode of action (Donn 1997), this approach has failed so far for glufosinate.

Even though it was possible to select an alfalfa cell line with an increased tolerance for phosphinothricin which was 20-fold higher than the tolerance level of the original cell line, all attempts to regenerate plants from the mutant cell line failed. In addition, all attempts to transfer the chromosome fragment conferring the locus of the amplified GS gene into mesophyll protoplasts of a highly regenerable alfalfa genotype were unsuccessful (Deak et al. 1988). The phosphinothricin-tolerant alfalfa cell line which became tolerant to the active ingredient, due to an amplification of a GS gene overexpressed the GS_1 enzyme three- to sevenfold (Donn et al. 1984). The full length GS_1 c-DNA isolated from the GS overproducing alfalfa line could complement a GS-deficient *E.coli* mutant (Dassarma et al. 1986). All attempts to mutagenize the alfalfa GS gene in vitro and to express the mutants in *E. coli* failed (H.M. Goodman, unpubl.). A few GS mutants were obtained which showed a reduced enzymatic activity and a lowered binding affinity for phosphinothricin. From these experiments and from the fact that after extended use of glufosinate in orchards, vineyards and plantations up to the present no glufosinate tolerant weed ecotypes were discovered, it was concluded that whenever a point mutation of a plant GS gene may lead to a mutant enzyme with lowered binding affinity for phosphinothricin, such mutants may not be viable because of the reduced binding affinity for glutamate and hence its reduced enzymatic activity; this may be as deleterious under atmospheric conditions as for the GS_2 mutants from barley.

The transgenic approach to overexpress the alfalfa GS_1 c-DNA under the control of the constitutive 35S promoter in tobacco cells led to tobacco plants with high amounts of alfalfa GS protein, elevated GS activity in shoots and leaves, but with only moderate levels of glufosinate tolerance (Eckes et al. 1989a). Apparently, the cytosolic expression of alfalfa GS_1 could not prevent partial inactivation of GS_2 in chloroplasts, leading to visible herbicidal symptoms.

4.7.2 Metabolic Inactivation of Glufosinate by Bar and Pat Enzymes

From the phosphinothricin producing *Streptomyces* species *S. hygroscopicus* and *S. viridochromogenes* the enzymes involved in the biosynthesis of bialaphos and their respective genes were investigated independently by two research groups. Amongst these genes, a gene conferring resistance against phosphinothricin was discovered in both species. The resistance gene from *S. hygroscopicus* was named bar (bialaphos resistance) gene (Thompson et al. 1987), whereas the respective gene from *S. viridochromogenes* was described as pat (phosphinothricin acetyl transferase) gene by Wohlleben et al. (1988). Both genes show a high degree of homology with 80% homology on the DNA level and 86% on the protein level.

Both genes code for a protein consisting of 184 amino acids. The enzymes have similar biochemical properties. They have a high substrate specificity. They accept only desmethyl phosphinothricin and phosphinothricin as substrates which they acetylate at the amino group, but none of the 20 proteinogenic amino acids are acetylated (Wehrmann et al. 1996). The acetylation of glutamate is 10,000-fold less efficient than acetylation of phosphinothricin. The biological function of these two enzymes in the bialaphos-producing strains is to protect the cells from accumulation of free phosphinothricin, because the GS of these *Streptomyces* species is sensitive to phosphinothricin. The final steps of bialaphos biosynthesis are as follows.

Desmethylphosphinothricin → *N*-acetyl-desmethylphosphinothricin → *N*-acetyldesmethyl-bialaphos → *N*-acetyl–bialaphos → bialaphos.

The cleavage of the protective acetyl group is the final step before the tripeptide is excreted.

Both research groups recognized the potential use of the bar and pat gene for the generation of glufosinate tolerant crop plants. Attempts to transfer and to express the resistance genes in plant cells were successful, if the genes were linked to a strong constitutive promoter. In both cases the 35S promoter from cauliflower mosaic virus was used (DeBlock et al. 1987; Wohlleben et al. 1988).

Initially, the bar and pat gene were linked with a neomycin-phosphotransferase gene as selectable marker, because it was not yet proven that these putative phosphinothricin resistance genes can be used as selectable markers per se. When the transgenic plants conferring glufosinate resistance were selected first for kanamycin resistance and then for glufosinate resistance, it became evident that the transgenic tissues tolerate, both in vitro as well as in planta, high doses of the active ingredient without any damage. This opened the opportunity to use these putative agronomic genes directly as selectable markers which can even be used for this purpose in breeding programs. A simple foliar glufosinate application on a segregating seedling population allows the recovery of the transgenic offspring. Such experiments showed the mendelian inheritance of the transferred gene due to its integration into the nuclear genome.

Because both *Streptomyces* genes are characterized by their high GC content, the group at Hoechst decided to create a synthetic pat gene in which the typical codon usage of plants was mimicked (Strauch et al. 1988; Eckes et al. 1989b). This synthetic pat gene codes for the same protein as the natural pat gene.

In retrospect, no significant differences in expression levels and expression stability could be observed between both versions of the pat gene, indicating that the high GC content of the natural gene did not lead to extensive gene inactivation due to methylation of the cytosine bases.

Even though pat and bar genes were used extensively as selectable marker genes for research purposes in a wide range of field crops, vegetables, fruit plants and even ornamentals, only a few crop species were chosen for commercial exploitation of the transgenic glufosinate-tolerant lines.

The costs for development and registration of glufosinate as a selective herbicide as well as the efforts for obtaining the release approvals for the respective transgenic events in a given crop have to be balanced against the market potential of glufosinate as a selective herbicide for a crop or even variety-specific indication. It is evident that only few indications for glufosinate as a selective herbicide in carefully chosen crop species will be commercially viable.

References

Acaster MA, Weitzmann PDJ (1985) Kinetic analysis of glutamine synthetases from various plants. FEBS Lett 189:241–244

Bayer E, Gugel KH, Haegele K, Hagenmaier H, Jessipow S, Koenig WA, Zaehner H (1972) Phosphinothricin und phosphinothricyl-alanyl-alanin. Helv Chim Acta 55:224–239

Dassarma S, Tisher E, Goodman HM (1986) Plant glutamine synthetase complements Glu A mutation in *Escherichia coli*. Science 232:1242–1244

Deak M, Donn G, Feher A, Dudits D (1988) Dominant expression of a gene amplification related herbicide resistance in *Medicago* cell hybrids. Plant Cell Rep 7:158–161

DeBlock M, Bottermann J, Vandewiele M, Dockx J, Thoen C, Gossele V, Movva N, Thompson C, VanMontagu M, Leemans J (1987) Engineering herbicide resistance in plants by expression of a detoxifying enzyme. EMBO J 6:2513–2518

Donn G (1997) Herbicide resistant crops generated by biotechnology. In: DePrado R, Jorrin J, Garcia-Torres L (eds) Weed and crop resistance to herbicides. Kluwer, Dordrecht, pp 217–227

Donn G, Tischer E, Smith J, Goodman H (1984) Herbicide resistant alfalfa cells: an example of gene amplification in plants. J Mol Appl Genet 2:621–635

Dorn E, Görlitz G, Heusel R, Stumpf K (1992) Verhalten von Glufosinat-ammonium in der Umwelt – Abbau im und Einfluβ auf das Ökosystem. Z Pflanzenkr Pflanzenschutz Sonderh 13:459–468

Eckes P, Schmitt P, Daub W, Wengenmayer F (1989a) Overproduction of alfalfa glutamine synthetase in transgenic tobacco plants. Mol Gen Genet 217:263–268

Eckes P, Uijtewaal B, Donn G (1989b) Synthetic gene confers resistance against the broad spectrum herbicide L-phosphinothricin in plants. J Cell Biochem 13D:334

Forde BG, Cullimore JV (1989) The molecular biology of glutamine synthetase in higher plants. In: Miflin BJ (ed) Oxford surveys of plant molecular and cell biology, vol 5. Oxford Univ Press, Oxford, pp 246–296

Hörlein G (1994) Glufosinate (phosphinothricin), a natural amino acid with unexpected herbicidal properties. Rev Environ Contam Toxicol 138:73–145

Ikeda M, Ogren WL, Hageman RH (1984) Effect of methionine sulfoximine on photosynthetic carbon metabolism in wheat leaves. Plant Cell Physiol 25:447–452

Jungk A (1984) Toxikologie der Pflanzenernährung/Düngerschäden. In: Hock B, Elstner EF (eds) Pflanzentoxikologie. BI Wissenschaftsverlag, Mannheim, pp 224–229

Keys AJ, Bird JF, Cornelius MJ, Lea PJ, Wallsgrove RM, Miflin BJ (1978) Photorespiratory nitrogen cycle. Nature 275:741–743

Kleiner D (1981) The transport of NH_3 and NH_4^+ across biological membranes. Biochim Biophys Acta 639:41–52

Köcher H (1983) Influence of the light factor on physiological effects of the herbicide Hoe 39866. Aspects Appl Biol 4:227–233

Köcher H, Lötzsch K (1985) Uptake, translocation and mode of action of the herbicide glufosinate-ammonium in warm climate weed species. Proc Asian-Pacific Weed Sci Soc 10th Conf:193–198

Lacuesta M, González-Moro B, González-Murua C, Aparicio-Tejo T, Monoz-Rueda A (1989) Effect of phosphinothricin (glufosinate) on activities of glutamine synthetase and glutamate dehydrogenase in *Medicago sativa* L. J Plant Physiol 1234:304–307

Lacuesta M, Monoz-Rueda A, González-Murua C, Sivak MN (1992) Effect of phosphinothricin (glufosinate) on photosynthesis and chlorophyll fluorescence emission by barley leaves illuminated under photorespiratory and non-photorespiratory conditions. J Exp Bot 43:159–165

Langston-Unkefer PL, Macy PA, Durbin RD (1984) Inactivation of glutamine synthetase by tabtoximine-β-lactam. Plant Physiol 76:71–74

Lea PJ (1991) The inhibition of ammonia assimilation: a mechanism of herbicide action. In: Baker NR, Percival MP (eds) Herbicides. Elsevier, Amsterdam, pp 267–298

Lea PJ (1993) Nitrogen metabolism. In: Lea PJ, Leegood RC (eds) Plant biochemistry and molecular biology. Wiley, Chichester, pp 155–180

Lea PJ, Ridley SM (1989) Glutamine synthetase and its inhibition. In: Dodge AD (ed) Herbicides and herbicide metabolism. Cambridge University Press, Cambridge, pp 137–170

Leason M, Cunliffe D, Parkin D, Lea PJ, Miflin B (1982) Inhibition of pea leaf glutamine synthetase by methioninesulfoximine, phosphinothricin and other glutamate analogs. J Phytochem 21: 855–857

Manderscheid R, Wild A (1986) Studies on the mechanism of inhibition by phosphinothricin of glutamine synthetase isolated from *Triticum aestivum* L. J Plant Physiol 123:135–142

Mase S (1984) Meiji Herbiace (UW 801, SF 1293, common name: bialaphos), a new herbicide. Jpn Pestic Information 45:27–30

Mastalerz P (1959) Inhibition of glutamine synthetase by phosphonic analogs of glutamic acid. Arch Immunol Terapii Doswiadczalnej 7:201–210

McNally SF, Hirel B (1983) Glutamine synthetase in higher plants. Physiol Vég 21:761–774

Miflin BJ, Lea PJ (1980) Ammonia assimilation. In: Miflin BJ (ed) The biochemistry of plants, vol 5: amino acids and derivatives. Academic Press, New York, pp 169–202

Niida T, Inouye S, Tsuruoka T, Shomura T, Kondo Y, Ogawa Y, Watanabe H, Sekizawa Y, Watanabe T, Igarashi H (1973) Antibiotic SF-1293 from *Streptomyces hygroscopicus*. German Offen DE 2 236 599 Meiji Seika Kaisha

Ogawa Y, Tsuruoka T, Inouye S, Niida T (1973a) Chemical structure of antibiotic SF-1293. Sci Rep Meiji Seika Kaisha 13:42–48

Ogawa Y, Tsuruoka T, Inouye S, Niida T (1973b) Chemical structure of antibiotic SF-1293. Sci Rep Meiji Seika Kaisha 13:49–53

Oliveira JC, Coruzzi GM (1999) Carbon and amino acids reciprocally modulate the expression of glutamine synthetase in *Arabidopsis*. Plant Physiol 121:301–309

Omura S, Hinotozawa K, Imanura N, Murata M (1984a) The structure of phosalacine, a new herbicidal antibiotic containing phosphinothricin. J Antibiot 37:939–940

Omura S, Murata M, Hanaki H, Hinotozawa K, Oiwa R, Fanaka H (1984b) Phosalacine, a new herbicidal antibiotic containing phosphinothricin, fermentation, isolation, biological activity and mechanism of action. J Antibiotic 37:829–835

Pace J, McDermott EE (1952) Methionine sulfoximine and some enzyme systems involving glutamine. Nature 169:413–416

Peterman TK, Goodman HM (1991) The glutamine synthetase gene family of *Arabidopsis thaliana*: light regulation and differential expression in leaves, roots and seeds. Mol Gen Genet 230:145–154

Ridley SM, McNally SF (1985) Effects of phosphinothricin on the isoenzymes of glutamine synthetase isolated from plant species which exhibit varying degrees of susceptibility to the herbicide. Plant Sci 39:31–36

Roberts JKM, Pang MKL (1992) Estimation of ammonium ion distribution between cytoplasm and vacuole using nuclear magnetic resonance spectroscopy. Plant Physiol 100:1571–1574

Ronzio RA, Meister A (1968) Phosphorylation of methionine sulfoximine by glutamine synthetase. Proc Natl Acad Sci 59:164–170

Sauer H, Wild A, Ruehle W (1987) The effect of phosphinothricin on photosynthesis II. The causes of inhibition of photosynthesis. Z Naturforsch 42c:270–278

Schwerdtle F, Bieringer H, Finke M (1981) Glufosinate-ammonium: ein neues nichtselektives Blattherbizid. Z Pflanzenkr Pflanzenschutz Sonderh 9:431–440

Shelp BJ, Swanton CJ, Mersey BG, Hall JC (1992) Glufosinate (phosphinothricin) inhibition of nitrogen metabolism in barley and green foxtail plants. J Plant Physiol 139:605–610

Strauch E, Arnold W, Alija R, Wohlleben W, Puehler A, Eckes P, Donn G, Uhlmann E, Hein F, Wengenmayer F (1988) Chemical synthesis and expression in plant cells and plants of phosphinothricin resistance gene with plant preferred codons. Eur Pat Appl EP275957 Hoechst AG

Takematsu T, Konnai M, Tachibana K, Tsurnoka T, Inouye S, Watanabe T (1979a) Antibiotic SF-1293 as herbicide. Jpn Kokai Tokky Koho JP79067026 Meiji Seika Kaisha; Germ Offen DE2858224

Takematsu T, Konnai M, Tachibana K, Tsurnoka T, Inouye S, Watanabe T (1979b) Herbicide for controlling weeds and bushes. Meiji Seika Kaisha Germ Offen DE 2856260

Thompson CJ, Movva NR, Tizard R, Crameri R, Davies JE, Lauwereys M, Botterman J (1987) Characterization of the herbicide resistance gene BAR from *Streptomyces hygroscopicus*. EMBO J 6:2519–2523

Wallsgrove RM, Turner JC, Hall NP, Kendall AC, Bright SW (1987) Barley mutants lacking chloroplast glutamine synthetase. Biochemical and genetic analysis. Plant Physiol 83:155–158

Wehrmann A, VanVliet A, Opsomer C, Botterman J, Schulz A (1996) The similarities of bar and pat gene products make them equally applicable for plant engineers. Nat Biotechnol 14:1274–1278

Wendler C, Barniske M, Wild A (1990) Effect of phosphinothricin (glufosinate) on photosynthesis and photorespiration in C_3 and C_4 plants. Photosyn Res 24:55–61

Wendler C, Putzer A, Wild A (1992) Effect of glufosinate (phosphinothricin) and inhibitors of photorespiration on activity. J Plant Physiol 139:666–671

Wild A, Wendler C (1990) Effect of glufosinate on amino acid content, photorespiration and photosynthesis. Pestic Sci 30:422–424

Wild A, Wendler C (1993) Inhibitory action of glufosinate on photosynthesis. Z Naturforsch 48c: 369–373

Wild A, Sauer H, Ruehle W (1987) The effect of phosphinothricin on photosynthesis. I. Inhibition of photosynthesis and accumulation of ammonia. Z Naturforsch 42c:263–269

Wohlleben W, Arnold W, Broer J, Hillmann D, Strauch E, Pühler A (1988) Nucleotide sequence of phosphinothricin-*N*-acetyl-transferase gene from *Streptomyces viridochromogenes*. Tue H94 and its expression in *Nicotiana tabacum*. Gene 70:25–37

5 Acetyl-CoA Carboxylase Inhibitors

Malcolm D. Devine

5.1 Introduction

Two important groups of herbicides, the cyclohexanediones (CHD) and aryloxyphenoxypropanoates (AOPP), inhibit the plastidic enzyme acetyl-CoA carboxylase (ACCase; E.C. 6.4.1.2). Representative compounds in these groups are shown in Fig. 1. A third class of inhibitor, based on a hybrid cyclic triketone structure, shows similar herbicidal activity (Rendina et al. 1995), but has not been developed commercially. CHD and AOPP herbicides are used to control a wide selection of grass weeds in both monocot and dicot crops. The basis of selectivity differs between dicot and grasses: in dicots, tolerance is based on the inherent insensitivity of dicot ACCase to these herbicides, whereas in certain cereal crops selectivity is based on higher rates of herbicide detoxification in the crop species (Devine and Shimabukuro 1994). This chapter will review the general activity of these herbicides, the biochemistry of the target enzyme, and the molecular basis of resistance in crops and weeds.

5.2 Symptoms of Herbicidal Activity

Injury symptoms tend to develop rather slowly in sensitive plants treated with CHD or AOPP herbicides. Growth (leaf elongation) stops within 24–48 h after herbicide application. Chlorosis is first observed on the youngest tissue, usually the emerging leaves. This reflects the fact that the initial phytotoxicity occurs primarily at the apical meristem, the major site of cell division and de novo fatty acid synthesis in these plants. In fact, 48–72 h after treatment the youngest emerged leaf can be quite easily separated from the rest of the plant by gently pulling it upwards; again, this reflects the tissue damage at the meristem. Chlorosis then spreads slowly through the rest of the plant, although it may take 7–10 days for the entire plant to be affected.

Phloem translocation of these herbicides through the plant is limited, resulting in relatively small amounts reaching the roots. For this reason, these herbicides seldom provide excellent control of perennial grass weeds. However, under certain conditions some control of perennials can be achieved.

P. Böger, K. Wakabayashi, K. Hirai (Eds.)
Herbicide Classes in Development

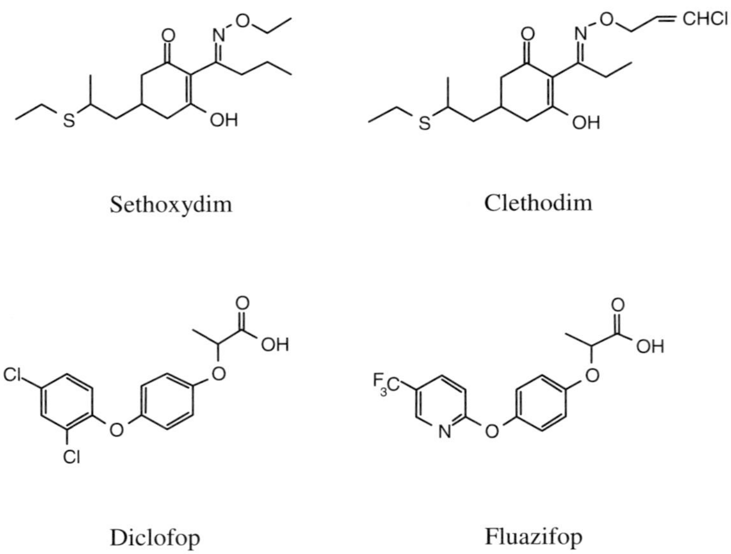

Fig. 1. Structures of two CHD herbicides, sethoxydim and clethodim, and two AOPP herbicides, diclofop and fluazifop. Note that diclofop and fluazifop are usually applied as the methyl- and butyl-esters, respectively, to facilitate penetration into the plant

No injury symptoms appear on dicot crops or weeds treated at typical use rates. Physiological injury can occur in cereal crops under certain conditions (e.g., low temperature at the time of application), presumably due to reduced rates of herbicide detoxification. However, most plants recover from this temporary injury within 7–10 days.

5.3 Biochemical Characteristics of the Target Enzyme

ACCase catalyzes the addition to CO_2 to acetyl-CoA to form malonyl-CoA, which is the initial product in the biosynthesis of acyl lipids (fatty acids). The chemical steps in the overall reaction can be represented as follows:

Enzyme-biotin + HCO_3^- + ATP $\rightarrow$ Enzyme-biotin-CO_2^- + ADP + Pi (catalyzed by biotin carboxylase)

Enzyme-biotin-CO_2^- + acetyl-CoA $\rightarrow$ Enzyme-biotin + malonyl-CoA (catalyzed by carboxyltransferase)

Malonyl-CoA is the substrate for fatty acid synthesis in the plastids, and also for fatty acid elongation and flavonoid and phytoalexin biosynthesis in the cytosol.

ACCase activity requires ATP and Mg^{2+}, and its activity is optimal under alkaline conditions (pH 8.0–8.2) (Herbert et al. 1996). ACCase is a biotinylated

Table 1. Organization of ACCase in higher plants. (Sasaki et al. 1995)

	Prokaryotic form	Eukaryotic form
Structure	Heterodimeric (separate BCC, BCase and CTase subunits)	Homodimer; single multifunctional polypeptide
Grasses	Absent	Plastids and cytosol
Dicots	Plastids	Cytosol
Sensitivity to CHD, AOPP	Insensitive	Sensitive (plastidic[a]) Insensitive (cytosolic)

[a] In a few grass species, the plastidic eukaryotic form of ACCase is insensitive to herbicides. See text for details.

enzyme that exists in two different forms in higher plants. The prokaryotic form is heterodimeric, and consists of four separate gene products: the biotin carboxyl carrier (BCC), biotin carboxylase (BCase), and carboxyltransferase (CTase; α and β subunits, See Konishi and Sasaki 1994; Sasaki et al. 1995; Ke et al. 2000). The genes for these subunits are coordinately expressed and the subunits are assembled to form the functional enzyme (Ke et al. 2000). The prokaryotic form of ACCase is relatively insensitive to inhibition by CHD and AOPP herbicides (see below). The eukaryotic, homodimeric form is a single polypeptide of around 220–230 kDa encompassing linked BCC, BCase, and CTase domains, and can be either sensitive (most plastidic forms) or resistant (cytosolic form) to herbicides. Some key elements of ACCase in grass and dicot plants are summarized in Table 1.

Egli et al. (1993) reported the presence of two isoforms of the eukaryotic ACCase in maize, which differed in sensitivity to the herbicides sethoxydim and haloxyfop. ACCase I, the plastidic form, was predominant and was sensitive to these compounds, whereas ACCase II, located in the cytosol, was relatively insensitive. The two forms of ACCase had similar molecular masses (ca. 220 kDa); no smaller polypeptides with ACCase activity were detected. A detailed examination of maize ACCase I and II by Herbert et al. (1996) showed similar results. In contrast, Incledon and Hall (1997) reported ACCase activity in maize associated with an 85-kDa protein, and suggested that the 220-kDa polypeptide in maize was composed of seven subunits. However, no genetic evidence has been proposed in support of smaller polypeptides with ACCase activity in grasses. It is possible that these smaller peptides exhibiting ACCase-like activity are subunits of related enzymes such as methylcrotonyl-CoA carboxylase (Ashton et al. 1994).

5.4 Mode of Action of Cyclohexanedione and Aryloxyphenoxypropanoate Herbicides

Although there were earlier indications that fatty acid synthesis was inhibited by CHD and AOPP herbicides (Hoppe and Zacher 1985), it was not until the late 1980s that the specific target site was identified as ACCase (Burton et al. 1987; Kobek et al. 1988; Rendina and Felts 1988; Secor and Cséke 1988). It was also shown that the stereoselectivity of AOPP herbicides [R(+) enantiomer is active, S(−) enantiomer inactive] reflected their inhibitory activity against ACCase (Hoppe and Zacher 1985; Walker et al. 1988; Secor et al. 1989). Both ACCase I and II are inhibited by CHD and AOPP herbicides, but ACCase II is up to 2000-fold less sensitive (Egli et al. 1993; Ashton et al. 1994; Herbert et al. 1996). Additional genetic evidence (reviewed below) adds support to ACCase as the primary target site of CHD and AOPP herbicides.

The kinetics of ACCase inhibition has been the subject of several detailed studies. Both CHD and AOPP herbicides are linear, noncompetitive inhibitors of ACCase with respect to the three enzyme substrates (Mg^{2+}-ATP, HCO_3^-, acetyl-CoA). However, the nearly competitive inhibition with respect to acetyl-CoA suggests that the herbicides most likely inhibit the transcarboxylase step of the reaction, and not the biotin carboxylation (Rendina et al. 1990; Burton et al. 1991). In addition, double inhibition studies have shown that binding of CHD and AOPP herbicides is mutually exclusive, suggesting that they share a common binding site (Rendina and Felts 1988; Rendina et al. 1990; Burton et al. 1991). However, the binding site(s) of these herbicides on ACCase have not yet been determined.

Another body of work has implicated disruption of membrane function as a component of the mode of action of AOPP herbicides (reviewed by Devine and Shimabukuro 1994). In particular, rapid depolarization of the plasma membrane electrogenic potential in sensitive species, the reversal of this in some resistant weed biotypes (Shimabukuro and Hoffer 1992), and the ability of 2,4-D to antagonize AOPP herbicides by blocking their effect on membrane potential have been cited as evidence of a specific membrane-related interaction. However, no target site associated with these activities has been identified, and no comprehensive explanation satisfactorily accounts for these intriguing results. In addition, more and more biochemical and genetic evidence is accumulating, in particular from CHD- and AOPP-resistant weeds, that whole-plant resistance and resistance at the level of ACCase are well correlated. Collectively, these results suggest that ACCase is the sole molecular target of CHD and AOPP herbicides.

5.5 Assays for Acetyl-CoA Carboxylase Activity

Several different assays have been used to measure ACCase activity in plant tissues. The most common method currently used is to make a crude ACCase preparation from young green leaf tissue (e.g., Shukla et al. 1997a), and to measure incorporation of ^{14}C from $H^{14}CO_3^-$ into heat- and acid-stable products. This assay lends itself easily to herbicide inhibition studies, in which various concentrations of herbicide are incorporated into the incubation medium prior to adding the $H^{14}CO_3^-$. Although the enzyme can be further purified for more detailed kinetic studies or fractionation of the different ACCase isoforms (Egli et al. 1993; Evenson et al. 1997; Incledon and Hall 1999), purification is not required to obtain a crude estimate of herbicide sensitivity. It has been shown, however, that different results on herbicide sensitivity can be obtained depending on how the enzyme preparation is handled (Shukla et al. 1997a). This points to the importance of working with clean enzyme preparations to generate reliable data.

A somewhat less refined version of this assay, often conducted with intact root or leaf tissue, measures the incorporation of ^{14}C into the tissue after feeding with ^{14}C-labeled acetate (Hoppe and Zacher 1985; Boldt and Barrett 1991; Di Tomaso et al. 1993). In general, this assay provides an approximate measure of lipid biosynthesis, but may overestimate it since the ^{14}C-acetate can be incorporated into products other than acyl lipids. However, this method provided some of the early evidence that fatty acid biosynthesis was the general target of CHD and AOPP herbicides (Hoppe and Zacher 1985).

5.6 Molecular Genetics of Resistance to Acetyl-CoA Carboxylase Inhibitors

Resistance to ACCase inhibitors in dicots is based on the insensitivity of the prokaryotic form of ACCase to CHD and AOPP herbicides, as described above (Rendina and Felts 1988; Konishi and Sasaki 1994). In most grasses the plastidic (ACCase I; eukaryotic) form of the enzyme is sensitive to herbicides (e.g., Burton et al. 1989). However, several exceptions exist, falling into three major categories: grasses with "natural" resistance, weed species in which resistance has evolved following repeated use of these herbicides, and crop genotypes selected in tissue culture for herbicide tolerance.

Several fescue species, including *Festuca rubra*, *F. ovina* and *F. amethystina* are tolerant of CHD and AOPP herbicides (Stoltenberg et al. 1989; Catanzaro et al. 1993). This tolerance is apparently based on insensitivity of the ACCase in these species to the herbicides. For example, ACCase from the fescue species was resistant to very high concentrations of fluazifop and sethoxydim (Catanzaro et al. 1993). It appears that even without herbicide selection, some grass species contain insensitive forms of ACCase I.

Maize is normally susceptible to ACCase inhibitors. However, maize mutants resistant to ACCase inhibitors have been isolated following selection of callus cultures on medium containing sethoxydim (Parker et al. 1990a; Marshall et al. 1992). This tolerance was stably inherited and was conferred by an altered form of ACCase with reduced sensitivity to CHD and AOPP herbicides. CHD-tolerant maize inbred lines have been developed from these initial selections. In related work, resistance in other maize lines was due to overexpression of the normal, herbicide-sensitive ACCase, not to the presence of a resistant form of ACCase (Parker et al. 1990b). In this case, a relatively small increase in expression of the target enzyme conferred a very high level of herbicide resistance, leaving open to question whether the resistance was entirely due to the upregulated ACCase activity. This tolerance was retained by some cell lines in the absence of herbicide, but lost after longer periods, indicating that it was not genetically stable. The mechanism of ACCase overexpression was not determined in this work.

Over the past 15 years resistance to CHD and AOPP herbicides has evolved in approximately 15 different grass weed species from Europe, the Americas, the far East and Australia (Heap 2001). Resistance can be conferred by either of two different mechanisms: altered forms of ACCase with reduced herbicide sensitivity, or enhanced rates of herbicide detoxification. These resistance mechanisms have been described in considerable detail in recent reviews (Devine and Shukla 2000; Devine and Preston 2000).

Reduced ACCase sensitivity to CHD and AOPP herbicides is the most common mechanism of resistance in resistant weed biotypes (Devine 1997). Many studies have been published showing reduced sensitivity of ACCase extracted from resistant compared to susceptible biotypes (Gronwald et al. 1992; Marles et al. 1993; Tardif and Powles 1993; Leach et al. 1995; Shukla et al. 1997a,b). The collected enzyme inhibition data from many of these resistant weed biotypes, when viewed in total, suggest that the resistant biotypes can be grouped into at least three or four different categories, each with a unique pattern of cross-resistance to different ACCase inhibitors. These groups include:

1. High level resistance to CHD, low or none to AOPP;
2. High level resistance to both CHD and AOPP;
3. High level resistance to fluazifop (AOPP), medium-low to other AOPP and CHD;
4. Medium to high level resistance to AOPP, none to CHD.

Similarly, Marshall et al. (1992) proposed that there were three to five different alleles of the major maize ACCase gene, each associated with different patterns of cross-resistance to ACCase inhibitors.

Detailed analyses of ACCase from resistant and susceptible biotypes have been described in several studies. Evenson et al. (1997) confirmed that resistance to diclofop in *Lolium multiflorum* was due to an altered form of ACCase I, the plastidic form (Table 2). The resistant and susceptible forms of ACCase of these *L. multiflorum* biotypes shared similar kinetic properties (Evenson et al. 1994). In other words, the mutation(s) causing resistance did not significantly change the catalytic function of the altered ACCase, other than the herbicide-binding properties. Incledon and Hall (1999) reported 5.5-fold higher V_{max} values for ACCase I from resistant maize compared to a susceptible line, but similar K_m values. Again, this indicates little overall change in the catalytic function of the enzyme. Several studies have shown no difference in growth and productivity between resistant and susceptible grass weeds with

Table 2. Herbicide sensitivity of ACCase I and II from diclofop-resistant or -susceptible *Lolium multiflorum* biotypes. (Reproduced from Evenson et al. 1997)

Biotype	Isoform	Source	Diclofop conc. (μM)	Inhibition (%)
Susceptible	ACCase I	Plastid	0.2	50
	ACCase II	Cytosol	125	42
Resistant	ACCase I	Plastid	7	50
	ACCase II	Cytosol	127	31

ACCase mutations, confirming that the change in ACCase does not significantly impair growth of the resistant biotypes (Wiederholt and Stoltenberg 1996a,b).

The results of Evenson et al. (1997) confirm that the cytosolic form, ACCase II, is relatively insensitive to diclofop in both susceptible and resistant biotypes. However, the sensitivity of ACCase I was greatly reduced in the resistant biotype, confirming that this is the mechanism of resistance. Somewhat different results have been reported from maize. These include changes to ACCase sensitivity of both forms of ACCase, and increased expression of the plastidic form (Incledon and Hall 1999). It is not clear how these results relate to the observed single-gene basis of resistance, or whether they involve some pleiotropic effects of the altered gene.

The molecular basis of resistance to herbicides that inhibit acetolactate synthase (ALS) and photosystem II electron transport has been well characterized, and specific gene mutations conferring different resistant phenotypes have been identified (Devine and Eberlein 1997; Devine and Preston 2000). By analogy, one can speculate that each of the above patterns of resistance to ACCase inhibitors may be attributed to a particular ACCase mutation, making the enzyme less susceptible to inhibition. Various studies have shown that ACCase resistance is controlled by a single, semi-dominant nuclear gene coding for the eukaryotic (plastidic) ACCase (Parker et al. 1990a; Betts et al. 1992). However, the mutations conferring these different patterns of resistance to ACCase inhibitors have not yet been identified.

In recent years, ACCase genes from various sources have been sequenced (Al-Feel et al. 1992; Gornicki et al. 1994; Podkowinski et al. 1996). For the eukaryotic ACCase, these sequences indicate an open reading frame of ca. 6700 base pairs, coding for a 2230-amino acid polypeptide of ca. 250 kDa. In a detailed molecular study involving complementation of a yeast ACCase null mutant with chimeric ACCase based partly on wheat ACCase, the "resistance determinant" was located to a 400 amino acid region corresponding to the CTase domain (Nikolskaya et al. 1999). Whether this corresponds to the putative herbicide binding site (see above) remains to be determined. More recently, we have obtained preliminary evidence that high-level resistance to sethoxydim in a biotype of *Setaria viridis* is associated with an A to C mutation at position 5582 of the *S. viridis* ACCase cDNA, coding for an Ile_{1806} to Leu substitution in the CTase domain (Zhang and Devine 2000 and unpubl. results). Further experiments are underway to identify mutations conferring other resistant phenotypes in grass weeds. Recently, two additional reports have been published confirming that an isoleucine to leucine substitution in the carboxyltransferase domain of the plastidic ACCase confers resistance to sethoxydim.

Resistance to CHD and AOPP herbicides can also be conferred by enhanced herbicide detoxification (Menendez and De Prado 1996; Preston et al. 1996; Hall et al. 1997; Hidayat and Preston 1997) or, in some cases, by a combination of two mechanisms (Maneechote et al. 1995). Some AOPP herbicides, such as diclofop-methyl, are metabolized by cytochrome P450 monooxygenases (CYP), followed by glycosylation or demethylation (Shimabukuro et al. 1979; Zimmerlin and Durst 1992; Barrett 2000). Others, such as fenoxaprop-ethyl, are metabolized by glutathione S-transferases (Edwards and Cole 1996). In principle, therefore, it is possible to create tolerance in crops by genetic transformation with the appropriate CYP or GST gene(s). However, to date there has been no economic incentive to develop herbicide-resistant crops by this approach.

In summary, ACCase has become an important target site for herbicide action, and many commercial ACCase inhibitors have been developed. Target site-based resistance has become common in grass weeds, particularly in crop rotations in which CHD and AOPP herbicides have been used repeatedly for grass weed control in alternating cereal and dicot crops. Although there are alternative weed control options for these resistant weeds, new herbicides that provide control of resistant and susceptible biotypes would be very useful.

References

Al-Feel W, Chirala SS, Wakil SJ (1992) Cloning of the yeast *FAS3* gene and primary structure of yeast acetyl-CoA carboxylase. Proc Natl Acad Sci USA 89:4534–4538

Ashton AR, Jenkins CLD, Whitfeld PR (1994) Molecular cloning of two different cDNAs for maize acetyl CoA carboxylase. Plant Mol Biol 24:35–49

Barrett M (2000) The role of cytochrome P450 enzymes in herbicide metabolism. In: Cobb AH, Kirkwood RC (eds) Herbicides and their mechanisms of action. Sheffield Academic Press, Sheffield, UK, pp 25–37

Betts KJ, Ehlke NJ, Wyse DL, Gronwald JW, Somers DA (1992) Mechanism of inheritance of diclofop resistance in a biotype of Italian ryegrass (*Lolium multiflorum*). Weed Sci 40:184–189

Boldt LD, Barrett M (1991) Effects of diclofop and haloxyfop on lipid synthesis in corn (*Zea mays*) and bean (*Phaseolus vulgaris*). Weed Sci 39:143–148

Burton JD, Gronwald JW, Somers DA, Connelly JA, Gengenbach BG, Wyse DL (1987) Inhibition of plant acetyl-CoA carboxylase by the herbicides sethoxydim and haloxyfop. Biochem Biophys Res Commun 148:1039–1044

Burton JD, Gronwald JW, Somers DA, Gengenbach BG, Wyse DL (1989) Inhibition of corn acetyl-CoA carboxylase by cyclohexanedione and aryloxyphenoxypropionate herbicides. Pestic Biochem Physiol 34:76–85

Burton JD, Gronwald JD, Keith RA, Somers DA, Gengenbach BG, Wyse DL (1991) Kinetics of inhibition of acetyl-coenzyme A carboxylase by sethoxydim and haloxyfop. Pestic Biochem Physiol 39:100–109

Catanzaro CJ, Burton JD, Skroch WA (1993) Graminicide resistance of acetyl-CoA carboxylase from ornamental grasses. Pestic Biochem Physiol 45:147–153

Délye C, Wang T, Darmency H (2002) An isoleucine-leucine substitution in chloroplastic acetyl-CoA carboxylase from green goxtail (*Setaria viridis* L. Beauv.) is responsible for resistance to the cyclohexanedione herbicide sethoxydim. Planta 214:421–427

Devine MD (1997) Mechanisms of resistance to acetyl-CoA carboxylase inhibitors: a review. Pestic Sci 51:259–264

Devine MD, Eberlein CV (1997) Physiological, biochemical and molecular aspects of herbicide resistance based on altered target sites. In: Roe RM, Burton JD, Kuhr RJ (eds) Herbicide activity: toxicology, biochemistry and molecular biology. IOS Press, Amsterdam, pp 159–185

Devine MD, Preston C (2000) The molecular basis of herbicide resistance. In: Cobb AH, Kirkwood RC (eds) Herbicides and their mechanisms of action. Sheffield Academic Press, Sheffield, pp 72–104

Devine MD, Shimabukuro RH (1994) Resistance to acetyl coenzyme A carboxylase inhibiting herbicides. In: Powles SB, Holtum JAM (eds) Herbicide resistance in plants: biology and biochemistry. Lewis, Boca Raton, pp 141–169

Devine MD, Shukla A (2000) Altered target sites as a mechanism of herbicide action. Crop Protect 19:881–891

Di Tomaso JM, Stowe AE, Brown PH (1993) Inhibition of lipid synthesis by diclofop-methyl is age dependent in roots of oat and corn. Pestic Biochem Physiol 45:210–219

Edwards R, Cole DJ (1996) Glutathione transferases in wheat (*Triticum*) species with activity toward fenoxaprop-ethyl and other herbicides. Pestic Biochem Physiol 54:96–104

Egli MA, Gengenbach BG, Gronwald JW, Somers DA, Wyse DL (1993) Characterization of maize acetyl-coenzyme A carboxylase. Plant Physiol 101:499–506

Evenson KJ, Gronwald JW, Wyse DL (1994) Purification and characterization of acetyl-coenzyme A carboxylase from diclofop-resistant and -susceptible *Lolium multiflorum*. Plant Physiol 105: 671–680

Evenson KJ, Gronwald JW, Wyse DL (1997) Isoforms of acetyl-coenzyme A carboxylase in *Lolium multiflorum*. Plant Physiol Biochem 35:265–272

Gornicki P, Podkowinski J, Scappino LA, DiMaio J, Ward E, Haselkorn R (1994) Wheat acetyl-coenzyme A carboxylase: cDNA and protein structure. Proc Natl Acad Sci USA 91:6860–6864

Gronwald JW, Eberlein CV, Betts KJ, Baerg RJ, Ehlke N J, Wyse DL (1992) Mechanism of diclofop resistance in an Italian ryegrass (*Lolium multiflorum* Lam.) biotype. Pestic Biochem Physiol 44:126–139

Hall LM, Moss SR, Powles SB (1997) Mechanisms of resistance to aryloxyphenoxypropionate herbicides in two resistant biotypes of *Alopecurus myosuroides* (blackgrass): herbicide metabolism as a cross-resistance mechanism. Pestic Biochem Physiol 57:87–98

Heap IM (2001) International survey of herbicide-resistant weeds. Internet. http://www.weedscience.com

Herbert D, Price LJ, Alban C, Dehaye L, Job D, Cole DJ, Pallett KE, Harwood JL (1996) Kinetic studies on two isoforms of acetyl-CoA carboxylase from maize leaves. Biochem J 318: 997–1006

Hidayat I, Preston C (1997) Enhanced metabolism of fluazifop acid in a biotype of *Digitaria sanguinalis* resistant to the herbicide fluazifop-P-butyl. Pestic Biochem Physiol 57:137–146

Hoppe HH, Zacher H (1985) Inhibition of fatty acid biosynthesis in isolated bean and maize chloroplasts by herbicidal phenoxy-phenoxypropionic acid derivatives and structurally related compounds. Pestic Biochem Physiol 24:298–305

Incledon BJ, Hall JC (1997) Evidence that maize acetyl-coenzyme A carboxylase does not function solely as a homodimer. J Agric Food Chem 45:4838–4844

Incledon BJ, Hall JC (1999) Inhibition of ACCase220 and ACCase240 isozymes from sethoxydim-resistant and -susceptible maize hybrids. J Agric Food Chem 47:299–304

Ke J, Wen T-N, Nikolau BJ, Wurtele ES (2000) Coordinate regulation of the nuclear and plastidic genes coding for subunits of the heteromeric acetyl-Coenzyme A carboxylase. Plant Physiol 122:1057–1071

Kobek K, Focke M, Lichtenthaler HK (1988) Fatty-acid biosynthesis and acetyl-CoA carboxylase as a target of diclofop, fenoxaprop and other aryloxy-phenoxypropionic acid herbicides. Z Naturforsch 43:47–54

Konishi T, Sasaki Y (1994) Compartmentalization of two forms of acetyl-CoA carboxylase in plants and the origin of their tolerance toward herbicides. Proc Natl Acad Sci USA 91: 3598–3601

Leach GE, Devine MD, Kirkwood RC, Marshall G (1995) Target enzyme-based resistance to acetyl-coenzyme A carboxylase inhibitors in *Eleusine indica*. Pestic Biochem Physiol 51:129–136

Maneechote C, Preston C, Powles SB (1995) A diclofop-methyl-resistant *Avena sterilis* biotype with a herbicide-resistant acetyl-coenzyme A carboxylase and enhanced metabolism of diclofop-methyl. Pestic Sci 49:105–114

Marles MAS, Devine MD, Hall JC (1993) Herbicide resistance in *Setaria viridis* conferred by a less sensitive form of acetyl coenzyme A carboxylase. Pestic Biochem Physiol 46:7–14

Marshall LC, Somers DA, Dotray PD, Gengenbach BG, Wyse DL, Gronwald JW (1992) Allelic mutations in acetyl-coenzyme A carboxylase confer herbicide tolerance in maize. Theor Appl Genet 83:435–442

Menendez J, De Prado R (1996) Diclofop-methyl cross-resistance in a chlorotoluron-resistant biotype of *Alopecurus myosuroides*. Pestic Biochem Physiol 56:123–133

Nikolskaya T, Zagnitko O, Tevzadze G, Haselkorn R, Gornicki P (1999) Herbicide sensitivity determinant of wheat plastid acetyl-CoA carboxylase is located in a 400-amino acid fragment of the carboxytransferase domain. Proc Natl Acad Sci USA 96:14647–14651

Parker WB, Marshall, LC, Burton JD, Somers DA, Wyse DL, Gronwald JW, Gengenbach BG (1990a) Dominant mutations causing alterations in acetyl-coenzyme A carboxylase confer tolerance to cyclohexanedione and aryloxyphenoxypropionate herbicides in maize. Proc Natl Acad Sci USA 87:7175–7179

Parker WB, Somers DA, Wyse DL, Keith RA, Burton JD, Gronwald JW, Gengenbach BG (1990b) Selection and characterization of sethoxydim-tolerant maize tissue cultures. Plant Physiol 92:1220–1225

Podkowinski J, Sroga GE, Haselkorn R, Gornicki P (1996) Structure of a gene encoding a cytosolic acetyl-CoA carboxylase of hexaploid wheat. Proc Natl Acad Sci USA 93:1870–1874

Preston C, Tardif FJ, Christopher JT, Powles SB (1996) Multiple resistance to dissimilar herbicide chemistries in a biotype of *Lolium rigidum* due to enhanced activity of several herbicide degrading enzymes. Pestic Biochem Physiol 54:123–134

Rendina AR, Felts JM (1988) Cyclohexanedione herbicides are potent inhibitors of acetyl-coA carboxylase from grasses. Plant Physiol 86:983–986

Rendina AR, Craig-Kennard AC, Beaudoin JD, Breen MK (1990) Inhibition of acetyl-coenzyme A carboxylase by two classes of grass-selective herbicides. J Agric Food Chem 38:1282–1287

Rendina AR, Campopiano O, Marsilii E, Hixon M, Chi H, Taylor WS (1995) Overlap between herbicidal inhibitors of acetyl-coenzyme A carboxylase: enhanced binding of cyclic triketones, a novel class of graminicide. Pestic Sci 43:368–371

Sasaki Y, Konishi T, Nagano Y (1995) The compartmentation of acetyl-coenzyme A carboxylase in plants. Plant Physiol 108:445–449

Secor J, Cséke C (1988) Inhibition of acetyl-coA carboxylase activity by haloxyfop and tralkoxydim. Plant Physiol 86:10–12

Secor J, Cséke C, Owen WJ (1989) Aryloxyphenoxypropanoate and cyclohexanedione herbicides. Inhibition of acetyl coenzyme A carboxylase. In: Whitaker JR, Sonnet PE (eds) Biocatalysis in agricultural biotechnology. ACS Symp Ser 389, Am Chem Soc, Washington, DC, pp 265–276

Shimabukuro RH, Hoffer BL (1992) Effect of diclofop on the membrane potentials of herbicide-resistant and susceptible annual ryegrass root tips. Plant Physiol 98:1415–1422

Shimabukuro RH, Walsh WC, Hoerauf RA (1979) Metabolism and selectivity of diclofop-methyl in wild oat and wheat. J Agric Food Chem 27:615–623

Shukla A, Dupont S, Devine MD (1997a) Resistance to ACCase-inhibitor herbicides in wild oat: evidence for target site-based resistance in two biotypes from Canada. Pestic Biochem Physiol 57:147–155

Shukla A, Leach GE, Devine MD (1997b) High-level resistance to sethoxydim conferred by acetyl-CoA carboxylase alterations in *Setaria faberi* and *Setaria viridis*. Plant Physiol Biochem 35: 803–807

Stoltenberg DE, Gronwald JW, Wyse DL, Burton JD, Somers DA, Gengenbach BG (1989) Effect of sethoxydim and haloxyfop on acetyl-coenzyme A carboxylase activity in *Festuca* species. Weed Sci 37:512–516

Tardif FJ, Powles SB (1993) Herbicide multiple resistance in a *Lolium rigidum* biotype is endowed by multiple mechanisms: isolation of a subset with a resistant acetyl-CoA carboxylase. Physiol Plant 91:488–494

Walker KA, Ridley SM, Harwood JL (1988) Effects of the selective herbicide fluazifop on fatty acid synthesis in pea and barley. Biochem J 254:811–817

Wiederholt RJ, Stoltenberg DE (1996a) Absence of differential fitness between giant foxtail (*Setaria faberi*) accessions resistant and susceptible to acetyl-coenzyme A carboxylase inhibitors. Weed Sci 44:18–24

Wiederholt RJ, Stoltenberg DE (1996b) Similar fitness between large crabgrass (*Digitaria sanguinalis*) accessions resistant or susceptible to acetyl-coenzyme A carboxylase inhibitors. Weed Technol 10:42–49

Zagnitko O, Jelenska J, Tevzadze G, Haselkorn R, Gornicki P (2001) An isoleucine/leucine residue in the carboxyltransferase domain of acetyl-CoA carboxylase is critical for interaction with aryloxyphenoxypropionate and cyclohexanedione inhibitors. Proc Natl Acad Sci USA 98:6617–6622

Zhang X-Q, Devine MD (2000) A possible mutation of plastidic ACCase gene conferring resistance to sethoxydim in green foxtail (*Setaria viridis*). Abstr Weed Sci Soc Am 40:33

Zimmerlin A, Durst F (1992) Aryl-hydroxylation of the herbicide diclofop by a wheat cytochrome P-450 monooxygenase: substrate specificity and physiological activity. Plant Physiol 100: 874–881

6 Inhibitors of Biosynthesis of Very-Long-Chain Fatty Acids

PETER BÖGER and BERND MATTHES

6.1 Introduction

Chloroacetamides have been used in maize, soybean or rice for about 50 years (Hamm 1974). During 1997/1998 in the USA, this class contributed to about 50% of the herbicides applied in corn and 11% in soybean (Anonymous 1999). Safeners have successfully broadened their use, and postemergence weed treatment has been improved by the concurrent application of chloroacetamides which are taken up via the soil. Their persistence ensures long-term weed control. Chloroacetamides are xylem-transported; they interfere with the early development of weeds. Germination generally takes place but growth is inhibited, and the seedlings do not emerge or remain stunted. Figure 1 demonstrates the latter effect for cucumber and barley seedlings. The first leaves emerging from the hypocotyl and the cotyledons of dicot plants are small and misformed, but the cotyledons and leaves are never bleached, showing a somewhat increased chlorophyll content. Cell division and enlargement are both inhibited (Deal and Hess 1980) which could also be shown with the microalgae *Chlamydomonas* (Fedtke 1982) and *Scenedesmus* (Weisshaar and Böger 1987). The latter authors assumed that an impaired membrane formation caused the halt of cell division.

The chloroacetamide mode of action has been a focus of research for many years and almost all components of basic plant metabolism were found to be affected. Protein formation was influenced (Sloan and Camper 1985; Zama and Hatzios 1987), and purine metabolism was modified (Narsaiah and Harvey 1977). Wilkinson (1981) reported reduced growth as being caused by a decreased terpenoid biosynthesis. Molin et al. (1986) pointed out that lignin and anthocyanin formation in sorghum was impaired. Leakage and an altered membrane permeability were found by Mellis et al. (1982) and Vavrina and Ashley (1983). Ebert (1980) found an impact on membrane formation as observed by electron microscopy. Generally, these effects were obtained by laboratory experiments with concentrations of 10^{-5} and 10^{-4} M. For a review of earlier findings on the mode of action and various physiological effects, see Fuerst (1987), LeBaron et al. (1988) and Sharp (1988). A recent review has been published by Böger et al. (2000).

Alachlor or metolachlor bind to nucleophiles like glutathione and cysteine in vitro (Leavitt and Penner 1979). These herbicides have been found to bind

P. Böger, K. Wakabayashi, K. Hirai (Eds.)
Herbicide Classes in Development

Fig. 1. *Above* and *center* Cucumber seedlings (*Cucumis sativus)* grown for 6–8 days on vermiculite with 80 μE m^{-2} s^{-1}, in a 16-h light/8-h dark regimen at 25°C with tap water in the presence of 1 μM (*middle*) and 10 μM metazachlor (*right*). *Left pots* show the controls. *Below* Single cucumber and barley seedlings treated with 1 μM metazachlor as above. Controls are shown on the *right*. Treated seedlings are small and stunted but never bleached

to many unidentified proteins (from oat) in both homogenates and the intact plant (LeBaron et al. 1988; McFarland and Hess 1985). Again, high concentrations of metolachlor or propachlor (10^{-4} M) were required and the quantities of alkylated proteins did not correlate with the growth inhibition observed. These compounds cannot be considered as general inhibitors of SH-containing enzymes because not all such enzymes are inhibited. Hence, a specificity of inhibition has to be assumed. Alkylation (e.g., of SH-groups) is an irreversible process and it should be noted that growth inhibition of the microalga *Scenedesmus* could only be restored after a 24-h cultivation time or longer in herbicide-free medium. This is in contrast to photosynthesis inhibitors (e.g., diuron or metribuzin) which allow for a recovery within minutes after washing off the herbicides. Chloroacetamides were not found to impair photosynthetic electron transport (Weisshaar and Böger 1987).

Three possibilities should be considered when investigating the chloroacetamide mode of action:

1. Higher concentrations used in both laboratory experiments and field applications may unspecifically alkylate various enzymes of different metabolic pathways. This may produce various phytotoxic effects and eventually death after long-term treatments.
2. Biosynthetic routes affected by these compounds may cause *secondary* effects such as modified terpenoid or hormone levels (see above). These effects are only *indirect* ones due to initial herbicide attack on one (or various?) enzyme(s).
3. Chloroacetamides in low concentrations may act on a highly sensitive target enzyme, possibly by a specific alkylation of the protein or by forming a tight enzyme-inhibitor complex. It was our objective to find such a specific target.

Although chloroacetamides are applied in relatively high doses (300 g a.i. pre-emergent and more per ha), the phytotoxic concentrations inside the cell have been found rather low, e.g., 0.1–0.7 μM metazachlor in corn (Fuerst et al. 1991) indicating an even smaller concentration at the target site. Growth of rice seedlings was halved by 50 nM metazachlor in aqueous medium (Couderchet et al. 1994), the I_{50} value for *Echinochloa* was estimated to about 10^{-8} M for thenylchor, and 10^{-8}–10^{-7} M for pretilachlor by inhibition of shoot growth (Asai and Yogo 1998 and pers. comm.). These in vivo data indicate a specific target for chloroacetamides which is inhibited by inhibitor concentrations below 10^{-6} M. Such a target is considered a *primary* one since it will be affected *first* when an inhibitor is entering the cell and its concentration is (still) low. When higher concentrations of the inhibitor have been accumulated in the cell, additional targets with a lower affinity may be impacted. Of course, inhibition of an assumed primary target should correlate with phytotoxicity exerted on the intact plant.

Accordingly, our search for the primary target was guided by the following strategy:

1. Changes of physiological and biochemical parameters should be measured which show up after a reasonable short time (minutes, hours, not days).
2. Changes against the control should become apparent with low inhibitor concentrations.
3. The metabolic changes observed should be specific for the phytotoxic compound class used.
4. Phytotoxicity in the glasshouse (Couderchet et al. 1997) and in the field as well as specific physiological and biochemical changes vs. control described below were exhibited by S-dimethenamid and S-metolachlor, not by the R-enantiomers. Studying the effects of S- vs. R-enantiomers was used as a means to obtain evidence that our biochemical findings obtained by different assays are related to phytotoxicity.

We assume that all parameter changes observed under these four conditions are due to a primary herbicide target whose inhibition triggers off metabolic changes with a lethal consequence. The stereospecificity provides strong evidence that the target is a key enzyme being specifically inhibited.

Many reports dealing with the mode of action referred to lipid metabolism. A role of acetyl-CoA was originally reported by Jaworski (1956) and lipid biosynthesis was subsequently found to be inhibited (Mann and Pu 1968). We could demonstrate with the microalga *Scenedesmus* that uptake of (labeled) acetate into acyl lipids was impaired by S-dimethenamid (Couderchet et al. 1997). This is a rapid effect showing up a couple of hours after addition of the herbicide while the uptake of amino acids (leucine or lysine) or sugars was not affected (Weisshaar and Böger 1987). When *Scenedesmus* was treated with several phytotoxic chloroacetamides, a strong accumulation of oleic acid (18:1) and a strong decrease in 18:2 and 18:3 fatty-acid species of the acyl lipids were observed (Couderchet and Böger 1993). Impairment of 18:2 desaturation was also found with embryoids of *Brassica napus* L. after treatment with 20 μM alachlor (Möllers and Albrecht 1994). Wu et al. (1999) did not observe a change of fatty acid composition by metolachlor or pretilachlor using monocotyledonous crop seedlings. Treatment with metolachlor (about 20 μM) changed the very long-chain constituents of waxes of primary leaves of sorghum including fatty acids, alcohols, aldehydes with a carbon number of 20–32 (Ebert and Ramsteiner 1984). In cucumber seedlings the same high metolachlor concentrations inhibited the formation of alkane homologues of waxes (Tevini and Steinmüller 1987). Thiocarbamates, or to be precise, the sulfoxide forms of *S*-ethyl dipropylthiocarbamate (EPTC) or triallate, obviously exert a similar mode of action as chloroacetamides. They have been found to inhibit wax biosynthesis and to change the wax composition (see Kern et al. 1997 for a recent paper).

Hence, findings of our laboratory and of others indicated that lipid biosynthesis may be the target domain. Accordingly, our subsequent investigations focused on fatty acid synthesis and further processing of long-chain (C_{18}) fatty acids.

6.2 The Model System

A substantial amount of [^{14}C]-labeled oleic or stearic acid was taken up by the green microalga *Scenedesmus acutus* when present in the autotrophic liquid culture medium. After separation of the lipids, part of these acids were found in a cellular fraction free of *soluble* lipids ("non-lipid-fraction", NLF). As demonstrated with different inhi-bitor concentrations (Table 1), 18:1 incorporation was severely inhibited by phytotoxically active chloroacetamides (I_{50} of metazachlor 8×10^{-8} M; for S-dimethenamid 10^{-7} M; Kring et al. 1995). Mitosis or photosynthesis inhibitors were found inactive; iodoacetamide exhibited a poor I_{50} value close to 10^{-5} M (Kring et al. 1995).

The NLF is a rather crude preparation. It consists of broken cell material and of sporopollenin (Wilmesmeier et al. 1993). This natural polymer is present in the cell wall of *Scenedesmus* and in pollen grains of higher plants. Apparently, part of the oleate applied is used for sporopollenin synthesis in the alga.

Only the S-form of dimethenamid inhibited oleate or stearate incorporation. A comparable growth inhibition of duckweed (*Lemna minor* L.) was found with an I_{50} value of about 3×10^{-8} M, which is close to the 10^{-7}-M figure of inhibition of the oleate label in crude NLF of *Scenedesmus acutus*. For thenylchlor we determined an I_{50} value of about 8×10^{-8} M, which again, is close to the approximate I_{50} value estimated by inhibition of rice shoot growth (Asai and Yogo 1998). The low I_{50} values are interpreted such that, firstly, the inhibition is a specific one. Secondly, the close agreement of the I_{50} values found with three different species is taken as evidence that the same target domain is attacked. Thirdly, we assume that the inhibition of oleate incorporation into the NLF of *Scenedesmus* is a measure of phytotoxicity. This is demonstrated by Fig. 2 where the growth inhibition of *Scenedesmus* was quantified either by a decrease of chlorophyll or by packed cell volume. All eight herbicides assayed were effective and cellular inhibition was positively correlated with inhibition of oleate incorporation into NLF (Böger et al. 2000). Butachlor was an outlier.

Since chloroacetamides are applied via the soil and prevent growth of the emerging seedlings, it is difficult to establish quantitatively reliable scores in the greenhouse. The specific and effective inhibition of the highly sensitive [^{14}C]-oleate incorporation into the non-lipid fraction of *Scenedesmus* allowed the development of a *quantitative* and *quick* activity assay for the first time (Couderchet et al. 1998). Table 1 demonstrates the inhibition of oleate incorporation into the NLF by several herbicides and fungicides. It is evident that only the first group – the herbicides – are active in our assay. Both enantiomers of metalaxyl, a fungicide chemically related to chloroacetamides, are inactive. EPTC is very weak. We assume that during this short-term assay with intact *Scenedesmus* EPTC could not be metabolically modified to the phytotoxic sulfoxide form.

Table 1. Inhibition of [^{14}C] oleate incorporation into the non-lipid fraction of *Scenedesmus acutus* in % of control. See Table 5 for chemical names of the tested compounds

Compounds	0.1 μM	1 μM	10 μM	100 μM
Metazachlor	51	64	90	–
rac-Dimethenamid	41	53	80	–
Mefenacet	15	41	80	–
Iodoacetamide	10	25	60	90
Piperophos	18	41	–	–
Anilofos	43	61	–	–
BAS 128682	–	0	0	40
Diuron	–	0	20	–
Oryzalin	–	0	10	20
Chlorsulfuron	–	1	3	–
R-Metalaxyl	2	5	–	–
S-Metalaxyl	0	1	–	–
Pyrazophos	3	3	6	–
EPTC	–	8	8	–

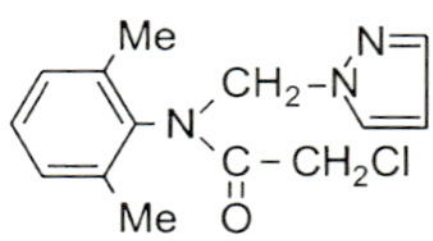

Metazachlor

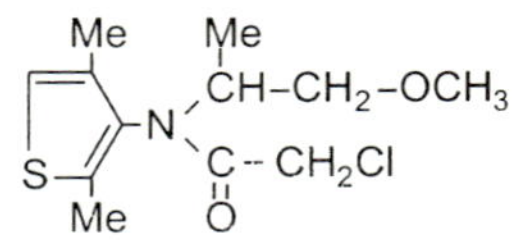

Dimethenamid

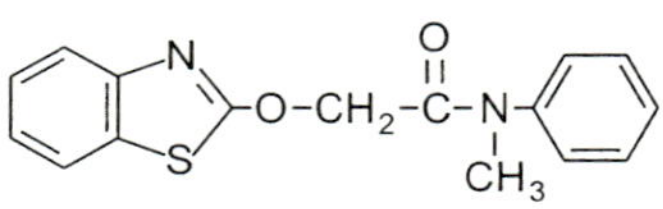

Mefenacet

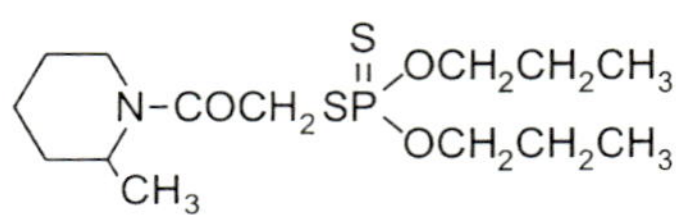

Piperophos

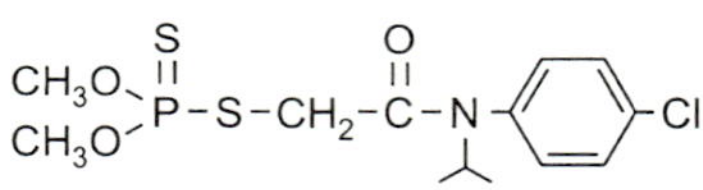

Anilofos

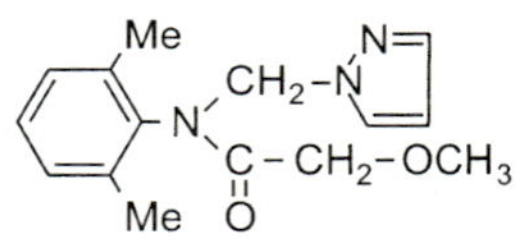

BAS 128682

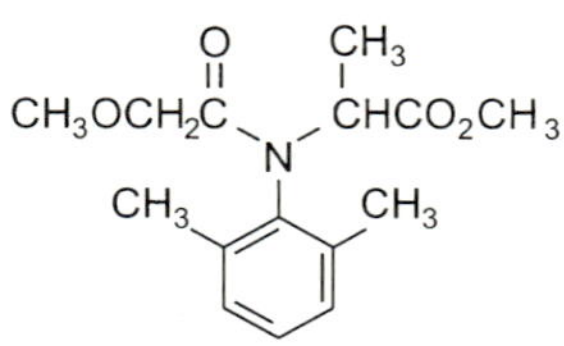

Metalaxyl

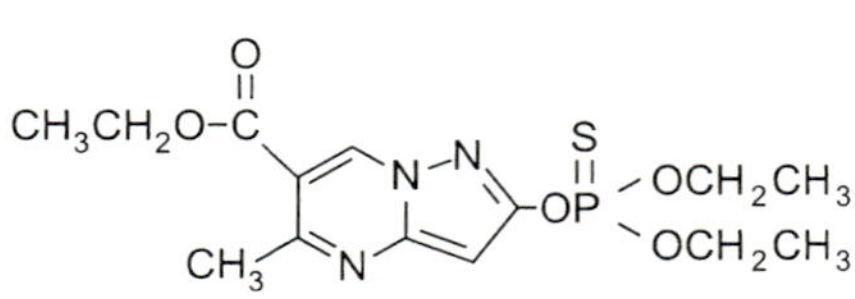

Pyrazophos

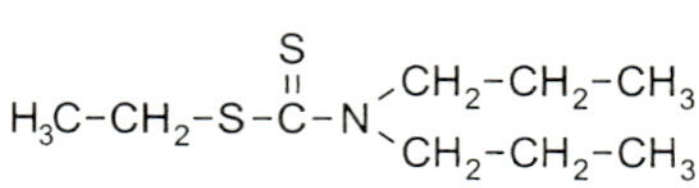

EPTC

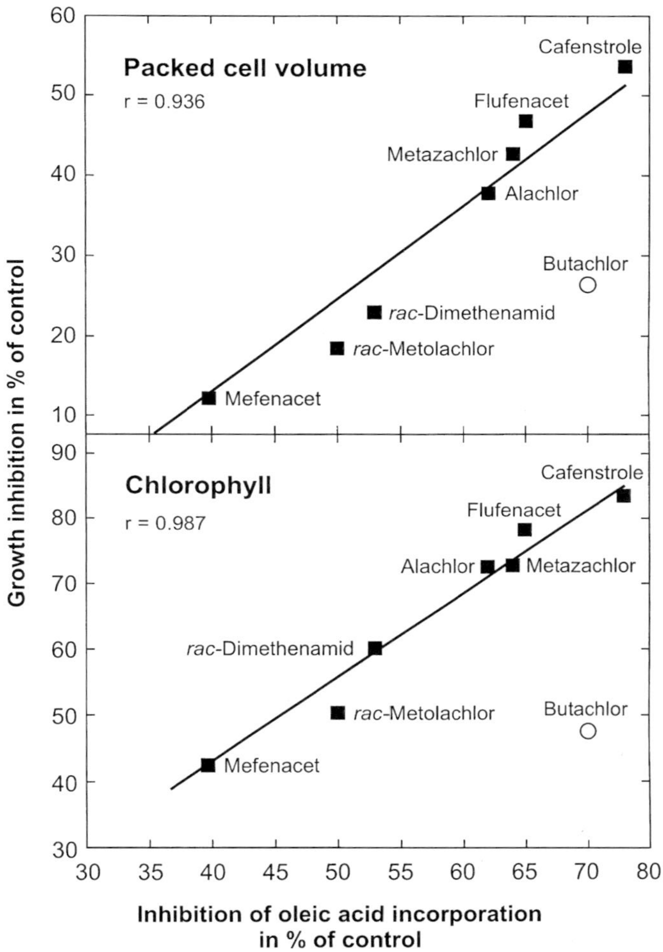

Fig. 2. Growth inhibition expressed by packed cell volume and chlorophyll content of the algal suspension of *Scenedesmus acutus* in liquid autotrophic culture. Inhibition by chloro-, oxyacetamides and cafenstrole is correlated with impaired [^{14}C]-oleate incorporation into a "non-lipid fraction" of that alga (Couderchet et al. 1998). This fraction includes very long-chain fatty acids which are produced from applied oleate (Kring et al. 1995, Schmalfuß et al. 1998)

The NLF was analyzed further. About 40% of the label, solubilized with acidified dioxane, could be extracted with n-hexane. Radio-HPLC showed several peaks of less polarity than oleic acid. These peaks were identified as 22:1, 24:1 and 26:1 monounsaturated *v*ery *l*ong-*c*hain *f*atty *a*cids (VLCFAs; Schmalfuß et al. 1998). A strong decrease in these peaks was observed with 1 μM S-metolachlor (Fig. 3, bottom). The oleate peak (18:1) increased vs. the control as expected, since VLCFAs are known to be formed by elongation of C16 and

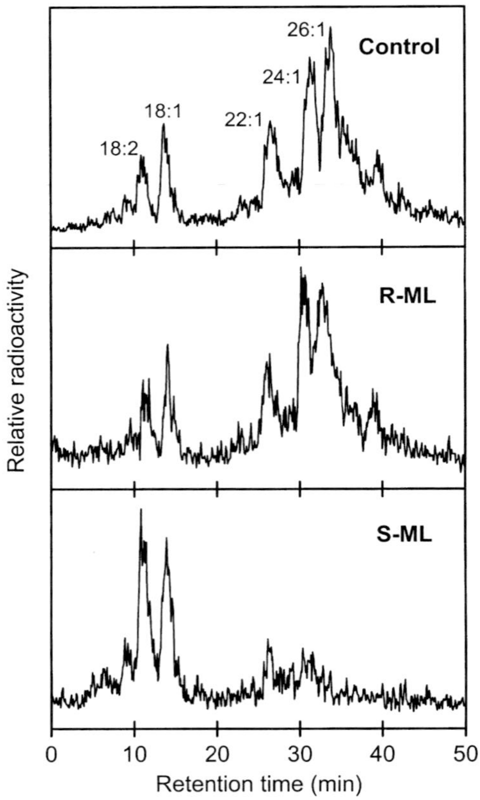

Fig. 3. HPLC-separation of the hexane extract from a dioxane/HCl subfraction of the non-lipid fraction obtained from *Scenedesmus*. Autotrophic cultures were grown for about 24 h in the presence of [^{14}C]-oleic acid and R- and S-metolachlor (*R-ML*, *S-ML*). The incorporation of the label into the fatty acid species indicated is expressed by relative radioactivity (see Schmalfuß et al. 1998 for details)

C18 fatty acids. It is noteworthy that the S-form was found active while the R-enantiomer was similar to the control. Inhibition of VLCFA biosynthesis in *Scenedesmus* was observed in the presence of all phytotoxic chloroacetamides as well as with new structures which exhibited a chloroacetamide-type activity in the glasshouse. We have assayed cafenstrole, phosphosulfonates (like RH-4641), or tetrazolinones (fentrazamide; see Fig. 4 for structures). These compounds showing high phytotoxicity in the glasshouse yielded I_{50} values generally below 1 μM in the *Scenedesmus* assay system as well.

Presumably, VLCFAs are required to stabilize the cell wall and/or the cell membrane of the alga. Cultivation of *Scenedesmus* with sublethal metazachlor concentrations produced large swollen cells. Obviously, the cell wall was

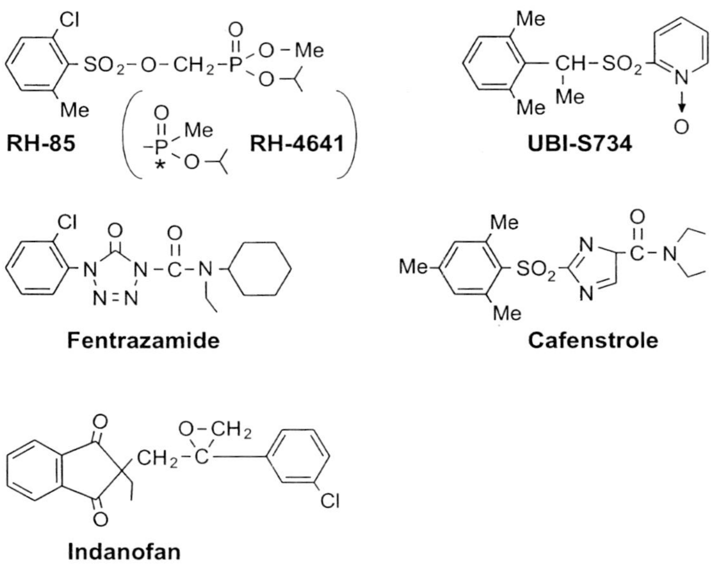

Fig. 4. New structures with chloroacetamide activity

weakened and cell division impaired. Correspondingly, a metazachlor-resistant *Scenedesmus* strain selected in our laboratory (Couderchet et al. 1995) has bigger and swollen cells. No processing of applied [^{14}C]-oleic acid could be observed with this mutant and only traces of VLCFAs were detected compared to about 24% of VLCFAs found in the wild-type NLF fraction. In contrast, 16:0 and 18:3 species were increased and 18:1 was more than tenfold higher in the mutant. Apparently, this resistance is neither caused by an insensitive target, nor by metabolic breakdown, but has its bearing on a markedly higher amount of C16 and C18 fatty acids replacing the missing VLCFAs in essential membranes or cell-wall components.

6.3 Very Long-Chain Fatty Acid Synthesis Inhibition in Intact Leaves

Figure 5 demonstrates the biosynthetic steps to produce the plant fatty acids dealt with herein. Fatty acids up to C16 or C18 are formed at the acyl-carrier protein of the plastid while further elongation takes place in the cytosol and at the endoplasmic reticulum. In contrast to the alga, VLCFAs of green plant tissue apparently consist of saturated fatty acids. As noted previously, higher concentrations of chloroacetamides as well as thiocarbamates inhibit formation of epicuticular waxes in intact higher plants (Ebert and Ramsteiner 1984;

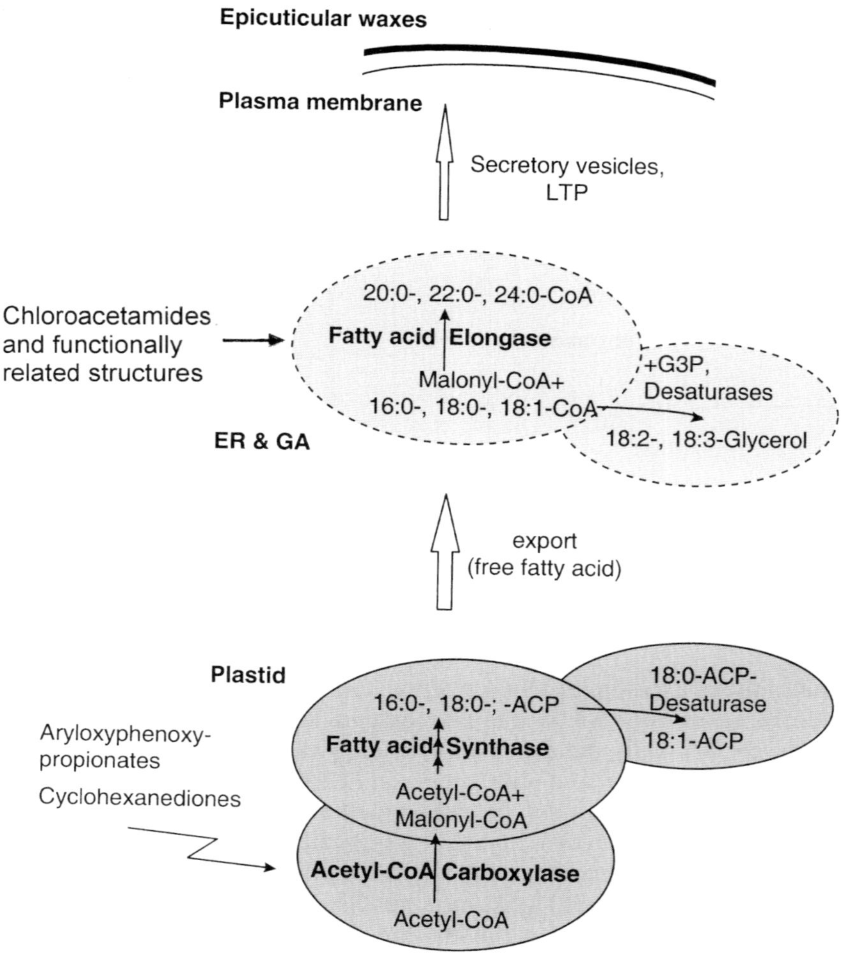

Fig. 5. Fatty-acid processing in plants. Fatty acids up to C16 and/or C18 chain length are produced in the plastid at the acyl-carrier protein (*ACP*) catalyzed by fatty-acid synthase. These fatty acids are exported into the cytosol where elongation to very long-chain species (VLCFAs) takes place at the endoplasmic reticulum (*ER*) and the Golgi apparatus (*GA*). The CoA-activated fatty acids are processed with malonyl-CoA by a VLCFA-elongase system which is strongly inhibited by chloroacetamides. VLCFAs are assumed to be transported by secretory lipid transport vesicles (*LTPs*), integrated into the plasma membrane and serve as precursors for epicuticular waxes. *G3P* glycerol-3-phosphate. Desaturations of fatty acid species occur in the plastid as well as at the ER-membrane

Tevini and Steinmüller 1987; Barrett and Harwood 1998). Some studies have documented that higher plants contain C20 to C26 fatty acids bound to phosphatidylserine (Murata et al. 1984). Plasma membranes contain cerebroside-linked VLCFAs as reported for *Secale cereale* L. (Cahoon and Lynch 1991),

Arabidopsis thaliana L. (Uemura et al. 1995) or *Avena fatua* L. (Renault et al. 1997). VLCFAs are minor, but presumably important components of plant cells but found enriched in epicuticular waxes and in the plasma membrane. More investigations on their role in the cell have been performed with animal systems (Poulos 1995) than with plants. Conceivably, lack of VLCFA biosynthesis will be phytotoxic. It should also be mentioned that wax-less *Arabidopsis* mutants are conditionally male-sterile since pollen development is disturbed (Millar et al. 1999). It may be concluded from our findings that this is due to decreased sporopollenin formation which needs VLCFA biosynthesis (see Sect. 6.2). It should be worthwhile to check for pollen production in plants treated with sublethal concentrations of chloroacetamides.

For analysis of intracellular VLCFAs, leaves were washed with chloroform to remove epicuticular waxes together with an excess of [^{14}C]-labeled precursors applied for fatty-acid biosynthesis. Then VLCFAs were extracted by a combined alkaline and acid hydrolysis. Leaf discs of *Cucumis sativus* L., *Hordeum vulgare* L. or *Zea mays* L. incorporated [^{14}C]-labeled stearic acid or [^{14}C]-malonate into 20:0, 22:0, 24:0 and 26:0 fatty acids (higher carbon numbers have not yet been found). A 2-h incubation time with metazachlor, metolachlor or butachlor inhibited formation of these VLCFAs with an I_{50} value in the range of 10–100 nM. A 1-μM concentration of metolachlor practically abolished this fatty-acid synthesis as shown by acyl lipid analysis of the plasma membrane isolated from *Cucumis* cotyledons (Fig. 6). Note that again the S-enantiomer was found to be active. The HPLC-separation pattern of the fatty acids looks like the *Scenedesmus* pattern of separation (Fig. 3). We conclude that both in algae and higher plants the same inhibition feature prevails. Higher plants exhibited a sensitivity comparable to that of *Scenedesmus*. The oxyacetamide flufenacet and other structures of Fig. 4 (Böger et al. 2000) gave similar results. Interestingly, maize leaf discs showed less sensitivity than, e.g., cucumber. Whether this reflects a (partial) dechlorination of metazachlor (by conjugation to glutathione) in maize or a less sensitive (enzyme) target remains to be studied.

For all higher plant species mentioned above the inhibition of [^{14}C]-malonate incorporation into VLCFAs depends on the fatty-acid species separated. The inhibition becomes stronger with increasing C-chain length. The formation of 22:0 and 24:0 species is markedly reduced with 10 nM metazachlor. The 20:0 formation requires 100 nM for substantial inhibition and even more is needed for inhibition of 18:0 (produced from 16:0; see Fig. 7). Formation of fatty acids with <C18 is not affected by chloroacetamides in the concentration range indicated (Matthes et al. 1998). Since malonate was found to actively perform the elongation, it can be safely concluded that the cytosolic *elongation* steps for VLCFAs, in principle, are identical to the fatty-acid build-up in the plastid. That implies a condensation of the activated acyl primer and malonyl-CoA followed by reduction of the ketoacyl-CoA, a subsequent dehydration and a second reduction. These four reactions have been characterized in part and are presumably organized as an "elongase"-enzyme complex (Domergue et al. 1998). It has been reported that the two reductases and the

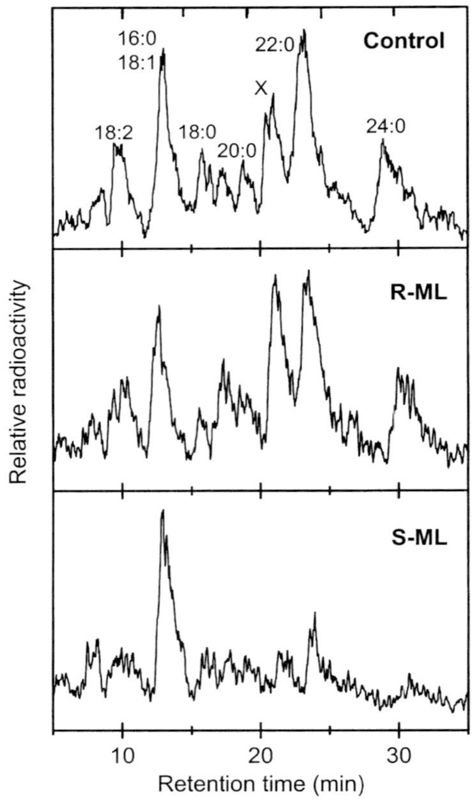

Fig. 6. Radio-HPLC analysis of [^{14}C]-labeled fatty acids of the isolated plasma membrane of *Cucumis sativus*. 2-[^{14}C]-malonate was applied for 24 h to 6-day-old seedlings treated either with 1 μM R- or S-metolachlor (*R-*, *S-ML*). Fatty acids were identified by standard fatty acid methyl esters. *X* not yet determined (probably a hydroxylated very long-chain fatty acid)

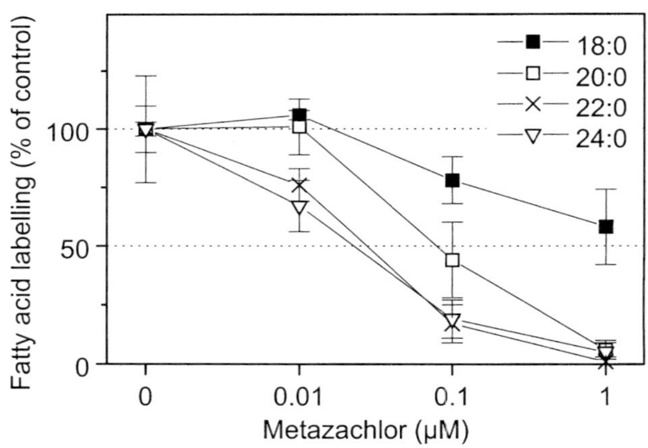

Fig. 7. Inhibition of 2-[^{14}C]-malonate incorporation into newly formed fatty acids of cucumber seedlings. The inhibitory effect of metazachlor increases with the chain length. The specific inhibition of elongation by chloroacetamides in cucumber starts with the processing of the 16:0 precursor into 18:0 (see text)

dehydratase involved are possibly shared with all elongase reactions while the *condensation step* is considered to be specific for the formation of VLCFAs (Miller and Kunst 1997). Since plastid-located fatty acid formation is not impaired by the inhibitors, obviously the condensation step in the *cytosol* is affected. This conclusion agrees with genetic studies. An elongase gene (*FAE1*), which apparently encodes a condensing β-ketoacyl-CoA synthase, was expressed in *Arabidopsis*, resulting in the formation of C20 and C22 VLCFAs that were not normally present in the wild type (Millar and Kunst 1997).

6.4 The Cell-Free Elongase System

Allium porrum seedlings also exhibited a similar sensitivity to chloroacetamides as was found with cucumber leaf discs. We chose leek because this plant has been intensively studied in relation to synthesis of VLCFAs (see, e.g., Cassagne et al. 1994; Domergue et al. 1998). Figure 8 demonstrates the four enzyme steps operating as the elongase system: (1) a 3-ketoacyl-CoA synthase, (2) a reductase to produce 3-hydroxyacyl-CoA with NAD(P)H, (3) a dehydratase to form a 2-enoyl-CoA, and eventually, and (4) a reductase to yield VLCFA-CoA again using reduced pyridine nucleotides. As indicated above in step (1), the condensing reaction of acyl-CoA and malonyl-CoA is apparently the limiting one. It may include at least two enzymes, one to elongate C18 (or C16) to C20 and C22 species, and another one for the elongation of C24-acyl primers (James et al. 1995; Lassner et al. 1996; Millar and Kunst 1997; Millar et al. 1999). Apparently, their elongation specificity depends on the plant species used, or whether seeds or vegetative tissues are investigated. The matter is still under investigation. Two elongases have been isolated from epidermal cell microsomes and distinguished by their preferential primers (18:0- or 20:0-CoA); no gene assignment has been published. The enzymes of acyl elongation are membrane-bound and thought to be associated with the endoplasmic reticulum and the Golgi apparatus of the cytosol and possibly with the plasma membrane (von Wettstein-Knowles 1993).

We have developed a short-term cell-free elongase assay (Schmalfuß et al. 2000) modified after Cassagne and coworkers (Agrawal et al. 1984; Cassagne et al. 1994) using malonyl-CoA and acyl-CoA of either 16:0, 18:0, 20:0 or 22:0 chain length. All elongation steps were strongly inhibited by metazachlor or cafenstrole; metazachlor yielded I_{50} values in the range from 0.5 to 1.7 μM. Once again acyl elongation was only inhibited by the S-enantiomer of metolachlor, not by the R-form. This could be shown with both a 20:0 acyl-CoA and an 18:0 acyl-CoA primer. Conclusively, the high sensitivity of the elongase system against chloroacetamides and functionally corresponding inhibitors is caused by a similar inhibition activity to *all* VLCFA-elongase steps.

Our findings suggest, by comparison of inhibition rates in vivo ($I_{50} \approx 10$–100 nM) and in vitro ($I_{50} \approx 1\,\mu$M) that inhibition is more sensitive under physio-

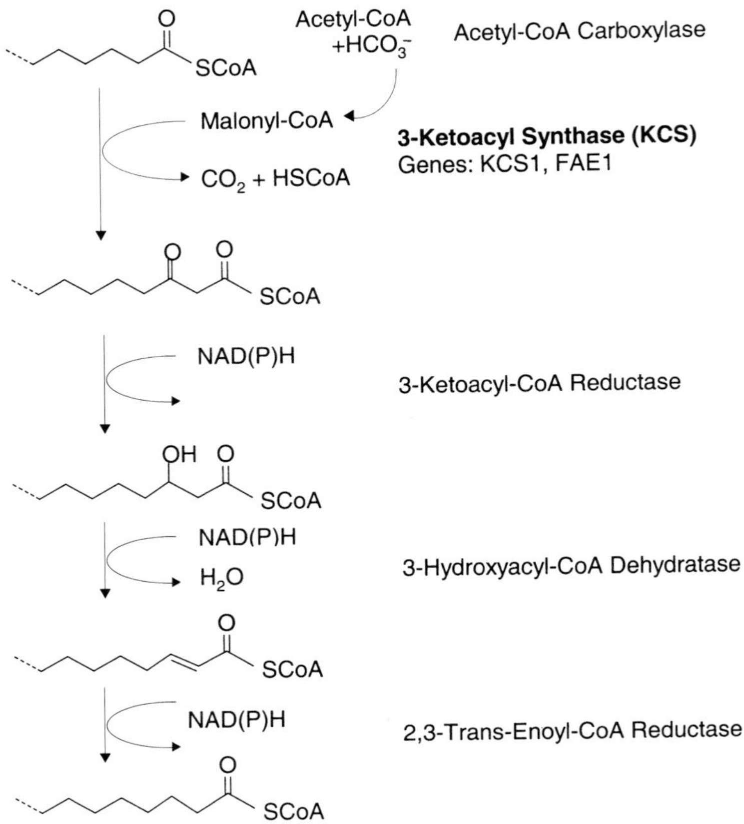

Fig. 8. The four-step reaction sequence for elongation of fatty acids at the endoplasmic reticulum. The first step catalyzed by 3-ketoacyl synthase, the condensing reaction, is considered to be rate-limiting and sensitive to chloroacetamides or functionally related structures like cafenstrole or fentrazamide. (Modified after Cook 1994)

logical conditions. In intact cucumber leaves or leek seedlings the formation of C18 → C20 was less inhibited than the formation of C20 → C22 and C22 → C24, whereas we found comparable inhibition for each elongation step in leek microsomes. This increase of inhibition can be explained as follows:

1. Fatty acid elongation proceeds sequentially. Since every elongation step is inhibited, 20:0- and 22:0-formation will limit substrate supply for the subsequent elongation steps and inhibition increases. Additionally, inhibition is time-dependent which determines elongation activity assayed in vitro (see Sect. 6.5).
2. The rate of inhibition does not only depend on inhibitor concentration, but also on the substrate concentration available. As shown in Fig. 9 (right

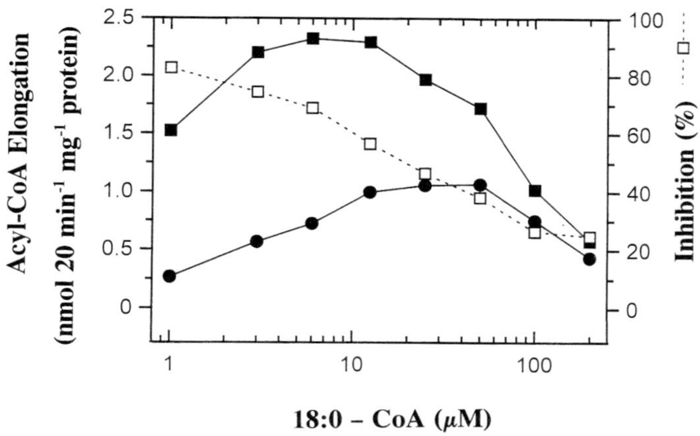

Fig. 9. Elongation of stearoyl (18:0)-CoA to 20:0 fatty acid by 2-[14C]-malonyl-CoA by isolated microsomes from leek (*Allium porrum*). Both the elongation reaction rate of the control (-■-) and the rate of the inhibited sample (1 μM metazachlor, -●-) depend on concentration of the 18:0-CoA substrate. Also, the degree of inhibition is influenced by the precursor concentration. (Modified after Schmalfuß et al. 2000)

axis), inhibition is most effective for very low (1 μM) substrate concentrations (Schmalfuß et al. 2000). Acyl-CoA concentrations in the living cell are assumed to be in the range of nM to μM (Ohlrogge and Jaworski 1997).

3. Higher concentrations of 18:0 CoAs are found to decrease the elongation activity (see Fig. 9, left axis). Presumably, inhibition of fatty acid elongation leads to accumulation of 18:0-CoA, which contributes to herbicidal inhibition of the elongation process.

Our preliminary data imply that there is a competition between acyl-CoA and the inhibitor at the active (substrate) site. This hypothesis is corroborated by the finding that iodoacetamide inhibition of acyl-CoA elongase from *Lunaria annua* L. could be alleviated, in part, by preincubation with oleoyl-CoA (Fehling et al. 1992). Future studies should prove whether inhibitor and acyl-CoA interact in a competitive manner.

Table 2 compiles the inhibition data found with intact organisms, the leek cell-free system and the findings with *Saccharomyces* which was transformed with the *Arabidopsis FAE*1-elongase gene encoding the 3-ketoacyl synthase of Fig. 8. The first four compounds are phytotoxic and they all inhibit VLCFA-formation albeit, to a somewhat different degree. It is again noteworthy that the phytotoxic S-enantiomer of metolachlor is active and the R-form is inactive in all four assay systems shown. The transgenic *Saccharomyces* strain exhibits a specific inhibition of VLCFA biosynthesis as is essentially found with *Cucumis* or *Allium*. This is direct evidence that the first step of the biosynthetic pathway, namely the condensing enzyme of the elongase system, is inhibited.

Table 2. Percent inhibition of VLCFA biosynthesis by herbicides in different species. Table assembled from data in the literature

Compounds tested	*Scenedesmus acutus* cells 18:1 elongation[a,b]	*Cucumis sativus*, cotyledons Total VLCFA[c]	PM-VLCFA[c]	*Allium porrum* Cell-free elongation[b]	*Saccharomyces cerevisiae* cells Recombinant *FAE1*[c,d]
Metazachlor	64	89	97	94	83
Fentrazamide	64	83	83	55	–
Cafenstrole	73	84	96	98	100
S-Metolachlor	68	64	69	22	31
R-Metolachlor	5	1	–5	0	1

Inhibition of VLCFA biosynthesis by 1 μM herbicide except for cafenstrole, 0.1 μM; *R*-metolachlor, 10 μM; PM, plasma membrane.
[a] Couderchet et al. (1998).
[b] Schmalfuß et al. (2000).
[c] Matthes (2000).
[d] *FAE1*, fatty acid elongase from *Arabidopsis* expressed in *Saccharomyces cerevisiae.* The strain was kindly provided by Dr. L. Kunst, Vancouver, Canada. This gene encodes the condensing enzyme specific for monounsaturated VLCFAs (see Sect. 6.4). The yeast strain used contains only minor amounts of saturated VLCFAs.

Table 3. Characteristics of intracellular membrane and plasma membrane fractions from *Cucumis sativus* seedlings

Characteristics	Intracellular[a] membranes	Plasma membrane
Protein (mg/g fresh wt.)	0.42 ± 0.14	0.11 ± 0.03
Presence of H^+-ATPase (%)[b]	<5	<95
Sterols (%)[b]	36 ± 11	64 ± 11
Chlorophyll (%)[b]	97 ± 3	3 ± 2
VLCFA content (nmol/g fresh wt.)	13 ± 3	30 ± 4

[a] Endoplasmic reticulum, thylakoid residues, Golgi apparatus, mitochondria, etc.
[b] Percent of data distribution between the two membrane fractions.

Recently we have isolated pure plasma membranes from *Cucumis sativus* (Matthes 2000), and found a high content of VLCFAs compared to the content in intracellular membranes present in the microsomal fraction (Table 3). As shown by Fig. 6, the VLCFAs of the plasma membrane (PM) strongly decrease under the influence of chloroacetamides. VLCFAs are bound to phosphatidylcholine, phosphatidylethanolamine or sphingolipids (cerebrosides). Seemingly, these lipids have a membrane-stabilizing function (Brown 1998; Moreau et al. 1998; Brown and London 2000). Depletion of the PM of VLCFAs will probably decrease its rigidity and permeability, associated with leakage and stalled cell division.

Table 4. Inhibition of the cell-free microsomal plant elongase from *Allium porrum* by metazachlor: Influence of preincubation; inhibition in percent of control

Herbicide conc.	Control (no preincubation)	Preincubation at 4°C		Preincubation at 30°C	
		60 min	180 min	10 min	30 min
0.1 μM	18	49	66	60	78
1.0 μM	51	92	99	93	98

After preincubation the assay was performed for 20 min at 30°C to determine the degree of inhibition shown above (Schmalfuß et al. 2000). The noninhibited sample did not show a decrease in activity during the experimental time.

6.5 Assumptions of the Reaction Mechanism

With the 20-min short-term in vitro assay, an approximately tenfold less sensitivity of elongation was observed than using the *intact* leaf disc or leek seedlings. The latter assay included a 6-h incubation with the inhibitor. The experiment of Table 4 explains this discrepancy (Table 5 lists the chemical names of compounds used). Pre-incubation of the elongase-containing microsomes for 60 and 180 min at 4 or 30°C increased the inhibition vs. the control. This is particularly obvious with a 0.1 μM metazachlor concentration (see also Schmalfuß et al. 2000). At 4°C the nonpreincubated sample yielded an 18% inhibition which increased to 66% after 180 min. The higher sensitivity found with *Scenedesmus* cells and intact leaves apparently is the result of a *prolonged* contact of the inhibitor with its target. An assay reaction mixture, which showed an almost complete inhibition by metazachlor, was subsequently diluted yielding a herbicide concentration too low to cause a substantial inhibition. Nevertheless, the elongase system remained inhibited. Once the inhibition is established after an appropriate incubation time, the inhibitor is tightly bound to the enzyme and cannot be removed (Schmalfuß et al. 2000).

In summary, the high sensitivity of the VLCFA-elongase system is obviously the result of three effects: *firstly,* there is a high affinity of the target to its inhibitor. *Secondly,* the inhibition increases along the subsequent elongation steps caused by the decreased concentration of CoA-activated acyl primers with each elongation step (see above and Schmalfuß et al. 2000). *Thirdly,* a longer incubation time together with an appropriate temperature produces an irreversible binding of the inhibitor to its target. Future studies should deal with the isolation of the enzyme-inhibitor complex and its characterization.

Table 1 and Fig. 4 show the structures of compounds which have been checked for elongase-inhibition activity in our laboratory. The active ones include several novel compounds to which a chloroacetamide-type activity has not been properly assigned before (anilofos, tridiphane, UBI-S734, RH-4641 or cafenstrole). At present, we assume that an alkylating reaction takes

Table 5. Chemical names of compounds dealt with

Alachlor	2-chloro-2′,6′-diethyl-*N*-(methoxymethyl)acetanilide
Anilofos	*S*-[4-chloro-*N*-(isopropyl)carbaniloyl]methyl *O*,*O*-dimethyl phosphorodithioate
Atrazine	2-chloro-4-ethylamino-6-isopropylamino-1,3,5-triazine
BAS-128682	2-methoxy-*N*-(2,6-dimethylphenyl)-*N*-(1*H*-pyrazol-1-ylmethyl)acetamide
Butachlor	*N*-(butoxymethyl)-2-chloro-2′,6′-diethylacetanilide
Cafenstrole	*N*,*N*-diethyl-3-mesitylsulfonyl-1*H*-1,2,4-triazole-1-carboxamide
Chlorsulfuron	1-(2-chlorophenylsulfonyl)-3-(4-methoxy-6-methyl-1,3,5-triazin-2-yl)urea
Dimethenamid	(*RS*)-2-chloro-N-(2,4-dimethyl-3-thienyl)-*N*-(2-methoxy-1-methylethyl)acetamide
Diuron	3-(3,4-dichlorophenyl)-1,1-dimethylurea
EPTC	*S*-ethyl dipropylthiocarbamate
Fentrazamide	1-(*N*-ethyl-*N*-cyclohexylcarbamoyl)-4-(2-chlorophenyl)-tetrazolinone
Flufenacet	*N*-(4′-fluorophenyl)-*N*-isopropyl)-2-(5-trifluoromethyl-1,3,4-thiadiazol-2-yloxy)acetanilide
Indanofan	(*RS*)-2-[2-(3-chlorophenyl)-2,3-epoxypropyl]-2-ethylindan-1,3-dione
Mefenacet	2-(1,3-benzothiazol-2-yloxy)-*N*-methylacetanilide
Metalaxyl	methyl (*RS*)-*N*-(2-methoxyacetyl)-*N*-(2,6-xylyl)-DL-alanine
Metazachlor	2-chloro-*N*-(pyrazol-1-ylmethyl)aceto-2′,6′-xylidide
Metolachlor	(*RS*)-2-chloro-6′-ethyl-*N*-(2-methoxy-1-methylethyl)aceto-2′-toluidide
Metribuzin	4-amino-6-(*tert*-butyl)-4,5-dihydro-3-methylthio-1,2,4-triazin-5-one
Oryzalin	3,5-dinitro-N^4, N^4-dipropylsulfanilamide
Piperophos	*S*-(2-methylpiperidino)carbonylmethyl *O*,*O*-dipropyl phosphorodithioate
Pretilachlor	2-chloro-2′,6′diethyl-*N*-(2-propoxyethyl)acetanilide
Propachlor	2-chloro-*N*-isopropylacetanilide
Pyrazophos	*O*,*O*-diethyl *O*-(6-ethoxycarbonyl-5-methylpyrazolo[1,5-a]pyrimidin-2-yl)phosphorothioate
RH-85	*O*,*O*-dimethyl (2-chloro-6-methylphenyl)sulfonyloxymethyl phosphonate
RH-4641	*O*-isopropyl (RS)-*P*-(2-ethyl-6-trifluoromethylphenyl)-sulfonyloxymethyl *P*-methyl phosphinate
Thenylchlor	2-chloro-*N*-(3-methoxy-2-thenyl)-2′,6′-dimethylacetanilide
Triallate	*S*-(2,3,3-trichloroallyl)-*N*,*N*-diisopropylthiocarbamate
UBI-S734	1-(2-ethyl-5-methylphenyl) 2-(*N*-oxidopyridyl) sulfone

place which is specific for the highly sensitive condensing enzyme of the plant microsomal elongase system. Formation of the covalent bond assumed between inhibitor and enzyme takes time and is temperature-dependent. The reaction is based on a nucleophilic attack of the β-ketoacyl synthase, most probably by its conserved *cysteine* of the reactive site. An active inhibitor should have an electrophilic C-atom made available by a leaving group. Chloro- or oxyacetamides have a reactive α-C atom since Cl or the heterocycle-oxy group probably splits off. Tetrazolinones or triazole amides (like cafenstrole) can react with the target enzyme through nucleophilic addition eliminating the tetrazolinone or the triazole moiety, respectively. The same reaction type may take place by opening the oxirane cycle of tridiphane. Phosphinosulfonates (RH-4641) or the phosphonosulfonate RH-85 apparently bind in a similar fashion by splitting off the 2,6-disubstituted benzenesulfonate. It is of interest that phytotoxic chloroacetamides, oxyacetamides, tetrazolinones and carbamoylsulfonyltriazoles all include a disubstituted amide nitrogen (having alkyl, substituted phenyl or hexyl substituents). As could be shown with appropriate analogs, an -N*H*-R group in the molecule yields an inactive compound. Seemingly, this also holds for active thiocarbamates (see Fig. 13 of Böger et al. 2000).

These assumed reaction mechanisms leading to a covalent binding between herbicide and target enzyme have to be verified by detailed analysis. Our hypothesis is corroborated by the structures of the *inactive* analogues. The metazachlor analog BAS-128682 (see Table 1) has a poor activity in our assay since by replacing Cl against a methoxy substituent (or a methyl group), no appropriate leaving group is present and the binding to the elongase cannot take place. A similar situation is evident with the inactive pyrazophos while the structurally comparable active anilofos contains an electrophilic C-atom with a leaving group.

The outlined enzyme-inhibitor binding mechanisms also hold for thiocarbamates like the well-known EPTC. This compound is practically inactive in our cell-free assay because it is not oxidized. In intact plants and by longer reaction times, the sulfur is assumed to be oxidized to -SO and/or SO_2. These groups can drag electrons off the neighboring carbonyl-C rendering it reactive.

6.6 Considerations on Resistance

Despite their widespread and repeated use, no serious resistance problems have shown up with chloroacetamides in corn and soybean fields. Four general explanations can be offered to explain the general situation (for exceptions see below).

1. The condensing enzymes sequenced so far include a highly conserved region with a *cysteine* at the reaction site within their catalytic domain (James et al. 1995; Millar et al. 1999; Todd et al. 1999). Inhibitor binding presumably requires this cysteine residue. An exchange of the cysteine will abolish both binding and catalytic function and will be lethal for the plant. Accordingly, no resistant weed mutants will occur. We believe that the findings outlined have strengthened this assumption.
2. An exchange of VLCFAs against "normal" fatty acids, as is the case in the resistant *Scenedesmus* mutant (Couderchet et al. 1995), appears to be a rare event since many reaction steps, like fatty-acid formation, transport and their incorporation into the right location of the cell have to be modified and regulated in concert.
3. It cannot be excluded as yet that chloroacetamides attack *additional* crucial targets leading to phytotoxicity (synthases of the condensing type operating in other biosynthetic pathways?), although, their inhibition may require

higher inhibitor concentrations than described above. If this holds true, target-enzyme resistance would require genetic modification of at least two (or even more) target enzymes in *one* mutant. This is an unlikely event. Out of about 10^8 individuals one weed individual with an altered β-ketoacyl-CoA synthase may show up. Genetic modification of two phytotoxic enzyme targets in a mutant would have a probability of $1:10^{16}$.

4. Metabolic resistance against metolachlor with a factor of 3–4 vs. the wild type has been achieved with transgenic tobacco by overexpression of a glutathione S-transferase (GST) gene from maize encoding the isoform GST-IV (Jepson et al. 1997). This (dimeric) isoform, including two GST-27 monomers (of 27 kDa), is instrumental for the conjugation of chloroacetamides with glutathione (see Sommer and Böger 1999 for details). By mutation, overexpression of an endogenous gene for the GST-27 subunit may occur resulting in resistance of a weed. However, this metabolic detoxification should neither be limited by the amount of glutathione (GSH) present, by the biosynthesis capacity to produce GSH, nor by the activity of GSH-reductase to keep glutathione reduced. Although no data are available it can be assumed that in many weeds, another component of this detoxification system, besides the GST-IV may be limiting. Then metabolic resistance requires increased expression of more than one gene. Again, a concurrent favorable alteration of two (or more) interactive cell constituents appears unlikely.

Resistance against metolachlor (also against propachlor, alachlor) was reported years ago for Australian rigid rye grass (*Lolium rigidum*) with resistance ratios of about 7 (Burnet et al. 1994). *Echinochloa crus-galli* with similar resistance factors against butachlor and thiobencarb has been found in rice fields in China (Huang and Gressel 1997). We obtained preliminary evidence that resistance of the *Echinochloa* strain is not due to the enzyme target. The mode of resistance is not known so the possibility of 4 (see above) cannot be excluded as yet. It is noteworthy that Moreland et al. (1995) have shown that isolated mung bean microsomes (including the cytochrome P450 monooxygenase system) can modify metolachlor or alachlor by O-demethylation. The rate, however, was found to be low. It is not known whether a P450 system in the resistant weeds is active enough to cope with the phytotoxic impact of these chloroacetamides, or whether these modified chloroacetamides are still phytotoxic. For metolachlor it has been shown (LeBaron et al. 1988) that in intact maize an O-demethylation does not occur with the free herbicide, but with the cysteine conjugate.

Acknowledgements. The authors are grateful to agrochemical companies which generously provided us with herbicides and analogs. These are BASF AG; Bayer AG, both Germany; DuPont, USA; Eiko Kasei, Tokuyama Corp., both Japan; Novartis (now Syngenta), Switzerland; Rohm and Haas, Uniroyal, both USA. The authors thank Dr. L. Kunst, Vancouver, Canada, for a *Saccharomyces* strain cloned with the FAE1 (elongase) gene.

References

Agrawal VP, Lessire R, Stumpf PK (1984) Biosynthesis of very-long-chain fatty acids in microsomes from epidermal cells of *Allium porrum* L. Arch Biochem Biophys 230:580–589

Anonymous (1999) Roundup usage doubles on US soybeans Agrow, no 330, pp 17–18

Asai M, Yogo Y (1998) Dose response analysis and estimation of I_{50} of paddy amide herbicides for prediction of duration of activity. Weed Sci Soc Am (WSSA) Abstr Book 38:65

Barrett PB, Harwood JL (1998) Naphthalic anhydride prevents inhibition of fatty acid elongation by thiocarbamates. Phytochemistry 49:1897–1903

Böger P, Matthes B, Schmalfuß J (2000) Towards the primary target of chloroacetamides – new findings pave the way. Pestic Manage Sci 56:497–508

Brown DA, London E (2000) Structure and function of sphingolipid- and cholesterol-rich membrane rafts. J Biol Chem 275:17221–17224

Brown RB (1998) Sphingolipid organization in biomembranes: what physical studies of model membranes reveal. J Cell Sci 111:1–9

Burnet MWM, Barr AR, Powles SB (1994) Chloroacetamide resistance in rigid ryegrass (*Lolium rigidum*). Weed Sci 42:153–157

Cahoon EB, Lynch DV (1991) Analysis of glucocerebrosides of rye (*Secale cereale* L. cv. Puma) leaf and plasma membrane. Plant Physiol 95:58–68

Cassagne C, Lessire R, Bessoule JJ, Moreau P, Creach A, Schneider F, Stubois B (1994) Biosynthesis of very-long-chain fatty acids in higher plants. Prog Lipid Res 33:55–69

Cook HW (1994) Fatty acid desaturation and chain elongation in eukaryotes. In: Vance DE, Vance JE (eds) Lipoproteins and membranes. Elsevier, Amsterdam, pp 129–152

Couderchet M, Böger P (1993) Chloroacetamide-induced reduction of fatty acid desaturation. Pestic Biochem Physiol 45:91–97

Couderchet M, Brozio B, Böger P (1994) Effect and metabolism of the chloroacetamide herbicide metazachlor: comparison of plant cell suspension cultures and seedlings. J Pestic Sci 19: 127–135

Couderchet M, Rumbolz J, Kring F, Böger P (1995) Characteristics of a metazachlor-resistant *Scenedesmus acutus* cell line. Pestic Biochem Physiol 52:222–233

Couderchet M, Bocion PF, Chollet R, Seckinger K, Böger P (1997) Biological activity of two stereoisomers of the *N*-thienyl chloroacetamide herbicide dimethenamide. Pestic Sci 50: 221–227

Couderchet M, Schmalfuß J, Böger P (1998) A specific and sensitive assay to quantify the herbicidal activity of chloroacetamides. Pestic Sci 52:381–387

Deal LM, Hess FD (1980) An analysis of the growth inhibitory characteristics of alachlor and metolachlor. Weed Sci 28:168–175

Domergue F, Besoule JJ, Moreau P, Lessire R, Cassagne C (1998) Recent advances in plant fatty acid elongation. In: Harwood JL (ed) Plant lipid biosynthesis: fundamentals and agricultural applications. Cambridge University Press, Cambridge, pp 185–222

Ebert E (1980) Herbicidal effects of metolachlor (2 chloro-N-[2-ethyl-6-methylphenyl]-*N*-[2-methoxy-1-methylethyl]acetamide) at a cellular level in sorghum. Pestic Biochem Physiol 13:227–236

Ebert E, Ramsteiner K (1984) Influence of metolachlor and the metolachlor protectant CGA 43089 on the biosynthesis of epicuticular waxes and the primary leaves of *Sorghum bicolor* Moench. Weed Res 24:383–389

Fedtke C (1982) Modes of herbicide action as determined with *Chlamydomonas reinhardii* and Coulter counting. In: Moreland DE, St John JB, Hess FD (eds) Biochemical responses induced by herbicides. ACS Ser 181, Am Chem Soc, Washington, DC, pp 231–250

Fehling E, Lessire R, Cassagne C, Mukherjee KD (1992) Solubilization and partial purification of constituents of acyl-CoA elongase from *Lunaria annua*. Biochim Biophys Acta 1126:88–94

Fuerst EP (1987) Understanding the mode of action of the chloroacetamide and thiocarbamate herbicides. Weed Technol 1:270–277

Fuerst EP, Lamoureux GL, Ahrens WH (1991) Mode of action of the dichloroacetamide antidote BAS 145-138 in corn. I. Growth responses and fate of metazachlor. Pestic Biochem Physiol 39:138–148

Hamm PC (1974) Discovery, development, and current status of the chloroacetamide herbicides. Weed Sci 22:541–545

Huang BQ, Gressel J (1997) Barnyardgrass (*Echinochloa crus-galli*) resistance to both butachlor and thiobencarb in China. Resist Pestic Manage 9:5–7

James Jr DW, Lim E, Keller J, Plooy I, Ralston E, Dooner HK (1995) Directed tagging of the *Arabidopsis* fatty acid elongation 1(*FAE1*) gene with the maize transposon activator. Plant Cell 7:309–319

Jaworski EJ (1956) biochemical action of CDAA, a new herbicide. Science 123:847–848

Jepson I, Holt DC, Roussel V, Wright SY, Greenland AJ (1997) Transgenic plant analysis as a tool for the study of maize glutathione S-transferases. In: Hatzios KK (ed) Regulation of enzymatic systems detoxifying xenobiotics in plants. Kluwer, Dordrecht, pp 313–323

Kern AJ, Jackson LL, Dyer WE (1997) Fatty acid and wax biosynthesis in susceptible and triallate-resistant *Avena fatua* L. Pestic Sci 51:21–26

Kring F, Couderchet M, Böger P (1995) Inhibition of oleic acid incorporation into a non-lipid fraction by chloroacetamide herbicides. Physiol Plant 95:551–558

Lassner MW, Lardizabal K, Metz JG (1996) A jojoba β-ketoacyl-CoA synthase cDNA complements the canola fatty acid elongation mutation in transgenic plants. Plant Cell 8:281–292

Leavitt JRC, Penner D (1979) In vitro conjugation of glutathione and other thiols with acetanilide herbicides and EPTC sulfoxide and the action of the herbicide antidote R-25788. J Agric Food Chem 27:533–536

LeBaron HM, McFarland JE, Simoneaux BJ (1988) Metolachlor. In: Kearney PC, Kaufman DD (eds) Herbicides – chemistry, degradation and mode of action. Dekker, New York, pp 335–382

Mann JD, Pu M (1968) Inhibition of lipid biosynthesis by certain herbicides. Weed Sci 22:197–198

Matthes B, Schmalfuß J, Böger P (1998) Chloroacetamide mode of action. II. Inhibition of very long chain fatty acid synthesis in higher plants. Z Naturforsch 53c:1004–1011

Matthes B (2000) Die Wirkungsweise herbizidaler Chloracetamide. PhD Thesis, University of Konstanz

McFarland JE, Hess FD (1986) Chloroacetamide herbicides alkylate plant proteins. Weed Sci Soc Am (WSSA) Abstr Book 26:81

Mellis JM, Pillai P, Davis DE, Truelove B (1982) Metolachlor and alachlor effects on membrane permeability and lipid synthesis. Weed Sci 30:399–404

Millar AA, Kunst L (1997) Very-long-chain fatty acid biosynthesis is controlled through the expression and specificity of the condensing enzyme. Plant J 12:121–131

Millar AA, Clemens S, Zachgo S, Giblin EM, Taylor DC, Kunst L (1999) *CUT1*, an *Arabidopsis* gene required for cuticular wax biosynthesis and pollen fertility, encodes a very-long-chain fatty acid condensing enzyme. Plant Cell 11:825–838

Möllers C, Albrecht, S (1994) Screening herbicide effects on lipid metabolism of storage lipids by in vitro culture of microspore-derived embryoids of *Brassica napus*. J Plant Physiol 144: 376–384

Molin WT, Anderson EJ, Porter CA (1986) Effects of alachlor on anthocyanin and lignin synthesis in etiolated sorghum (*Sorghum bicolor* (L.) Moench) mesocotyls. Pestic Biochem Physiol 25:105–111

Moreau P, Bessoule S, Mongrand S, Testet P, Cassagne C (1998) Lipid trafficking in plant cells. Prog Lipid Res 37:371–391

Moreland DE, Corbin FT, Fleischmann, TJ, McFarland JE (1995) Partial characterization of microsomes isolated from mung bean cotyledons. Pestic Biochem Physiol 52:98–108

Murata N, Sato N, Takahashi N (1984) Very-long chain saturated fatty acids in phosphatidyl-serine from higher plant tissues. Biochim Biophys Acta 795:147–150

Narsaiah DB, Harvey RG (1977) Alachlor placement in the soil as related to phytotoxicity to maize (*Zea mays* L.) and soybean (*Glycine max*. L.) seedlings. Weed Res 17:163–168

Ohlrogge JB, Jaworski JG (1997) Regulation of fatty acid synthesis. Annu Rev Plant Physiol Plant Mol Biol 48:109–136

Poulos A (1995) Very long chain fatty acids in higher animals – a review. Lipids 30:1–14

Renault S, Shukla A, Giblin M, MacKenzie SL, Devine MD (1997) Plasma membrane lipid composition and herbicide effects on lipoxygenase activity do not contribute to differential membrane responses in herbicide-resistant and -susceptible wild oat (*Avena fatua* L.) biotypes. J Agric Food Chem 45:3269–3275

Schmalfuß J, Matthes B, Mayer P, Böger P (1998) Chloroacetamide mode of action. I. Inhibition of very long chain fatty acid synthesis in *Scenedesmus acutus.* Z Naturforsch 53c:995–1003

Schmalfuß J, Matthes B, Knuth K, Böger P (2000) Inhibition of acyl-CoA elongation by chloroacetamide herbicides in microsomes from leek seedlings. Pestic Biochem Physiol 67:25–35

Sharp DB (1988) Alachlor. In: Kearney PC, Kaufman DD (eds) Herbicides – chemistry, degradation and mode of action. Dekker, New York, pp 301–333

Sloan ME, Camper ND (1985) Effects of alachlor and metolachlor on cucumber seedlings. Environ Exp Bot 26:1–7

Sommer A, Böger P (1999) Characterization of recombinant corn glutathione S-transferase isoforms I, II, III, and IV. Pestic Biochem Physiol 63:127–138

Tevini M, Steinmüller D (1987) Influence of light, UV-B radiation, and herbicides on wax biosynthesis of cucumber seedlings. J Plant Physiol 131:111–121

Todd J, Post-Beittenmiller D, Jaworski JG (1999) *KCS1*encodes a fatty acid elongase 3-ketoacyl-CoA synthase affecting wax biosynthesis in *Arabidopsis thaliana.* Plant J 17:119–130

Uemura M, Joseph RA, Steponkus PL (1995) Cold acclimation of *Arabidopsis thaliana.* Plant Physiol 109:15–30

Vavrina CS, Ashley RA (1983) Effect of alachlor on PEG_{6000} uptake, root osmotic potential, and root leakage. Weed Sci 31:600–603

Weisshaar H, Böger P (1987) Primary effects of chloroacetamides. Pestic Biochem Physiol 28: 286–293

Wettstein-Knowles PM von (1993) Waxes, cutin and suberin. In: Moore TS (ed) Lipid metabolism in plants. CRC Press, Boca Raton, pp 127–166

Wilkinson RE (1981) Metolachor influence on growth and terpenoid synthesis. Pestic Biochem Physiol 16:63–71

Wilmesmeier S, Steuernagel S, Wiermann R (1993) Comparative FTIR and ^{13}C CP/MAS NMR spectroscopic investigations on sporopollenin of different systematic origin. Z Naturforsch 48c:697–701

Wu J, Hwang IT, Hatzios KK (1999) Effects of chloroacetanilide herbicides on membrane fatty acid desaturation and lipid composition in rice, maize and sorghum. Pestic Biochem Physiol 66:161–169

Zama P, Hatzios KK (1987) Interaction between the herbicide metolachlor and the safener CGA-92194 at the levels of uptake and macromolecular synthesis in sorghum leaf protoplasts. Pestic Biochem Physiol 27:86–96

7 Cellulose Biosynthesis Inhibitor Herbicides

KEVIN C. VAUGHN

7.1 Introduction

Cellulose biosynthesis inhibitor (CBI) herbicides are a small group of chemically unrelated compounds including the herbicides dichlobenil, isoxaben and flupoxam (Sabba and Vaughn 1999; Fig. 1). In addition, the auxinic-type herbicide quinclorac may have a second site of action in monocots that affects cellulose biosynthesis (Koo et al. 1996, 1997). Although they are a rather small group of compounds in relation to all of the herbicides, they have several qualities that make them quite important. For example, the lack of field resistance to these compounds and a site of action not shared by mammals make them an important group in terms of resistance management and approval by government agencies concerned with toxicological issues. Moreover, they appear to be useful tools to unravel the complexities of the plant cell wall and, more specifically, the production of cellulose.

CBI herbicides induce the same sort of symptomatology as the mitotic disrupter herbicides: swollen root tips after growth of the seedlings in the presence of the herbicide (Roberts 1990; Heim et al. 1998; Sabba and Vaughn 1999), so-called radial root swelling (Baskin et al. 1992). Because many assume that this symptom is associated only with mitotic disrupter herbicides, this single symptom may be misinterpreted, resulting in classification of the herbicide in the wrong group. Such was the case with flupoxam, a herbicide that induces extensive radial swelling of the root tips (O'Keefe and Klevorn 1991). However, subsequent investigations have not indicated any mitotic or microtubule effect of this herbicide (Hoffman and Vaughn 1996), yet substantial effects are noted on new cell wall synthesis (Heim et al. 1998; Vaughn and Turley 2001). Thus, although mitotic disrupter herbicides have been described classically as causing this effect, a herbicide should certainly not be placed in this group without further verification.

Another important distinction between CBI and mitotic disrupter herbicides lies in the selectivity differences of these groups. Mitotic disrupter herbicides are most effective on small seeded monocots (Vaughn 2000). The CBI herbicides are most effective on dicots, with either less dichlobenil (DCB) or no (isoxaben) effects on the monocots (Sabba and Vaughn 1999). Recently, Heim et al. (1998) have described subtle differences in root appearance that

P. Böger, K. Wakabayashi, K. Hirai (Eds.)
Herbicide Classes in Development

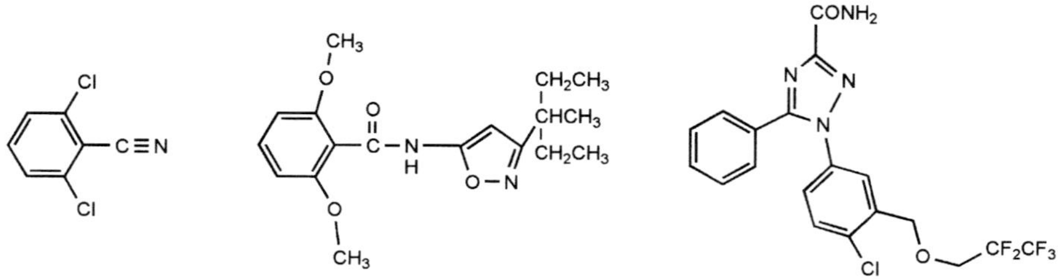

Fig. 1. Chemical structures of the three major cellulose biosynthesis inhibitor herbicides: dichlobenil (*right*), isoxaben (*center*) and flupoxam (*right*)

might be useful in diagnosis of CBI symptomatology compared with mitotic disrupter herbicides.

7.2 Mode of Action Studies

Each of the CBI herbicides has been shown to inhibit the incorporation of radiolabeled glucose into an acid insoluble product that is assumed to be cellulose (Hogetsu et al. 1974; Heim et al. 1990a,b, 1998). Even though cellulose is the most abundant biopolymer in the world, attempts to obtain consistent cellulose synthase activity in vitro have been very difficult (see reviews in Delmer and Amor 1995; Brown et al. 1996). Instead, callose is often produced in these in vitro assays with a limited amount of cellulose. This makes such in vitro assay systems relatively useless in studying effects of these herbicides. Thus, certain in vivo systems that allow one to study de novo formation of cell walls or cellulose have been utilized to investigate CBI herbicides.

7.2.1 Cell Plates

Cell plates, which are cell walls formed to separate the daughter nuclei after cell division, are an attractive system for studying wall biosynthesis in that the wall is formed de novo in about 20–90 min, depending upon the species (Samuels et al. 1995). Moreover, recent advances in cell synchronization and cell plate ontogenesis have improved our understanding of how the various polysaccharides are assembled into this new wall (Samuels et al. 1995). These data indicate a prominent role of callose in causing cell plate spreading, whereas cellulose plays a role later in the development associated with cell plate stiffening.

DCB has been the most intensively studied in terms of cell plate formation. In the numerous light and electron microscopic studies, a consistent observation has been that the plates have been much more undulated and thicker than noted in the untreated cell plates (Fig. 2; Gonzalez-Reyes et al. 1986; Mineyuki

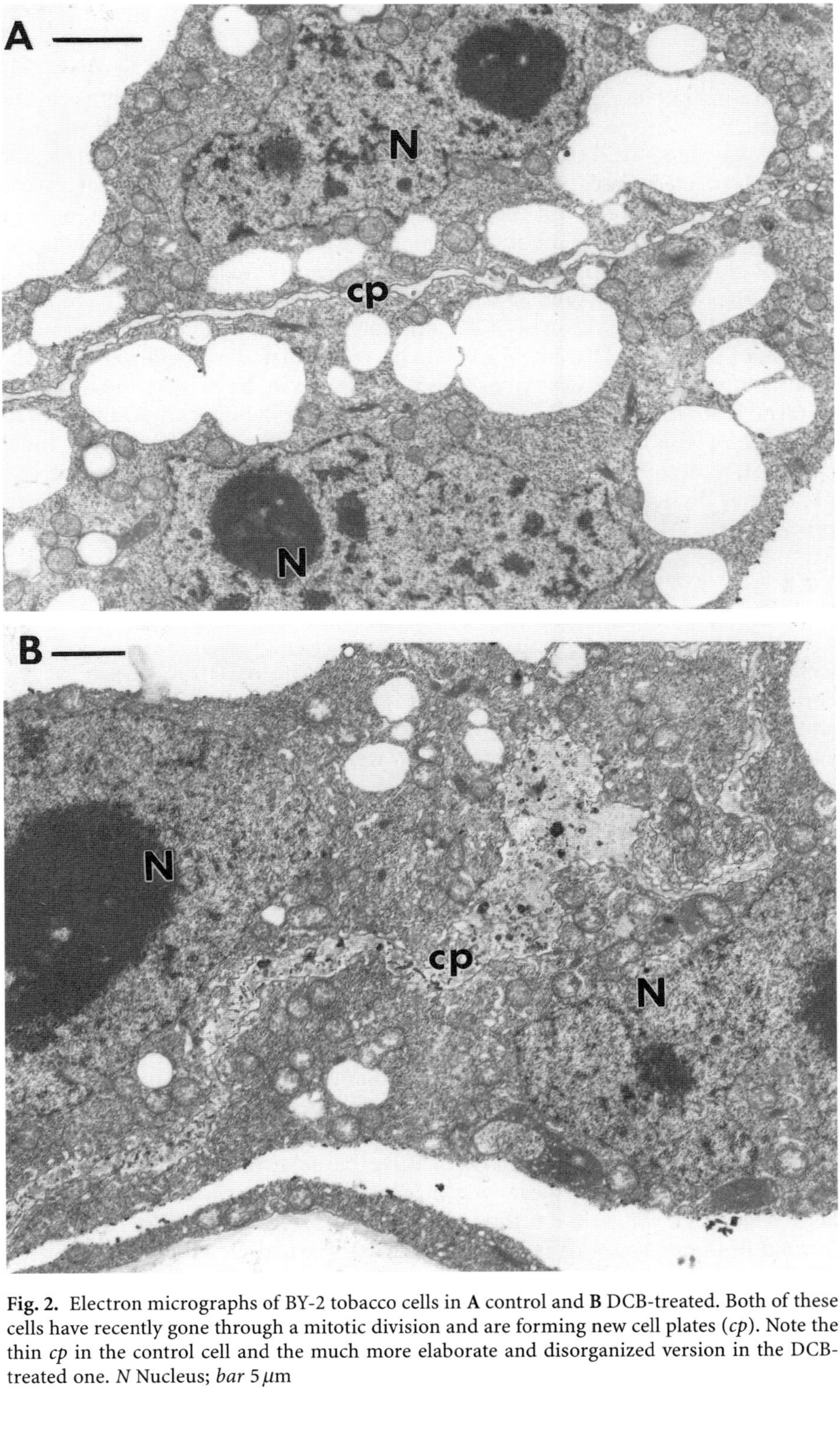

Fig. 2. Electron micrographs of BY-2 tobacco cells in **A** control and **B** DCB-treated. Both of these cells have recently gone through a mitotic division and are forming new cell plates (*cp*). Note the thin *cp* in the control cell and the much more elaborate and disorganized version in the DCB-treated one. *N* Nucleus; *bar* 5 μm

and Gunning 1990; Vaughn et al. 1996). Often the plates and the resulting cell walls are incomplete or attached only to one parental wall. Inclusions in the cell plate resemble plasmodesmata. In accord with the increased spreading of the cell plate after DCB treatment, Vaughn et al. (1996) have demonstrated that the DCB-affected plates are highly enriched in callose (and to a lesser extent xyloglucan) compared with the controls (Fig. 3). A simple model to explain this is that inhibition of cellulose biosynthesis by DCB increases the amount of UDP-glucose available for callose or xyloglucan synthesis.

The effects of flupoxam and isoxaben on cell plate formation have been investigated less than those of DCB, although both herbicides are effective disrupters of the process. Unlike the wider, more undulated wall induced by DCB, treatment with either isoxaben or flupoxam results in the production of a very thin cell plate that is enriched in neither callose nor xyloglucan. Antibodies to esterified pectins do react with these thin cell plates, however. These data indicate that, although these two herbicides block cellulose synthesis in the cell plates, they do not result in the diversion of the UDP-glucose pools into callose production.

7.2.2 Developing Cotton Fibers

Cotton fibers are elongated single cell hairs arising from the surface of the cotton ovule. Although the primary wall is similar in composition to other trichome walls (Vaughn and Turley 1999), the secondary wall is composed almost exclusively of cellulose. The extent of new wall produced is so great compared to the unexpanded epidermal cell that these essentially represent a completely de novo cell wall. An added advantage of this system is that the ovules may be cultured in vitro, allowing manipulation of the media and culture conditions to examine the effects on fiber production (Montezinos and Delmer 1980). Treatment of developing cotton fibers with any of the CBI herbicides results in severe inhibition of fiber development (Vaughn and Turley 2001). In all cases, the fibers produced are much more isodiametric than the elongated cells produced in the untreated fibers (Fig. 4A, B). When these fibers were analyzed by immunocytochemistry, two basic types of fiber alterations were noted. Control fibers contained a cell wall with two layers, an outer wall area enriched in pectins and an inner layer enriched in cellulose-xyloglucan. The ratio of the two wall components was about two thirds to three fourths in the inner layer and one third to one fourth in the outer layer (Vaughn and Turley 1999). After DCB treatment, the outer layer was of the same relative proportion and composition, but now the inner layer was highly enriched in callose. This new inner layer exhibited the same ultrastructural characteristics as the cell plates formed in the presence of DCB, including the unusual electron opaque inclusions. Only very weak labeling was obtained with a cellulase-gold probe, indicating that these fibers had virtually no cellulose present. A different effect was noted after treatment with flupoxam and isoxaben. With these herbicide treatments, the outer layer of the fiber was apparently increased in relative

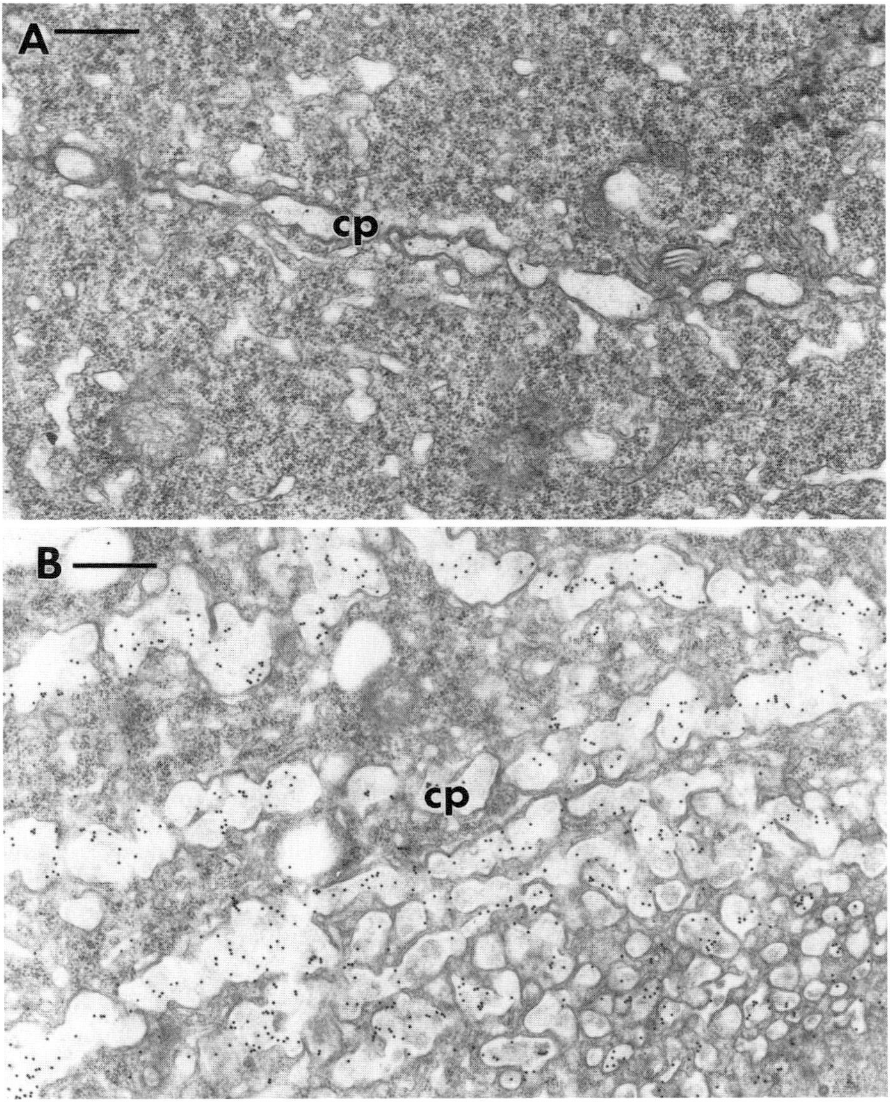

Fig. 3. Electron micrographs of BY-2 tobacco cells in **A** control and **B** DCB-treated and labeled with antibodies to callose and secondary-antibody gold complexes. Note the label along the thin cell plate (*cp*) of the control cell and the very strong labeling in the elaborate cell plate formed in the presence of DCB. *Bar* 0.5 μm

proportion to the fiber wall and the inner layer had a similar structural organization to the outer wall. When probed immunocytochemically, the fiber wall was highly enriched in de-esterified pectin, even in the inner layers of the wall where de-esterified pectins are not usually noted. When probed with the cellulase-gold probe very little labeling was obtained, with the exception of

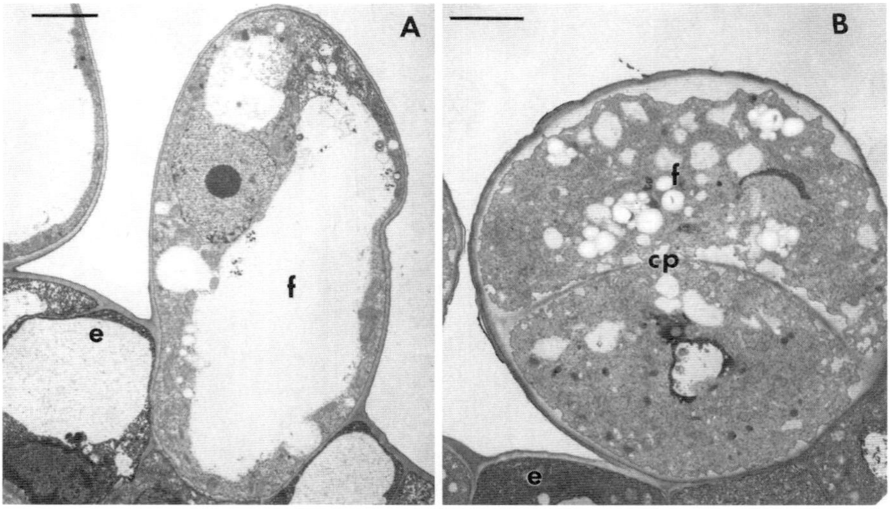

Fig. 4. **A** Electron micrograph of a control 1-day-old cotton fiber cell and **B** that treated with the herbicide isoxaben. Instead of the elongated fibers noted in control cells (**A**), the isoxaben-treated fiber is distinctly spherical and possess walls composed primarily of pectin rather than cellulose. Note also the presence of a new cell wall transecting the fiber (*cp*) that occurs rarely in control fibers but frequently after isoxaben treatment. *Bar* 2.0 μm

small patches that occurred at the plasma membrane-wall interface. Unlike the DCB treatment, the fiber cells treated with isoxaben and flupoxam also initiated cell divisions during fiber growth, with ~60% of the fibers containing cell plates (Fig. 4B).

These two quite different model systems, cell plate formation and biogenesis of cotton fibers, are both useful in investigating the effects of the CBI herbicides. Despite the difference in wall formation, in terms of rapidity and mechanism, the CBI herbicides caused similar sorts of effects in both systems. DCB caused a diversion of cellulose biosynthesis into callose. Treatment with flupoxam and isoxaben stopped the production of both callose and cellulose in both plates and fibers.

7.2.3 Azido-DCB Derivative

Delmer and colleagues showed that an azido-derivative of DCB bound to an 18-kDa polypeptide. Because this polypeptide is much lower in molecular mass than the cellulose synthase (62 kDa), these workers assumed that this might be a regulatory component of the cellulose synthase complex. Wang et al. (1997) developed a similar DCB analog and used it to identify a similar molecular mass polypeptide in a marine diatom, although the diatom protein fractionated with the soluble fraction rather than the membrane fractions. Certainly the increase in production of callose instead of cellulose in both the cell plates (Vaughn et al. 1996) and cotton fibers (Vaughn and Turley 2001) is a strong indicator that DCB might be causing a shift in the production of cellulose into callose, rather than just a simple inhibition of cellulose production. Cloning of this gene product and further identification will be necessary to determine the molecular mechanism by which DCB regulates the flow of carbon in cellulose biosynthesis.

7.3 Resistant Biotypes

No examples of field-selected resistance to the CBI herbicides have been reported, with the possible exception of a smooth crabgrass biotype resistant to quinclorac (Koo et al. 1996, 1997). Similarly, the generation of laboratory mutants resistant to these herbicides in *Arabidopsis* from a mutagenized seed population also proved difficult (Heim et al. 1989, 1990a,b). Selection of two isoxaben-resistant mutants and one DCB-resistant mutant were isolated from very extensive mutant screens with over two million mutagenized seedlings screened. The isoxaben resistant lines were 50- to 100-fold resistant to isoxaben than the wild type, whereas the DCB-resistant line was much less so, about 5- to 7-fold based upon growth only. The isoxaben-resistant mutants are not cross-resistant to either DCB or flupoxam, and the DCB mutant is not cross-resistant to either flupoxam or isoxaben (Vaughn, unpubl.). These data indicate that the three herbicides act at three different sites or have three different binding sites on the same protein.

Microscope investigations on cell plate formation in the resistant lines reveal that the characteristic cell plate abnormalities associated with that particular herbicide to which the resistant biotypes were selected were absent or reduced in the resistant biotype (Vaughn, unpubl.). However, effects of the other CBI herbicides were typical of those noted in the normally susceptible wild type. These data agree with the relative effects of these herbicides on radiolabeled glucose incorporation into cellulose in these lines. None of the mutants exhibited enhanced herbicide metabolism (Heim et al. 1989, 1990a).

With the advent of sequencing technology and the identification of the cellulose synthase gene family, Sommerville and colleagues (pers. comm.) have confirmed that at least one of the isoxaben-resistant mutants carries an alteration in one of the cellulose synthase genes. This is the first molecular confirmation that any CBI herbicide interacts directly with cellulose synthase. Similar DNA sequencing of the alteration(s) in the DCB-resistant line may allow us to determine if the polypeptide labeled by azido-DCB is altered in the DCB-resistant biotype.

7.4 Habituation

Normally, studies of herbicide action involve relatively narrow windows of herbicide treatment (1–24h) and then the effects at the herbicide active site or some physiological parameter are monitored. However, in the real use of herbicides, the plant is confronted with the herbicide over a relatively long period. Thus, the plant might respond to the herbicide in one way in the short term, but differently after continuing exposure.

Delmer and her colleagues (Delmer et al. 1987; Shedletzky et al. 1990, 1992) were the first to investigate this sort of effect with cultures exposed to the herbicide DCB. Although the expectation is that the cells might somehow adapt to the presence of DCB by producing cellulose despite the presence of herbicide, something quite different occurred. An early indicator of the difference was the shift in morphology of the cultures. Normal suspension cultures are relatively homogeneous, with small clumps or strands of 5–20 cells (Fig. 5). After DCB treatment, the cells were much more clumped than previously noted, with huge aggregates of cells rather than the finely dispersed suspension. Biochemical analysis revealed the near absence of cellulose in these DCB-treated cultures, but rather a great enrichment in uronic acids (pectins) was

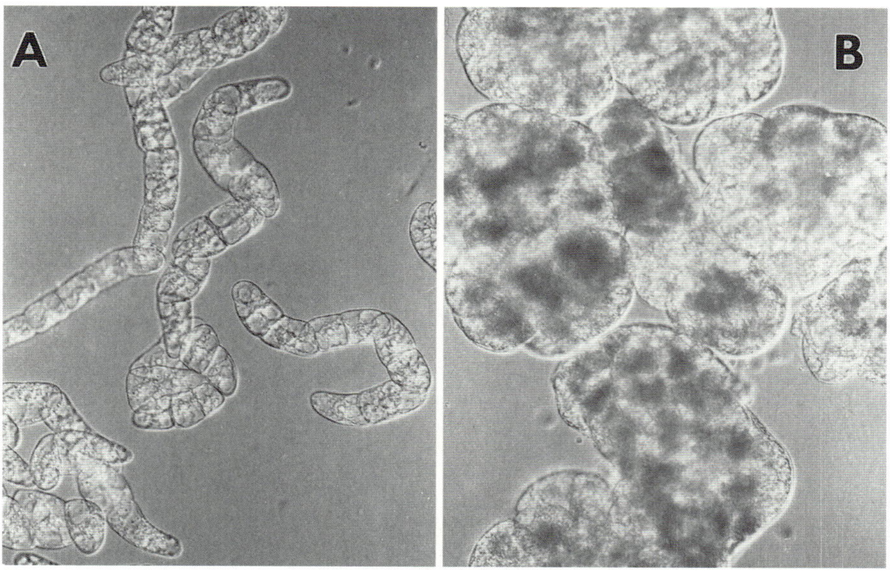

Fig. 5. Nomarski differential interference micrographs of BY-2 **A** cultures and **B** those habituated on isoxaben. Although the control BY-2 cells form nonclumping filaments, those habituated to isoxaben produce dense clumps of cells because of the increase in pectins in these walls

observed. Delmer and colleagues showed that both dicots and monocots responded similarly to the DCB habituation, although there were subtle differences between the dicot species and more substantial differences in wall composition in the one monocot examined.

Sabba et al. (1999) examined the habituation of the rapidly growing BY-2 cells to DCB and found similar shifts in the presence of pectins and absence of cellulose in the cell walls of the habituated line. The wall morphology was shown to consist of lamellate sheets of primarily de-esterified pectins, with an increase in the wall protein extensin that is involved in wall strengthening. Although callose accumulation is an early and short-term symptom of DCB treatment, the walls did not accumulate massive amounts of callose. Rather, the callose was present as short strands that paralleled the distribution of microtubules in these cells. One interpretation of these data is that the cellulose synthase components were still being directed by the microtubules, but the presence of DCB diverted the production to callose. Presumably callose was then metabolized into glucose units so that the entire wall was not callose-enriched.

Habituation of cultures to isoxaben followed a similar pattern to that observed for DCB except that the response of the cultures was much quicker (Sabba and Vaughn 1999; Sabba and Vaughn, in prep.). Cells treated with DCB seemed to go through a phase when there was an initial buildup of callose (as would be expected of the short-term effects seen in cell plate and cotton fibers), whereas cells treated with isoxaben immediately started to clump. Within several weeks, the cells were in small clumps and exhibited many properties of the DCB-habituated lines, with the exception of callose accumulation. Because no callose accumulates in isoxaben-treated cell plates in BY-2 cells (Vaughn, unpubl.), the habituation of the cells to isoxaben goes directly into pectin production without an intervening period of adaptation.

All of the habituation studies described above were performed on suspension cultures so that the cells are constantly bathed in a herbicide-containing medium. Alvarez's group has shown that it is possible to select for similar sorts of habituated cells by simply growing callus cultures on media supplemented with DCB (Encina et al. 2001) or isoxaben (Diaz-Cacho et al. 1999). These habituated callus cultures contained the same sorts of pectin alterations as those described for suspension cells described above, although the severity of the decreases in cellulose and increases in pectin were less than those found in suspension cultures. It is possible that as the callus becomes larger and more separated from the source of herbicide, the effective herbicide concentrations also become less.

Another interesting use of the habituated cell lines has been the isolation and characterization of cellulose synthase itself. Nakagawa and Sakurai (1998) found that the DCB-habituated BY-2 cells produced increased amounts of the celA-1 protein (cellulose synthase). Several reasons could be evoked to explain this increase. One may be related to the plant sensing the need to make more cellulose and increasing the amount of cellulose synthase to compensate for the lack of cellulose being produced, or that DCB somehow stabilizes the cellulose synthase protein.

7.5 The Unusual Case of Quinclorac

Although the other CBI herbicides seem to have only cellulose biosynthesis inhibition as a primary mechanism of action, the auxinic herbicide quinclorac may act as a CBI herbicide in some monocot species. Work of Koo et al. (1996, 1997) revealed that radiolabeled glucose incorporation into the cellulosic fraction was inhibited by quinclorac in grass species in which the auxinic-type effects are not noted. These data indicated that quinclorac might have a second site of action unrelated to its primary mechanism in the dicots. A quinclorac-resistant biotype of smooth crabgrass required higher concentrations of quinclorac to inhibit glucose incorporation into cellulose.

7.6 Conspectus

CBI herbicides are a small group of herbicides that are effective inhibitors of cellulose biosynthesis. Because this site of action is not shared with mammals, the toxicity of these compounds makes them ideal choices in terms of registration. Moreover, the lack of field-selected herbicide resistance makes them a powerful tool in weed resistance management. Aside from their use as herbicides, the CBI herbicides have been extremely useful in terms of understanding the nature of the plant cell wall and cell plate formation in plants. The role of callose in cell plate spreading and the ability of cells to form walls composed essentially totally of pectins are two important discoveries related directly to the CBI herbicides.

Acknowledgments. Portions of this work were supported by a NRI grant to Kevin C. Vaughn. Work of postdoctoral scientists John C. Hoffman, Neil A. Durso, and Robert P. Sabba are gratefully acknowledged as is the technical assistance of Ms. Lynn Libous-Bailey.

References

Baskin TI, Betzner AS, Hoggart R, Cork A, Williamson RE (1992) Root morphology mutants in *Arabidopsis thaliana*. Aust J Plant Physiol 279:717–720

Brown RM, Saxena IM, Kudlicka K (1996) Cellulose biosynthesis in higher plants. Trends Plant Sci 1:149–156

Delmer DP, Amor Y (1995) Cellulose biosynthesis. Plant Cell 7:987–1000

Delmer DP, Read SM, Cooper G (1987) Identification of a receptor protein in cotton fibers for the herbicide 2,6-dichlorobenzonitrile. Plant Physiol 84:415–420

Diaz-Cacho P, Moral R, Encina A, Acebas JL, Alvarez J (1999) Cell wall modifications in bean (*Phaseolus vulgaris*) callus cultures tolerant to isoxaben. Physiol Plant 107:54–59

Durso NA, Vaughn KC (1997) The herbicidal manipulation of callose levels in plants radically affects cell plate structure. Plant Physiol 114S:87

Encina AE, Moral RM, Acebas JL, Alvarez JM (2001) Characterization of cell walls in bean (*Phaseolus vulgaris* L.) callus cultures tolerant to dichlobenil. Plant Sci 160:3312–339

Gonzalez-Reyes JA, Navas P, Garcia-Herdugo G (1986) An ultrastructural study of cell plate modifications induced by 2,5-dichlorobenzonitrile. Protoplasma 132:172–178

Heim DR, Roberts JL, Pike RD, Larrinua IM (1989) Mutation of a locus of *Arabidopsis thaliana* confers resistance to the herbicide isoxaben. Plant Physiol 90:858–861

Heim DR, Roberts JL, Pike PD, Larrinua IM (1990a) A second locus, *Ixr* B1 in *Arabidopsis thaliana*, that confers resistance to the herbicide isoxaben. Plant Physiol 92:858–861

Heim DR, Skomp JR, Waldron C, Larrinua IM (1990b) Isoxaben inhibits the synthesis of acid insoluble cell wall materials in *Arabidopsis thaliana*. Plant Physiol 93:93–99

Heim DR, Larrinua IM, Murdoch MG, Roberts JL (1998) Triazofenamide is a cellulose biosynthesis inhibitor. Pestic Biochem Physiol 59:163–168

Hoffman JC, Vaughn KC (1996) Flupoxam induces classic club root morphology but is not a mitotic disrupter herbicide. Pestic Biochem Physiol 55:49–53

Hogetsu T, Shibaoka H, Shimokoriyama M (1974) Involvement of cellulose synthesis in actions of gibberellin and kinetin on cell expansion. 2,6-dichlorobenzonitrile as a new cellulose synthesis inhibitor. Plant Cell Physiol 15:389–393

Koo SJ, Neal JC, DiTomaso JM (1996) 3,7-Dichloroquinolinecarboxylic acid inhibits cell wall biosynthesis in maize roots. Plant Physiol 112:1383–1389

Koo SJ, Neal JC, DiTomaso JM (1997) Mechanism of action of quinclorac in grass roots. Pestic Biochem Physiol 57:44–53

Mineyuki Y, Gunning BES (1990) A role of preprophase bands of microtubules in maturation of new cell walls and a general proposal on the function of preprophase band sites in cell division in higher plants. J Cell Sci 97:527–537

Montezinos D, Delmer DP (1980) Characterization of inhibitors of cellulose synthesis in cotton fibers. Planta 148:305–311

Nakagawa N, Sakurai N (1998) Increase in the amount of celA-1 protein in tobacco BY-2 cells by a cellulose biosynthesis inhibitor, 2,6-dichlorobenzonitrile. Plant Cell Physiol 39:779–785

O'Keefe MG, Klevorn TB (1991) A new pre- and post-emergence herbicide for broad-leaf weed control in winter cereals. Brighton Crop Prot Conf Weeds 1:63–68

Roberts JL (1990) Root tips and leaf protoplasts respond to the herbicide isoxaben with cell expansion and wall deformation. Plant Physiol 93S:107

Sabba RP, Vaughn KC (1999) Tobacco BY-2 cells habituated to the cellulose-inhibiting herbicide isoxaben produce cell walls devoid of cellulose and enriched in pectin. Weed Sci Soc Am Abstr 39:132

Sabba RP, Durso NA, Vaughn KC (1999) Structural and immunocytochemical characterization of dichlobenil-habituated tobacco BY-2 cells. Int J Plant Sci 160:275–290

Samuels AL, Giddings TH, Staehelin LA (1995) Cytokinesis in tobacco BY-2 and root tip cells: a new model of cell plate formation in higher plants. J Cell Biol 130:1345–1357

Shedletzky E, Shmuel M, Delmer D, Lamport DTA (1990) Adaption and growth of tomato cells on the herbicide 2,6-dichlorobenzonitrile leads to production of unique cell walls virtually lacking a cellulose-xyloglucan network. Plant Physiol 94:980–987

Shedletzky E, Shmuel M, Trainin T, Kalman S, Delmer D (1992) Cell wall structure in cells adapted to growth on the cellulose synthesis inhibitor 2,6-dichlorobenzonitrile. Plant Physiol 100: 120–130

Vaughn KC (2000) Anticytoskeletal herbicides. In: Nick P (ed) Plant microtubules, potential for biotechnology. Springer, Berlin Heidelberg New York, pp 193–205

Vaughn KC, Turley RB (1999) The primary walls of cotton fibers contain an ensheathing pectin layer. Protoplasma 209:226–237

Vaughn KC, Turley RB (2001) Ultrastructural effects of cellulose biosynthesis inhibitor herbicides on developing cotton fibers. Protoplasma 216:80–93

Vaughn KC, Hoffman JC, Hahn MG, Staehelin LA (1996) The herbicide dichlobenil disrupts cell plate formation: immunogold characterization. Protoplasma 194:117–132

Wang Y, Lu J, Mollet JC, Gretz MR, Hoagland KD (1997) Extracellular matrix assembly in diatoms (Bacillariphycea). II. 2,5-dichlorobenzonitrile inhibition of motility and stalk production in the marine diatom *Achnanthus longipes*. Plant Physiol 113:1071–1080

Wells B, McCann MC, Shedletzky E, Delmer D, Roberts K (1994) Structural features of cell walls from tomato cells adapted to grow on the herbicide 2,6-dichlorobenzonitrile. J Microsc 173:155–164

8 Inhibitors of Protoporphyrinogen Oxidase: A Brief Update

HIROSHI MATSUMOTO

This contribution is a brief overview on recent findings. Further details have been published in *Peroxidizing Herbicides* (ed. by P. Böger and K. Wakabayashi) issued in 1999 by Springer Publ., Heidelberg, Germany.

8.1 Introduction

Chlorophyll and heme are synthesized by the ubiquitous and highly conserved tetrapyrrole pathway in photosynthetic organisms. The first committed precursor of the pathway is 5-aminolevulinic acid (ALA) which is made from glutamate via the C_5 pathway and is converted by a series of reactions into protoporphyrin IX, the last common intermediate between heme and chlorophyll (Fig. 1). Insertion of Fe^{2+} into protoporphyrin IX produces protoheme, whereas chelation of Mg^{2+} is the first step of the chlorophyll branch. Earlier studies revealed that plastids contained the entire pathway for the production of chlorophyll and heme. Parallel to the plastidic tetrapyrrolic pathway, activities of the last two enzymes of the heme synthesizing pathway, protoporphyrinogen oxidase and ferrochelatase, were found in mitochondria (Jacobs et al. 1991; Smith et al. 1993). Thus, protoporphyrinogen IX is distributed between the plastidic pathway and the mitochondrial heme synthesis pathway. Although the final steps of plant heme synthesis occur parallel in mitochondria and plastids, the control mechanism of the intercompartmental allocation is still under investigation.

Protoporphyrinogen oxidase (E.C. 1.3.3.4) is the last enzyme in the common tetrapyrrole biosynthesis pathway before the pathway branches toward chlorophyll and heme synthesis. The enzyme is the target of many class of herbicides including diphenyl ethers, cyclic imides, and thiadiazolidines. The application of protoporphyrinogen oxidase-inhibiting herbicides to plant leads to the peroxidative destruction of cellular membrane and bleaching of tissues in the presence of light. From these characteristics, the herbicides are also designated as photobleaching or peroxidizing herbicides. This brief chapter summarizes fundamental and recently obtained knowledge on protoporphyrinogen oxidase and its inhibitors.

P. Böger, K. Wakabayashi, K. Hirai (Eds.)
Herbicide Classes in Development

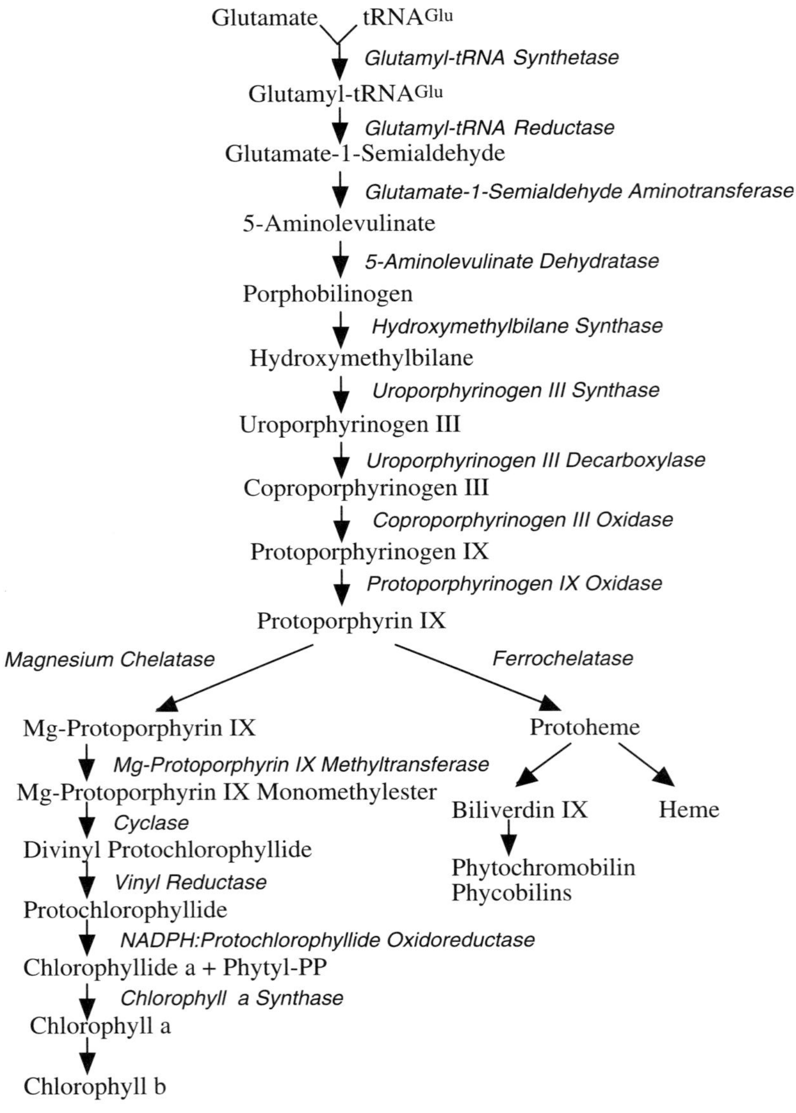

Fig. 1. The tetrapyrrole pathway in plants

8.2 Protoporphyrinogen Oxidase Inhibitors and Their Mode of Action

Protoporphyrinogen oxidase catalyses the six-electron oxidation of nonfluorescent protoporphyrinogen IX to fluorescent protoporphyrin IX and in plastids it is associated with the envelope and the thylakoid membrane (Matringe et al. 1992a; Böger and Wakabayashi 1999). In mitochondria, the enzyme is associated with the inner membrane facing with its active center, the cytoplasmic side (Deybach et al. 1985). The bicyclic structure of the herbicides allows a competitive inhibition by filling the complementary space of the binding site for the natural substrate, protoporphyrinogen IX (Matringe et al. 1992b; Nandihalli et al. 1992). Although the primary target of the herbicides was known, the subsequent mechanisms have not yet been thoroughly elucidated. In treated tissues, these herbicides cause the accumulation of protoporphyrin IX. It is assumed that excess protoporphyrinogen leaks out of the plastid to cytoplasm, not to be metabolized by herbicide-inhibited mitochondrial protoporphyrinogen oxidase, and oxidized to protoporphyrin IX in the cytoplasm (exterior of the plastids). The accumulation of protoporphyrin IX can proceed by a nonenzymatic or by an enzymatic oxidation. The latter may be mediated by nonbiosynthetic protoporphyrinogen oxidase activity which is insensitive against the herbicides (Duke et al. 1994). Such enzyme activities have been found in the endoplasmic reticulum and plasmalemma, and their oxidative activity with protoporphyrinogen IX may be a side reaction. Furthermore, a peroxidase has been reported oxidizing protoporphyrinogen (Yamato et al. 1994). Protoporphyrin IX accumulation is a very rapid process with a concurrent halt of chlorophyll and heme biosynthesis. Protoporphyrin IX is known as a potent photosensitizer. In the presence of light and molecular oxygen, it generates high levels of singlet oxygen and toxic oxygen radicals which attack the unsaturated fatty acids of the cell membrane. Peroxidation of fatty acids results in leakage of membranes (loss of water and ions), pigment breakdown and eventually necrosis of the leaf. The deleterious effects of protoporphyrin IX occur because it cannot be rechanneled into the plastid-located pathway (Jacobs and Jacobs 1993; Lee et al. 1993). Recent reviews explain well the action mechanism of the protoporphyrinogen oxidase-inhibiting herbicides (Böger and Sandmann 1998; Hess 2000).

The protoporphyrinogen oxidase-inhibiting herbicides are applied pre- or post-emergence. Some of the herbicides have been used for more than 30 years in agriculture. Most of them are poorly translocated within the leaf tissue; they often do not reach the growing points which is a disadvantage to achieve efficient weed control. Furthermore, poor selectivity may be a problem. Cinidon-ethyl, an isoindoldione herbicide, which has been commercially introduced, however, selectively controls broadleaf weeds in cereals. This inhibits protoporphyrinogen oxidase and its selectivity to wheat is due to rapid metabolism in the plants (Grossmann and Schiffer 1999). More recently the tricyclic protoporphyrinogen oxidase inhibitor benzfendizone was introduced which is considered to act by mimicking three of the pyrrole rings of protoporphyrinogen IX (Theodoridis et al. 2000).

8.3 Biochemical Characterization of Protoporphyrinogen Oxidase

Studies on structure and function of protoporphyrinogen oxidase were stimulated by those of herbicide binding to the target. The herbicides undergo rapid, reversible specific binding and compete with protoporphyrinogen IX for the substrate binding site on the enzyme. Kinetic studies with mammals and plants confirm that herbicide inhibition of enzyme activity is competitive with the substrate (Camadro et al. 1991). Mixed-type inhibition for the diphenyl ether acifluorfen has been reported for the yeast membrane. However, this mixed-type inhibition in yeast has been shown to be due to partial hydrolysis of the enzyme and the mode of inhibition is competitive (Camadro et al. 1994).

An important component of the enzyme catalytic system is the flavin associated with the enzyme. Yeast and mammalian protoporphyrinogen oxidase has been shown to be FAD-containing enzymes (Camadro et al. 1994, Dailey et al. 1995), and sequence analysis of the cloned eukaryotic protoporphyrinogen oxidase shows that all share a common $\beta\alpha\beta$-ADP(Rossmann)-binding fold characteristic of flavoproteins (Nishimura et al. 1995; Camadro and Labbe 1996; Dailey and Dailey 1996). Moreover, the flavoproteins are revealed to have common motif shortly after the $\beta\alpha\beta$-dinucleotide binding motif (Dailey and Dailey 1998; Vallon 2000). The role of the flavin in catalysis was investigated by studying the reactivity of yeast protoporphyrinogen oxidase toward a potential inhibitor of flavoproteins, the diphenyleneiodonium cation (DPI; Arnould et al. 1997, 1998). The cofactor is probably essential for stabilizing the protein structure, providing a rigid skeleton for the polypeptide chain.

The binding site of herbicides to protoporphyrinogen oxidase was examined with a photoaffinity radioligand. Birchfiled et al. (1998) suggested that there was one herbicide-binding site per FAD and the site may not be proximal to FAD. From the determination of the specific binding site for a diazoketone derivative of [^{3}H]-acifluorfen, Arnould and Camadro (1998) concluded that the active site of the enzyme was in the C-terminal domain of the protein, at the interface between the C- and N-terminal domains.

Protoporphyrinogen oxidase is highly resistant to proteases. Arnould et al. (1999) reported that palmitoylation of the enzyme occurred in vivo both in yeast cells and in a heterologous bacterial expression systems and suggested that this acylation stabilizes a protease-resistant conformation of the enzyme.

8.4 Protoporphyrinogen Oxidase Genes and Transgenic Herbicide-Resistant Plants

Genes involved in protoporphyrinogen oxidase activity have been identified first from *Escherichia coli* (Sasarman et al. 1993) and *Bacillus subtilis* (Hansson and Hederstedt 1992) and are designated *hemG* and *hemY*, respectively. Both genes encode peptides that do not share any sequence similarity. They represent two distinct protoporphyrinogen-oxidizing systems, the HemY-type oxygen-dependent and the bacterial multicomponent systems. The purified yeast protoporphyrinogen oxidase is a 55-kDa protein (Camadro et al. 1994), and its gene was identified by functional complementation of a *hem14-1* yeast mutant that is deficient in enzyme activity and resembles the HemY protein. The first plant protoporphyrinogen oxidase gene was isolated from *Arabidopsis thaliana* by functional complementation (Narita et al. 1996). It encoded a protein of 537 amino acid residues showing only little homology to the enzymes encoded from human or mouse. Subsequently, Lermontva et al. (1997) identified two different cDNA sequences for protoporphyrinogen oxidase in tobacco. The deduced protein sequences designated as protoporphyrinogen oxidase I and II have a molecular mass of 60 and 55 kDa, respectively, and share only 27.3% similarity. Translocation studies and immunological analysis proved that the proteins are imported either exclusively into plastids (protoporphyrinogen oxidase I) or into mitochondria (protoporphyrinogen oxidase II). These isoforms of protoporphyrinogen oxidase closely resemble known protoporphyrinogen oxidase sequences, e.g., from *Arabidopsis* (Narita et al. 1996), *Chlamydomonas reinharditii* (Randolph-Anderson et al. 1998), human (Nishimura et al. 1995) or *Bacillus subtilis* (Hansson and Hederstedt 1992). Recently, isolation and cloning of the cDNA for protoporphyrinogen oxidase of several plant species have been reported. Adomat and Böger (2000) isolated and sequenced the cDNA of the plastidic protoporphyrinogen oxidase from chicory (*Cichorium foliosum*). The cDNA of the plastidic isoform was also isolated from spinach (*Spinacia oleracea*) (Che et al. 2000)

Several strategies have been used for obtaining plants resistant to protoporphyrinogen oxidase-inhibiting herbicides. Resistance can be obtained by an alteration of the herbicide-binding site of the enzyme, preventing stable binding of specific herbicides. Screens for resistant spontaneous and induced mutants have been employed. Mutant cell cultures have been selected by protoporphyrinogen oxidase-inhibitor-containing medium with the aim of understanding the molecular mechanism of herbicide resistance and identifying the gene that confers this resistance (Pornprom et al. 1994; Ichinose et al. 1995). Horikoshi and Hirooka (1999) isolated a tobacco cell lines resistant to protoporphyrinogen oxidase-inhibiting herbicide pyraflufen-ethyl. Then they isolated cDNA of the enzyme by gene complementation of an *E. coli* strain defective in *hemG* gene (Horikoshi et al. 1999). The sequence of the plastidic protoporphyrinogen oxidase cDNA from a resistant cell line revealed a single point mutation. A point mutation, Val389Met, of the plastidic isoform also conferred the resistance of *C. reinhardtii rs-3* mutant to a N-phenyl heterocyclic protoporphyrinogen oxidase inhibitor S-23142 (Randolph-Anderson et al. 1998).

Protoporphyrinogen oxidase originating from the aerobic bacterium *Bacillus subtilis* is poorly inhibited by diphenyl ether-type herbicides (Corrigall et al. 1998). Expression of *B. subtilis* protoporphyrinogen oxidase (HemY) in the cytoplasm or in the plastid of tobacco and rice plants leads to a higher resistance against the herbicide oxyfluorfen (Choi et al. 1998; Lee et al. 2000). Transgenic tobacco plants that overexpress an *Arabidopsis* plastidic isoform of protoporphyrinogen oxidase was resistant to acifluorfen (Lermontova and Grimm 2000). Overproduction of mitochondrial protoporphyrinogen oxidase in a mutant tobacco cell culture also gave rise to higher resistance to the inhibitors (Watanabe et al. 1998; Warabi et al. 2001)

8.5 Recent Advances in QSAR Studies

Quantitative structure-activity relationship (QSAR) methods have been successful to characterize the activity of structurally related compounds because biological activity is correlated with structural features of ligands. Steric and electrostatic parameters are most commonly used in QSAR because interactions between herbicides and the receptor binding sites usually involve non-covalent binding. Structure-activity relationship (SAR) studies on diphenyl ether herbicides published before 1995 mainly used whole-plant biological activity (see e.g., Fujita and Nakayama 1999 for details). However, in vivo biological activity data are not useful for investigation of the relationship between structural characteristics of inhibitors and their molecular binding site. At the whole-plant level, other factors such as uptake, translocation and metabolism contribute to the overall activity of protoporphyrinogen oxidase inhibitors. As a result, the most herbicidally active compounds may not necessarily be the most potent protoporphyrinogen oxidase inhibitors (Dayan et al. 1997; Ishida et al. 2000). Conversely, strong inhibitors of the enzyme may have poor herbicidal activity (Dayan and Duke 1997).

Protoporphyrinogen oxidase is known to be the target of numerous compounds, including many hetero-bicyclic structures. Several two-dimensional (2-D) QSAR studies have correlated diphenyl ether inhibition with the enzyme inhibition. However, these studies are limited in that the models could not be used to predict the activity of structurally related derivatives. Recently developed 3-D QSAR methods have markedly enhanced the molecular analyses that were previously possible with classical QSAR methods. When the shape of a ligand binding pocket is not yet characterized, as in the case with protoporphyrinogen oxidase, we must rely on 3-D QSAR information using inhibitory data to glimpse at the ligand-receptor interaction. Durst (1998), using a set of 24 diphenyl ethers, demonstrated the benefit of 3-D techniques such as Comparative Molecular Field Analysis (CoMFA; Cramer et al. 1988) in determining QSAR among the compounds. Subsequently, Dayan and Allen (2000) presented a 3-D approach that described the spatial characteristics of 31 diphenyl ether analogues required for the enzyme-inhibiting activity at the molecular level. Increased predictability was achieved by aligning the diphenyl ether compounds along with the trifluoromethyl phenyl-substituted benzene ring (q^2: the value evaluating the overall predictiveness of the analysis =0.70) rather than along the nitro-substituted benzene ring ($q^2 = 0.65$) or along the centroids ($q^2 = 0.69$). Prediction of protoporphyrinogen oxidase inhibition was best approximated with the data set aligned on the trifluoromethyl phenyl rings suggesting the orientation of the plane of the trifluoromethyl phenyl ring plays an important role in binding to the enzyme.

As part of the process to optimize the herbicidal activity of 6-membered heterocyclic benzoxazinone-type protoporphyrinogen oxidase inhibitors, QSAR was developed for the benzoxazinone replacements using molecular properties, and these were compared to published QSAR (Theodoridis 1997) for acyclic 2,4,5-trisubstituted phenyl-heterocyclic inhibitors (Lyga et al. 1999). Since the molecular properties of the acyclic and bicyclic inhibitors are similar they may interact with the same binding site.

8.6 Antioxidative Stress Responses of Plants to Protoporphyrinogen Oxidase Inhibitors

Molecular oxygen is very unusual in that its two least strongly bound electrons are unpaired and have parallel spins. This property means that ground-state O_2 is triplet but that its excitation leads to the formation of the highly reactive singlet state. Generation of the reactive oxygen species (ROS, e.g., singlet oxygen, superoxide anion radical, hydrogen peroxide) in plant cells is inevitable in metabolism even under normal environmental condition. Generation of ROS increases with the accumulation of tetrapyrrole intermediates. Therefore, the tetrapyrroles biosynthesis is highly regulated in part to avoid their abnormal accumulation. The detrimental effects of accumulated tetrapyrroles are evident in plants treated with a variety of herbicides that act via inhibition of protoporphyrinogen oxidase. Light-activated protoporphyrin IX generates ROS including singlet oxygen and superoxide anion (Yanagida et al. 2000).

Plants must protect themselves against the toxic effect of ROS. ROS in cells are detoxified by the antioxidative system consisting of a set of enzymes and low molecular weight antioxidants. However, greater amount of ROS are often generated exceeding the capacity of the scavenging mechanism. There is considerable evidence for the involvement of antioxidative systems in plant tolerance to diphenyl ether-type herbicides. Recently we reexamined the involvement of the ROS-scavenging systems in plant tolerance to oxyfluorfen (Yanagida et al. 1999). Among the tested species, rice showed the highest tolerance to the herbicide, while buckwheat showed the lowest. However, both plants rapidly accumulated comparable amounts of protoporphyrin IX. The measurement of antioxidative enzymes and antioxidants revealed that superoxide dismutase and catalase activities and glutathione content were much higher in rice than in the other species. These enzyme activities in the herbicide-treated rice plants were significantly stimulated before protoporphyrin IX accumulation reached its maximum. The antioxidative ability of buckwheat was lowest and was not increased with herbicide treatment. Furthermore, we found common chickweed (*Stellaria media* Vill.) plants have some natural tolerance to a singlet oxygen generator, rose bengal (Matsumoto et al. 1999). These results strongly suggest that antioxidative ability is one of the critical factors determining the tolerance to oxyfluorfen in some plant species. The induction of antioxidative enzyme activities and mRNA levels were also demonstrated in oxyfluorfen-treated tobacco plants (Lederer et al. 1999).

The antioxidative defense responses of transgenic tobacco plants expressing antisense RNA for uroporphyrinogen decarboxylase of coproporphyrinogen oxidase were investigated (Mock et al. 1998). Compared with control plants, the transformants had an increased level of mRNAs particularly those encoding superoxide dismutase, catalase, and glutathione peroxidase. *Arabidopsis* plants expressing an antisense protoporphyrinogen oxidase gene showed necrotic lesion on leaves and had a high endogenous salicylic acid level. Treatment of wild-type plants with sublethal concentrations of protoporphyrinogen oxidase-inhibiting herbicides also induced defense responses that conferred enhanced resistance to *Peronospora paracritica* infection (Molina et al. 1999). These results demonstrate that genetic or chemical disruption of the metabolic pathway can lead to the induction of a set of defense responses. Treatment of soybean suspension cultures with salicylic acid, an inducer of pathogen defense, strongly stimulated the specific activities of the antioxidative enzymes and enhanced protecting activity against oxyfluorfen-induced lipid peroxidation (Lederer et al. 1999). This indicates that salicylic acid may be involved in regulating this oxidative stress response.

References

Adomat C, Böger P (2000) Cloning, sequence, expression, and characterization of protoporphyrinogen IX oxidase from chicory. Pestic Biochem Physiol 66:49–62

Arnould S, Camadro J-M (1998) The domain structure of protoporphyrinogen oxidase, the molecular target of diphenyl ether-type herbicides. Proc Natl Acad Sci USA 95:10553–10558

Arnould S, Berthon JL, Hubert C, Dias M, Cibert C, Monet R, Camadro J-M (1997) Kinetics of protoporphyrinogen oxidase inhibition by diphenyleneiodonium derivatives. Biochemistry 36:10178–10184

Arnould S, Takahashi M, Camadro J-M (1998) Stability of recombinant yeast protoporphyrinogen oxidase: effect of diphenyl ether-type herbicides and diphenyleneiodonium. Biochemistry 37:12818–12828

Arnould S, Takahashi M, Camadro J-M (1999) Acylation stabilizes a protease-resistant conformation of protoporphyrinogen oxidase, the molecular target of diphenyl ether-type herbicides. Proc Natl Acad Sci USA 96:14825–14830

Birchfield NB, Latli B, Casida JE (1998) Human protoporphyrinogen oxidase: relation between the herbicide binding site and the flavin cofactor. Biochemistry 37:6905–6910

Böger P, Sandmann G (1998) Action of modern herbicides. In: Raghavendra AS (ed) Photosynthesis: a comprehensive treatise. Cambridge University Press, Cambridge, pp 337–351

Böger P, Wakabayashi K (1999) General physiological characteristics and mode of action of peroxidizing herbicides. In: Böger P, Wakabayashi K (eds) Peroxidizing herbicides. Springer, Berlin Heidelberg New York, pp 163–190

Camadoro J-M, Labbe P (1996) Cloning and characterization of the yeast HEM14 gene coding for protoporphyrinogen oxidase, the molecular target of diphenyl ether-type herbicides. J Biol Chem 271:9120–9128

Camadro J-M, Matringe M, Scalla R, Labbe P (1991) Kinetic studies on protoporphyrinogen oxidase inhibition by diphenyl ether herbicides. Biochem J 277:17–21

Camadro J-M, Thome F, Brouillet N, Labbe P (1994) Purification and properties protoporphyrinogen oxidase from the yeast *Saccharomyces cerevisiae.* Mitochondrial location and evidence for a precursor form of the protein. J Biol Chem 269:32085–32091

Che F-S, Watanabe N, Iwano M, Inokuchi H, Takayama S, Yoshida S, Isogai A (2000) Molecular characterization and subcellular localization of protoporphyrinogen oxidase in spinach chloroplasts. Plant Physiol 124:59–70

Choi KW, Han OH, Lee HJ, Yun YH, Moon YH, Kim M, Kuk YI, Han SU, Guh JO (1998) Differential susceptibilities of wheat and barley to diphenyl ether herbicide oxyfluorfen. Biosci Biotechnol Biochem 62:558–560

Corrigall AV, Siziba KB, Maneli MH, Shephard EG, Ziman M, Dailey TA, Dailey HA, Kirsch RE, Meissner PN (1998) Purification of and kinetic studies on a cloned protoporphyrinogen oxidase from the aerobic bacterium *Bacillus subtilis.* Arch Biochem Biophys 358:251–256

Cramer RD III, Patterson DE, Bunce JD (1988) Comparative molecular field analysis (CoMFA): effect of shape of binding of steroids to carrier proteins. J Am Chem Soc 110:5959–5967

Dailey HA, Dailey TA (1996) Protoporphyrinogen oxidase of *Myxococcus xanthus*: expression, purification, and characterization of cloned enzyme. J Biol Chem 271:8714–8718

Dailey TA, Dailey HA (1998) Identification of an FAD superfamily containing protoporphyrinogen oxidases, monoamine oxidases, and phytoene desaturase. J Biol Chem 273:13658–13662

Dailey TA, Dailey HA, Meissner P, Prasad AR (1995) Cloning, sequence, and expression of mouse protoporphyrinogen oxidase. Arch Biochem Biophys 324:379–384

Dayan FE, Allen SN (2000) Predicting the activity of the natural phytotoxic diphenyl ether cyperine using comparative molecular field analysis. Pestic Manage Sci 56:717–722

Dayan FE, Duke SO (1997) Phytotoxicity of protoporphyrinogen oxidase inhibitors: phenomenology, mode of action and mechanisms of resistance. In: Roe RM, Burton JD, Kuhr RJ (eds) Herbicides activity: toxicology, biochemistry and molecular biology. ISO Press Inc, Amsterdam, pp 11–35

Dayan FE, Duke SO, Reddy KN, Hamper BC, Leschinski KL (1997) Effect of isoxazole herbicides on protoporphyrinogen oxidase and porphyrin physiology. J Agric Food Chem 45:967–975

Deybach JC, Da Silva V, Grandchamp B, Nordmann Y (1985) The mitochondrial location of protoporphyrinogen oxidase. Eur J Biochem 149:431–435

Duke SO, Nandihalli UB, Lee HJ, Duke MV (1994) Protoporphyrinogen oxidase as the optimal herbicide site in the porphyrin pathway. In: Duke SO, Rebeiz CA (eds) Porphyric pesticides. ACS Symp Series 559, American Chemical Society, Washington, DC, pp 191–204

Durst GL (1998) Comparative molecular field analysis (CoMFA) of herbicidal protoporphyrinogen oxidase inhibitors using a standard steric and electrostatic fields and an alternative LUMO field. Quant Struct Act Relat 17:419–426

Fujita T, Nakayama A (1999) Structure-activity relationship and molecular design of peroxidizing herbicides with cyclic imide structures and their relatives. In: Böger P, Wakabayashi K (eds) Peroxidizing herbicides. Springer, Berlin Heidelberg New York, pp 91–139

Grossmann K, Schiffer H (1999) Protoporphyrinogen oxidase-inhibiting activity of the new, wheat-selective isoindolinone herbicide, cinidon ethyl. Pestic Sci 55:687–695

Hansson M, Hederstedt L (1992) Cloning and characterization of the *Bacillus subtilis* hemEHY gene cluster, with encodes protoheme IX biosynthetic enzymes. J Bacteriol 174:8081–8093

Hess FD (2000) Light-dependent herbicides: an overview. Weed Sci 48:160–170

Horikoshi M, Hirooka T (1999) Selection of tobacco cell lines resistant to photobleaching herbicides. J Pestic Sci 24:13–16

Horikoshi M, Mametsuka K, Hirooka T (1999) Molecular basis of photobleaching herbicide resistance in tobacco. J Pestic Sci 24:17–22

Ichinose K, Che F-S, Kimura Y, Matsunobu A, Sato F, Yoshida S (1995) Selection and characterization of protoporphyrinogen oxidase inhibiting herbicide (S23142) resistant photomixotrophic cultured cells of *Nicotiana tabaccum*. J Plant Physiol 146:693–698

Ishida S, Miller-Sulger R, Kohno H, Böger P, Wakabayashi K (2000) Enzymatic activity of protoporphyrinogen-IX oxidase from various plant species: Its sensitivity to peroxidizing herbicides. J Pestic Sci 25:18–23

Jacobs JM, Jacobs NJ (1993) Porphyrin accumulation and export by isolated barley (*Horduem vulgare*) plastids. Plant Physiol 101:1181–1187

Jacobs JM, Jacobs NJ, Sherman TD, Duke SO (1991) Effect of diphenyl ether herbicides on oxidation of protoporphyrinogen to protoporphyrin in organellar and plasma membrane enriched fractions of barley. Plant Physiol 97:197–203

Lederer B, Knörzer OC, Böger P (1999) Differential gene expression in plants stressed by the peroxidizing herbicide oxyfluorfen. Z Naturforsch 54c:764–770

Lee HJ, Duke MV, Duke SO (1993) Cellular localization of protoporphyrinogen-oxidizing activities of etiolated barley *Hordeum vulgare* L. leaves: relationship to mechanism of action of protoporphyrinogen oxidase-inhibiting herbicides. Plant Physiol 102:881–889

Lee HJ, Lee SB, Chung JS, Han SU, Han O, Guh JO, Jeon JS, An G, Back K (2000) Transgenic rice plants expressing a *Bacillus subtilis* protoporphyrinogen oxidase gene are resistant to diphenyl ether herbicide oxyfluorfen. Plant Cell Physiol 41:743–749

Lermontova I, Grimm B (2000) Overexpression of plastidic protoporphyrinogen IX oxidase leads to resistance to the diphenyl-ether herbicide acifluorfen. Plant Physiol 122:75–83

Lermontova I, Kruse E, Mock N-P, Grimm B (1997) Cloning and characterization of a plastidal and mitochondrial isoform of tobacco protoporphyrinogen IX oxidase. Proc Natl Acad Sci USA 94:8895–8900

Lyga JW, Chang JH, Theodoridis G, Baum JS (1999) Structural replacements for the benzoxazinone protox inhibitor. Pestic Sci 55:281–287

Matringe M, Camadro J-M, Block MA, Joyard J, Scalla R, Labbe P, Douce R (1992a) Localization within the chloroplasts of protoporphyrinogen oxidase, the target enzyme for diphenyl etherlike herbicides. J Biol Chem 267:4646–4651

Matringe M, Mornet R, Scalla R (1992b) Characterization of [^{3}H] acifluorfen binding to purified pea etioplasts, and evidence that protoporphyrinogen oxidase specifically binds acifluorfen. Eur J Biochem 209:861–868

Matsumoto H, Kashimoto Y, Warabi E (1999) Basis for common chickweed (*Stellaria media*) tolerance to oxyfluorfen. Pestic Biochem Physiol 64:47–53

Mock H-P, Keetman U, Kruse E, Rank B, Grimm B (1998) Defense responses to tetrapyrrole-induced oxidative stress in transgenic plants with reduced uroporphyrinogen decarboxylase or coproporphyrinogen oxidase activity. Plant Physiol 116:107–116

Molina A, Volrath S, Guyer D, Maleck K, Ryals J, Ward E (1999) Inhibition of protoporphyrinogen oxidase expression in *Arabidopsis* causes a lesion-mimic phenotype that induces systemic acquired resistance. Plant J 17:667–678

Nandihalli UB, Duke MV, Duke SO (1992) Quantitative structure-activity relationships of protoporphyrinogen oxidase-inhibiting diphenyl ether herbicides. Pestic Biochem Physiol 43: 193–211

Narita S, Tanaka R, Ito T, Osada K, Taketani S, Inokuchi H (1996) Molecular cloning and characterization of a cDNA that encodes protoporphyrinogen oxidase of *Arabidopsis thaliana*. Gene 182:169–175

Nishimura K, Taketani S, Inokuchi H (1995) Cloning of a human cDNA for protoporphyrinogen oxidase by complementation in vivo of a *hemG* mutant of *Escherichia coli*. J Biol Chem 270: 8076–8080

Pornprom T, Matsumoto H, Usui K, Ishizuka K (1994) Characterization of oxyfluorfen tolerance in a selected soybean cell line. Pestic Biochem Physiol 50:107–114

Randolph-Anderson BL, Sato R, Johnson AM, Harris EH, Hauser CR, Oeda K, Ishige F, Nishio S, Gillham NW, Boynton JE (1998) Isolation and characterization of a mutant protoporphyrinogen oxidase gene from *Chlamydomonas reinhardtii* conferring resistance to porphyric herbicides. Plant Mol Biol 38:839–859

Sasarman A, Letowski J, Czaika G, Ramirez V, Nead MA, Jacobs JM, Morais R (1993) Nucleotide sequence of the *hemG* gene involved in the protoporphyrinogen oxidase activity of *Escherichia coli* K12. Can J Microbiol 39:1155–1161

Smith AG, Marsh O, Elder GH (1993) Investigation of the subcellular location of the tetrapyrrole-biosynthesis enzyme coproporphyrinogen oxidase in higher plants. Biochem J 292:503–508

Theodoridis G (1997) Structure-activity relationships of herbicidal aryltriazolinones. Pestic Sci 50:283–290

Theodoridis G, Bahr JT, Hotzman FW, Sehgel S, Suarez DP (2000) New generation of protox-inhibiting herbicides. Crop Protection 19:533–535

Vallon O (2000) New sequence motif in flavoproteins: evidence for common ancestry and tools to predict structure. Proteins Struct Funct Genet 38:95–114

Warabi E, Usui K, Tanaka Y, Matsumoto H (2001) Resistance of a soybean cell line to oxyfluorfen by overproduction of mitochondrial protoporphyrinogen oxidase. Pestic Manage Sci 57:743–748

Watanabe N, Che F-S, Iwano M, Takayama S, Nakano Y, Yoshida S, Isogai A (1998) Molecular characterization of photomixotrophic tobacco cells resistant to protoporphyrinogen oxidase-inhibiting herbicides. Plant Physiol 118:751–758

Yamato S, Katagiri M, Ohkawa H (1994) Purification and characterization of protoporphyrinogen-oxidizing enzyme with peroxidase activity and light-dependent herbicide resistance in tobacco cultured cells. Pestic Biochem Physiol 50:72–82

Yanagida M, Matsumoto H, Usui K (1999) Responses of antioxidative systems to oxyfluorfen and their role in herbicidal tolerance of plants. J Weed Sci Tech 44:67–76

Yanagida M, Matsumoto H, Usui K (2000) Determination of porphyrin-induced peroxidation of plant lipids by oxygen electrode. J Weed Sci Tech 45:26–33

9 Genetic Engineering of Herbicide-Resistant Plants

MAMORU HORIKOSHI

9.1 Introduction

Since the marketing of FlavrSavr in 1994 by Calgene as the first commercial introduction of genetically modified (GM) crops (James 1996), the area of GM crops planted has increased drastically. In 2000, the global area of GM crops was estimated to be 44.2 million ha which was more than 25-fold compared with that in 1996. It was also estimated that 16% of the global area of the four major crops (soybean, cotton, canola and corn) was occupied by GM crops. Although many kinds of GM crops have been commercially approved, for example, herbicide-, insect-, disease-resistant crops and crops with improved quality, herbicide-resistant (HR) crops are dominant, occupying 74% of the global area of GM crops (James 2000). Commercialized HR crops are glyphosate-resistant crops sold as Roundup Ready, glufosinate-resistant crops sold as Liberty Link and bromoxynil-resistant crops sold as BXN crops. Imidazolinone- or sulfonylurea-resistant crops sold as IMI or STS crops, respectively, are not transgenic ones, but bred by conventional selection and crossing (Copping 1998).

The first objective for the production of HR crops by global agrochemical companies might be to recover the cost spent for the discovery, development, and registration of new herbicides, which have increased annually (Duke 1996). The use of a herbicide would increase with the production of HR crops, because crops which could be applied with the herbicide could be expanded. Moreover, some benefit would be acquired from seed sales. HR genes could also be used as the selection marker gene for the transformation of crops to introduce various traits into plants (Buchanan-Wallaston et al. 1992; Rathore et al. 1993).

There are many risks and benefits of HR crops to growers and consumers such as water contamination, herbicide use, food contamination, accelerated loss of small farms, cost of farm production, consumer costs and herbicide resistance (Duke 1996). For more detailed information about these problems, see reviews in Burnside (1996) and Kuiper et al. (2000).

This chapter technically outlines the methods with which a plant is transformed and a resistant crop is developed, describing strategy, cloning of the genes, gene transfer, and vector constructs.

P. Böger, K. Wakabayashi, K. Hirai (Eds.)
Herbicide Classes in Development

9.2 Strategy

HR plants were mainly produced by the selection of tissue-cultured cells and the regeneration to plants at an early stage of the research. However, the progress in gene cloning and gene transfer, especially using *Agrobacterium tumefaciens*, has made genetic engineering the most popular at present. One of the reasons must be that the resistant gene could be transferred to many plant species, once the gene was cloned (Dale et al. 1993). However, some commercial HR crops were produced by conventional selection and breeding to avoid the problem of public acceptance. There are several strategies for the production of HR plants by genetic engineering; the introduction of the gene encoding the herbicide-inactivating enzyme, the introduction of the mutant, or foreign gene, encoding the target enzyme with low affinity to the herbicide, and the overexpression of the target enzyme (Mazur and Falco 1989). The first two strategies have been practically applied to commercial HR crops and described in detail below. The third one was applied especially in the early stage of transgenic research, because the resistance in cultured plant cells and microorganisms were often conferred by the overexpression of the target enzyme. However, this strategy is in decline at present, possibly because the level of resistance was not high enough for practical use.

9.2.1 The Gene Encoding the Herbicide-Inactivating Enzyme

This strategy has been most widely applied for the production of HR crops. The key step is to clone the gene encoding a herbicide-inactivating or detoxifying enzyme with high specificity and efficiency.

Glufosinate-resistant crops sold as Liberty Link were produced by the introduction of the *bar* gene encoding the glufosinate-inactivating enzyme (Rasche 1997). The *bar* gene was cloned from *Streptomyces hygroscopicus* which produced bialaphos, the precursor of glufosinate (phosphinothricin). The *bar* gene encodes phosphinothricin-*N*-acetyl-transferase (PAT) which acetylates bialaphos to an inactivated form and prevents autotoxicity of bialaphos in the bacterium.

Bromoxynil-resistant crops sold as BXN were produced by the introduction of the *bxn* gene encoding the bromoxynil-inactivating enzyme (Stalker et al. 1996). Bromoxynil-contaminated soil was screened for the selection of bacterium with high inactivating activity, because bromoxynil is rapidly inactivated in soil. *Klebsiella ozaenae* which utilized bromoxynil as the sole source of carbon, was selected. The *bxn* gene coding for nitrilase with high specificity to bromoxynil was cloned from the bacterium as the resistant gene. The nitrilase converts bromoxynil to nontoxic 3,5-dibromo 4-hydroxybenzoic acid.

Glyphosate-resistant crops sold as Roundup Ready were produced by the introduction of the foreign gene encoding the target enzyme with low affinity to the herbicide (Padgette et al. 1996). However, the *gox* gene encoding the glyphosate-inactivating enzyme was also introduced into some plant species, possibly to enhance the level of resistance (Wells 1995). Glyphosate is known to be readily inactivated in soil and the bacterium with high glyphosate-inactivating activity was screened from industrial-activated sludge from a glyphosate waste-stream facility. *Achromobacter* sp. strain LBAA, thus selected, utilized glyphosate as the sole source of carbon and phosphorus by the *gox* gene encoding glyphosate oxidoreductase (GOX) which converted glyphosate to glyoxylate and aminomethylphosphonate.

The gene encoding the 2,4-D (2,4-dichlorophenoxyacetic acid)-inactivating enzyme was also cloned from a soil bacterium because 2,4-D was readily inactivated in soil (Llewellyn and Last 1996). *Alcaligenes eutrophus*, thus selected, utilized 2,4-D as the sole source of carbon. The *tfdA* gene from the bacterium converts 2,4-D (2,4-dichlorophenoxyacetic acid) to 2,4-dichlorophenol, the inactivated form.

The examples described above utilize the gene encoding the enzyme specific to a herbicide. An alternative strategy is to utilize the gene encoding the enzyme involved in the resistance to more than one herbicide such as glutathione *S*-transferase and cytochrome P-450. The introduction of the human gene encoding cytochrome P-450 conferred resistance to several herbicides (Inui et al. 1999).

9.2.2 Mutant or Foreign Gene Encoding the Target Enzyme with Low Affinity to the Herbicide

Although theoretically this strategy is applicable to the production of any HR crops, the application to commercial HR crops has been restricted to glyphosate-resistant crops sold as Roundup Ready.

As described above, glyphosate-resistant crops sold as Roundup Ready were produced by the introduction of a foreign gene encoding EPSPS (5-enolpyruvyl-shikimate-3-phosphate synthase) with a low affinity to the herbicide (Padgette et al. 1996). Resistant EPSPS genes from *Salmonella typhimurium* (Comai et al. 1983), *E. coli* (Kishore et al. 1986) and petunia (Padgette 1991; Ruff et al. 1991) were introduced into plants in early experiments. However, the levels of resistance were not enough. Generally, when an enzyme with a high herbicide-binding constant was produced by a mutant gene, its enzymological characteristics were found to be unfavorable for the maximal enzyme activity leading to decreased growth and fitness of the plants transformed with this gene. It was known that EPSPSs from some bacteria were naturally resistant to glyphosate. EPSPS from *Agrobacterium* sp. strain CP4 was selected with high glyphosate-resistance and catalytic efficiency in the presence of glyphosate. The CP4 EPSPS gene was cloned from the bacterium and used for the production of Roundup Ready crops such as soybean, canola, cotton, maize and sugar beet (Moll 1997).

The target enzyme of sulfonylurea, imidazolinone and triazolopyrimidine herbicides is acetolactate synthase (ALS). Various resistant ALS genes were cloned from tobacco (Lee et al. 1988) and *Arabidopsis thaliana* (Haughn et al. 1988). These gene products showed different levels of resistance to sulfonylureas, imidazolinones and triazolopyrimidines (Devine and Shukla 2000). These resistant ALS genes were introduced into plants individually, or in combination, and conferred resistance to these herbicides. Though some of these genes conferred resistance even at field trials, these genes have not been used for commercialization (Saari and Mauvais 1996).

As for protoporphyrinogen oxidase (Protox), the target enzyme of photobleaching or peroxidizing herbicides such as diphenyl ethers, resistant Protox genes were cloned from *Chlamydomonas reinhardtii* resistant to S-23142 (Randolph-Anderson et al. 1998). Horikoshi and Hirooka (1999) selected tobacco cell lines resistant to pyraflufen-ethyl by subculturing tobacco calli regenerated from protoplasts in the presence of a derivative of pyraflufen-ethyl. The mechanisms of resistance were analyzed by evaluating the cross-resistance to other herbicides, measuring Proto IX accumulation and enzyme activity in vitro. The resistance was conferred by the alteration of Protox to the herbicide-resistant form in a cell line. Further analysis of the cell line revealed that a single-point mutation causing one amino acid substitution in the chloroplastic Protox gene was the molecular basis of the resistance (Fig. 1; Horikoshi et al. 1999). The plants transformed with the resistant Protox gene were tolerant to the application of pyraflufen-ethyl at the practical dose rate (Horikoshi 1999). Another approach by Syngenta was to clone resistant Protox genes mutagenized in *E. coli* mutator strain XL-1 red in which random mutations were generated spontaneously (Ward 1995; Volrath et al. 1997). Recently, Syngenta announced the planning of marketing crops resistant to Protox-inhibiting herbicides in 2003 using the "Acuron" gene (Agrow 1999). The gene might be cloned by the method described above. The Protox gene from *Bacillus subtilis*, which was naturally resistant to the herbicides, was also used for the production of resistant rice plants at an experimental level (Lee et al. 2000).

The trials to produce glufosinate-resistant plants by this strategy were not successful. One reason might be the existence of multiple nuclear genes of glutamine synthetase (GS), the target enzyme of the herbicide, in plants. Therefore, mutations should be introduced into each gene (Mazur and Falco 1989).

9.3 Cloning of the Genes

9.3.1 Genetic Resource

The first step in the cloning of the gene to be introduced into plants is to choose the genetic resource for the gene. In general, genes are cloned from microorganisms, cultured plant cells and mutant plants. In some cases, the

Nuc.(717) C→T ⇒ Codon(231) GCT→GTT ⇒ Amino acid(231) Ala→Val

```
  1  MTTTPIANHP NIFTHQSSSS PLAFLNRTSF IPFSSISKRN SVNCNGWRTR CSVAKDYTVP
 61  SSAVDGGPAA ELDCVIVGAG ISGLCIAQVM SANYPNLMVT EARDRAGGNI TTVERDGYLW
121  EEGPNSFQPS DPMLTMAVDC GLKDDLVLGD PNAPRFVLWK GKLRPVPSKL TDLPFFDLMS
181  IPGKLRAGFG AIGLRPSPPG HEESVEQFVR RNLGGEVFER LIEPFCSGVY AGDPSKLSMK
241  AAFGKVWKLE ETGGSIIGGT FKAIKERSST PKAPRDPRLP KPKGQTVGSF RKGLRMLPDA
301  ISARLGSKLK LSWKLSSITK SEKGGYHLTY ETPEGVVSLQ SRSIVMTVPS YVASNILRPL
361  SVAAADALSN FYYPPVGAVT ISYPQEAIRD ERLVDGELKG FGQLHPRTQG VETLGTIYSS
421  SLFPNRAPKG RVLLLNYIGG AKNPEILSKT ESQLVEVVDR DLRKMLIKPK AQDPLVVGVR
481  VWPQAIPQFL VGHLDTLSTA KAAMNDNGLE GLFLGGNYVS GVALGRCVEG AYEVASEVTG
541  FLSRYAYK
```

Amino acid sequence of chloroplastic Protox from the wild-type cell line

Fig. 1. Mutation in the resistant Protox cDNA from a resistant cell line

wild-type gene encoding the target enzyme sensitive to a herbicide is mutagenized to the resistant gene by molecular biological techniques such as site-directed mutagenesis and polymerase chain reaction (PCR) (Sambrook and Russell 2000).

9.3.1.1 Microorganism

Microorganisms can be the genetic resource of the genes which encode the herbicide-inactivating enzyme as in the case of *bar*, *bxn*, *gox* and *tfdA* which inactivate bialaphos, bromoxynil, glyphosate and 2,4-D, respectively (Sect. 9.2.1). Some bacteria have the target enzyme which is naturally resistant to a herbicide. The EPSPS gene resistant to glyphosate and the Protox gene resistant to diphenyl ether herbicide were cloned from *Agrobacterium* CP4 and *Bacillus subtilis*, respectively (Sect. 9.2.2). Moreover, mutagenesis of bacteria by chemical mutagens can be a powerful tool to obtain the mutant strain in which the target enzyme is converted to the resistant form. The resistant EPSPS genes were cloned from resistant strains of *Salmonella typhimurium* (Comai et al. 1983) and *E. coli* (Kishore et al. 1986) mutagenized by chemical mutagens.

9.3.1.2 Plant Tissue Culture

Plant cells are rich sources of genetic variability called somaclonal variation. Therefore, the cell line resistant to a herbicide can be obtained by the selection of calli regenerated from protoplasts. Alternatively, cultured plant cells treated with chemical mutagens to enhance mutagenesis can be used for the selection. In this case, the resistant mechanism should be verified, because various resistance mechanisms are known in cultured cells such as reduced uptake, enhanced metabolism, overexpression of the target enzyme and mutation of the target enzyme gene to the HR form. Resistant ALS genes were cloned from resistant cell lines (Lee et al. 1988) treated with chemical mutagens (Chaleff and Ray 1984). A resistant Protox gene was cloned from a resistant cell line (Horikoshi et al. 1999) without mutagen treatment (Horikoshi and Hirooka 1999).

9.3.1.3 Mutant Plants

In some cases, seedlings from plant seeds treated with a mutagen could be used for the selection of resistant lines. Soybean lines resistant to sulfonylurea were selected from seedlings of seeds treated with chemical mutagens (Sebastian et al. 1989). The ALS enzyme in the line was resistant to the herbicide.

Arabidopsis thaliana is known as a model plant for plant molecular biology. Seeds treated with chemical or radioactive mutagens are commercially available. A mutant line resistant to sulfonylurea was selected from seedlings (Haughn and Somerville 1986) and the resistant ALS gene was cloned (Haughn et al. 1988). The complete genome sequence of the plant was analyzed recently (The *Arabidopsis* genome initiative, Nature 408:796–815, 2000).

9.3.2 Cloning Methods

9.3.2.1 The Information of Protein

The method which utilizes information of the protein requires purification of the target enzyme. Then, the expression library from the genetic resource is screened with an antibody raised against the protein. Alternatively, the amino acid sequence of the protein should be analyzed, even partially. Depending on the information of the nucleotide sequence revealed from the amino acid sequence, the cDNA library is screened with oligonucleotide probes. The EPSPS gene was cloned from a *Petunia* cell line which overproduced the enzyme, using oligonucleotide probes designed from the amino acid sequence revealed by microsequencing of purified enzyme preparations (Shah et al. 1986).

Purification of a protein is rather difficult at present. However, recent progress on proteomics (Ziv and de Vienne 1999) will facilitate protein analysis.

9.3.2.2 The Information of Nucleic Acid

The method which utilizes the information of the nucleic acid depends on the gene from another organism. The gene from another organism is used as the probe for the library screening. The ALS genes were isolated from tobacco and *Arabidopsis thaliana* using the yeast ALS gene as a probe (Mazur et al. 1987). Subsequently, the plant ALS genes were used as the probes to clone the resistant ALS gene from tobacco (Lee et al. 1988) and *Arabidopsis thaliana* (Haughn et al. 1988). Alternatively, if the complete nucleotide sequence of the target gene is already analyzed, the entire gene can be amplified directly by PCR. The resistant Protox gene was amplified by PCR with primers designed from the nucleotide sequence of the wild-type Protox gene (Horikoshi et al. 1999).

9.3.2.3 Bacterial Genetics

Bacterial genetics is one of the most progressive area in molecular biology. Therefore, many techniques, such as mutagenesis, transposon tagging, and genetic complementation can be used and the target gene can be cloned rather easily.

Wild-type plant genes sensitive to a herbicide could be mutagenized in *E. coli* by chemical mutagens or in *E. coli* mutator strains such as XL-1 red, as in the case of the *Petunia* EPSPS gene (Ruff et al. 1991) or the Protox genes of *Arabidopsis thaliana* and maize (Ward 1995), respectively. Site-directed mutagenesis was used for the *Petunia* EPSPS gene (Padgette et al. 1991). Transposon tagging is to isolate a mutant by gene disruption caused by the transposon insertion and clone the gene using transposon sequences as the tag. The *tfdA* gene which inactivated 2,4-D was isolated from *Alcaligenes eutrophus* by this method (Streber et al. 1987). Genetic complementation is to complement a mutant of microorganisms by the expression of the functionally equivalent gene from another organism. The wild-type Protox gene was cloned from tobacco using an *E. coli hemG* mutant (Horikoshi et al. 1999). The alfalfa GS gene complemented the growth of *E coli glnA* mutant (DasSarma et al. 1986) and was used for the selection of the resistant GS gene (Mazur and Falco 1989).

9.4 Gene Transfer

The methods for gene transfer are classified into two classes; the direct and indirect gene transfer. The direct gene transfer includes polyethylene-glycol (PEG)-mediated gene transfer, electroporation and particle bombardment. The indirect gene transfer uses the *Agrobacterium*-mediated gene transfer. The most effective one is *Agrobacterium*-mediated gene transfer, though restricted to dicots in the past. However, the method for an effective transformation of rice was developed recently (Hiei et al. 1997). Types of vectors differ between the direct and indirect gene transfer (Sect. 9.5.2).

9.4.1 Polyethylene-Glycol-Mediated Gene Transfer and Electroporation

These methods cause the uptake of naked DNA into protoplasts. PEG-mediated gene transfer uses the chemical treatment of protoplasts by PEG. Electroporation is achieved by the application of a high electric field to protoplasts between two electrodes. DNA is then integrated into chromosomal DNA randomly. These methods require the regeneration to plants from protoplasts to produce transgenic plants. Because the *Agrobacterium*-mediated gene transfer was restricted to dicots, as described above, these methods were effective for rice, in which the regeneration system was established using embryogenic callus cultures (Shimamoto et al. 1989; Datta et al. 1990). However, these methods declined after development of the method for the effective *Agrobacterium*-mediated gene transfer of rice. Electroporation is still one of the most efficient method to transform *E. coli.*

9.4.2 Particle bombardment

Particle bombardment is a mechanical method to introduce naked DNA into plant tissues. DNA is delivered into plant cells with particles accelerated by bombardment using a ballistic device. This method is most versatile, being applicable to any type of plant cells, tissues and organs, and to any plant species. The only limitation is the regeneration to plants from treated tissues. Therefore, this method is valuable to recalcitrant plant species in which the *Agrobacterium*-mediated gene transfer or the regeneration of plants from protoplasts is ineffective. Roundup Ready soybeans were produced by this method (Padgette et al. 1996). Transgenic wheat and barley resistant to glufosinate were produced by the introduction of the *bar* gene with this method (Vasil et al. 1992; Wan and Lemaux 1994). The only disadvantage is that regenerated plants are sometimes chimeras.

Particle bombardment was also used for plastid transformation which resulted in transgenic plants resistant to glyphosate (Daniell et al. 1998). The plastid transformation prevents gene escape through pollen, because the introduced genes are maternally inherited. Gene escape through pollen is a the serious problem in HR crops (Daniell et al. 1998).

9.4.3 *Agrobacterium*-Mediated Gene Transfer

Agrobacterium tumefaciens is a soilborne Gram-negative bacterium which causes tumors called crown galls in host plants by the transfer of T-DNA (transferred DNA) in the Ti plasmid (tumor-inducing plasmid) of the bacterium into plant cells. *Agrobacterium*-mediated gene transfer utilizes this naturally occurring transformation event. *Agrobacterium*-mediated gene transfer is the most effective method for gene transfer into dicotyledonous host plants. Many of HR dicotyledonous plants were produced by the method. The molecular mechanism of the gene transfer by the bacterium has been reviewed (Hooykaas and Schilperoort 1992). Briefly, T-DNA in the Ti plasmid flanked by the left and right 25-base-pair border sequences is introduced into the plant genome by the function of virulence (*vir*) genes in the Ti plasmid and chromosomal DNA of the bacterium, which encode proteins required for the T-DNA transfer.

The methods for the efficient transformation of rice (Hiei et al. 1997) and maize (Ishida et al. 1996) were developed recently using embryogenic calli, super-binary vector and acetosyringone which was a potent inducer of *vir* genes.

9.5 Vector Constructs

9.5.1 Expression Cassettes

In order to efficiently express the gene introduced into plant chromosomal DNA, the promoter sequence for the transcription of mRNA by DNA-dependent RNA polymerase and the terminator sequence for the termination of the transcription must be fused to the gene to form expression cassettes. A selection marker gene is also necessary for the selection of transformants regenerated. In some cases, enhancer sequences are also used to enhance the transcription. Transit peptide sequences are necessary when gene products should be imported into organelles.

9.5.1.1 Promoter and Terminator

The most commonly used promoter is CaMV (Cauliflower mosaic virus) 35S promoter. The promoter is known to be a strong and constitutive one. The promoter was originally used by CaMV, a double-stranded DNA virus, for the transcription of 35S mRNA in infected plant cells. Other promoters include CaMV 19S promoter and nopaline synthase (NOS) promoter from the Ti plasmid (Fraley et al. 1983). The RbcS (ribulose-1,5-bisphosphate carboxylase, small subunit) promoter and the PR-1a (pathogenesis related-1a) promoter are inducible by light and stress, respectively (Broglie et al. 1984; Uknes et al. 1993).

The NOS terminator is most commonly used. Mitsuhara et al. (1996) reported that the CaMV 35S terminator was more effective than the NOS terminator.

9.5.1.2
Selection Marker Gene

The first widely used selection marker gene was the neomycin phosphotransferase (NPTII) gene from transposon Tn5 (Velten and Schell 1985) which conferred resistance to aminoglycoside antibiotics such as kanamycin and G-418 in transformed plant cells. Some plants, such as wheat and rice, are not so sensitive to these antibiotics. In these cases, the hygromycin phosphotransferase gene (Gritz and Davies 1983) which conferred resistance to hygromycin was used for the efficient selection of transformants. The *bar* gene which conferred resistance to bialaphos was also used as a selection marker gene for rice (Rathore et al. 1993). Recently, the *pmi* gene encoding phosphomannose-isomerase was used as the selection marker gene for the positive selection of transgenic maize in the presence of mannose (Negrotto et al. 2000).

9.5.1.3
Enhancer Sequence

In order to enhance the expression of the introduced gene, DNA fragments from various origins have been used as the enhancer sequence. The omega sequence from TMV (tobacco mosaic virus), intron sequence from a gene for phaseolin and 5′-upstream sequence of the CaMV 35S promoter are known as the enhancer sequences (Mitsuhara et al. 1996).

9.5.1.4
Transit Peptide Sequence

Most of the target enzymes for herbicides are localized in chloroplasts (Berg et al. 1999). Therefore, resistant genes should be imported into chloroplasts. Roundup Ready soybeans were produced by the introduction of the CP4 EPSPS gene fused with the transit peptide sequence for chloroplasts from *Petunia* (Padgette et al. 1996). Transit peptide sequences for mitochondria are also known.

9.5.2
Type of Vectors

9.5.2.1
Vectors for Direct Gene Transfer

No specialized sequence is required for the vectors used for direct gene transfer except expression cassettes described above. In general, high copy number plasmids such as pUC or pBluescript series are used because relatively high plasmid amounts are required for these methods.

9.5.2.2
Vectors for *Agrobacterium*-Mediated Gene Transfer

For *Agrobacterium*-mediated gene transfer, oncogenic genes in T-DNA must be removed, at first, resulting in a disarmed Ti plasmid. Two types of vectors are known for *Agrobacterium*-mediated gene transfer; cointegrate vectors and binary vectors. Cointegrate vectors require homologous recombination with the disarmed Ti plasmid to transfer expression cassettes into T-DNA. The binary vector system consists of two autonomously replicating plasmids in *A. tumefaciens*. The binary vector, which contains expression cassettes between border sequences and can be replicated in *E. coli*, and the disarmed Ti plasmid without T-DNA to assist gene transfer into the plant genome. Binary vectors are now more widely used than cointegrate vectors because they are easy to handle. Some binary vectors such as the pBI series (Jefferson et al. 1987) based on pBIN19 (Bevan 1984) are commercially available. Super-binary vectors such as pSB131, derived from *A. tumefaciens* strain A281 which was highly efficient in the transformation of higher plants (Komari 1990), were used for the transformation of rice and maize (Ishida et al. 1996; Hiei et al. 1997).

9.5.2.3
Other Vectors

Vectors for plastid transformation have been developed for high expression and gene containment (Zoubenko et al. 1994). Transferred genes are flanked with plastid genome sequences and transferred by homologous recombination into the plastid genome.

The MAT (multi-auto-transformation) vector system was developed for the production of marker-free transgenic plants and the repeated transformation (Ebinuma et al. 1997; Sugita et al. 2000). The MAT vector is a binary vector which comprises the *ipt* gene from *A. tumefaciens* as a dominant selection marker gene and, in addition, a removable DNA element. The *ipt* gene induces abnormal transgenic shoots in a hormone-free medium and is then omitted by the removable DNA element resulting in normal shoots. The risk of an antibiotics-resistant gene used as the selection marker gene (Kuiper et al. 2000) is eliminated by this method.

9.6
Conclusions

Although GM crops have been accepted rather favorably in the USA, they have not been accepted so favorably in the EU and Japan. Experimental data suggesting the risks of GM crops such as environmental safety and low yield have been presented, even in the USA. The techniques for the production of GM crops seem to be the only ones which can improve crop yield drastically (James 1996) to feed the increasing global population and is indispensable for the future of the human race. Because HR crops dominate GM crops and HR genes can be used as the selection marker gene to introduce various traits into plants, HR crops should lead to public acceptance by reducing the risks. Fortunately, new technologies to reduce risks such as plastid transformation and marker-free transformation have been developed. These techniques will facilitate the production of HR crops of the next generation.

HR crops are dominated by crops resistant to nonselective herbicides such as glyphosate and glufosinate. Therefore, the development of new nonselective herbicides is desired in order to increase options for farmers and avoid resistance by herbicide rotation. New technologies such as genomics, proteomics, DNA chips (microarrays), high-throughput screening and combinatorial chemistry (Berg et al. 1999; Cole et al. 2000) will facilitate the development of new herbicides and crops resistant to a certain herbicide.

References

Agrow (1999) Novartis plans herbicide-tolerant crops. Agrow 323, PJB Publ Ltd, Surrey, UK, p 21

Berg D, Tietjen K, Wollweber D, Hain R (1999) From genes to targets: impact of functional genomics on herbicide discovery. Brighton Crop Protection Conf Weeds, pp 491–500

Bevan M (1984) Binary *Agrobacterium* vectors for plant transformation. Nucleic Acids Res 12: 8711–8721

Broglie R, Coruzzi G, Fraley RT, Rogers SG, Horsch RB, Niedermeyer JG, Fink CL, Chua NH (1984) Light-regulated expression of a pea ribulose-1,5-bisphosphate carboxylase small subunit gene in transformed plant cells. Science 224:838–843

Buchanan-Wallaston V, Snape A, Cannon F (1992) A plant selectable marker gene based on the detoxification of the herbicide dalapon. Plant Cell Rep 11:627–634

Burnside OC (1996) An agriculturalist's viewpoint of risks and benefits of herbicide-resistant cultivars. In: Duke SO (ed) Herbicide-resistant crops. Lewis, Boca Raton, pp 391–408

Chaleff RS, Ray TB (1984) Herbicide-resistant mutants from tobacco cell cultures. Science 223:1148–1151

Cole D, Pallett K, Rodgers M (2000) Discovering new modes of action for herbicides and the impact of genomics. Pestic Outlook 11:223–229

Comai L, Sen L, Stalker DM (1983) An altered *aroA* gene products confers resistance to the herbicide glyphosate. Science 221:370–371

Copping LG (1998) Genetically modified crops II-genetic engineering for herbicide tolerance. Agrow reports, PJB Publ Ltd, Surrey

Dale PJ, Irwin JA, Scheffler JA (1993) The experiment and commercial release of transgenic crop plants. Plant Breed 111:1–8

Daniell H, Datta, R, Varma S, Gray S, Lee SB (1998) Containment of herbicide resistance through genetic engineering of the chloroplast genome. Nat Biotechnol 16:345–348

DasSarma S, Tischer E, Goodman HM (1986) Plant glutamine synthetase complements a *glnA* mutation in *Escherichia coli*. Science 232:1242–1244

Datta SK, Peterhans A, Datta K, Potrykus I (1990) Genetically engineered fertile *indica*-rice recovered from protoplasts. Bio/technology 10:736–740

Devine MD, Shukla A (2000) Altered target sites as a mechanism of herbicide resistance. Crop Prot 19:881–889

Duke SO (1996) Herbicide-resistant crops – background and perspectives. In: Duke SO (ed) Herbicide-resistant crops. Lewis, Boca Raton, pp 1–12

Ebinuma H, Sugita K, Matsunaga E, Yamakado M (1997) Selection of marker-free transgenic plants using the isopentenyl transferase gene. Proc Natl Acad Sci USA 94:2117–2121

Fraley RT, Rogers SG, Horsch RB, Sanders PR, Flick JS, Adams SP, Bittner ML, Brand LA, Fink CL, Fry JS, Galluppi GR, Goldberg SB, Hoffmann NL, Woo SC (1983) Expression of bacterial genes in plant cells. Proc Natl Acad Sci USA 80:4803–4807

Gritz L, Davies J (1983) Plasmid-encoded hygromycin B resistance: the sequence of hygromycin B phosphotransferase gene and its expression in *Escherichia coli* and *Saccharomyces cerevisiae*. Gene 25:179–188

Haughn GW, Somerville C (1986) Sulfonylurea-resistant mutant of *Arabidopsis thaliana*. Mol Gen Genet 204:430–434

Haughn GW, Smith J, Mazur B, Somerville C (1988) Transformation with a mutant *Arabidopsis* acetolactate synthase genes renders tobacco resistant to sulfonylurea herbicides. Mol Gen Genet 211:266–271

Hiei Y, Komari T, Kubo T (1997) Transformation of rice mediated by *Agrobacterium tumefaciens*. Plant Mol Biol 35:205–218

Hooykaas PJJ, Schilperoort RA (1992) *Agrobacterium* and genetic engineering. Plant Mol Biol 19:15–38

Horikoshi M (1999) Fundamental studies on the production of plants resistant to Protox inhibiting herbicides. J Pestic Sci 24:328–335

Horikoshi M, Hirooka T (1999) Selection of tobacco cell lines resistant to photobleaching herbicides. J Pestic Sci 24:13–16

Horikoshi M, Memetsuka K, Hirooka T (1999) Molecular basis of photobleaching herbicide resistance in tobacco. J Pestic Sci 24:17–22

Inui H, Ueyama Y, Shiota N, Ohkawa Y, Ohkawa H (1999) Herbicide metabolism and cross-tolerance in transgenic potato plants expressing human CYP1A1. Pestic Biochem Physiol 64: 33–46

Ishida Y, Saito H, Ohta S, Hiei H, Komari T, Kumashiro T (1996) High efficiency transformation of maize (*Zea Mays* L) mediated by *Agrobacterium tumefaciens*. Nat Biotechnol 14:745–750

James C (1996) Global review of the field testing and commercialization of transgenic plants:1986 to 1995 the first decade of crop biotechnology. International Service for the Acquisition of Agri-biotech Applications, Ithaca, USA http://www.isaaa.org/

James C (2000) Global status of commercialized transgenic crops: 2000. International Service for the Acquisition of Agri-biotech Applications, Ithaca, USA. http://www.isaaa.org/

Jefferson RA, Kavanagh TA, Bevan MW (1987) GUS fusions: beta-glucuronidase as a sensitive and versatile gene fusion marker in higher plants. EMBO J 6:3901-3907

Kishore GM, Brundage L, Kolk K, Padgette SR, Rochester D, Huynh QK, Della-Cioppa G (1986) Isolation, purification, and characterization of a glyphosate-tolerant mutant *E. coli* EPSP synthase. Fed Proc 45:1506

Komari T (1990) Transformation of cultured cells of *Chenopodium quinoa* by binary vectors that carry a fragment of DNA from the virulent regions of pTiBo542. Plant Cell Rep 9:303–306

Kuiper HA, Kleter GA, Noordam MY (2000) Risks of the release of transgenic herbicide-resistant plants with respect to humans, animals, and the environment. Crop Prot 19:773–738

Lee HJ, Lee SB, Chung JS, Han SU, Han O, Guh JO, Jeon JS, An G, Back K (2000) Transgenic rice plants expressing a *Bacillus subtilis* protoporphyrinogen oxidase gene are resistant to diphenyl ether herbicide oxyfluorfen. Plant Cell Physiol 41:743–749

Lee KY, Townsend J, Tepperman J, Black M, Chui CF, Mazur B, Dunsmuir P, Bedbrook J (1988) The molecular basis of sulfonylurea herbicide resistance in tobacco. EMBO J 7:1241–1248

Llewellyn D, Last D (1996) Genetic engineering of crops for tolerance to 2,4-D. In: Duke SO (ed) Herbicide-resistant Crops. Lewis, Boca Raton, pp 159–174

Mazur BJ, Falco SC (1989) The development of herbicide resistant crops. Annu Rev Plant Physiol Plant Mol Biol 40:441–470

Mazur BJ, Chui CF, Smith JK (1987) Isolation and characterization of plant genes coding for acetolactate synthase, the target enzyme for two classes of herbicides. Plant Physiol 85: 1110–1117

Mitsuhara I, Ugaki M, Hirochika H, Oshima M, Murakami T, Gotoh Y, Katayose Y, Nakamura S, Honkura R, Nishimiya S, Ueno K, Mochizuki A, Tanimoto H, Tsugawa H, Otsuki Y, Ohashi Y

(1996) Efficient promoter cassettes for enhanced expression of foreign genes in dicotyledonous and monocotyledonous plants. Plant Cell Physiol 37:49–59

Moll S (1997) Commercial experience and benefits from glyphosate tolerant crops. Brighton Crop Protection Conf Weeds:931–940

Negrotto D, Jolley M, Beer S, Wenck AR, Hansen G (2000) The use of phosphomannose-isomerase as a selectable marker to recover transgenic maize plants (*Zea mays* L.) via *Agrobacterium* transformation. Plant Cell Rep 19:798–803

Padgette SR, Re DB, Gasser CS, Eichholtz DA, Frazier RB, Hironaka CM, Levine EB, Shah DM, Fraley RT, Kishore GM (1991) Site-directed mutagenesis of a conserved region of the 5-enolpyruvylshikimate-3-phosphate synthase active site. J Biol Chem 266:22364-22369

Padgette SR, Re DB, Barry GF, Eichholtz DE, Delannay X, Fuchs RL, Kishore GM, Fraley RT (1996) New weed control opportunities: development of soybean with Roundup Ready gene. In: Duke SO (ed) Herbicide-resistant crops. Lewis, Boca Raton, pp 53–84

Randolph-Anderson BL, Sato R, Johnson AM, Harris EH, Hauser CR, Oeda K, Ishige F, Nishio S, Gillham NW, Boynton JE (1998) Isolation and characterization of a mutant protoporphyrinogen oxidase gene from *Chlamydomonas reinhardtii* conferring resistance to porphyric herbicides. Plant Mol Biol 38:839–59

Rasche E (1997) Glufosinate ammonium tolerant crops – international commercial developments and experiences. Brighton Crop Protection Conf Weeds, pp 941–946

Rathore KS, Chowdhury VK, Hodges TK (1993) Use of *bar* as a selectable marker gene and for the production of herbicide-resistant rice plants from protoplasts. Plant Mol Biol 21:871–884

Ruff T, Eichholtz D, Re D, Padgette SR, Kishore G (1991) Effects of amino acid substitutions on glyphosate tolerance and activity of EPSPS. Plant Physiol Suppl 96:94

Saari LL, Mauvais CJ (1996) Sulfonylurea herbicide resistant crops. In: Duke SO (ed) Herbicide-resistant crops. Lewis, Boca Raton, pp 127–142

Sambrook J, Russell DW (2000) Molecular cloning, a laboratory manual, 3rd edn. CSHL Press, Cold Spring Harbor, NY

Sebastian SA, Fader GM, Ulrich JF, Forney DR, Chaleff RS (1989) Semidominant soybean mutations for resistance to sulfonylurea herbicides. Crop Sci 29:1403–1408

Shah DM, Horsh RB, Klee HJ, Kishore GM, Winter JA, Tumer NE, Hironaka CM, Sanders PR, Gasser CS, Aykent S, Siegel NR, Rogers SG, Fraley RT (1986) Engineering herbicide tolerance in transgenic plants. Science 233:478–481

Shimamoto K, Terada R, Izawa T, Fujimoto H (1989) Fertile transgenic rice plants regenerated from transformed protoplasts. Nature 338:274–276

Stalker DM, Kiser JA, Baldwin G, Coulombe B, Houck CM (1996) Cotton weed control using the BXN system. In: Duke SO (ed) Herbicide-resistant crops. Lewis, Boca Raton, pp 93–105

Streber WR, Timmis KN, Zenk MH (1987) Analysis, cloning, and high-level expression of 2,4-dichlorophenoxyacetate monooxygenase gene *tfdA* of *Alcaligenes eutrophus* JMP134. J Bacteriol 169:2950–2955

Sugita K, Kasahara T, Matsunaga E, Ebinuma H (2000) A transformation vector for the production of marker-free transgenic plants containing a single copy transgene at high frequency. Plant J 22:461–469

Uknes S, Dincher S, Friedrich L, Negrotto D, Williams S, Thompson-Taylor H, Potter S, Ward E, Ryals J (1993) Regulation of pathogenesis-related protein-1a gene expression in tobacco. Plant Cell 5:159–169

Vasil V, Castillo AM, Fromm ME, Vasil IK (1992) Herbicide resistant fertile transgenic plant obtained by microprojectile bombardment of regenerable embryogenic callus. Bio/technology 10:667–674

Velten J, Schell J (1985) Selection-expression plasmid vectors for use in genetic transformation of higher plants. Nucleic Acids Res 13:6981-6998

Volrath SL, Jhonson MA, Potter SL, Ward ER, Heifetz PB (1997) DNA molecules encoding plant protoporphyrinogen oxidase and inhibitor-resistant mutant thereof. WO 97/32011

Wan Y, Lemaux PG (1994) Generation of large numbers of independently transformed fertile barley plants. Plant Physiol 104:37–48

Ward ER (1995) Manipulation of protoporphyrinogen oxidase enzyme activity in eukaryotic organisms. WO 95/34659

Wells BH (1995) Development of glyphosate tolerant crops into the market. Brighton Crop Protection Conf Weeds:787–790

Ziv M, de Vienne D (1999) Proteomics: a link between genomics, genetics and physiology. Plant Mol Biol 44:575–580

Zoubenko OV, Allison LA, Svab Z, Maliga P (1994) Efficient targeting of foreign genes into the tobacco plastid genome. Nucleic Acids Res 22:3819–3824

10 Major Synthetic Routes for Modern Herbicide Classes and Agrochemical Characteristics

KENJI HIRAI, ATSUSHI UCHIDA, and RYUTA OHNO

10.1 Introduction

Practical herbicides are classified into several groups according to their mode of action by the Herbicide Resistance Action Committee (HRAC) in co-operation with the Weed Science Society of America (WSSA). The actual classification is composed of 15 groups with 9 subgroups and currently contains a total of 269 kinds of herbicides. Chapters 1–8 deal with representative herbicides belonging to 8 HRAC groups; B (acetolactate synthase inhibitors), F_1–F_3 (carotenoid biosynthesis inhibitors), G (EPSP synthase inhibitors), H (glutamine synthetase inhibitors), A (ACCase inhibitors), K_3 (very long-chain fatty acid, VLCFAs biosynthesis inhibitors), L (cellulose biosynthesis inhibitors) and E (protoporphyrinogen-IX oxidase, PPO inhibitors). Each group contains 50, 19, 2, 2, 16, 21, 4 and 26 kinds of practical herbicides, respectively. This chapter briefly summarizes agrochemical characteristics and major synthetic routes for these modern herbicides and chronologically reviews the structural evolution of related compounds (more than 900) disclosed since 1990 except for the PPO inhibitors since 1995.

10.2 Acetolactate Synthase Inhibitors

Acetolactate synthase (ALS) inhibitors act by inhibiting biosynthesis of the essential branched amino acids such as valine and isoleucine (cf. Chap. 1). ALS inhibitors are nontoxic for man, animals, fish and shellfish because these amino acids are only formed in plants. Moreover, ALS inhibitors exhibit superior efficacy against weeds at the low application rate of less than 100 g/ha, and some of them show potent herbicidal activity even at rates of less than 10 g/ha. For these reasons, ALS inhibitors have been aggressively investigated by many companies since the 1970s and many practical ALS inhibitors, which have contributed greatly to modern agricultural production, have been launched until today.

Almost all ALS inhibitors can be roughly grouped into three classes, sulfonylureas, pyrimidyl(thio)oxybenzoates and imidazolinones by their similarity of chemical structures. Additionally, triazolinone and triazolopyrimidine

P. Böger, K. Wakabayashi, K. Hirai (Eds.)
Herbicide Classes in Development

classes derived from sulfonylureas are also ALS inhibitors. Group B of the HRAC classification includes 50 kinds of ALS inhibitors. All ALS inhibitors disclosed since 1990 are documented in this chapter and classified by their chemical structure, although some of them are not regarded as ALS inhibitors. Also, agricultural properties and several major synthetic routes of practical ALS inhibitors are briefly summarized. Moreover, the structural evolution of each class is chronologically reviewed relating to their chemical structures.

10.2.1 Sulfonylurea Acetolactate Synthase Inhibitors

The most important key to the market launch of sulfonylurea herbicides was the discovery of chlorsulfuron and sulfometuron-methyl by DuPont in the early 1980s, since then, 27 compounds have been commercialized and 3 compounds are under development. These practical sulfonylureas are shown in Table 1. Sulfonylureas have been subjected to structural modifications in order to reduce the dosage dramatically, widen the weed spectrum and enhance selectivity for crops. Hence, modern sulfonylureas exhibit extremely strong activity against numerous weeds, including grass and broadleaf weeds, and have played important roles in increasing the yield of primary crops such as rice, wheat, barley, soybeans, corn and so on. Additionally, some sulfonylureas such as metsulfuron-methyl are also used in non-crops.

10.2.1.1 Practical Sulfonylurea Acetolactate Synthase Inhibitors

Chlorsulfuron was the first sulfonylurea class ALS inhibitor to be commercialized by DuPont in 1984. It consists of a 2-chlorophenylsulfonyl group and a 4,6-disubstituted triazine ring, which are connected together through the urea moiety. Chlorsulfuron is the origin of the term of "sulfonylurea". Since the discovery of chlorsulfuron, explosive developments of the sulfonylurea class have been carried out by many companies. Chlorsulfuron has good selectivity for wheat, barley, oats and rye and shows good herbicidal activity against broadleaf weeds at as low as 4–30 g/ha with pre- or post-emergence application. Moreover, it controls different varieties of grass weeds and the water hyacinth. In particular, chlorsulfuron plays a most important role in spring wheat in North America by controlling harmful broadleaf weeds at rates of 20–30 g/ha by post-emergence application. Since the discovery of chlorsulfuron, introduction of a variety of substituents at the benzene and triazine rings has been actively accomplished because a slight difference in the chemical structure considerably affects herbicidal activity and crop safety.

Metsulfuron-methyl bearing a methoxycarbonyl group at the *ortho*-position of the benzene ring has been used as either a pre- or post-emergence herbicide at 2–8 g/ha in wheat, barley, oats and turf. It controls a wide range of annual and perennial broadleaf weeds. The herbicide is superior to

Table 1. Practical sulfonylurea ALS inhibitors

Chemical structure	ISO name Code No. Company	Dose Appl. method Target crops	Patent No.
<Sulfonylureas with triazine ring>			
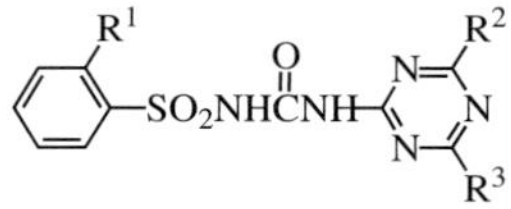	chlorsulfuron DPX-W-4189/DuPont (R^1=Cl, R^2=MeO, R^3=Me)	4~30 g/ha pre, post cereals, turf	US4127405
	metsulfuron-methyl DPX-6376/DuPont (R^1=MeOOC, R^2=MeO, R^3=Me)	2~8 g/ha pre, post non-crop, cereals	US4370480
	triasulfuron CGA-131036/Ciba-Geigy (R^1=$Cl(CH_2)_2O$, R^2=MeO, R^3=Me)	5~15 g/ha pre, post cereals	US4514212 EP44808
	ethametsulfuron-methyl DPX-A-7881/DuPont (R^1=MeOOC, R^2=EtO, R^3=MeNH)	10~120 g/ha pre, post rice, rape, turf	EP136061
	cinosulfuron CGA-142264, CG-148 Ciba-Geigy (R^1=$MeO(CH_2)_2O$, R^2=R^3=MeO)	10~40 g/ha post rape	US4479821 EP44807
	prosulfuron CGA-152005, CG-205 Ciba-Geigy (R^1=$CF_3(CH_2)_2$, R^2=MeO, R^3=Me)	12~40 g/ha post maize, turf, corn	EP120814
	tritosulfuron BASF (R^1=R^2=CF_3, R^3=MeO)		DE4038430
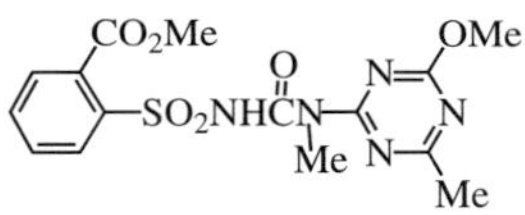	tribenuron-methyl DPX-L-5300, DPX-53 DuPont	5~30 g/ha post cereals, citrus	WO88/02599
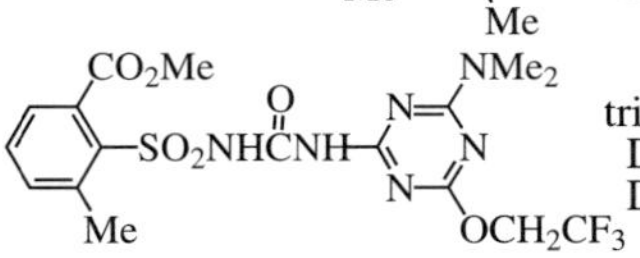	triflusulfuron-methyl DPX-66037 DuPont	10~25 g/ha post sugar beet	WO89/9214
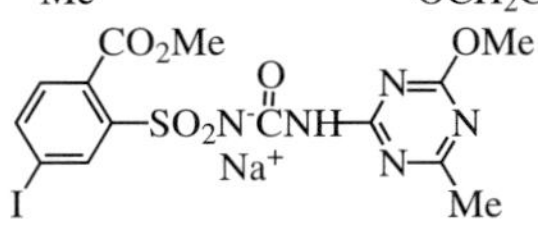	iodosulfuron-methyl-sodium AE F115008 AgrEvo	post cereals	DE19520839 WO92/13845
<Sulfonylureas with pyrimidine ring>			
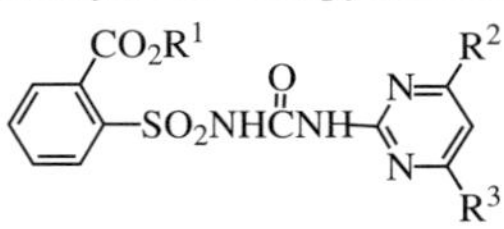	sulfometuron-methyl DPX-T-F-5648, MB-13/DuPont (R^1=R^2=R^3=Me)	70~840 g/ha pre, post non-crop, turf	US4127405
	chlorimuron-ethyl DPX-F-6025, DPX-25/DuPont (R^1=Et, R^2=MeO, R^3=Cl)	9~13 g/ha pre, post soybeans, turf	US4394506 US4547215
	primisulfuron-methyl CGA-136872/Ciba-Geigy (R^1=Me, R^2=R^3=CHF_2O)	20~40 g/ha post soybeans, maize	EP84020 US4478635
	oxasulfuron CGA-277476/Novartis (R^1=3-oxetanyl, R^2=R^3=Me)	45~80 g/ha post soybeans	EP496701 US5209771

Table 1. *Continued*

Chemical structure	ISO name Code No. Company	Dose Appl. method Target crops	Patent No.
<Sulfonylureas with pyrimidine ring>			
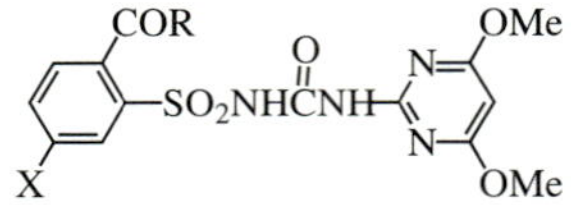	foramsulfuron Hoechst-Schering-AgrEvo (R=Me_2N, X=OHCNH)		DE4335297
	mesosulfuron Hoechst-Schering-AgrEvo (R=MeO, X=$MeSO_2NHCH_2$)		DE4415049
<Pyridylsulfonylureas>			
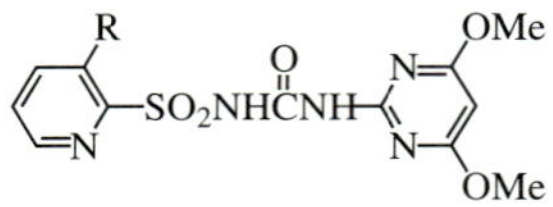	nicosulfuron DPX-V-9360/DuPont (R=Me_2NCO)	40~60 g/ha post maize	US4789393 WO90/05728
	rimsulfuron DPX-E-9636/DuPont (R=$EtSO_2$)	5~15 g/ha post corn, turf, potatoes	EP3411011
	flazasulfuron SL-160, OK-116/Ishihara (R=CF_3)	25~100 g/ha post sugarcane, turf	WO94/23063
	trifloxysulfuron CGA-292230/Novartis (R=CF_3CH_2O)	cereals	WO92/16522
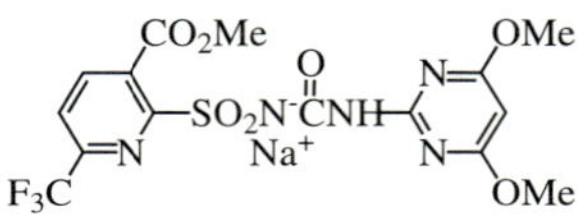	flupyrsulfuron-methyl-Na DPX-KE-459, IN-KE-459, JE-183/DuPont	10 g/ha post soybeans, cereals	WO88/04297 EP327251 EP502740
<Sulfonylureas with heterocycles>			
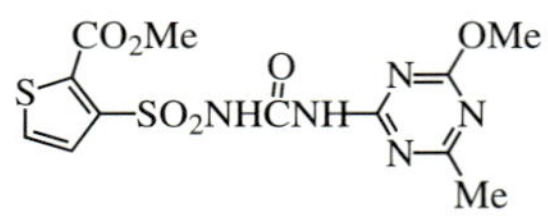	thifensulfuron-methyl DPX-M-6316, DPX-16 DuPont	17~35 g/ha post wheat, barley, citrus, soybeans	US4484029
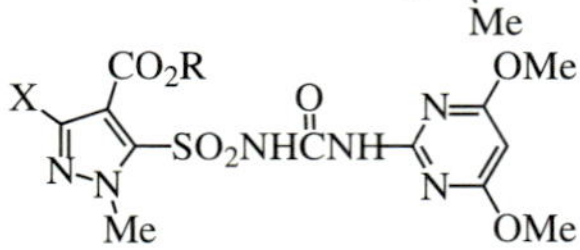	pyrazosulfuron-ethyl NCA-256, NC-311/Nissan (R=Et, X=H)	14~30 g/ha pre, post rice, turf	JP0656792
	halosulfuron-methyl NC-319/Nissan (R=Me, X=Cl)	70~140 g/ha, pre 18~35 g/ha, post corn, turf, sugarcane	JP60208977
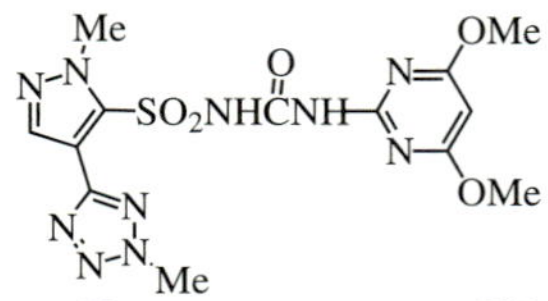	azimsulfuron DPX-A-8947, DPX-47 DuPont	8~20 g/ha post rice	US4746353 US4786311
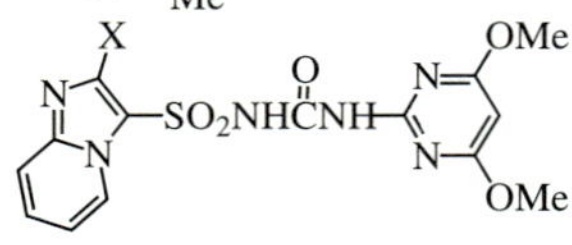	imazosulfuron TH-913/Takeda (X=Cl)	75~100 g/ha pre, post rice, turf	JP6438091
	sulfosulfuron MON-37500, TKM-19 Takeda (X=$EtSO_2$)	10~30 g/ha post cereals	JP05-9102 EP477808

Table 1. *Continued*

Chemical structure	ISO name Code No. Company	Dose Appl. method Target crops	Patent No.
<Benzyl-, Anilino- and Phenoxy-sulfonylureas>			
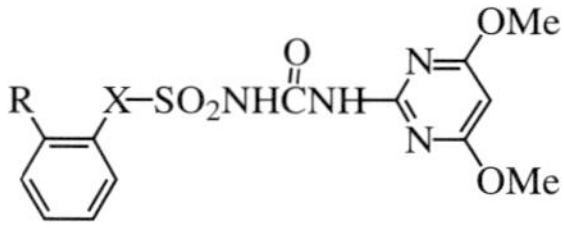	bensulfuron-methyl DPX-F-5384, DPX-84/DuPont (R=MeOOC, X=CH_2)	20~100 g/ha pre, post rice	US4420325
	cyclosulfamuron AC-322140, AC-140/ACC (R=c-PrCO, X=NH)	25~60 g/ha pre, post rice, cereals, turf	US5009699 EP613618 WO92/00952
	ethoxysulfuron Hoe-095404, Hoe-404/AgrEvo (R=EtO, X=O)	10~120 g/ha pre, post rice, cereals, turf	EP342569 EP507093 EP560178
<Sulfonamidosulfonylurea>			
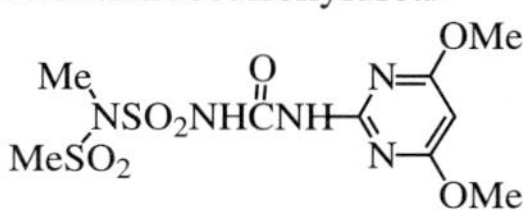	amidosulfuron Hoe-032 Hoechst	15~60 g/ha post cereals	EP298901

chlorsulfuron in controlling violets and turfed knotweeds except for activities against catchweed bedstraw. Ciba-Geigy's sulfonylurea, triasulfuron was developed as a wheat and barley herbicide. It shows excellent herbicidal activity against *Apera spica-venti* and some *Lolium* sp., violets, catchweed bedstraw and broadleaf weeds at 5–15 g/ha with pre- or post-emergence application. Ethametsulfuron-methyl having a 6-ethoxy-4-methylamino-1,3,5-triazine ring was launched in 1993 by DuPont as an oilseed rape herbicide. Pre- and post-emergence treatment provides good herbicidal activity at 10–40 g/ha in spring rape and 80–120 g/ha in winter rape. It controls important weeds such as wild chamomile, henbit, ladysthumb, wild mustard, common purslane, common chickweed and *Polygonum tomentosum*. Cinosulfuron was developed as a post-emergence herbicide. It shows strong activity against annual and perennial weeds including needle spikerush, Japanese bulrush, *Sagittaria pygmaea*, *Alisma canaliculatum* and *Cyperus serotinus* except for barnyard grass at 10–40 g/ha in transplanted and direct seeded rice. Prosulfuron was commercialized as a post-emergence corn herbicide. It is effective against broadleaf weeds such as velvetleaf, common lambsquarters, common purslane, common cocklebur, common ragweed and morningglory at rates of 12–40 g/ha. Tritosulfuron is characterized by two trifluoromethyl groups at the 2- and 6-positions of the benzene and triazine rings, respectively. It is currently developed by BASF.

Tribenuron-methyl is a unique sulfonylurea substituted with a methyl group at the ureido-nitrogen atom. It was launched as a post-emergence herbicide in wheat. It is active on various kinds of perennial broadleaf weeds at 5–30 g/ha and shows high safety in crop rotation of rape or beans because the introduction of the methyl group at the ureido-nitrogen atom promotes rapid decomposition in soil. Triflusulfuron-methyl was DuPont's first sugar beet herbicide with post-emergence application and was launched in 1993. It controls annual and perennial broadleaf weeds and annual grasses such as catchweed bedstraw, slender amaranth, black nightshade, prostrate knotweed, wild mustard, saltmarsh aster, and water foxtail at rates of 10–25 g/ha in two applications. AgrEvo launched a new sulfonylurea herbicide, iodosulfuron-methyl-sodium for use on European cereals in 1999. It is effective against grass weeds including blackgrass, ryegrass and meadow grass and has a wide spectrum of broadleaf weeds by post-emergence application.

The first sulfonylurea with a pyrimidine ring was sulfometuron-methyl and its first registration was approved in 1985. It has been used in non-crop fields and grass fields, and exhibits superior efficacy at 70–840 g/ha against grass weeds, especially johnsongrass and broadleaf weeds. Structural modifications at the pyrimidine ring of sulfometuron-methyl led to chlorimuron-ethyl, which has dramatically improved the crop safety. It is used as a post-emergence herbicide in soybeans and controls common cocklebur, slender amaranth and morningglory at rates of 9–13 g/ha. Moreover, chlorimuron-ethyl has been used in combination with metribuzin as a pre-emergence herbicide.

Introduction of a difluoromethoxy group at the pyrimidine ring gives good selectivity for crops. For example, primisulfuron-methyl is a post-emergence herbicide that controls grass weeds such as johnsongrass and quackgrass as well as broad leaf weeds in maize at 20–40 g/ha. Application of less than 20 g/ha has no influence on crop rotations. Oxasulfuron is a post-emergence herbicide that controls a range of broadleaf and grass weeds such as velvetleaf, common cocklebur, common purslane, common ragweed, barnyardgrass, morningglory and johnsongrass at application rates of 45–80 g/ha in soybeans. Foramsulfuron and mesosulfuron, having additional substituents at the benzene rings, are under development.

Replacements of the usual benzene rings against hetero rings such as pyridine, thiophene, pyrazole or imidazole produced a new class of sulfonylurea herbicides. Nicosulfuron is the first example. It can control annual and perennial grass weeds such as barnyardgrass, fall panicum, quackgrass and johnsongrass as well as broadleaf weeds at 40–60 g/ha. Rimsulfuron was commercialized as a post-emergence herbicide. It exhibits strong herbicidal activity against quackgrass, johnsongrass, *Breea setosa*, purple nutsedge and atrazine-resistant broadleaf weeds at the extremely low rates of 5–15 g/ha. Green foxtail, *Panicum bisulcatum*, and southern crabgrass are sensitive for rimsulfuron. Also, it is used as a post-emergence herbicide in potato fields. Flazasulfuron is a pre- or a post-emergence herbicide active on broadleaf and grass weeds including umbrella plant at 25–100 g/ha in sugarcane. It is also useful as a turf herbicide. Trifloxysulfuron is now under development as a cereal herbicide by Syngenta. Flupyrsulfuron-methyl-sodium, with a trifluoromethyl group at 6-positon on the pyridine ring, controls annual grass weeds such as blackgrass and broadleaf weeds in cereals at a rate of 10 g/ha.

DuPont launched thifensulfuron-methyl in 1988. It is positioned as the first sulfonylurea herbicide with a five-membered heterocycle instead of the phenyl group and controls important broadleaf weeds and *Apera spica-venti* by post-emergence application to wheat or barley at 17–35 g/ha. Nissan's pyrazosulfuron-ethyl with a pyrazole ring has been developed as a pre- or post-emergence herbicide to control annual and perennial grass weeds and many broadleaf weeds in direct-sown and transplanted rice at 14–30 g/ha. In particular, it exhibits good efficacy against *Eleocharis kuroguwai*, *Cyperus serotinus*, arrowhead, *Sagittaria pygmaea*, *Oenanthe javanica* and *Potamogeton distinctus* in rice fields. Halosulfuron-methyl is selective for corn, sugarcane and turf, and can be used for controlling velvetleaf, common cocklebur and purple nutsedges with pre- or post-emergence application of 70–140 and 18–35 g/ha, respectively. DuPont's rice herbicide, azimsulfuron, was first launched in Malaysia in 1996. It controls grass weeds such as *Cyperus serotinus* and *Eleocharis kuroguwai* at the rates of 8–20 g/ha. Azimsulfuron is much more active than DuPont's older rice herbicide, bensulfuron-methyl; however, the activity against broadleaf weeds is weaker. Takeda developed two sulfonylurea herbicides with imidazopyridine rings, namely, imazosulfuron and sulfosulfuron. The former was first launched in 1993 in Japan. It controls annual and perennial broadleaf weeds, except for barnyardgrass in rice, and several combination products are now available. The latter is a post-emergence herbicide for use in wheat. It controls some grass weeds and many broadleaf weeds at rates as low as 10–30 g/ha. It is recommended for use at 27 g/ha to control wild oat, redroot pigweed, chickweed, wild mustard and stinkgrass. It also suppresses green foxtail, quackgrass and dandelion.

In the aforementioned sulfonylurea herbicides, the arylsulfonyl moiety (Aryl-SO_2) is a common structure, which is important in providing strong herbicidal activity against a wide range of weeds. In the course of structural modifications of this moiety, it was shown that an aryl-X-SO_2 – moiety (X = CH_2, O or NH) was remarkably effective for increasing crop safety even at the sacrifice of powerful activity against weeds. Benzyl-modified sulfonylurea, bensulfuron-methyl is a useful rice herbicide that controls broadleaf weeds and umbrella plant except for barnyardgrass at rates of 20–100 g/ha with pre- or post-emergence application. Cyclosulfamuron, decorated with a substituted anilino group at the sulfonylurea moiety, was launched by the American Cyanamid Comp. (ACC) group in 1997. It controls annual and perennial broadleaf weeds at application rates of 25–60 g/ha in the paddy field. Cyclosulfamuron is also active on catchweed bedstraw, wild chamomile, Persian speedwell and wild mustard, and can be applied pre- and post-emergence in autumn wheat and post-emergence in spring wheat. The development, however, has been discontinued. Ethoxysulfuron was first developed as a rice herbicide and then it was recognized as a pre- or post-emergence herbicide in small-grain cereals to control *Cyperus* sp. and cleavers. It shows excellent activity against important broadleaf weeds at 10–120 g/ha in rice.

Amidosulfuron has a unique chemical structure in the sulfonylurea classes, because it includes an *N*-methyl-*N*-methylsulfonylamino group instead of the usual aryl group at the sulfonylurea moiety. It was launched as a post-emergence herbicide by Hoechst in 1990. The active ingredient controls important broadleaf weeds including wild mustard and shepherd's purse in wheat and barley at rates of 15–60 g/ha.

10.2.1.2
Structural Evolution of Sulfonylurea Acetolactate Synthase Inhibitors

Structural modifications of triazine and pyrimidine sulfonylureas disclosed since 1990 are chronologically demonstrated in Figs. 1 and 2, respectively. With regard to the substituents at *ortho*-position of the benzene ring, slightly bulky and hydrophobic forms are essential for potent herbicidal activity. Almost all sulfonylureas cited in these figures are substituted by alkoxycarbonyl [**1–6, 13, 14, 19, 23–44**], sulfonyl [**49, 50, 53**], sulfenyl [**9, 47, 48, 52**], fluorinated alkyl [**10, 17, 54, 58**], trifluoromethoxy [**12, 63**], alkylthio groups [**46, 51**] or a chlorine atom [**10, 16, 18, 62**] at *ortho*-position. Additional substitution at the benzene ring often seems to improve herbicidal activity and crop safety. Sulfonylureas, for example triflusulfuron-methyl with *ortho*-disubstituted phenyl rings, tend to be highly active and several compounds [**10, 16, 42, 43, 62**] have been produced since 1990. Moreover, sulfonylureas bearing *ortho*- and *meta*-disubstituted phenyl rings keep strong activity, although it depends on the substituent at *meta*-position. Especially, sulfonylureas [**6, 11, 28–32, 34–39, 51, 53**] with 2,5-disubstituted phenyl rings have been studied, and iodosulfuron-methyl-sodium [**5**], mesosulfuron [**33**] and foramsulfuron [**45**] are under development. Also, sulfonylureas with a 2,3-disubstituted phenyl ring have been continued [**40, 41, 44**].

In the course of optimization of substituents at 1,3,5-triazine and pyrimidine rings, introduction of methoxy groups at the 4- and 6-positions gave good results. Sulfonylureas with the 4,6-dimethoxy-1,3,5-triazine or 4,6-dimethoxypyrimidine ring were easily decomposed, so that carryover, a serious problem in crop rotation, was overcome. Further structural modifications are observed in the investigations of some sulfonylureas [**12–20**], of which urea moieties are modified with a methyl, ethoxymethyl, or butyl and other substituents. Additionally, an amidine or guanidine moiety is introduced as a mimic of the urea group. Especially sulfonylguanidine is a prototype lead compound for triazolopyrimidine ALS inhibitors mentioned later.

Structural evolution of pyridyl-sulfonylureas is chronologically depicted in Fig. 3. Practical pyridyl-sulfonylureas, namely, flazasulfuron, rimsulfuron, nicosulfuron and flupyrsulfuron-methyl-sodium have electron-withdrawing groups such as trifluoromethyl, ethanesulfonyl and dimethylcarbamoyl groups at the neighboring position of the sulfonyl group at the pyridine ring. Pyridyl-sulfonylureas proposed in the early 1990s possess electron-donating groups such as amino [**76–82**], alkoxy [**88–93**] and substituted alkyl groups [**94, 95**]. Alternatively, new pyridyl-sulfonylureas [**83–87**] with substituted amino groups only at 6-position of the pyridine ring have been suggested.

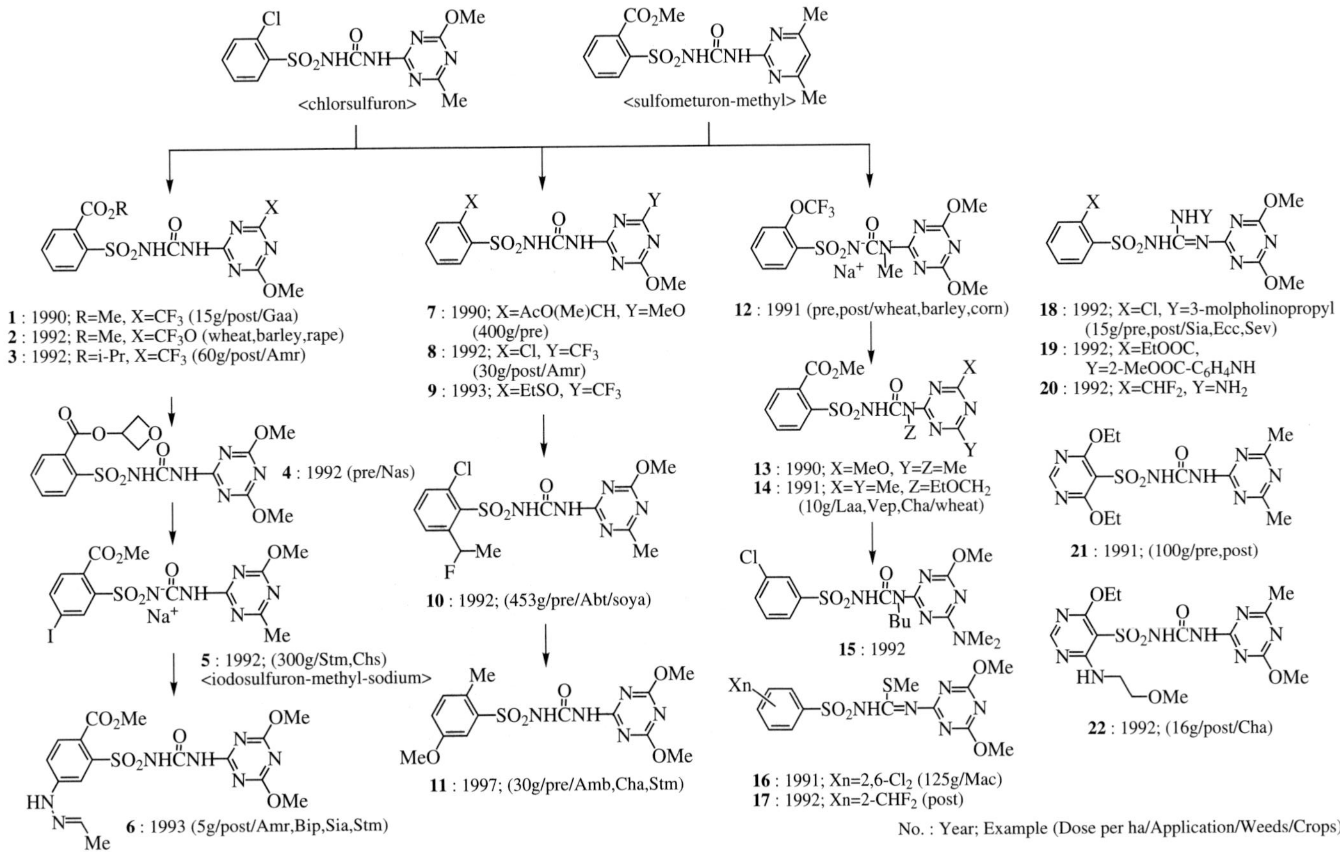

Fig. 1. Structural evolution of sulfonylurea ALS inhibitors with triazine ring since 1990

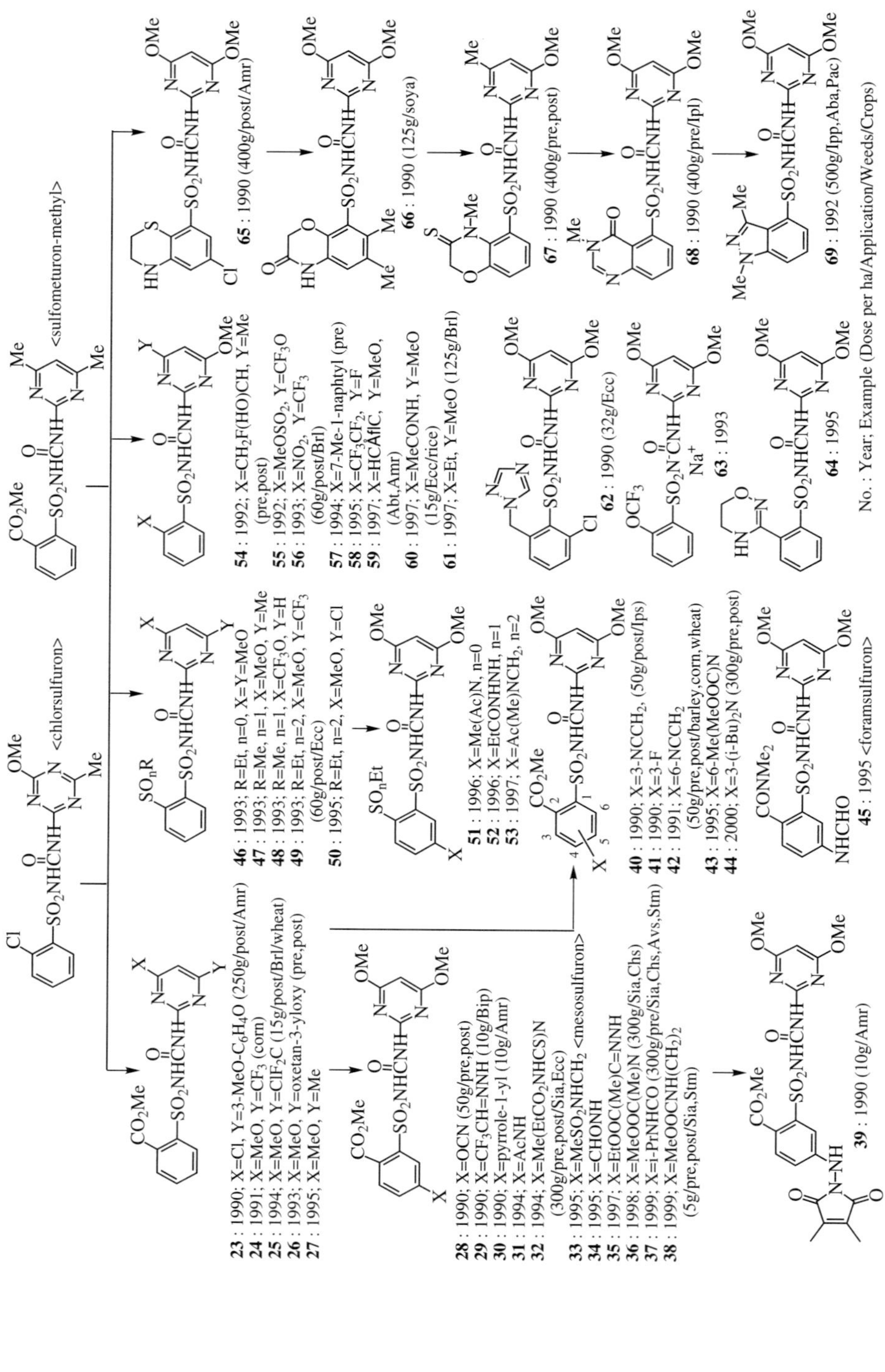

Fig. 2. Structural evolution of sulfonylurea ALS inhibitors with pyrimidine ring since 1990

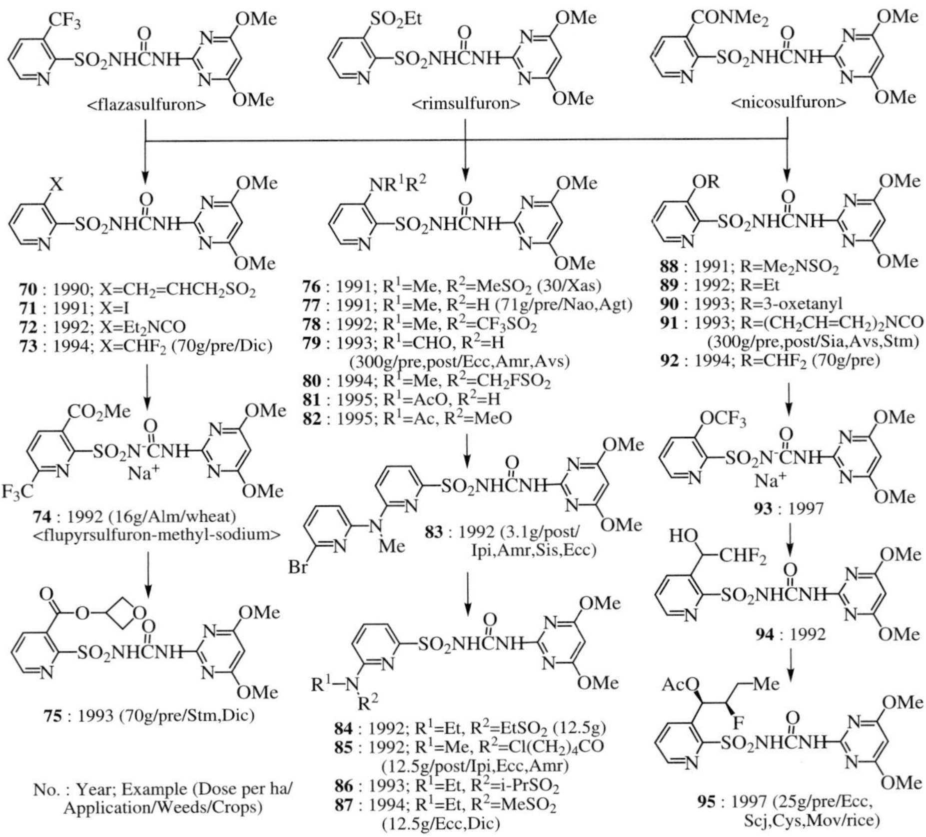

Fig. 3. Structural evolution of pyridylsulfonylurea ALS inhibitors since 1990

Various kinds of sulfonylureas with five-membered heterocycles have been aggressively developed, as shown in Fig. 4. Thifensulfuron-methyl was the first sulfonylurea herbicide having a thiophen-3-yl group. Other sulfur-containing heterocycles such as thiazole and thiadiazole rings have been proposed since the 1990s.

Structural modifications of sulfonylureas with nitrogen-containing heterocycles such as pyrrole and pyrazole rings are shown chronologically in Fig. 5. Pyrazosulfuron-ethyl and halosulfuron-methyl are leading compounds for these sulfonylureas. Additionally, sulfonylureas with bicyclic heterocycles such as imidazopyridine [**125**] and benzopyrazole [**126**] have been studied continuously. Among them, imazosulfuron and sulfosulfuron are practical ALS inhibitors developed by Takeda.

Other sulfonylureas possessing substituted benzyl, sulfonamido, anilino and phenoxy groups are depicted in Fig. 6. In spite of good biological properties investigations of almost all derivatives have been discontinued.

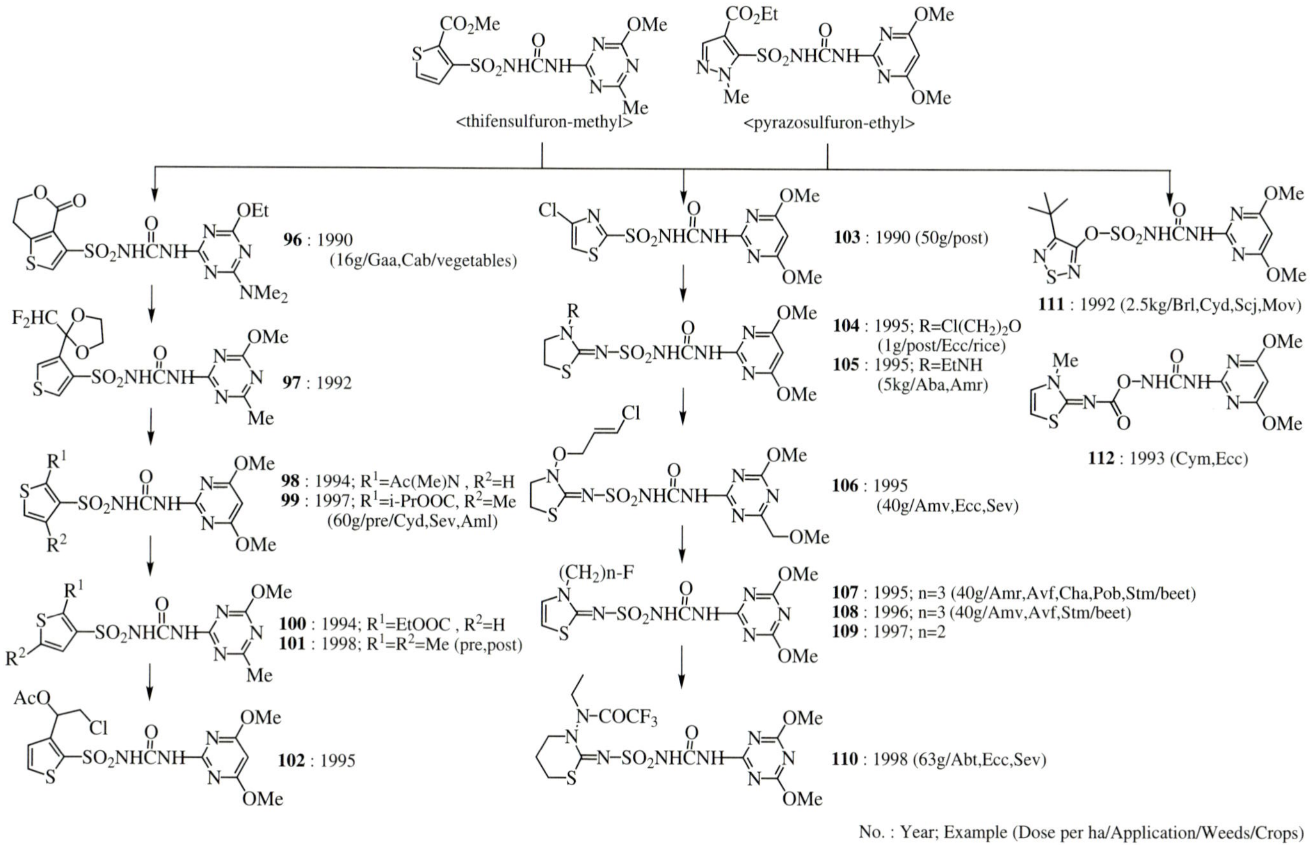

Fig. 4. Structural evolution of sulfonylurea ALS inhibitors with sulfur-containing heterocycles since 1990

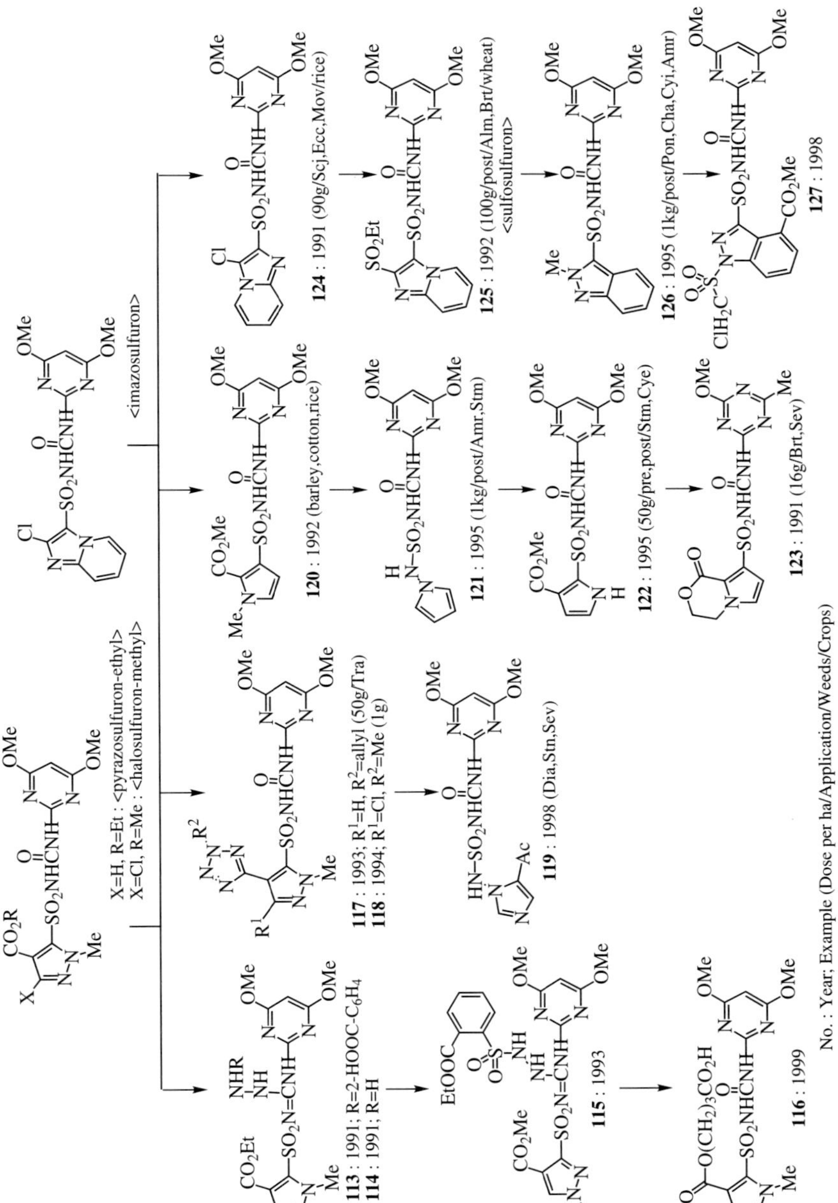

Fig. 5. Structural evolution of sulfonylurea ALS inhibitors with nitrogen-containing heterocycles since 1990

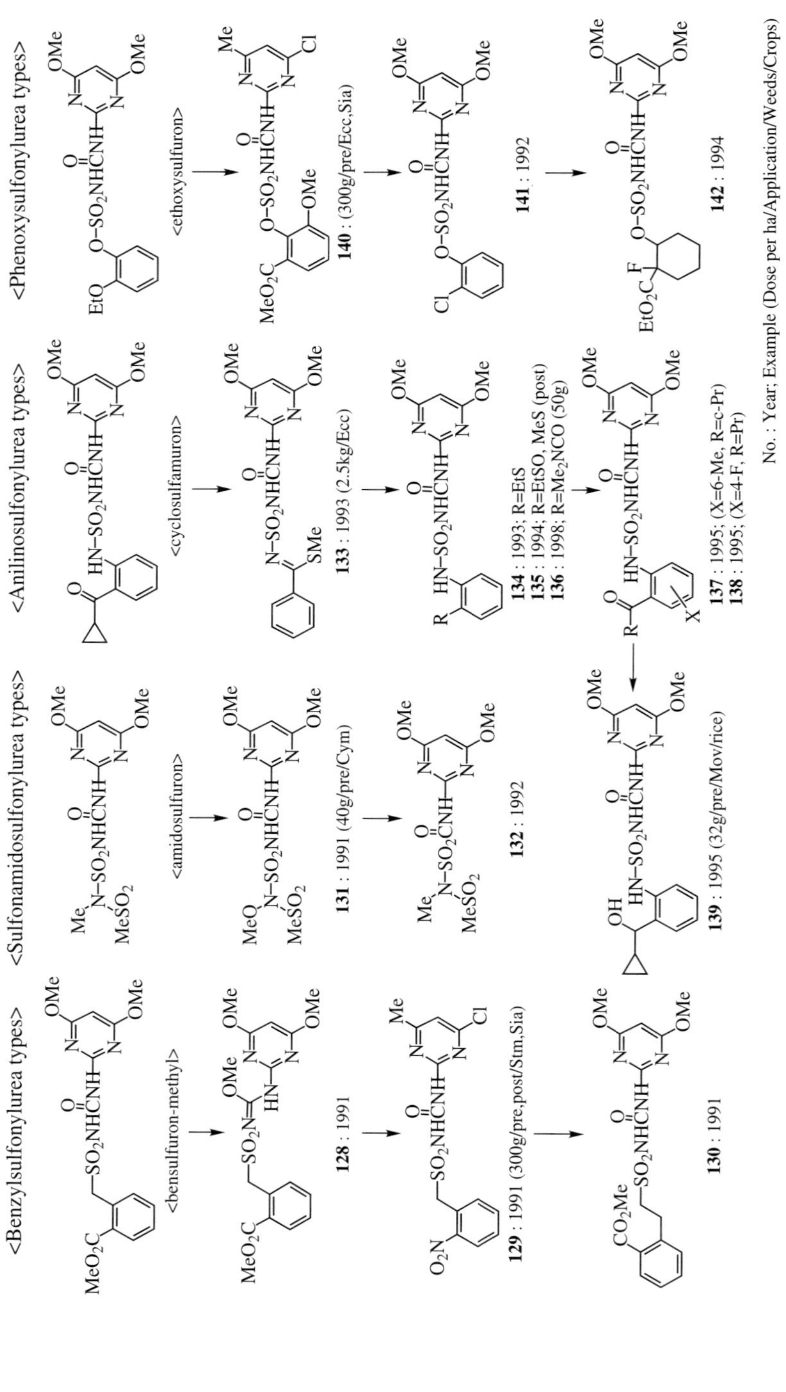

Fig. 6. Structural evolution of benzyl-, sulfonamido-, anilino- and phenoxy-sulfonylurea ALS inhibitors since 1990

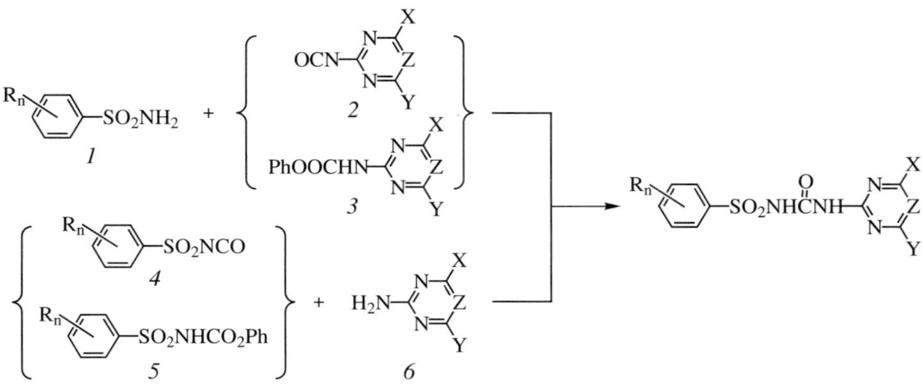

Scheme 1. General synthetic routes for sulfonylurea moiety

10.2.1.3
Major Synthetic Routes for Sulfonylureas

There are generally two methods for preparing sulfonylurea (-SO_2NHCONH-) herbicides as shown in Scheme 1. One is the reaction of arylsulfonamides (*1*) with isocyanates (*2*) or phenyl carbamates (*3*), the other one is the reaction of arylsulfonyl isocyanates (*4*) or phenyl *N*-phenylsulfonyl carbamates (*5*) with arylamine (*6*). Major synthetic pathways for modern sulfonylurea ALS inhibitors are described briefly below.

The synthetic route for chlorsulfuron is shown in Scheme 2. Here, the key intermediate is sulfonamide (*8*) that is prepared from 2-chloroaniline (*7*) by three subsequent steps, diazotization, chlorosulfonation and amination. Phosgenation of sulfonamide (*8*) is accelerated by the addition of a lower alkyl isocyanate such as butyl isocyanate to yield 2-chlorophenylsulfonyl isocyanate (*9*), which is condensed with 2-amino-1,3,5-triazine (*10*) affording chlorsulfuron.

The synthesis of flupyrsulfuron-methyl-sodium is also depicted in Scheme 2. Cyclocondensation of 4-butoxy-3-buten-2-one (*11*) with ester (*12*) followed by bromination with phosphoryl bromide gives pyridine-3-carboxylate (*13*). After benzylthiolation of *13*, treatments with hypochlorite and *tert*-butylamine give *N*-*tert*-butylpyridine-2-sulfonamide (*14*). Deprotection of the *tert*-butyl group by trifluoroacetic acid affords sulfonamide (*15*), which reacts with phenyl *N*-(4,6-dimethoxypyrimidin-2-yl)carbamate (*16*) to give flupyrsulfuron-methyl-sodium.

Scheme 2 also shows the synthetic procedure for thifensulfuron-methyl. The key intermediate, methyl 3-aminothiophene-2-carboxylate (*18*) is easily synthesized by cycloaddition reaction with chloroacrylonitrile (*17*) and methyl thioglycolate. After diazotization of the amino group of (*18*), treatment with sulfur dioxide in the presence of copper chloride and then ammonia gas gives

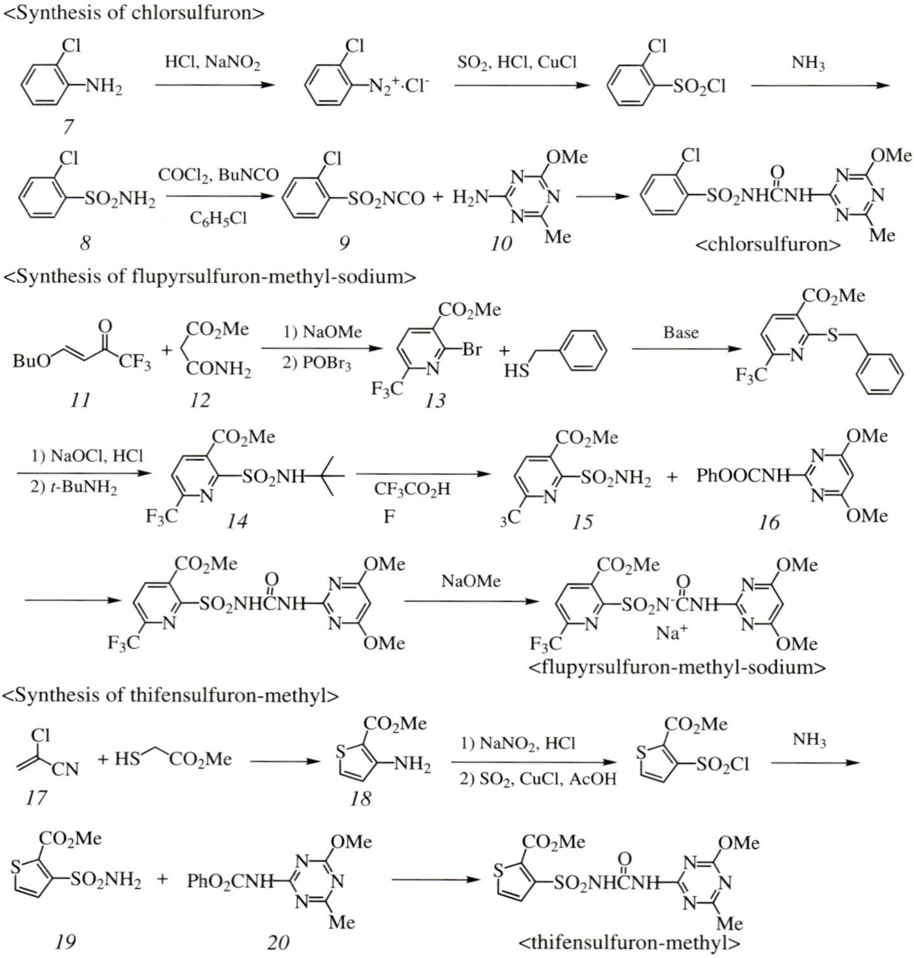

Scheme 2. Major synthetic routes for chlorsulfuron, flupyrsulfuron-methyl-sodium and thifensulfuron-methyl

thiophene-3-sulfonamide (*19*). Reaction of sulfonamide (*19*) with phenyl *N*-(4-methoxy-6-methyl-1,3,5-triazin-2-yl)carbamate (*20*) easily proceeds to yield thifensulfuron-methyl.

For the synthesis of halosulfuron-methyl, methyl 5-aminosulfonyl-3-chloro-1-methylpyrazole-4-carboxylate (*24*) is an important key precursor. There are two synthetic pathways for *24* as depicted in Scheme 3. The first method is characterized by using lithium reagent. Methyl 3-amino-1-methylpyrazole-4-carboxylate (*21*) is selectively chlorinated at 3-position *via* diazonium salt affording 3-chloropyrazole (*22*). The compound (*22*) is treated with sulfur

<Synthesis of halosulfuron-methyl (Method-1)>

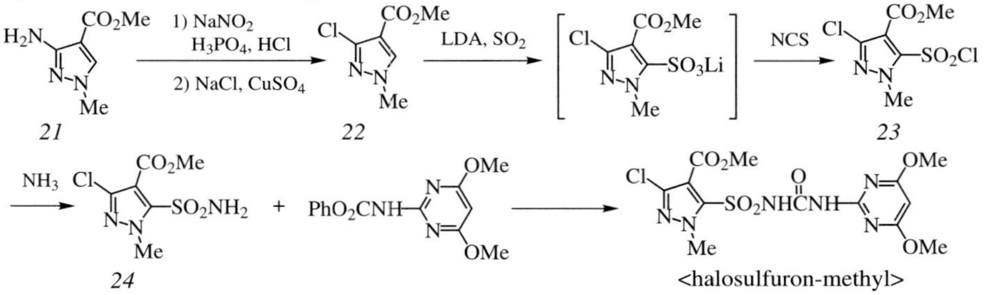

<Synthesis of halosulfuron-methyl (Method-2)>

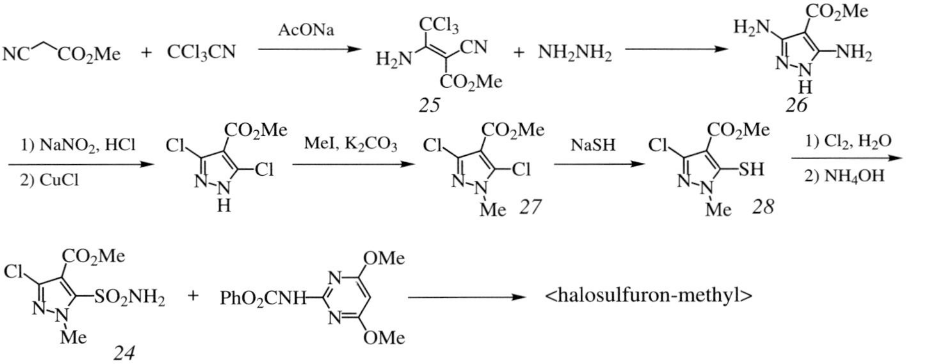

<Synthesis of cyclosulfamuron>

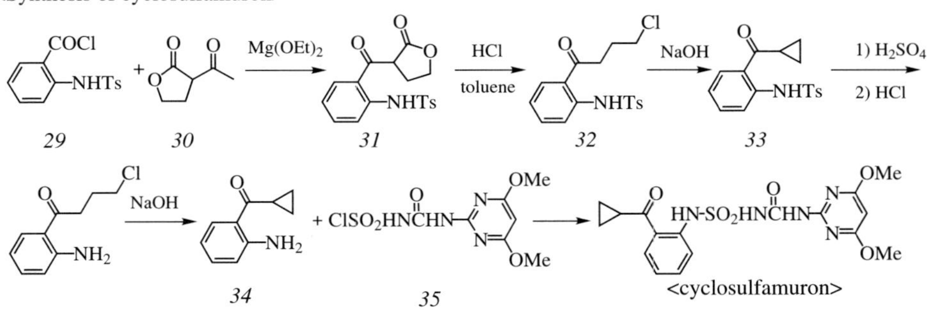

<Synthesis of ethoxysulfuron>

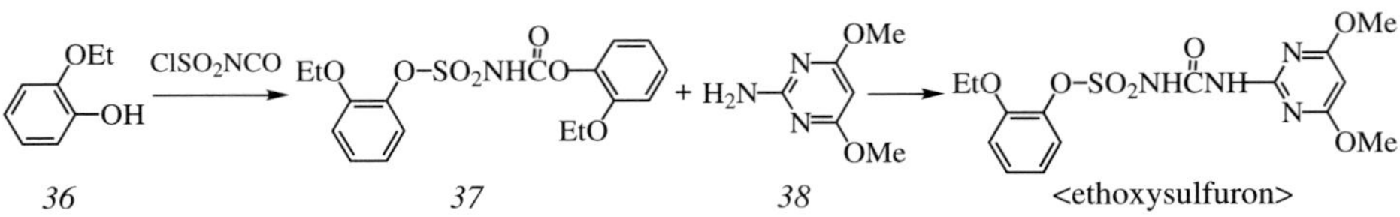

Scheme 3. Major synthetic routes for halosulfuron-methyl, cyclosulfamuron and ethoxysulfuron

dioxide after lithiation by lithium diisopropylamide (LDA) and chlorinated by *N*-chlorosuccinimide (NCS) to give pyrazole (*23*). The key intermediate (*24*) is readily obtained by amination of *23*. The other is the method without lithium reagent in spite of somewhat long steps. Methyl cyanoacetate condenses with trichloroacetonitrile to obtain acrylate (*25*). Cycloaddition reaction of *25* with hydrazine yields 3,5-diaminopyrazole-4-carboxylate (*26*), which is dichlorinated at the 3- and 5-positions via a diazonium salt and methylated at the 1-position. Selective thiolation at the 5-position of *27* by sodium hydrosulfide is accomplished to give methyl 3-chloro-5-mercapto-1-methylpyrazole-4-carboxylate (*28*), followed by treatment with a chlorine gas in water and amination subsequently giving the desired intermediate (*24*). Pyrazole-4-carboxylate (*24*) thus obtained is easily converted to halosulfuron-methyl.

Scheme 3 also shows the synthetic scheme for cyclosulfamuron including a unique method for introduction of a cyclopropylcarbonyl group. Condensation of anthraniloyl chloride (*29*) with 2-acetyl-γ-butyrolactone (*30*) in the presence of magnesium diethoxide gives anilide (*31*), which is heated together with concentrated hydrochloric acid in toluene to yield 4-chloro-1-(2-*N*-tosylaminophenyl)-1-butanone (*32*). Ring closure of *32* by treating with aqueous sodium hydroxide gives cyclopropyl ketone (*33*), followed by hydrolysis of the tosyl group and recyclization yielding *ortho*-aminophenyl cyclopropyl ketone (*34*). The ketone (*34*) reacts with chlorosulfonylurea (*35*) to yield cyclosulfamuron.

An efficient synthetic approach of ethoxysulfuron is shown in Scheme 3. It is readily synthesized by the reaction of pyrimidine (*38*) with 2-ethoxyphenylcarbamate (*37*), which is derived by condensation of two equivalents of phenol (*36*) and chlorosulfonyl isocyanate.

10.2.2 Triazolinone Acetolactate Synthase Inhibitors

10.2.2.1 Practical Triazolinone Acetolactate Synthase Inhibitors

Triazolinone herbicides are classified as second-category ALS inhibitors designed on the basis of the usual sulfonylurea ALS inhibitors. Most triazolinone ALS inhibitors, in general, consist of substituted phenyl groups and 4,5-disubstituted triazolinone rings bridged together by a sulfonylaminocarbonyl (-SO_2NHCO-) moiety. Investigations have been actively continued since the 1990s and two practical herbicides, flucarbazone-sodium and procarbazone-sodium, have been commercialized and are under development (Table 2). Flucarbazone-sodium is a post-emergence cereal graminicide and controls wild oat and green foxtail at 30 g/ha. Procarbazone-sodium is being developed as a wheat herbicide.

Table 2. Practical triazolinone ALS inhibitors

Chemical structure	ISO name Code No. Company	Dose Appl. method Target crops	Patent No.
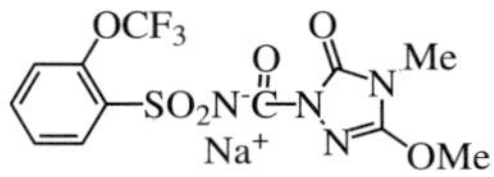	flucarbazone-sodium BAY-MKH-6562 Bayer	30 g/ha post cereals	US5541337 EP507171
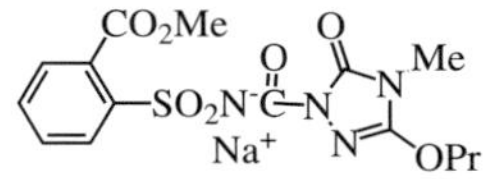	procarbazone-sodium BAY-MKH-6561 Bayer	30~70 g/ha wheat	US5541337 US6147221 US6147222

10.2.2.2 Structural Evolution of Triazolinone Acetolactate Synthase Inhibitors

Structural modifications of orthodox triazolinone ALS inhibitors are summarized in the left column of Fig. 7. *Ortho*-positions of the benzene rings are modified by a variety of substituents, which are not always limited to electron-withdrawing groups such as the alkoxycarbonyl group. On the other hand, the substituents on the triazolinone rings are restricted to electron-donating groups except for the incipient compound [**144**].

10.2.2.3 Major Synthetic Routes for Triazolinone Acetolactate Synthase Inhibitors

As an example of synthetic routes for triazolinone ALS inhibitors, the major synthetic pathway of flucarbazone-sodium is illustrated in Scheme 4. Cycloaddition reaction of phenyl carbazate (*39*) and trimethyliminocarbonate (*40*) together with elimination of phenoxide gives 3-methoxytriazolinone (*41*), which reacts with 2-trifluoromethoxyphenylsulfonyl isocyanate (*42*) to yield flucarbazone-sodium (*43*).

10.2.3 Triazolopyrimidine Acetolactate Synthase Inhibitors

Triazolopyrimidine herbicides are grouped as the third-category ALS inhibitors that are distinguished by their triazolopyrimidine rings. Dow Agrosciences disclosed the characteristic triazolopyrimidine in the 1980s. The synthetic strategy for the construction of the triazolopyrimidine ring is of some interest to herbicide researchers. At first, 3-(arylsulfonylamino)triazolopyrimidine derivatives were designed and synthesis was attempted; however, further development was interrupted because of their low solubility. Further structural modifications resulted in inversion of -SO_2 and -NH moieties, which improved

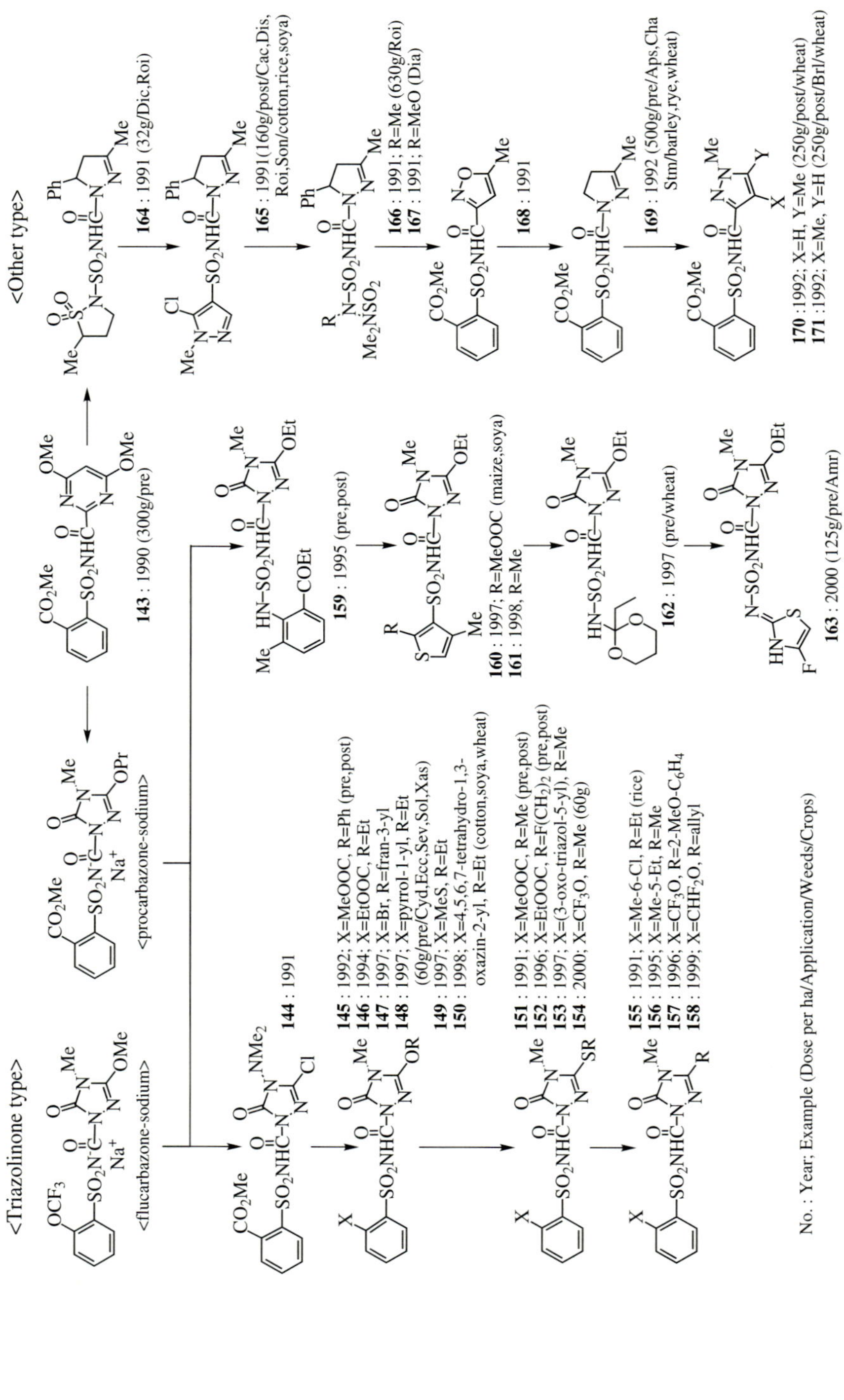

No. : Year; Example (Dose per ha/Application/Weeds/Crops)

Fig. 7. Structural evolution of triazolinone ALS inhibitors and related compounds since 1990

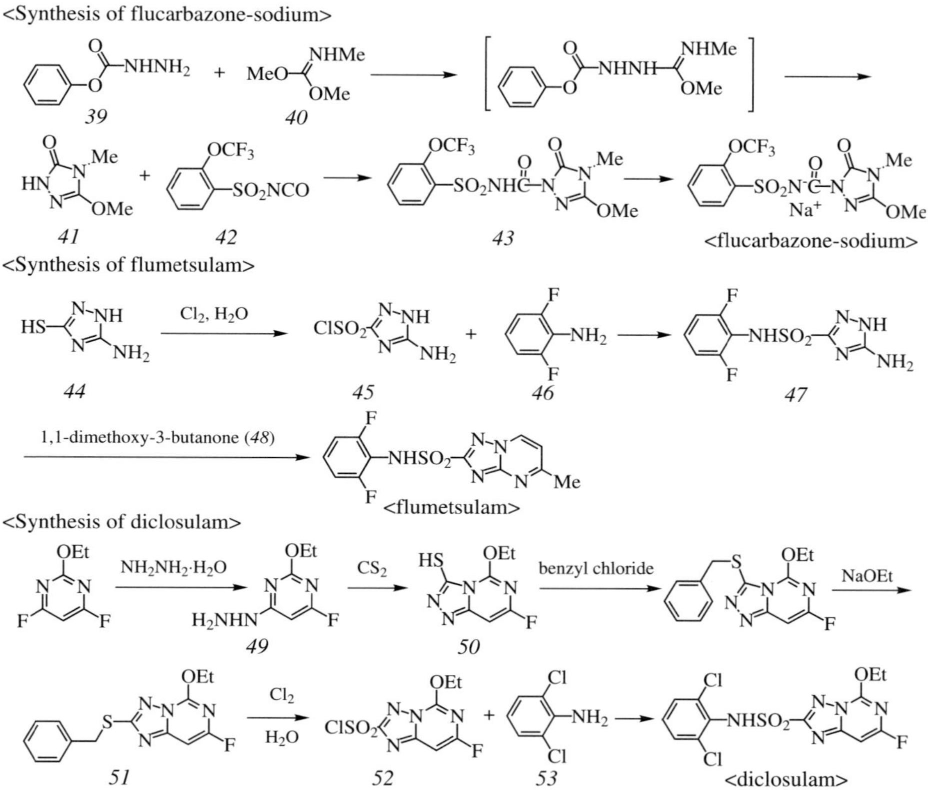

Scheme 4. Major synthetic routes for flucarbazone-sodium, flumetsulam and diclosulam

their solubility. Optimization of thus obtained 3-(arylaminosulfonyl)triazolopyrimidine derivatives produced a novel 1,2,4-triazolo[1,5-*a*]pyrimidine ALS inhibitor, flumetsulam. Subsequent unique structural modifications led to a regioisomer, namely, 1,2,4-triazolo[1,5-*c*]pyrimidine, which brought the further discoveries of several practical herbicides, cloransulam-methyl, diclosulam, or florasulam.

10.2.3.1 Practical Triazolopyrimidine Acetolactate Synthase Inhibitors

Flumetsulam, launched in 1994, shows strong herbicidal activity against broadleaf weeds including velvetleaf by pre- and post-emergence applications, and exhibits excellent safety for soybeans, wheat and corn (Table 3). Metosulam is used as a post-emergence herbicide for corn, wheat and barley. Application of 20–30 g/ha is suitable for corn, and 5–15 g/ha for wheat and

Table 3. Practical triazolopyrimidine ALS inhibitors

Chemical structure	ISO name Code No. Campany	Dose Appl. method Target weeds	Patent No.
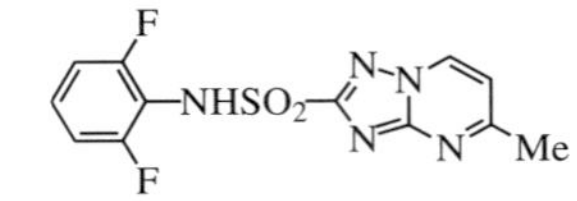	flumetsulam DE-498 Dow Elanco	17~70 g/ha pre, post corn, soybeans, wheat	US4988812 JP03153689
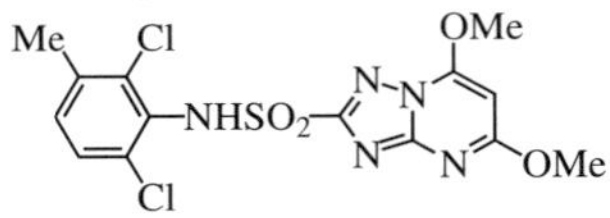	metosulam DE-511 Dow Elanco	5~15 g/ha (wheat) 20~30 g/ha (corn) pre, post	EP343752
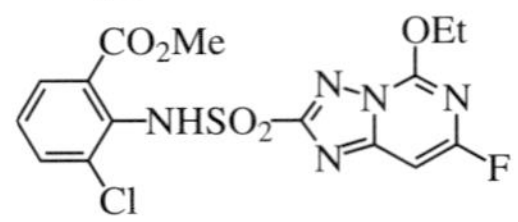	cloransulam-methyl XDE-565 Dow Elanco	35~44 g/ha, pre 17.5~35g/ha, post soybeans	US5163995
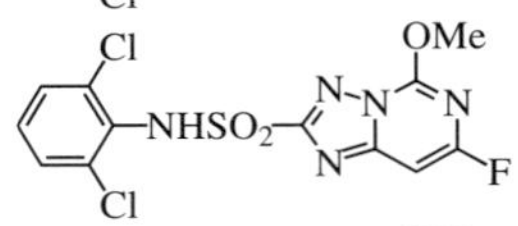	diclosulam XDE-564 Dow Elanco	26~35 g/ha pre, post soybeans, peanuts	EP343752 US5163995
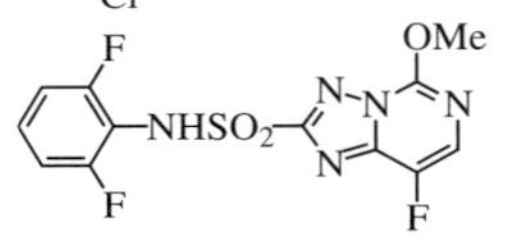	florasulam DE-570 Dow Elanco	5~10 g/ha pre, post turf, corn, cereals	EP343752 US5163995
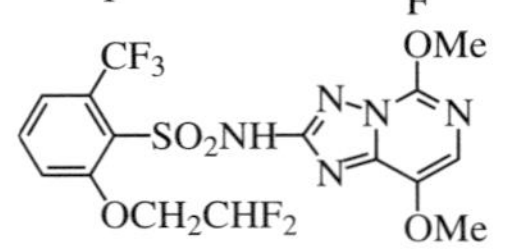	penoxsulam Dow Elanco		US5858924

barley. The herbicide controls many broadleaf weeds including catchweed bedstraw, common chickweed, redroot pigweed, black nightshade and *Polygonum persicaria*.

Cloransulam-methyl is a broadleaf herbicide for use on soybeans by either pre- or post-emergence. It controls cocklebur, velvetleaf, morningglory, ragweed, purslane, lambsquarters, sunflower, *Polygonum lapathifolium* and prickly sida with the use rate of 35–44 g/ha pre-emergence application, and cocklebur, velvetleaf, morningglory, common ragweed, sunflower at 17.5 g/ha post-emergence application. Diclosulam gives pre-emergence control of broadleaf weeds in soybeans and pre- and post-emergence control in peanuts. It is also a soil-applied herbicide with activity against a wide spectrum of broadleaf weeds and some grass suppression. Florasulam launched in 1999 was developed to control broadleaf weeds, particularly cleavers, in small-grain cereals and corn. The *N*-triazolopyrimidylbenzene sulfonamide derivative, penoxsulam, is under development.

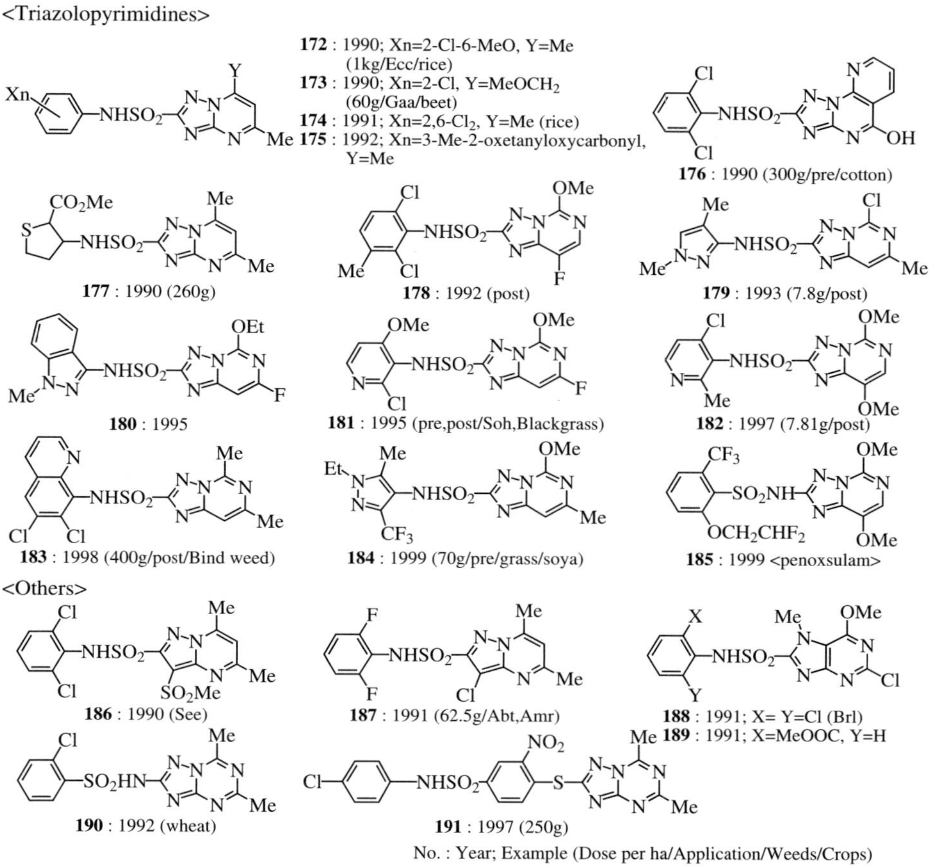

Fig. 8. Structural evolution of triazolopyrimidine ALS inhibitors and related compounds since 1990

10.2.3.2
Structural Evolution of Triazolopyrimidine Acetolactate Synthase Inhibitors

Research on triazolopyrimidine ALS inhibitors has been actively pursued since the late 1980s. Triazolopyrimidine derivatives disclosed since 1990 are illustrated in Fig. 8. Investigations of 1,2,4-triazolo[1,5-*a*]pyrimidines, however, have declined since 1992.

On the other hand, 1,2,4-triazolo[1,5-*c*]pyrimidines have been continuously studied and various derivatives have been proposed. In addition to *N*-phenyl derivatives [**178, 183**], *N*-pyrazolyl [**179, 180, 184**] and *N*-pyridyl derivatives [**181, 182**] are also shown in Fig. 8. *N*-triazolopyrimidyl benzenesulfonamide derivative, penoxsulam [**185**], is being developed. Triazolopyrimidine rings are converted to other bicyclic hetero rings such as pyrazopyrimidine [**186, 187**], imidazopyrimidine [**188, 189**] and triazolotriazine [**190, 191**].

10.2.3.3
Major Synthetic Routes for Triazolopyrimidine Acetolactate Synthase Inhibitors

A synthetic route for flumetsulam is shown in Scheme 4. Treatment of 3-amino-5-mercapto-1,2,4-triazole (*44*) with chlorine gas in water gives chlorosulfone (*45*), which reacts with 2,6-difluoroaniline (*46*) to yield 1,2,4-triazolesulfonamide (*47*). Then, a cyclization reaction of *47* with 1,1-dimethoxy-3-butanone (*48*) provides flumetsulam.

Synthetic procedure of diclosulam is also illustrated in Scheme 4. 3-Mercapto triazolopyrimidine (*50*) is easily prepared by treating hydrazinopyrimidine (*49*) with carbon disulfide. After benzylation at the sulfur atom, an abnormal rearrangement proceeds in the presence of sodium ethoxide to give 1,2,4-triazolo[1,5-*c*]pyrimidine (*51*), which is converted to sulfonylchloride (*52*). Finally, condensation of *52* with 2,6-dichloroaniline (*53*) gives diclosulam.

10.2.4
Acetolactate Synthase Inhibitor-Like Miscellaneous Pyrimidines and Related Compounds

Miscellaneous sulfonylurea-like ALS inhibitors, which can neither be grouped into sulfonylurea, triazolinone nor triazolopyrimidine classes, are summarized in Fig. 9.

10.2.5
Pyrimidyl(thio)oxybenzoate Acetolactate Synthase Inhibitors

It has been revealed that pyrimidyl(thio)oxybenzoates without a sulfonylurea moiety exhibit ALS inhibitory activity. These pyrimidyl(thio)oxybenzoates are regarded as the analogues of the usual sulfonylurea ALS inhibitors in which the sulfonylurea moiety is entirely replaced by oxygen and sulfur atoms. A pair of methoxy groups is suitable for substituents at the pyrimidine ring, and at *ortho*-position of the benzene ring, electron-withdrawing groups such as methoxycarbonyl and carboxy groups are appropriate. These compounds are called pyrimidyl(thio)oxybenzoate ALS inhibitors and several practical herbicides have been developed until today.

10.2.5.1
Practical Pyrimidyl(thio)oxybenzoate Acetolactate Synthase Inhibitors

Practical pyrimidyl(thio)oxybenzoate ALS inhibitors are shown in Table 4. Pyrithiobac-sodium is the first post-emergence herbicide to be approved. It

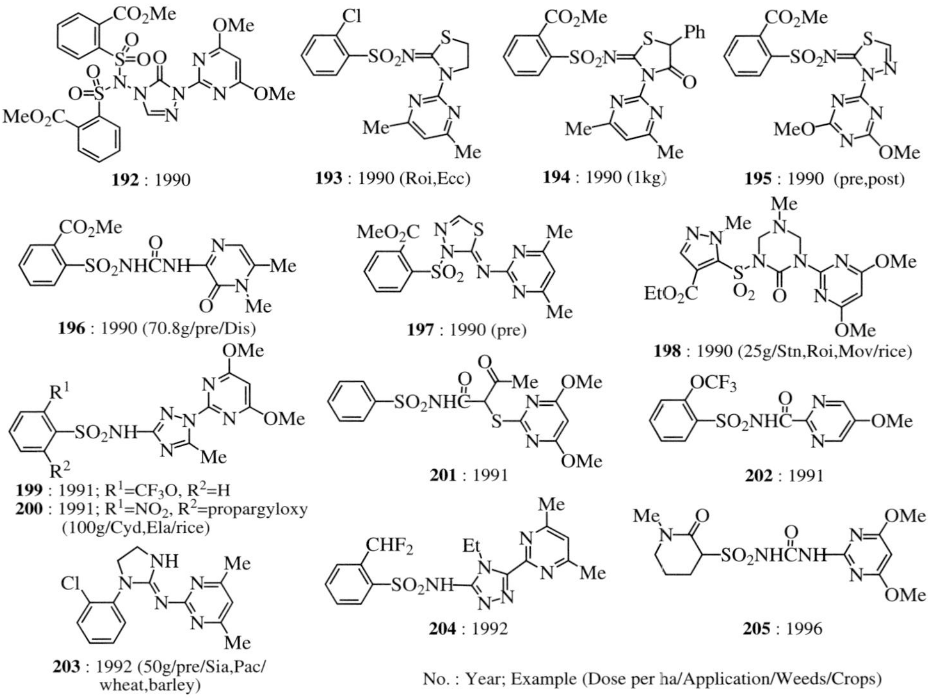

Fig. 9. ALS inhibitor-like miscellaneous pyrimidines and related compounds disclosed since 1990

controls broadleaf weeds including morningglory, common cocklebur, and slender amaranth at rates of 30–100 g/ha in cotton.

Pyriminobac-methyl, which was first launched in Japan in 1996, is selective for rice and specific for barnyardgrass control from pre-emergence to the four-leaf stage of growth. It is effective at 30–90 g/ha. Its high crop safety is considered to be conferred by introduction of a 1-(methoxyimino)ethyl group at the 3-position of the benzene ring. A pyrimidylthiobenzofuranone derivative, pyriftalid, is under development as a rice herbicide by Syngenta. Bispyribac-sodium commercialized in 1997 is a post-emergence herbicide that controls annual and perennial grass, broadleaf and sedge weeds at rates of 15–300 g/ha in dry or water-seeded rice. Pyribenzoxim is a broad-spectrum rice herbicide, and shows excellent post-emergence activity against various grass and broadleaf weeds such as barnyardgrass, blackgrass and *Polygonum* sp. at 30–50 g/ha in rice, wheat, barley or turf.

Table 4. Practical pyrimidy(thio)oxybenzoate ALS inhibitors

Chemical structure	ISO name Code No. Company	Dose Appl. method Target crops	Patent No.
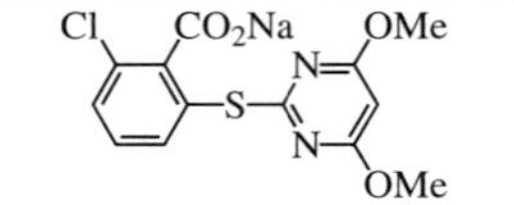	pyrithiobac-sodium KIH-2031,KIH-8921 Kumiai	30~100 g/ha pre, post cotton	JP01230561 EP315889 US5149357
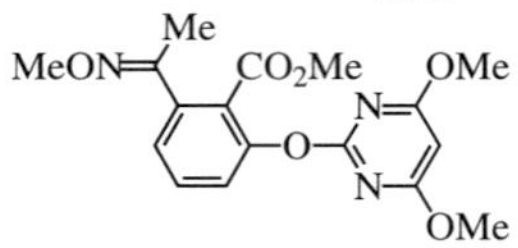	pyriminobac-methyl KIH-6127, KUH-920 Kumiai	30~90 g/ha pre, post rice	EP435170 JP04134080 US5118339
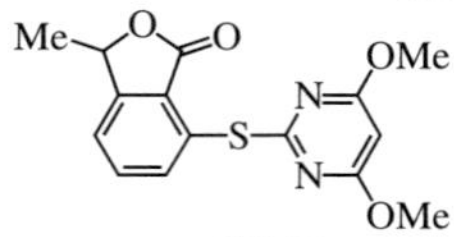	pyriftalid Novartis	180 g/ha rice	JP1143409
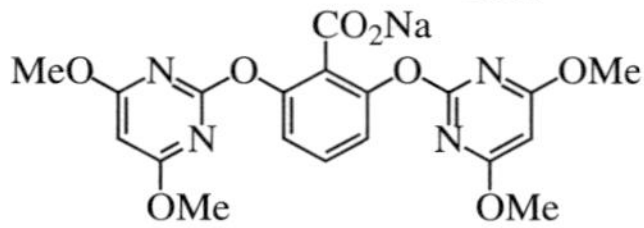	bispyribac-sodium KUH-911 Kumiai	15~300 g/ha pre, post rice	JP01250365 EP321846 US4906285
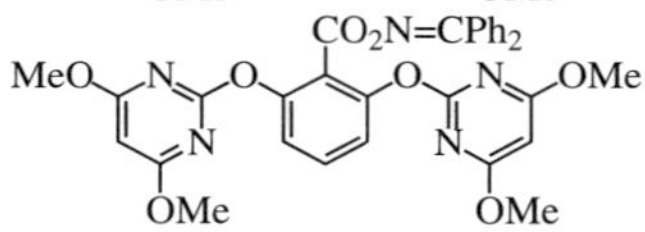	pyribenzoxim LGC-40863 LG Chemical	30~50 g/ha post rice, turf, cereals	JP07196629 EP658549

10.2.5.2 Structural Evolution of Pyrimidyl(thio)oxybenzoate Acetolactate Synthase Inhibitors

Pyrimidyl(thio)oxybenzoate ALS inhibitors released since 1990 are summarized in Fig. 10. Almost all compounds have a variety of substituents at 2-position of the benzene ring; for example, ester [**221, 229, 230, 240–242, 248, 250**], amide [**222, 236, 237**], iminomethyl [**253–255, 257**], hydrazonomethyl [**256, 258, 259**], substituted alkyl groups [**260–267**] and so on. There are many ALS inhibitors modified by acetal moieties [**261, 262**], cyanohydroxymethyl groups [**266, 267**] or oxazole rings [**251, 252**] at *ortho*-position, because these substituents are easily metabolized to the active species with the carboxy group in plants and soil. Furthermore, various substituents have been introduced at the 3-position of the benzene ring to enhance their crop safety.

Since the discovery of pyrithiobac-sodium with practical herbicidal activity and the introduction of various heterocycles at the 2-position of the pyrimidine ring, great attempts have been made to find more potent herbicides. Structural modifications of pyrimidyl(thio)oxy-substituted benzologues and -heterocycles are chronologically shown in Fig. 11. Among them, pyriftalid [**276**] is a practical ALS inhibitor. Several 2-pyrimidyloxycyclopentane-1-

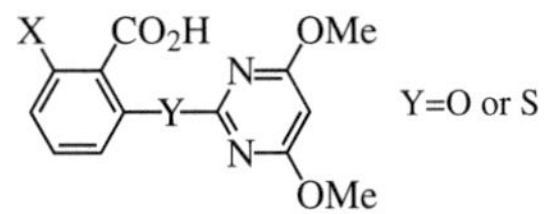

206 : 1990; X=Cl, sodium salt (63g/Ipp,Xas/cotton) <pyrithiobac-sodium>
207 : 1990; X=4-Cl-C_6H_4 (500g)
208 : 1991; X=C_6H_5S
209 : 1991; X=Ph (125g/post/Soh,Gaa,Blackgrass)
210 : 1992; X=EtON=CH (250g/post/Brl)
211 : 1992; X=2-Me-thiazol-5-yl (15.6g/post/Amr)
212 : 1992; X=Ac (1kg/pre/Eco,Mov,Scj)
213 : 1992; X=CH_2=CH (60g/post/wheat)
214 : 1993; X=CH_3CH=CH (1kg/Ecc)
215 : 1995; X=allyl (1kg/Ecc,Mov/rice)
216 : 1995; X=(2-MeO-pyridin-6-yl)
217 : 1996; X=morpholino (63g/post/Sef,Xap,Abt,Amr)
218 : 1997; X=$MeOOCCH_2$ (100g)
219 : 1997; X=3-$MeSO_2C_6H_4$ (62.5g/post/Avf,Ecc,Alm)
220 : 1997; X=2-$MeSO_2$-pyridin-6-yl

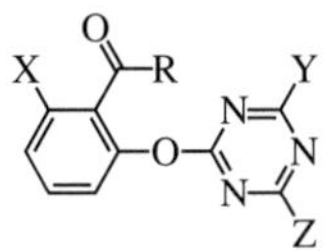

221 : 1990; R=MeO, X=H, Y=Me, Z=MeNH (1kg/See)
222 : 1993: R=Me_2N, X=Cl, Y=Z=MeO (125g/Ips)

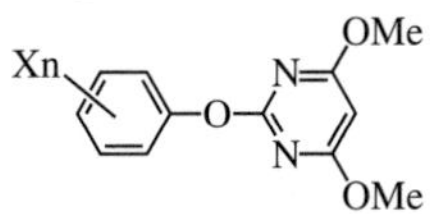

223 : 1990; Xn=2-CF_3SO_2NH-3-Cl (pre,post/Gaa,Abt)
224 : 1991; Xn=2-Cl (100g/post/Eco,Sap/rice)
225 : 1994; Xn=4-(3-CF_3-C_6H_4O)-C_6H_4 (4kg/Mov)
226 : 1994; Xn=2-(oxazol-5-yl) (Dic)

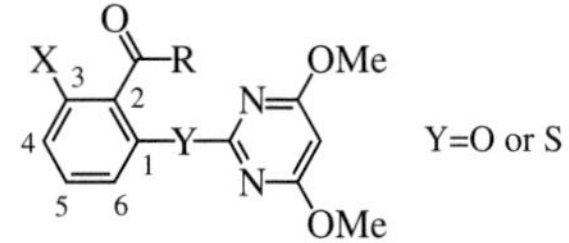

227 : 1991; R=Me_2NNH, X=H (15g/Cha/rice)
228 : 1991; R=Me_2NO, X=H (31g/pre/Aba)
229 : 1991; R=MeO, X=MeON=(Me)C (1kg/pre/Ecc) <pyriminobac-methyl>
230 : 1991; R=MeO, X=(2-MeOOC-C_6H_4)OOC (500g/Ecc)
231 : 1991; R=(1,3-dioxan-2-yl)ethyl, X=H (125g/Ecf,Avf/upland)
232 : 1992; R=Me_2NCH=(Me)C, X=H
233 : 1992; R=$MeSO_2$(Me)NO, X=H (500g/pre/Ecc/rice)
234 : 1992; R=NH_2NHCO, X=H (250g/rice)
235 : 1992; R=$PhCH_2O$, X=Ph (1kg/Ecc,Mov/rice)
236 : 1992; R=HOCO(Me)CHNH, X=H (3kg/Ecc)
237 : 1992; R=i-$PrSO_2NH$, X=H (2kg/pre/Soh/soya)
238 : 1992; R=$PhSO_2NH$, X=H
239 : 1993; R=Me_2N, X=pylazol-1-yl
240 : 1993; R=MeO, X=F (125g/Sev/soya)
241 : 1993; R=MeO, X=(1,3-dioxolan-2-yl)ethyl (1kg)
242 : 1993; R=i-PrCOO(Me)CHO, X=Cl (1kg/Ecc,Mov,Scj/rice)
243 : 1993; R=MeO, X=PhCH=CH (pre,post)
244 : 1994; R=Me_2NO, X=Ph
245 : 1994; R=MeO(Me)C=N, X=Cl
246 : 1995; R=HOOC (300g/pre,post/Sia,Chs,Lom, Pam)
247 : 1995; R=Me_2CNO, X=2-MeO-pyridin-6-yl
248 : 1995; R=t-BuO, X=1-(dioxolan-2-yl)propyl (320g/Ecc/rice)
249 : 1995; R=Me_2(MeO)C, X=Ac (32g/pre/Ecc,Brl,Scj/rice)
250 : 1995; R=MeO, X=t-Bu-C_6H_4ON=(Me)C (30g/Ecc/rice)

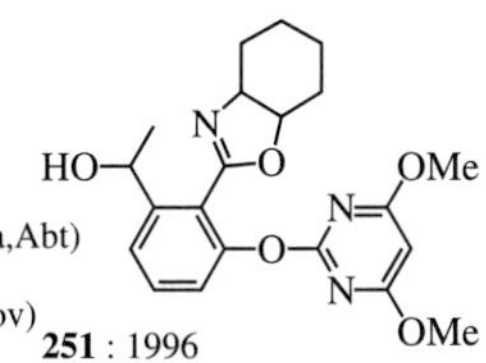

251 : 1996 (250g/pre/Sef,Abt,Ama)

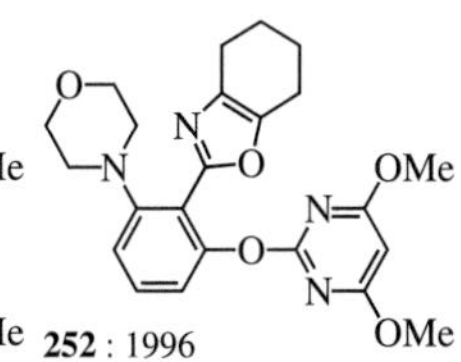

252 : 1996 (250g/pre/Ecc,Abt,Sef)

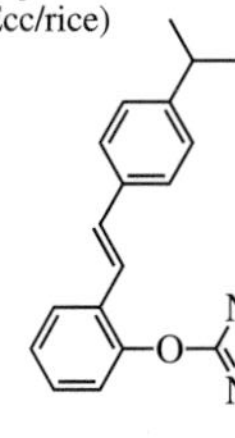

268 : 1994 (320g/Ecc/rice)

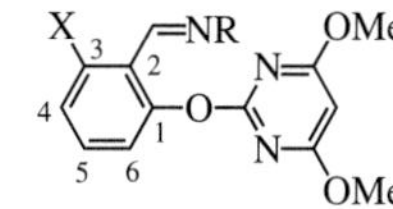

253 : 1990; R=C_6H_5CH=CHO, X=H (300g/post/Ecc,Sap/rice)
254 : 1991; R=EtCOO(Me)CHO, X=H (300g/pre/Ecc,Sch/soya,cotton,corn)
255 : 1991; R=PhCH=CHO, X=H (300g/Amr)
256 : 1991; R=PhNH, X=F
257 : 1992; R=Ph, X=H (3kg/pre/Avf)
258 : 1992; R=NH_2, X=H (300g/pre/Sev,Soh/cotton,soya)
259 : 1992; $C_6H_5CH_2NH$, X=H (300g/pre/Soh,Sev/soya,cotton,corn)

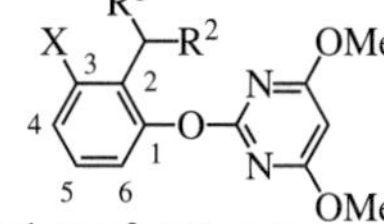

260 : 1991; R^1=H, R^2=HO, X=H (500g/Pac,Scj,Sat,Mov/rice)
261 : 1992; R^1=R^2=-O$(CH_2)_2$O-, X=H (31.3g/Pac)
262 : 1992; R^1=R^2=-O$(CH_2)_2$O-, X=Br (500g/Pac,Dia)
263 : 1992; R^1=R^2=$C_8H_{17}O$, X=H (500g/pre/Cha,Ecc,Sev,Dia)
264 : 1993; R^1=5-Me-(1,3,4-thiadiazol-2-yl), X=H (200g/pre/Ecc)
265 : 1994; R^1=H, R^2=N_3, X=Cl (1kg/Ecc/rice)
266 : 1995; R^1=AcO, R^2=NC (1kg/pre)
267 : 1995; R^1=HO, R^2=NC (50g/post/Amr,Ecc/soya,corn,rice,cotton)

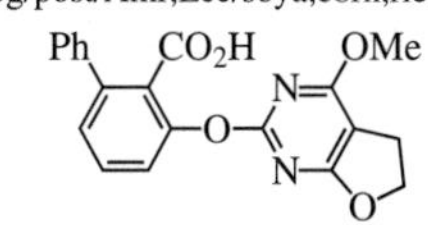

269 : 1992

No. : Year; Example (Dose per ha/Application/Weeds/Crops)

Fig. 10. Structural evolution of pyrimidyl(thio)oxybenzoate ALS inhibitors and related compounds since 1990 (1)

<Pyrimidyl(thio)oxy-substituted benzologues>

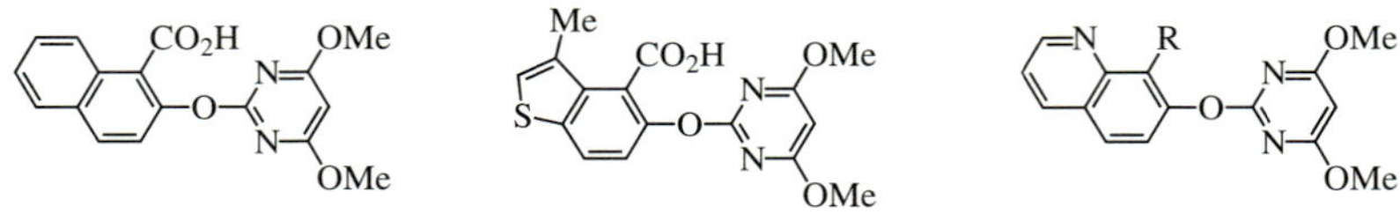

270 : 1990 (250g/Dia,Amr,Cha)

275 : 1994 (1kg/Mov,Ecc,Scj)

271 : 1991; R=PhNHN=CH
272 : 1994; R=Me_2NCO (Gaa/wheat)

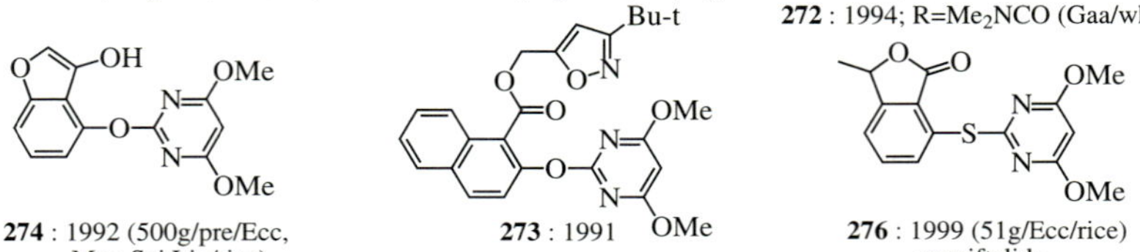

274 : 1992 (500g/pre/Ecc, Mov,Scj,Lip/rice)

273 : 1991

276 : 1999 (51g/Ecc/rice) <pyriftalid>

<Pyrimidyl(thio)oxy-substituted heterocycles>

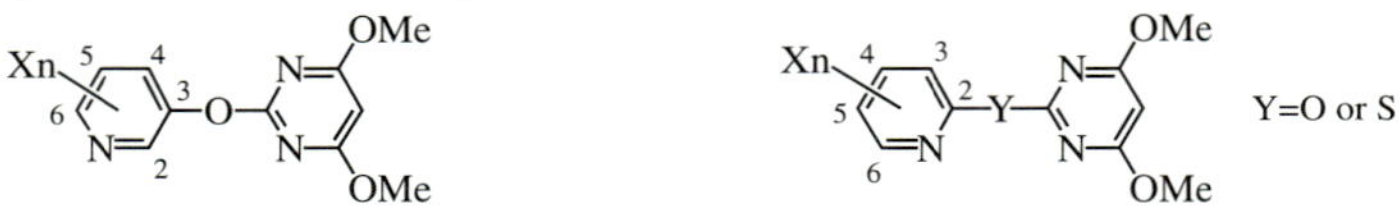

277 : 1990; Xn=2-PhC(Me)=NOCO
278 : 1990; Xn=2-Me_2C=NOCO
279 : 1991; Xn=2-$(MeO)_2$CH (3kg/Ecc,Sev,Amr/soya,cotton)
280 : 1992; Xn=4-EtOOC$(CH_2)_2$ (500g/Pac,Mov/rice)
281 : 1992; Xn=2-MeOOC (3kg/Amr,Ala/soya,cotton,corn)
282 : 1992; Xn=2-$HOCH_2$ (Amr,Ala/soya,cotton,corn)
283 : 1993-95; Xn=2-$MeCO_2$-6-Me_2N (100g/Mov,Cyd,Ecc)
284 : 1994; Xn=2-$PhCH_2$S (1kg/Ecc)
285 : 1995; Xn=2-MeOOC-6-pyrrolidino (Ecc,Mov,Bul)

286 : 1992; Xn=3-PhNHN=CH
287 : 1992; Xn=3-MeOOC-4-Ph (1kg/Ecc,Mov,Scj)
288 : 1992; Xn=3-HOOC (pre,post)
289 : 1994; Xn=3-MeOOC-6-Me (100g/Ecc)
290 : 1993,1994; Xn=3-MeOOC-5-pyrrolidino (1kg/pre/Ecc,Mov,Scj/rice)
291 : 1994; Xn=3-HOOC-4-thienyl-5-Cl (100g,Ecc)
292 : 1994; Xn=6-$PhCH_2$S (5kg/Ecc)
293 : 2000; Xn=3-H_2C=CMe (630g/Stm)

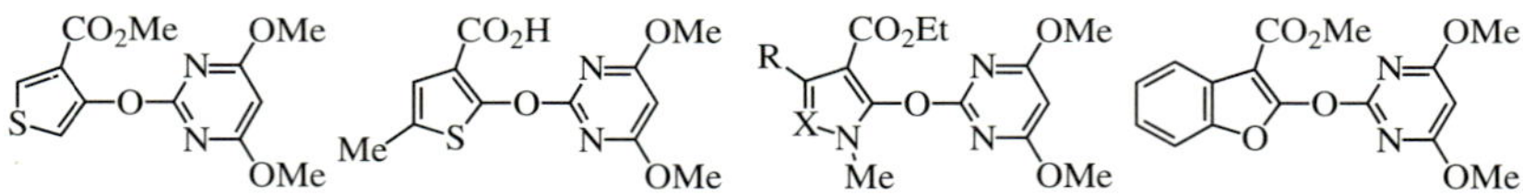

294 : 1990

295 : 1991 (1kg/Ecc,Cyd, Mov,Scj)

296 : 1991; R=H, X=CH
297 : 1992; R=Ph, X=N (1kg/Mov,Pac,Scj/rice)

298 : 1993 (4kg/pre/Ecc, Mov,Scj)

No. : Year; Example (Dose per ha/Application/Weeds/Crops)

Fig. 11. Structural evolution of pyrimidyl(thio)oxybenzoate ALS inhibitors and related compounds since 1990 (2)

carboxylate and cyclohexane-1-carboxylate derivatives are illustrated in Fig. 12. There are many compounds with a fluorine atom introduced at 1-position of the cycloalkane-1-carboxylates.

Pyrimidyl(thio)oxyacetate derivatives without the ring system have been actively studied (Fig. 13). More than ten kinds of pyrimidyl(thio)oxy-acetates [**334–349, 353–356**] and -acetamides [**357–359**] were synthesized in the early 1990s. These compounds have various substituents such as alkyl, phenoxy and alkylidene groups at the α-position of carbonyl groups. Apparently, the alkoxycarbonyl groups are not always necessary for herbicidal activity, because several compounds [**361–365**] without such groups have also been proposed. The right column of Fig. 13 shows unusual ALS inhibitor-like pyrimidine

<Cyclopentyloxypyrimidine type>

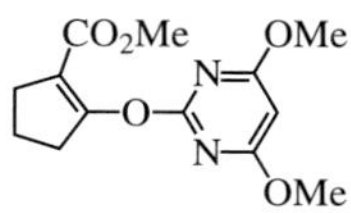

299 : 1990 (63g/post/Son,Roi)

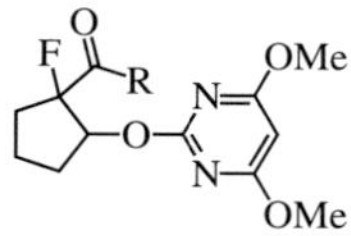

302 : 1993; R=EtO (1kg/post/Dia,Ecc,Sev,Abt)
303 : 1994; R=HO (1kg/Dic)
304 : 1994; R=$PhSO_2NH$
305 : 1994; R=(3-Me-C_6H_4)SO_2NH (1kg/Ecc)
306 : 1994; R=$MeSO_2N(Me)NH$ (1kg/Dic)
307 : 1994; R=$PhSO_2NHNH$ (1kg/Ecc)

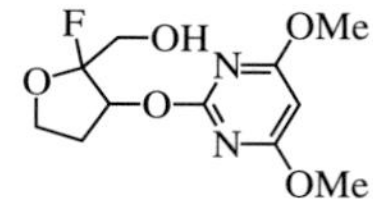

309 : 1993 (soya,corn,wheat)

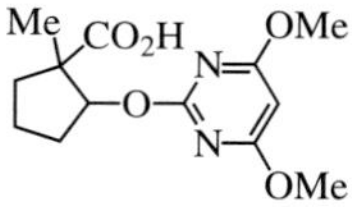

300 : 1992 (4kg/Ecc,Scj)

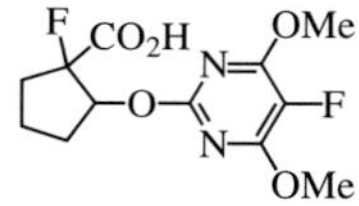

310: 1994 (1kg/Ecc)

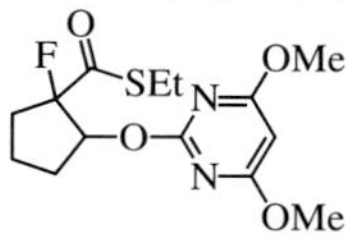

301 : 1992 (100g/pre/Ecc,Abt, Sev/soya,cotton)

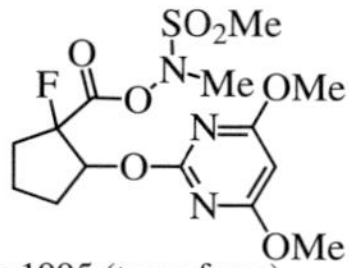

308 : 1995 (trans form) (1kg/pre/Ecc,Scj,Mov,Lip)

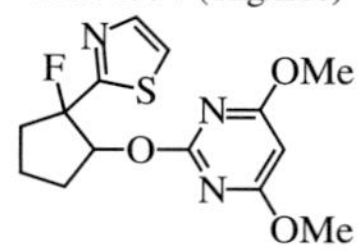

311 : 1995 (1kg/pre/Ecc,Lip,Scj/rice)

<Cyclohexyloxypyrimidine type>

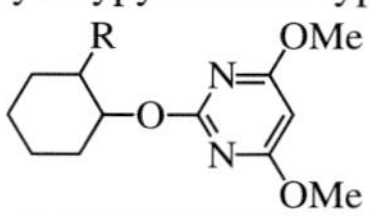

312 : 1991; R=MeOOC (300g/post/Abt,See)
313 : 1991; R=HOOC (200g/post/Ecc,Dic, Abt/rape,cotton)
314 : 1991; R=CHO (200g/post/Ecc,Dic,Abt)
315 : 1992; R=$HOCH_2$ (100g/post)
316 : 1994; R=Me_2C=NNHCO (1kg/post/Ecc, Dic,Abt,Sev,Amr,Xap)
317 : 1994; R=$MeSO_2NHCO$ (160g/post/Ecc, Dic,Abt/cotton)
318 : 1995; R=HOOC (cis form) (pre/Ecc,Sia)

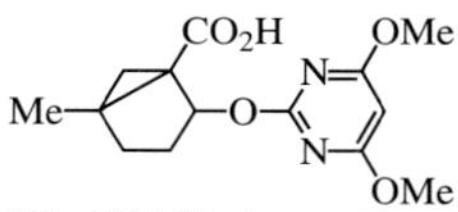

321 : 1994 (5kg/pre,post/soya)

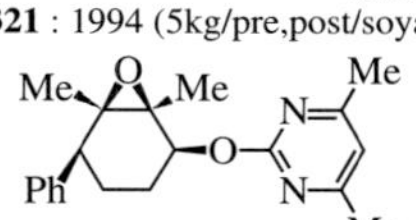

322 : 1998 (300g/Ecc/wheat)

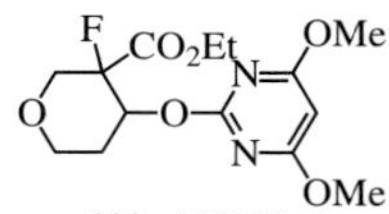

329 : 1993(1kg)

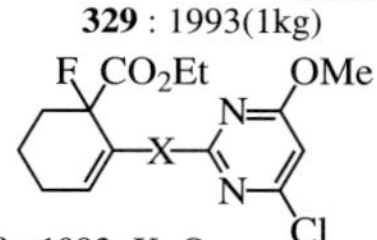

330 : 1993; X=O (5kg/pre,post/Ecc,Dic,Sev)
331 : 1993; X=S (1kg/pre,post/Ecc,Dic,Sev)

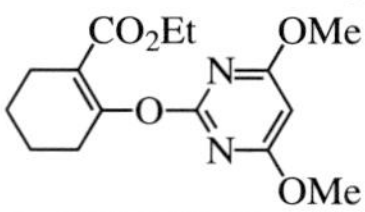

319 : 1992 (200g/Amr,Pon/ corn,sunflower,wheat)

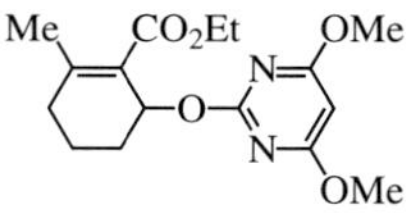

320 : 1993

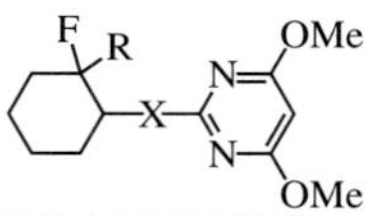

323 : 1993; R=MeOOC, X=S (5kg/pre/Dic,Ecc,Sev,Amv)
324 : 1993; R=EtSCO, X=O
325 : 1993; R=$HOCH_2$, X=O (Ecc,Dia,Sev)
326 : 1994; R=$(MeO)_3P$=NCO, X=O
327 : 1994; R=AcNHCO, X=O (1kg/Cyd)
328 : 1994; R=MeNHCOMHCO, X=O (1kg/post/Ecc,Abt,Dic,Sev)

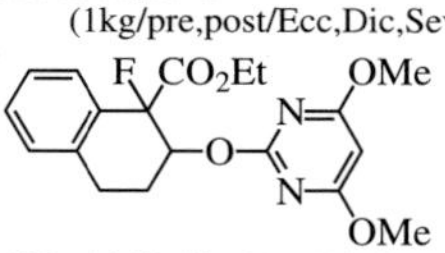

332 : 1993 (5kg/Ecc,Dia,Sev)

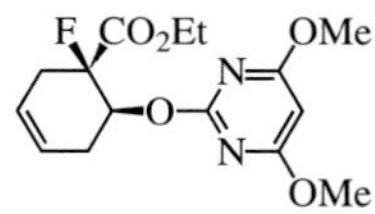

333 : 1995 (1kg/Ecc,Mov,Scj)

No. : Year; Example (Dose per ha/Application/Weeds/Crops)

Fig. 12. Structural evolution of cycloalkyl(thio)oxypyrimidine ALS inhibitors since 1990

derivatives [**366–381**], which have neither an ether nor a thioether moiety. Among the compounds in Fig. 13, α-hydroxybenzylpyrimidine derivative [**371**] shows a strong herbicidal activity in paddy rice.

Only five kinds of bis(pyrimidyloxy)benzoate ALS inhibitors [**382–386**] have been known since 1990, as illustrated in Fig. 14. Miscellaneous compounds that may be classified as ALS inhibitors are also demonstrated.

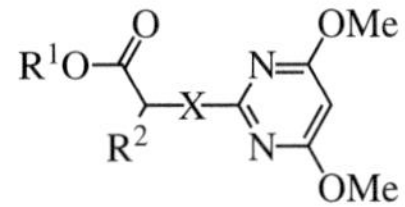

334 : 1991; R^1=Me, R^2=allyl, X=O (1kg/post/Abt)
335 : 1991; R^1=allyl, R^2=Pr, X=S (2kg/pre/Ecc,Dia)
336 : 1992; R^1=Me, R^2=PhO, X=S (100g/Vep/corn,soya,wheat)
337 : 1992; R^1=H, R^2=i-Pr, X=O (2kg/Ecc)
338 : 1992; R^1=Et, R^2=$PhCH_2O(Me)CH$, X=O (2kg/Ecc,Scj)
339 : 1993; R^1=$PhCH_2OCOCH_2$, R^2=i-Pr, X=O (300g/Sia,Ecc,Lom)
340 : 1993; R^1=Me, R^2=t-Bu, X=O
341 : 1993; R^1=Me, R^2=$(MeO)_2CH$, X=S (1kg/Ecc,Sev,Amr)
342 : 1993; R^1=H, R^2=i-Pr, X=S (2kg/post/Sia,Stm)
343 : 1994; R^1=H, R^2=$PhCH_2O(Me)(Ph)C$, X=O (125g/Goh)
344 : 1994; R^1=Me, R^2=$(MeO)_2PO$, X=O (1kg/rice)
345 : 1995; R^1=Me, R^2=PhC(AcO)H, X=S
346 : 1995; R^1=Me, R^2=$Ph(Me)(N_3)C$, X=O (3kg/Ecc,Lom)
347 : 1996; R^1=Me, R^2=1-OH-c-pentyl, X=O (2.5kg/post/Ecc)

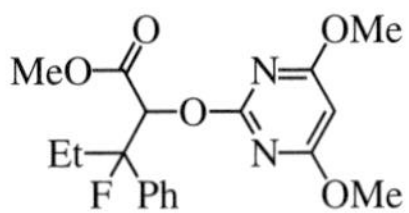

348 : 1992 (100g/pre/Avf,Sev/ sugar beet,soya,corn,rice)

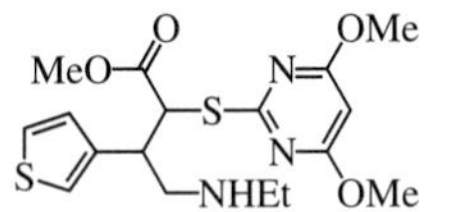

349 : 1992

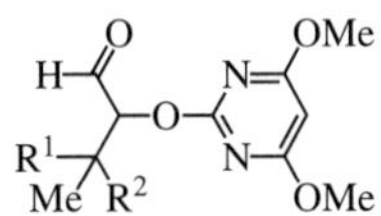

350 : 1993; R^1=Me, R^2=H
351 : 1994; R^1=thienyl, R^2=F

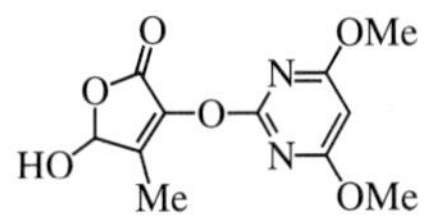

352 : 1994

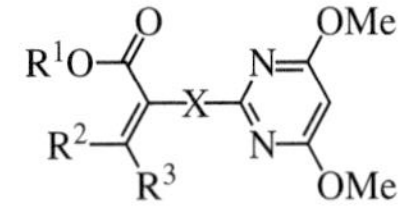

353 : 1991; R^1=Me, R^2=Ph, R^3=H, X=S (100g/post/Gaa,See,Sol/corn,wheat)
354 : 1991; R^1=Me, R^2=3-thienyl, R^3=H, X=S (125g)
355 : 1993; R^1=Me, R^2=OH, R^3=H, X=O (pre,post)
356 : 1993; R^1=Et, R^2=R^3=Me, X=O

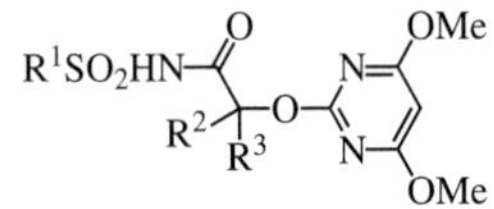

357 : 1993; R^1=Me, R^2=t-Bu, R^3=H
358 : 1993; R^1=Me_2N, R^2=i-Pr, R^3=MeO (soya)
359 : 1994; R^1=c-Pr, R^2=$Me_2(MeO)C$, R^3=H (corn,soya)

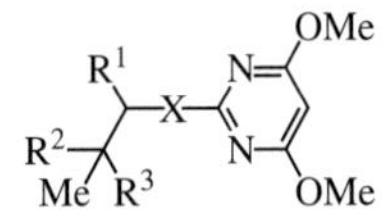

360 : 1991; R^1=Me, R^2=EtOOC, R^3=H, X=O (2.5kg/Ecc,Dic)
361 : 1993; R^1=MeSSC, R^2=R^3=Me, X=O
362 : 1994; R^1=CH_2Cl, R^2=Me, R^3=H, X=S
363 : 1994; R^1=$HOCH_2$, R^2=Me, R^3=H, X=O
364 : 1994; R^1=HOC(NC)H, R^2=R^3=Me, X=S

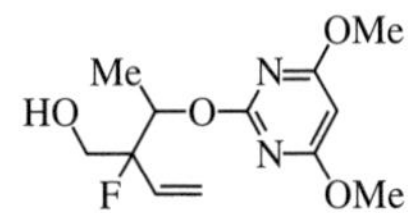

365 : 1994

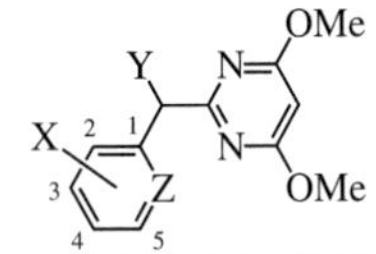

366 : 1991; X=2-NaOOC-3-Me, Y=H, Z=CH (400g/Ecc/cotton)
367 : 1991; X=2-MeO_2C-3-thienyl, Y=H, Z=N
368 : 1992; X=2-MeOOC, Y=H, Z=N (100g/pre/Gaa)
369 : 1996; X=4-Cl, Y=tetrazol-5-yl, Z=N (1kg/post/Sia,Avs,Stm)
370 : 1999; X=2-CH_2FSO_2NH, Y=H, Z=CH
371 : 2000; X=2-F_2HCSO_2NH-3-$MeOCH_2$, Y=HO, Z=CH (16g/pre/Ecc,Mov,Scj/rice)

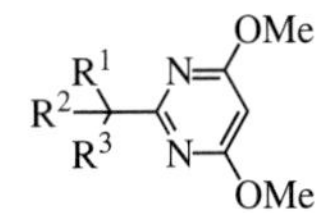

372 : 1991; R^1=HO, R^2=3-Pr-C_6H_4, R^3=H (1kg/pre/Son,Cye,Blackgrass)
373 : 1991; R^1=MeO, R^2=Pr, R^3=H (300g/pre/Brt)
374 : 1992; R^1=HO, R^2=$Me_2(EtO)C$, R^3=H (200g/pre/Dic/cotton)
375 : 1992; R^1=i-Pr, R^2=$MeSO_2NHCO$, R^3=H (1kg/Ecc,Mov,Scj)
376 : 1992; R^1=$MeSO_2NH$, R^2=i-Pr, R^3=H (250g/Mov,Ecc,Scj)
377 : 1992; R^1=MeOOC, R^2=Ph, R^3=F (125g/pre,post/Avf,Vep)
378 : 1994; R^1=$MeSO_2NH$, R^2=t-Bu, R^3=H (250g/pre/Ecc,Amr,Cha)
379 : 1994; R^1=MeOOC, R^2=Ph, R^3=MeO
380 : 1995; R^1=$MeSO_2NH$, R^2=i-Pr, R^3=H (Ecc,Mob)
381 : 1995; R^1=$MeOSO_2$, R^2=R^3=H

No. : Year; Example (Dose per ha/Application/Weeds/Crops)

Fig. 13. Structural evolution of 2-substituted dimethoxypyrimidine ALS inhibitors since 1990

<Bis(pyrimidyloxy)benzoate type>

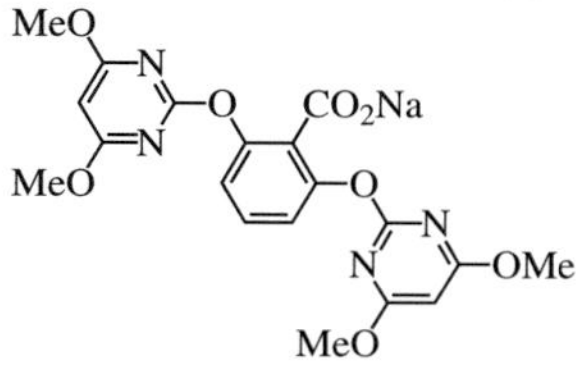

382 : 1992 <bispyribac-sodium> (40g/Stm,Gas,Vim,Mai,Ala/wheat,barley)

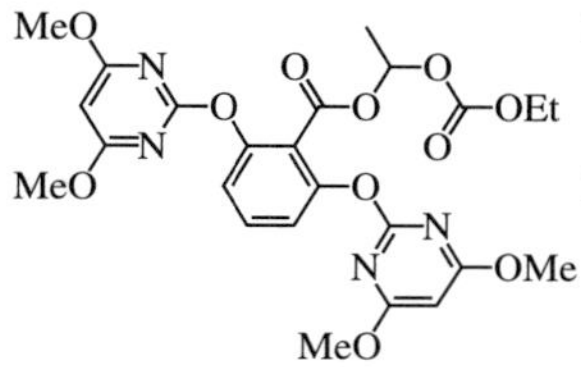

383 : 1994 (4kg/Amr)

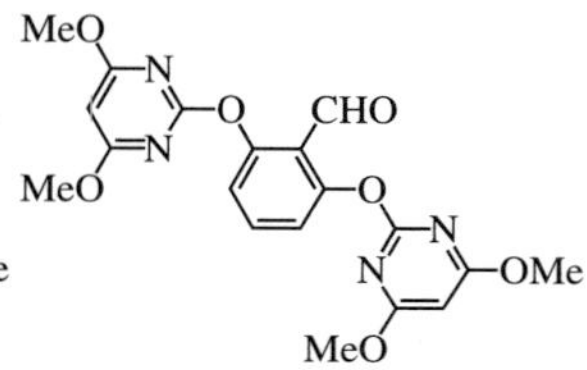

384 : 1992

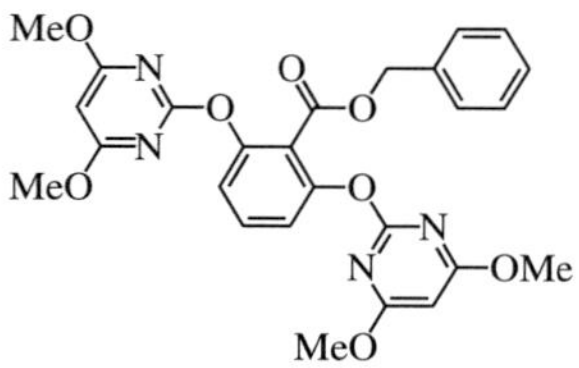

385 : 1993 (1kg/Ecc)

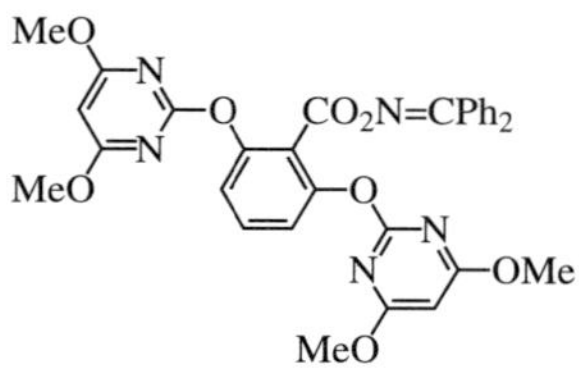

386 : 1993 <pyribenzoxim>

<Miscellaneous compounds>

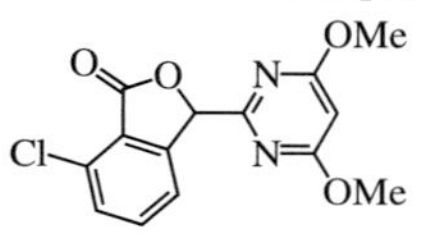

387 : 1991

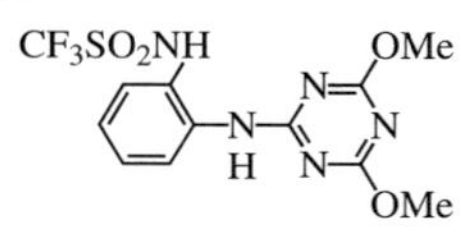

388 : 1993 (250g/pre)

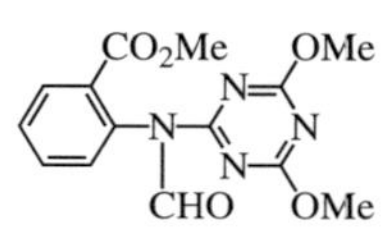

389 : 1993 (1kg/Ecc,Mov,Scj)

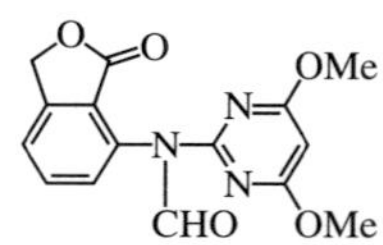

390 : 1993 (1kg/pre,post/Ecc, Pel,Amv,Cha,Cyi)

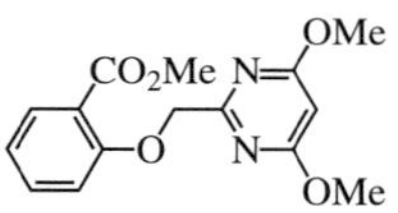

391 : 1992 (1kg/Scj,Ecc)

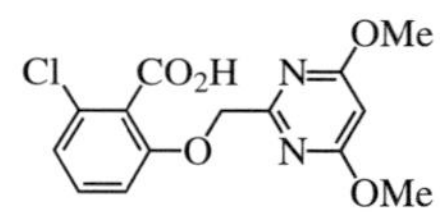

392 : 1995

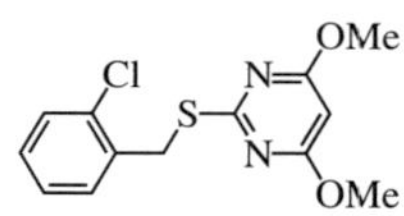

393 : 1993

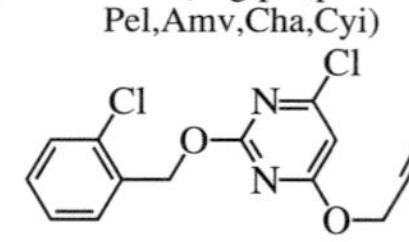

394 : 1995 (50g/Amr)

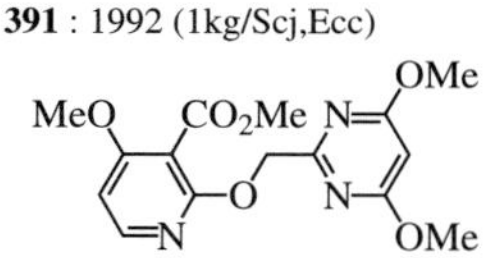

395 : 1998 (10g/pre/Gas,Vep,Lom, Cha,Pob,Avf/wheat,barley)

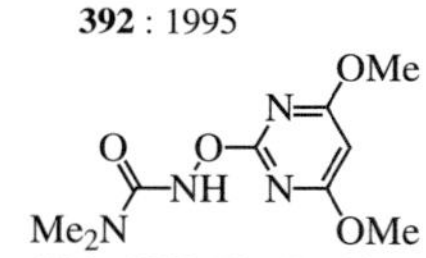

396 : 1993 (1kg/Ecc/rice)

No. : Year; Example (Dose per ha/ Application/Weeds/Crops)

Fig. 14. Bis(pyrimidyloxy)benzoate ALS inhibitors and miscellaneous compounds disclosed since 1990

10.2.5.3 Major Synthetic Routes for Pyrimidyl(thio)oxybenzoate Acetolactate Synthase Inhibitors

A synthetic route for pyriminobac-methyl is shown in Scheme 5. 3-Benzyloxybenzophenone ketal (*54*) is regioselectively carboxylated at the 2-position with carbon dioxide *via* lithiated benzene prepared by butyl lithium or LDA. After esterification and deprotection of the ketal, the benzyl group is removed by hydrogenation to yield methyl 2-acetyl-6-hydroxybenzoate (*55*). Condensation of *55* with 4,6-dimethoxy-2-methanesulfonylpyrimidine

<Synthesis of pyriminobac-methyl>

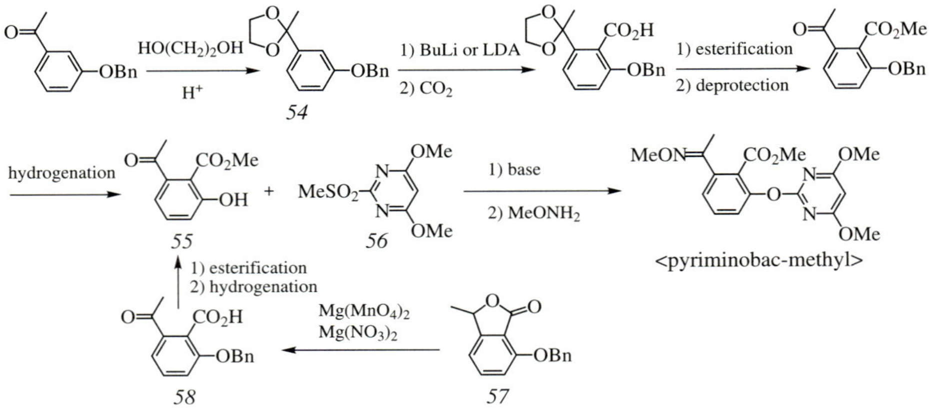

<Synthesis of cyclopentyloxypyrimidine, **302**>

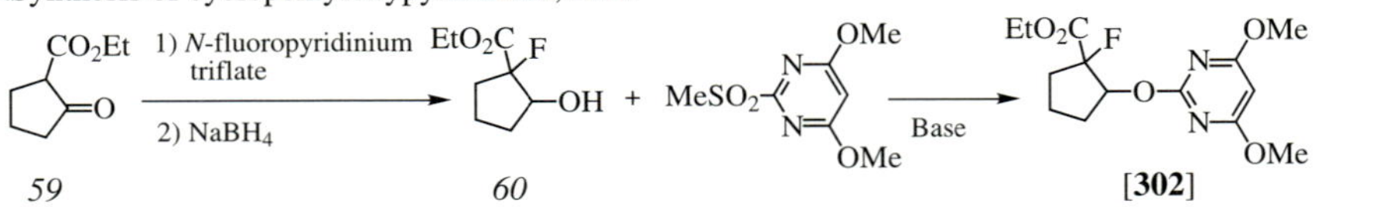

Scheme 5. Major synthetic routes for pyriminobac-methyl and cyclopentyloxypyrimidine, **302**

(*56*) and subsequent methoxyimination of the acetyl group gives pyriminobac-methyl. Moreover, there is an alternative synthetic pathway for the key intermediate (*55*), which consists of three steps: oxidative hydrolysis of 7-benzyloxy-3-methylphthalide (*57*) in the presence of magnesium permanganate and magnesium nitrate, esterification and deprotection of *58*.

Scheme 5 also shows a synthetic scheme for compound **302** as an example of 1-fluoro-2-(4,6-dimethoxypyrimidyloxy)cycloalkane-1-carboxylate derivative depicted in Fig. 12. Ethyl 2-oxocyclopentane-1-carboxylate (*59*) is fluorinated at 1-position by *N*-fluoropyridinium triflate and the carbonyl group is reduced by sodium borohydride to give ethyl 1-fluoro-2-hydroxycyclopentane-1-carboxylate (*60*), which reacts with 4,6-dimethoxy-2-methylsulfonylpyrimidine to yield the desired compound **302**.

10.2.6 Imidazolinone Acetolactate Synthase Inhibitors

10.2.6.1 Practical Imidazolinone Acetolactate Synthase Inhibitors

Imidazolinone ALS inhibitors include six practical herbicides, imazamethabenz-methyl, imazapyr, imazethapyr, imazapic, imazamox and

Table 5. Practical imidazolinone ALS inhibitors

Chemical structure	ISO name Code No. Company	Dose Appl. method Target crops	Patent No.
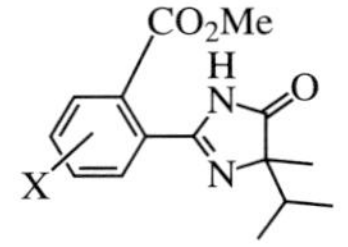	imazamethabenz-methyl AC-222293, AC-293 ACC (X=4- and 5-Me 50% mixture)	350~1000 g/ha post cereals	US4608437 US4798619
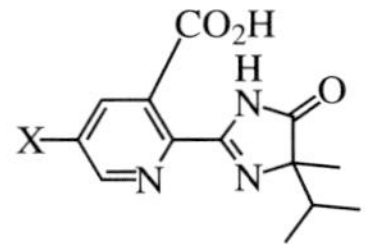	imazapyr AC-252925, AC-925 ACC (X=H)	280~1680 g/ha pre, post non-crop	US4658030
	imazethapyr AC-263499 ACC (X=Et)	35~45 g/ha pre, post soybean, turf, potato	US4816588 US2964529
	imazapic AC263222, AC222 ACC (X=Me)	50~105 g/ha pre, post soybeans, peanuts	GB2174395
	imazamox AC-263 ACC (X=MeOCH$_2$)	35~45 g/ha post soybeans, peanuts, maize, corn	JP08337571
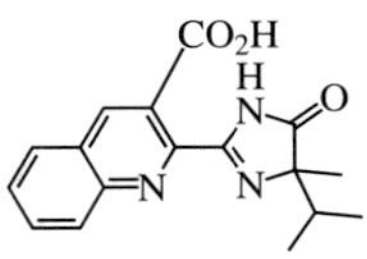	imazaquin AC-252214 ACC	140~150 g/ha pre, post soybeans, turf	US4656283

imazaquin. Development of imidazolinone ALS inhibitors was the unrivaled research area of ACC in spite of several proposals from other companies. All practical herbicides have the same imidazolinone ring system with methyl and isopropyl groups at the 5-position. Additionally, the carboxy group or its mimic is introduced at *ortho*-position of the aromatic rings as well as being present in other ALS inhibitors as shown in Table 5.

Imazamethabenz-methyl is applied at rates of 350–1000 g/ha with post-emergence in wheat and barley. It controls *Apera spica-venti* and annual bluegrass, despite weak activity against water foxtail, catchweed bedstraw and wild buckwheat. The non-selective herbicide, imazapyr, is used for pre- or post-emergence control of annual and perennial grasses, sedges, and broadleaf weeds as well as for bush and deciduous tree species at rates of 280–1680 g/ha. Imazethapyr was launched in 1987 for pre- and post-emergence control of grass and broadleaf weeds in soybeans and other legumes. It shows effective herbicidal activity against southern crabgrass, livid amaranth, fall panicum, giant foxtail and yellow foxtail at the low rates of 35–45 g/ha. Imazapic is an early post-emergence herbicide for control of a wide range of weeds including

Euphorbia esula, *Cassia obtusifolia*, *Desmodium tortuosum* and *Panicum texanum* at 50–105 g/ha in soybeans, peanuts and sugarcane. Imazamox is a broad-spectrum imidazolinone herbicide that gives contact and residual control of all major soybean weeds such as wild mustard, *Polygonum lapathifolium*, common lambsquarters, velvetleaf, slender amaranth, fall panicum, giant foxtail and green foxtail with post-emergence at 35–45 g/ha. Imazaquin is a quinoline-3-carboxylic acid derivative and used for pre-plant, pre-emergence or early post-emergence control of broadleaf weeds in soybeans, applied at 140–150 g/ha, the 5-chlorinated derivative of imazaquin is more active in spite of inferior safety for soybeans.

10.2.6.2 Structural Evolution of Imidazolinone ALS Inhibitors

Development of imidazolinone herbicides started with a random screening test, in which 2,3-dimethyl-2-(phthalimido)butanamide was found to have sufficient activity to warrant investigations. The phthalimide was cyclized to a tricyclic imidazoisoindole, which was a significant lead compound for further evolution leading to imidazolinone chemistry. As a result, the base-catalyzed ring-opening reaction of the imidazoisoindole resulted in the first candidate, 2-(4-isopropyl-4-methyl-5-oxo-2-imidazolin-2-yl)benzoate, for novel imidazolinone ALS inhibitors. Detailed structural optimization of the benzene ring was performed to discover imazamethabenz-methyl as a practical herbicide. Structural modifications of imidazolinone ALS inhibitors are depicted in Fig. 15. Investigation of these class herbicides declined in the 1990s.

10.2.6.3 Major Synthetic Routes for Imidazolinone Acetolactate Synthase Inhibitors

The synthetic route for imazamethabenz-methyl is depicted in Scheme 6. 5-Methylphthalic anhydride (*61*) reacts with 2-amino-2,3-dimethylbutylonitrile (*62*), which is prepared by condensation of methyl(isopropyl)ketone and potassium cyanide in ammonia water in high yield, to give a regio-isomeric mixture of phthalamic acids (*63*, *63′*). Ring closure of the mixture by thionyl chloride produces a single isomer (*64*), of which hydrolysis in sulfuric acid and subsequent cyclization with aqueous sodium hydroxide give a mixture of two isomers (*65*, *65′*). Finally, non-regioselective hydrolysis by sodium methoxide yields a mixture of 4-methyl- and 5-methyl-2-(4-isopropyl-4-methyl-5-oxoimidazolin-2-yl)benzoates, namely, imazamethabenz-methyl. A key intermediate for production of imazaquin is quinoline-2,3-dicarboxylate (*68*); see Scheme 6 for a representative method. The additional reaction of aniline (*66*) to acetylene dicarboxylate or condensation reaction of aniline with 2-oxosuccinate gives 2-anilinomaleate (*67*), which is cyclized to *68* using Vilsmeier reagent. Quinoline dicarboxylate thus obtained is readily converted to imazaquin similar to the method for imazamethabenz-methyl.

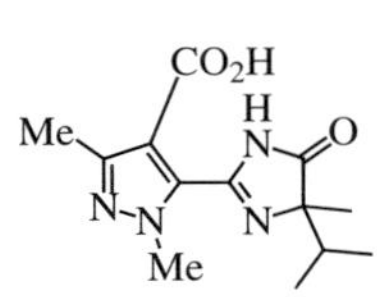

397 : 1990

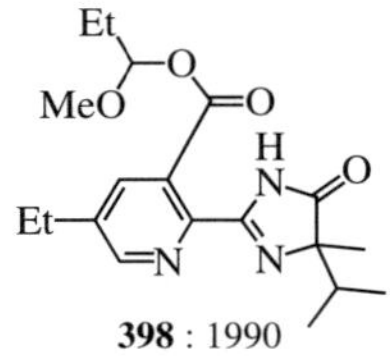

398 : 1990

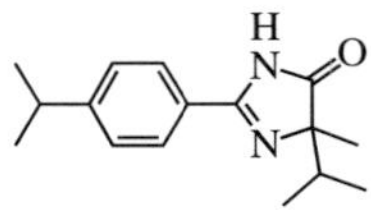

399 : 1990 (1kg/Ecc)

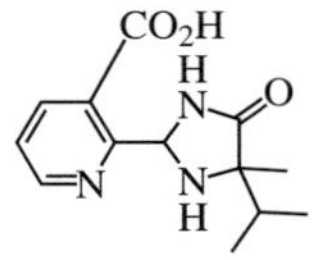

400 : 1990 (100g)

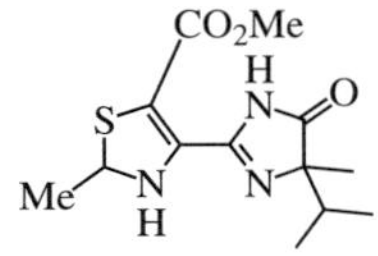

401 : 1990 (3kg/Stm)

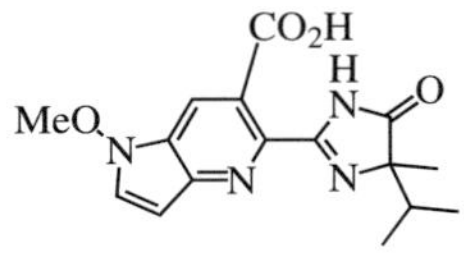

402 : 1991

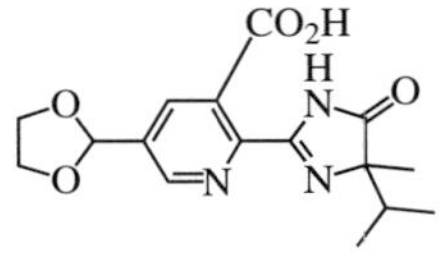

403 : 1991 (Ecc,Amr/corn, cotton,sugar beet,soya)

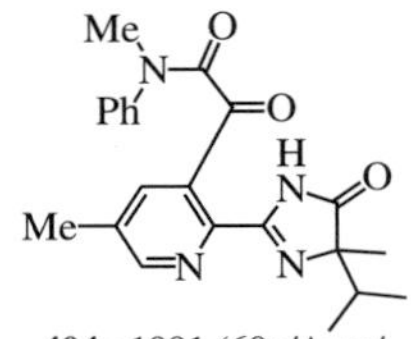

404 : 1991 (60g/Amv/ corn,wheat)

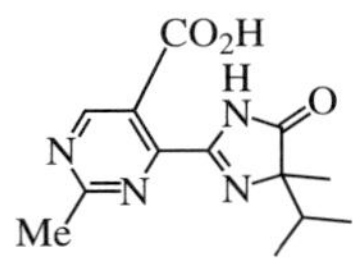

405 : 1992 (pre,post/rice)

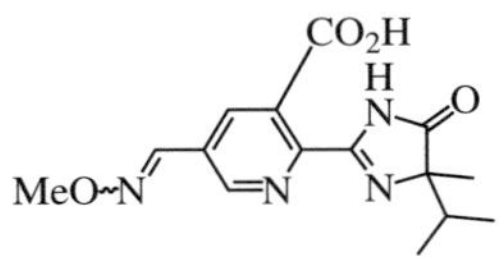

406 : 1993 (250g/pre)

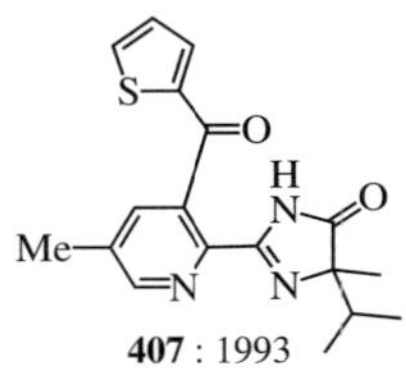

407 : 1993

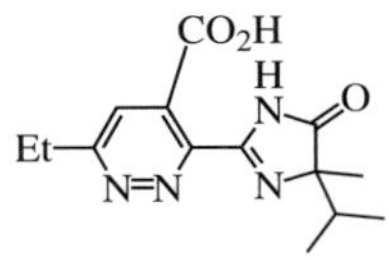

408 : 1994

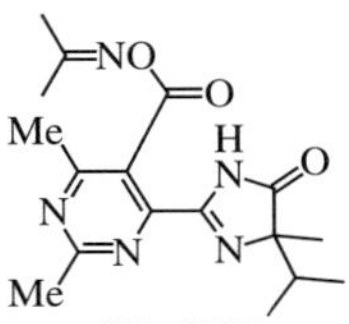

409 : 1994

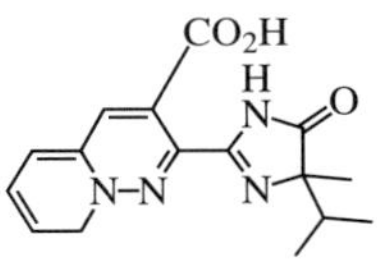

410 : 1994

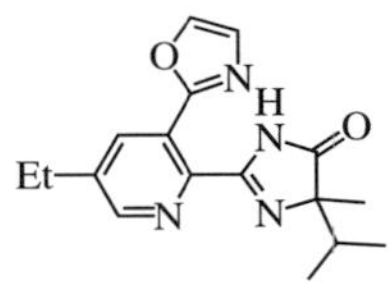

411 : 1995 (100g/Aba)

No. : Year; Example
(Dose per ha/Application/
Weeds/Crops)

Fig. 15. Imidazolinone ALS inhibitors disclosed since 1990

10.3 Carotenogenesis Inhibitors

Prenyltransferase, phytoene desaturase (PDS), ζ-carotene desaturase, and 4-hydroxyphenylpyruvate dioxygenase (4-HPPD) are well-known target enzymes for carotenoid biosynthesis inhibitors (cf. Chap. 2). This section deals with agrochemical characteristics and major synthetic pathways for each enzyme inhibitor and also describes structural evolution of all new compounds released as carotenoid biosynthesis inhibitors since 1990.

10.3.1 Phytoene Desaturase Inhibitors

10.3.1.1 Practical Phytoene Desaturase Inhibitors

PDS, which catalyzes dehydrogenation of phytoene to zetacarotene, is the major target enzyme of herbicides in the carotenoid biosynthesis pathway. Up to now,

<Synthesis of imazamethabenz-methyl>

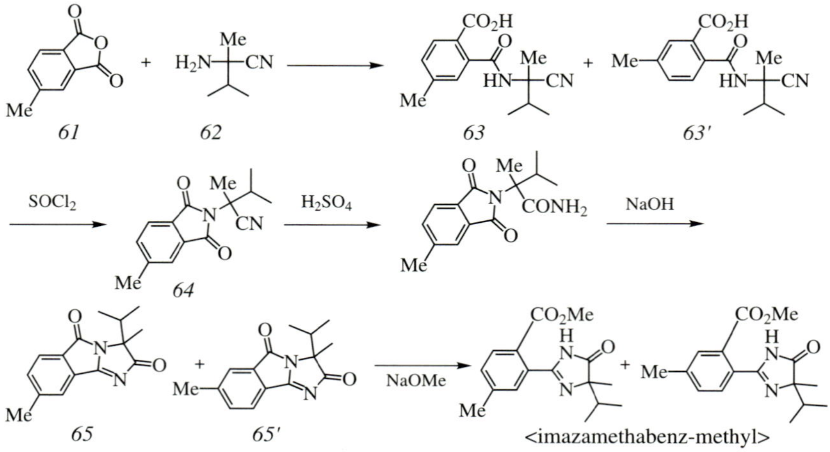

<Synthesis of imazaquin *via* quinoline-2,3-dicarboxylate>

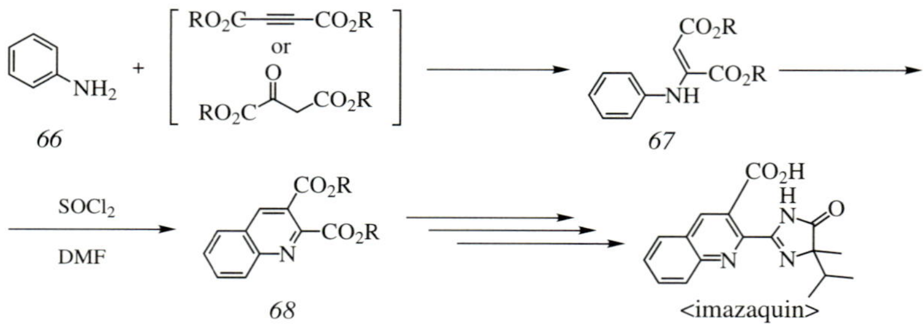

Scheme 6. Major synthetic routes for imazamethabenz-methyl and imazaquin

norflurazon, diflufenican, fluridone, flurochloridone and flurtamone have been launched as PDS inhibitors, and currently, picolinafen and beflubutamid are being actively developed as practical herbicides. While chemical structures of these PDS inhibitors are very different, as shown in Table 6, interestingly, a 3-trifluoromethylphenyl group is a common structure in all compounds.

Norflurazon, a pyridazinone PDS inhibitor developed by Sandoz, is used as a pre-emergence herbicide in upland fields and orchards. It controls annual grasses, broadleaf weeds and some perennial grasses in cotton, nuts, soybeans, peanuts, citrus, vines, cranberries and hops at 1–8 kg/ha. Flurochloridone is a pyrrolidinone herbicide developed by Stauffer Chemical. It is used to control chickweed, ivyleaf speedwell and field violet in winter wheat and rye; redroot pigweed, purslane and black nightshade in cotton; catchweed bedstraw, black nightshade and birdseye speedwell in potatoes; and a wide range of weeds in sunflowers at 0.5–2 kg/ha by pre-emergence application. Fluridone is a 4-pyridone PDS inhibitor developed by Elanco. In the beginning, it was used in

Table 6. Practical PDS inhibitors with 3-trifluoromethylphenyl

Chemical structure	ISO name Code No. Company	Dose Appl. method Target crops	Patent No.
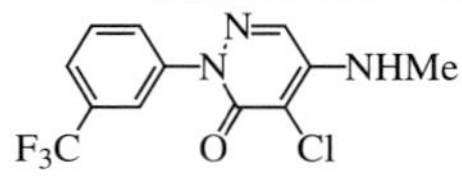	norflurazon SAN-978938 Sandoz	1~8 kg/ha pre cotton, soybeans, orchard	CH482684 US3644355 US3834889
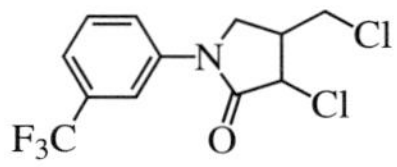	flurochloridone YH-44, R-40244 Stauffer Chemical	0.5~2 kg/ha pre cotton, potatoes, cereals	DE2612731
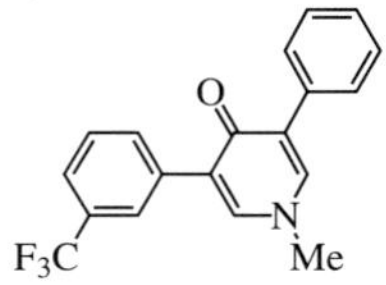	fluridone=fluoridone HOK-854, EL-171 Elanco	0.25~4 kg/ha pre, post cotton, corn, rice	DE2537753
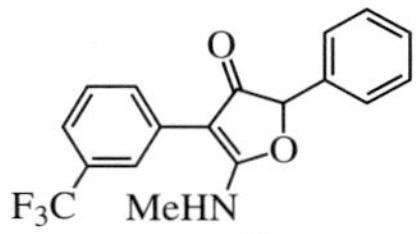	flurtamone RE-40885 Chevron	250~500 g/ha pre, post cotton, sorghum cereals, peanuts	DE3422346
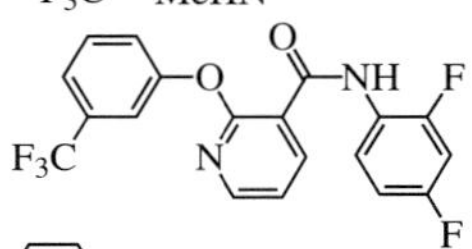	diflufenican MB-38544 Rhône Poulenc	125~250 g/ha pre cereals, sunflower	EP53011
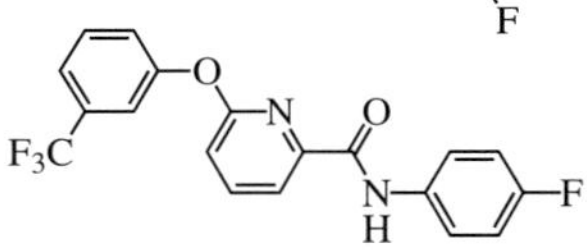	picolinafen AC 90001 ACC	50 g/ha pre, post cereals, lupines	EP447004
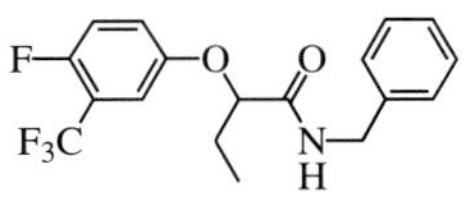	beflubutamid UBH-820 Ube	170~255 g/ha early post wheat, barley, rye	JP6310749 JP01268658

cotton to control annual broadleaf weeds and grasses by pre-emergence treatment at 300–600 g/ha and perennial weeds such as johnsongrass, purple nutsedges, and bermudagrass at 0.8–1.2 kg/ha. Currently, it is applied as an aquatic herbicide for control of most submerged and emerged aquatic weeds such as bladderwort, contrail, watermilfoil, southern naiad, pondweed and paragrass in ponds, lakes, reservoirs, irrigation ditches and rivers. Flurtamone, a pre- and early post-emergence herbicide, was discovered by Chevron and launched by Rhône-Poulenc in 1990. The substituents at the dihydrofuranone ring are the same as fluridone. It controls annual broadleaf weeds and some grasses in wheat, peas and sunflowers at 250–500 g/ha.

Diflufenican; a pyridinecarboxamide herbicide, developed by Rhône-Poulenc, is used at 125–250 g/ha pre- or early post-emergence application in autumn-sown wheat and barley to control grass and broadleaf weeds such as catchweed bedstraw, Persian speedwell and violet. Picolinafen is an early post-emergence herbicide reported by ACC in 1999. It selectively controls important annual broadleaf weeds such as Persian speedwell, catchweed bedstraw, henbit, wild mustard and field violet in winter wheat, barley, durum and others at 50 g/ha. Picolinafen is a regioisomer of diflufenican, i.e., the position of the carboxamide moiety is different from that of diflufenican.

Beflubutamid is a new PDS inhibitor developed by Ube Industries. It shows efficient control of a wide range of broadleaf weeds such as birdseye speedwell, henbit and field violet without injury to wheat, barley, rye and triticale at rates of 170–255 g/ha. It has pre- to post-emergence activity and is effective by early post-emergence application.

10.3.1.2
Structural Evolution of Phytoene Desaturase Inhibitors

Structural modifications of pyrrolidinone PDS inhibitors have been carried out mainly by ICI, Mitsui, Bayer and Ciba-Geigy (Fig. 16). In the early 1990s, modifications at the 3- and 4-positions of the pyrrolidinone ring were actively investigated. Many proposed derivatives were decorated particularly at the 3-position with substituted phenyl groups. Furthermore, the 3-trifluoromethylphenyl group at the 1-position of flurochloridone has been changed to benzyl, cumyl and other groups. Flurochloridone has two asymmetric carbons in the pyrrolidinone ring, a 3,4-*trans* stereochemistry gives better herbicidal activity than the *cis* form. Pyrrolidinone derivatives **435**, **436** and **437** in Fig. 16 are thought to be derived from flurochloridone, however, 3,5-dichlorocumyl and similar groups on the nitrogen atom suggest that these compounds are analogues of oxaziclomefone (MY-100), of which target enzyme has not been confirmed.

Although the hetero ring systems of fluridone, flurtamone and NTN-28621 are completely different, their three-dimensional structures are thought to be very similar due to two phenyl groups at the neighboring atoms of the carbonyl groups. Therefore, this class of compounds may be called "*diaryl-heterocycle PDS inhibitors*". From this point of view, several novel compounds released since 1990 are summarized in Fig. 17. In almost all compounds, one phenyl ring has the trifluoromethyl group at *meta*-position peculiar to PDS inhibitors and the other phenyl ring is substituted by electron-withdrawing groups at *para*-position. The latest compound [**448**] disclosed by DuPont has a 3-trifluoromethylbenzyl group instead of the usual phenyl group.

Figure 18 shows pyridinecarboxamide PDS inhibitors represented by diflufenican and picolinafen. These derivatives are also characterized by the 3-trifluoromethylphenyl group at the hetero rings. Presumably compounds [**454, 455, 456**] were derived by structural modifications of diflufenican and compounds [**459–467**] from picolinafen. In the first group, the trifluoromethylphenoxy and carbamoyl groups are placed next to each other like diflufenican, while the latter group has these groups on both sides of the nitrogen atom like picolinafen.

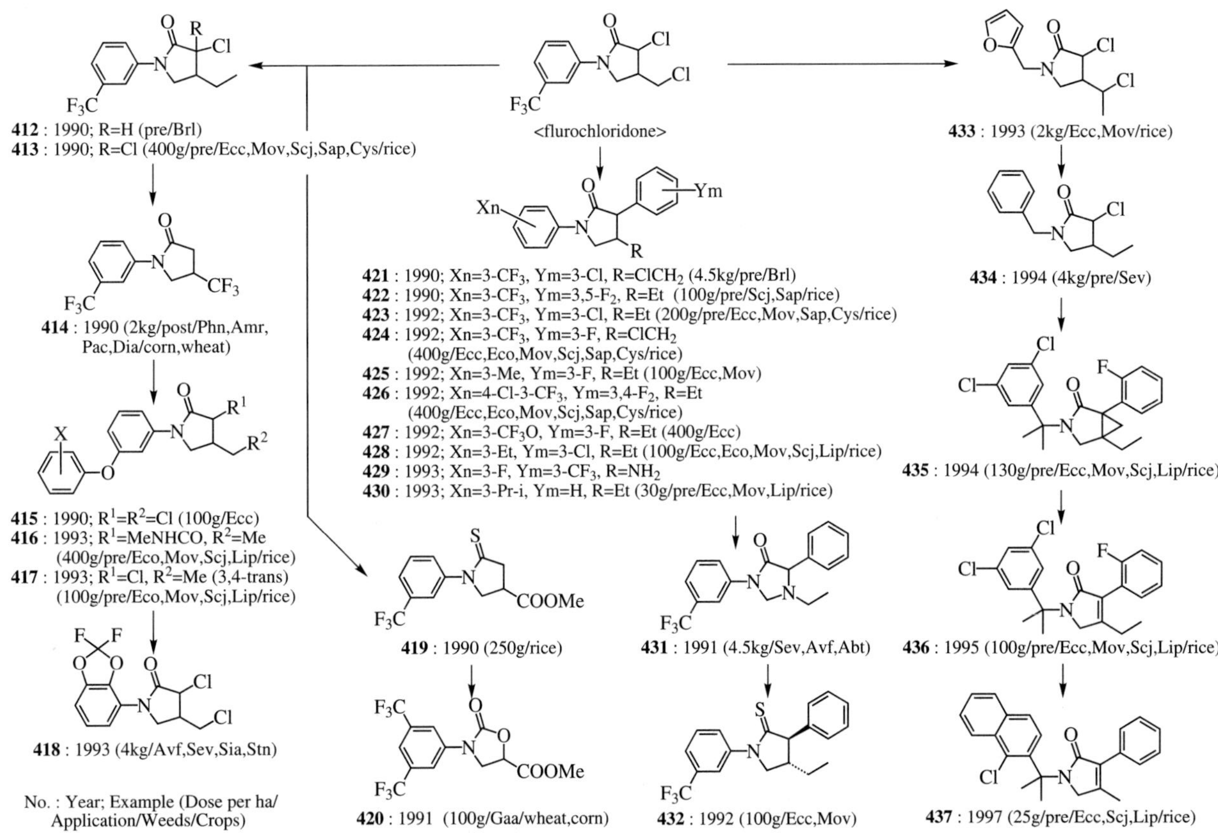

Fig. 16. Structural evolution of pyrrolidinone PDS inhibitors and related compounds since 1990

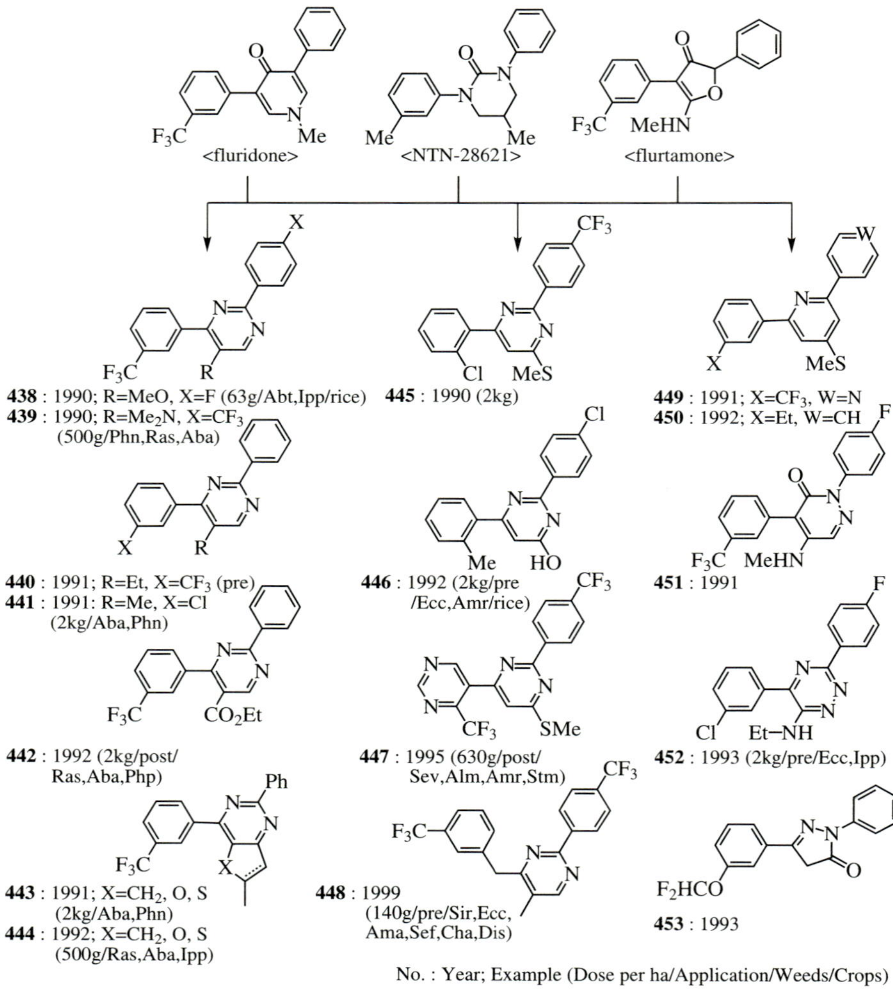

Fig. 17. Structural evolution of diaryl-heterocycle PDS inhibitors and related compounds since 1990

10.3.1.3
Major Synthetic Routes for Phytoene Desaturase Inhibitors

Major synthetic routes for norflurazon, flurochloridone, picolinafen, fluridone, flurtamone and beflubutamid are demonstrated in Scheme 7.

The pyridazinone ring of norflurazon is constructed by the addition reaction of 3-trifluoromethylphenylhydrazine and mucochloric acid followed by cyclization with acetic anhydride to yield 4,5-dichloro-2-(3-

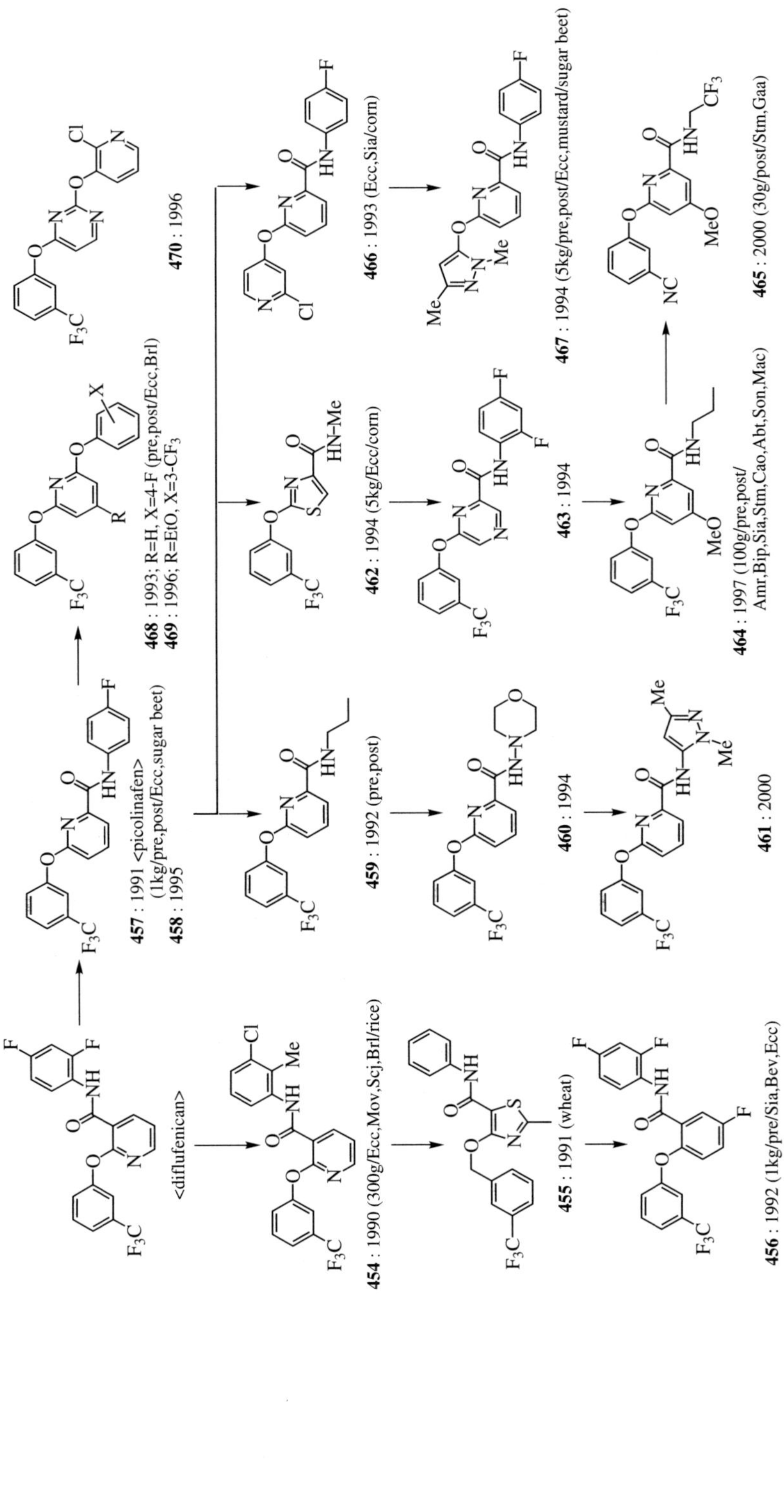

No. : Year; Example (Dose per ha/Application/Weeds/Crops)

Fig. 18. Structural evolution of pyridinecarboxamide PDS inhibitors and related compounds since 1990

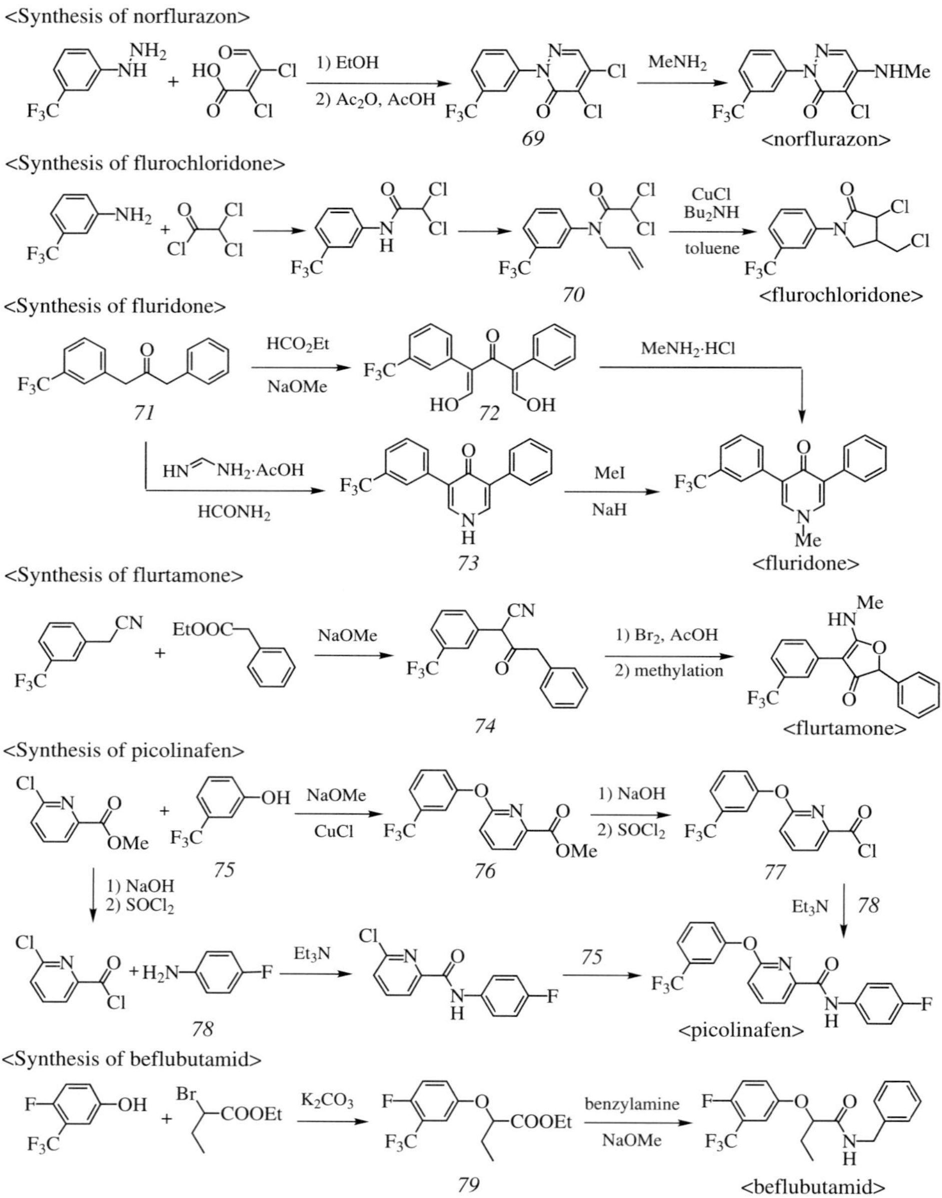

Scheme 7. Major synthetic routes for norflurazon, flurochloridone, picolinafen, fluridone, flurtamone and beflubutamid

trifluoromethylphenyl)pyridazin-3-one (*69*). Subsequent nucleophilic substitution of *69* with methylamine gives norflurazon. Flurochloridone is readily synthesized by cyclocondensation of *N*-allyl-*N*-(3-trifluoromethylphenyl)dichloroacetamide (*70*), which is prepared by reacting 3-trifluoromethylaniline with allyl chloride and dichloroacetyl chloride.

There are two pathways for the synthesis of fluridone. 1-(3-Trifluoromethylphenyl)-3-phenyl-2-propanone (*71*) can react with ethyl formate in the presence of a base to yield a di-formylated intermediate (*72*), which is cyclized with methylamine to give fluridone. It is also synthesized by methylation of 3-(3-trifluoromethylphenyl)-5-phenyl-4(1*H*)-pyridone (*73*) that is prepared by the reaction of 1-(3-trifluoromethylphenyl)-3-phenyl-2-propanone (*71*) with formamidine and formamide. The synthetic pathway for flurtamone is very simple. 3-Trifluoromethylbenzyl cyanide reacts with phenylacetate in the presence of a base to give 4-phenyl-2-(3-trifluoromethylphenyl)-3-oxo-butyronitrile (*74*), which is treated with bromine in acetic acid and methylated to yield flurtamone.

3-Trifluoromethylphenol (*75*) reacts with methyl 6-chloropicolinate in the presence of cuprous chloride to obtain methyl 6-(3-trifluoromethylphenoxy)picolinate (*76*). The ester is hydrolyzed and the resulting carboxylic acid is treated with thionyl chloride to give acid chloride (*77*), which condenses with 4-fluoroaniline (*78*) affording picolinafen. After amidation of picolyl chloride with *78*, introduction of the 3-trifluoromethylphenoxy group gives picolinafen. Synthesis of beflubutamid is also illustrated in Scheme 7. 4-Fluoro-3-trifluoromethylphenol reacts with 2-bromobutanoate to yield 2-(4-fluoro-3-trifluoromethylphenoxy)butanoate (*79*), followed by amidation with benzylamine in the presence of a base affording good yields of beflubutamid.

10.3.2
4-Hydroxyphenylpyruvate Dioxygenase Inhibitors

10.3.2.1
Practical 4-Hydroxyphenylpyruvate Dioxygenase Inhibitors

4-Hydroxyphenylpyruvate dioxygenase (4-HPPD) catalyzes the conversion of *p*-hydroxyphenylpyruvate to homogentisate in the biosynthesis pathway of plastoquinone and α-tocopherol. When 4-HPPD inhibitors block the hydroxylation of 4-hydroxyphenylpyruvate, homogentisate is not produced and no plastoquinone formed. Plastoquinone is thought to be an acceptor of hydrogen from phytoene, therefore the absence of plastoquinone leads to an accumulation of phytoene and carotenoid biosynthesis is impaired.

Investigations on 4-HPPD inhibitors, which were classified as bleaching herbicides, have been aggressively carried out since the 1990s, because these inhibitors exhibit excellent herbicidal activity even at low dosage like ALS and PPO inhibitors. Table 7 shows pyrazolone, triketone and isoxazole herbicides belonging to 4-HPPD inhibitors. All compounds have similar structural

Table 7. Practical 4-HPPD inhibitors

Chemical structure	ISO name Code No. Company	Dose Appl. method Target crops	Patent No.
<Pyrazoles>			
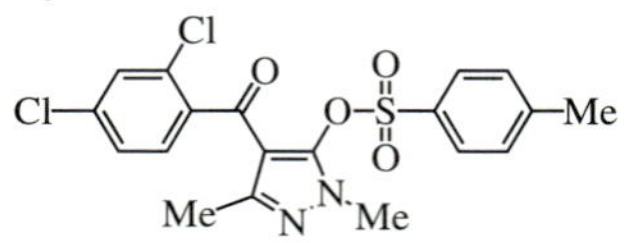	pyrazolate = pyrazolynate SW-751 Sankyo	2~3 kg/ha pre rice	JP101829 GB1463473
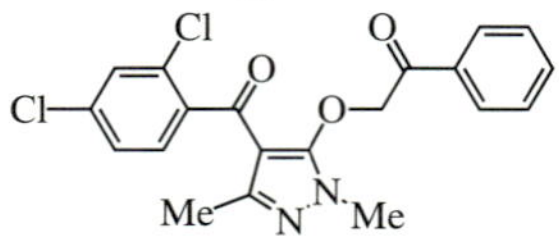	pyrazoxyfen SL-49 Ishihara	3 kg/ha pre, early post rice	JP5470269
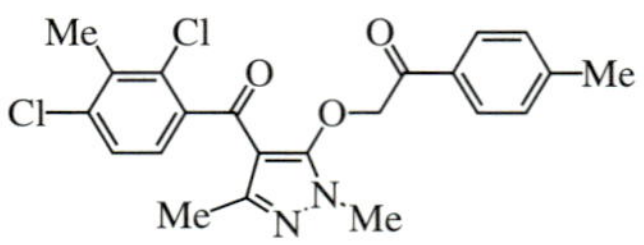	benzofenap MY-98 Mitsubishi-Petrochemical	1.2~2.4 kg/ha pre, early post rice	JP57072903
<Triketones>			
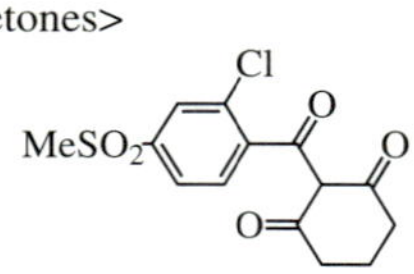	sulcotrione SC-0051 ICI-A-0051 Zeneca	0.25~1 kg/ha pre, post corn, sugarcane, cerials	EP249150
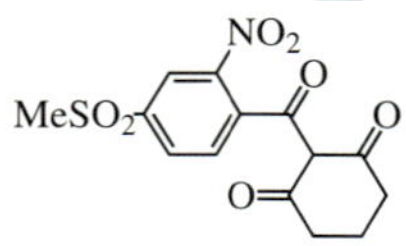	mesotrione ZA-1296 Zeneca	75~150 g/ha pre, post corn	EP186118 US4780127
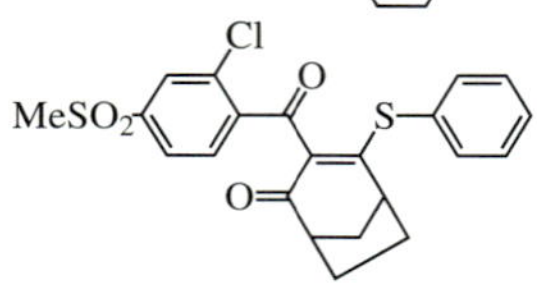	benzobicyclon SB-500 SAN-1315H SDS Bioteck	300 g/ha rice	JP0625144
<Isoxazoles>			
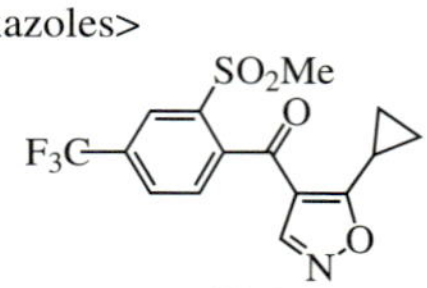	isoxaflutole RP-201772 EXP-30953 Rhône-Poulenc	75~150 g/ha pre corn, sugarcane	EP470856
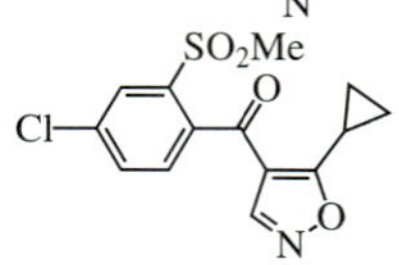	isoxachlortole RPA-201735 Rhône-Poulenc	pre	EP470856

characteristics, i.e., their benzoyl groups are substituted with various electron-withdrawing groups such as chlorine atom or trifluoromethyl, nitro and methanesulfonyl groups at *ortho*- and *para*-positions of the benzene rings.

In a series of pyrazole 4-HPPD inhibitors, Sankyo's pyrazolate (pyrazolynate) was first launched for use on rice. It is a pre-emergence herbicide that shows excellent efficacy against harmful weeds such as pondweed, arrowhead and waterplantain, and good safety for rice even at 2–3 kg/ha. Pyrazoxyfen developed by Ishihara Sangyo is a pre- or early post-emergence herbicide for use in rice fields. It is distinguished from pyrazolate by changing the tosyl group into a phenacyl group at the 5-position of the pyrazole ring. This modification led to an improvement of the efficacy against perennial weeds and of the weed spectrum. The herbicide is effective on annual weeds as well as perennial weeds in rice fields at 3 kg/ha. Benzofenap developed by Mitsubishi Petrochemical, is an active herbicide for rice like pyrazolate and pyrazoxyfen. The herbicide effectively controls annual and perennial weeds pre- and post-emergence at 1.2–2.4 kg/ha. A methyl group at the 3-position of the benzoyl moiety enhances herbicidal activity against broadleaf weeds, but was found ineffective on grass weeds and umbrella plants; however, its safety for rice is moderately improved.

In a series of triketone 4-HPPD inhibitors, sulcotrione was first developed by Zeneca and has already been sold as a pre- or post-emergence herbicide for use on corn. It controls broadleaf and grass weeds at 300–450 g/ha, while its efficacy against green foxtail is slightly inferior. Mesotrione, a nitro analogue of sulcotrione, was developed by Zeneca as a pre- and post-emergence herbicide for soybeans. Substitution by the nitro group extremely enhances its activity to control annual broadleaf weeds at 75 g/ha and annual grass weeds at 150 g/ha. The herbicide is also effective on sulfonylurea-resistant weeds. The first approvals in the USA and France are expected in 2001. Benzobicyclon was disclosed by SDS Biotech in 1994 and it has been developed for pre- and early post-emergence herbicides in rice fields. It exhibits potent activity at 300 g/ha against annual broadleaf and grass weeds and perennial weeds except arrowhead. Especially, the herbicide gives long residual control.

Isoxaflutole is a pre-emergence herbicide in corn and sugarcane. It controls many broadleaf and grass weeds including lambsquarters, pigweed, velvetleaf, waterhemp, barnyardgrass, *Kochia*, foxtails and wild proso millet at 75–150 g/ha. It also controls acetolactate synthase- and triazine-resistant weeds. Isoxaflutole has already been launched in Jamaica and France. Isoxachlortole is a chloro analogue of isoxaflutole and is effective against broadleaf and grass weeds at a lower rate of use.

10.3.2.2
Structural Evolution of 4-Hydroxyphenylpyruvate Dioxygenase Inhibitors

Structural modifications of pyrazole 4-HPPD inhibitors are demonstrated in Fig. 19. The benzoyl group at the 4-position of the pyrazole ring is a common structure in almost all derivatives. Furthermore, as a combination of the substituents at the 2- and 4-positions at the benzoyl group, chloro-chloro, chloro-methylsulfonyl or methyl-methylsulfonyl groups seem to be more effective for

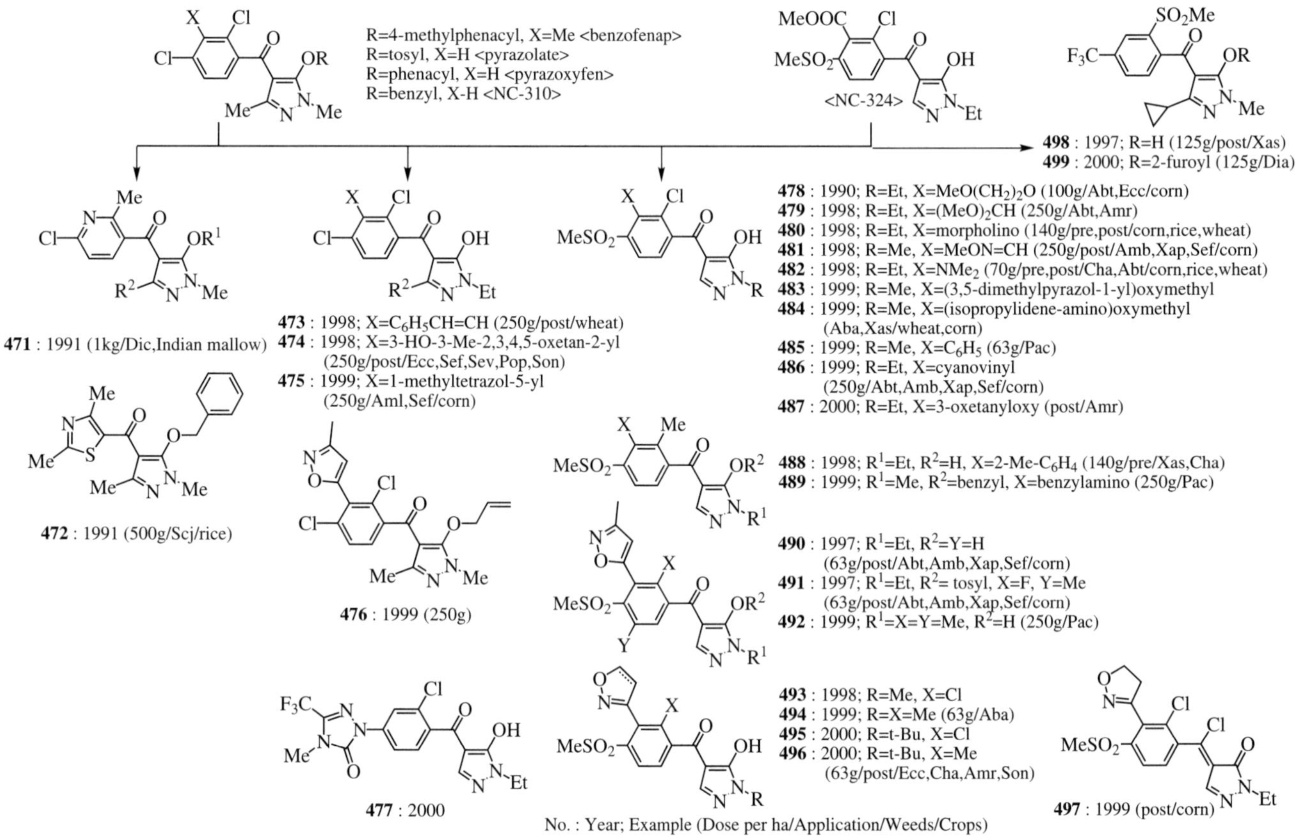

Fig. 19. Structural evolution of pyrazole 4-HPPD inhibitors since 1990

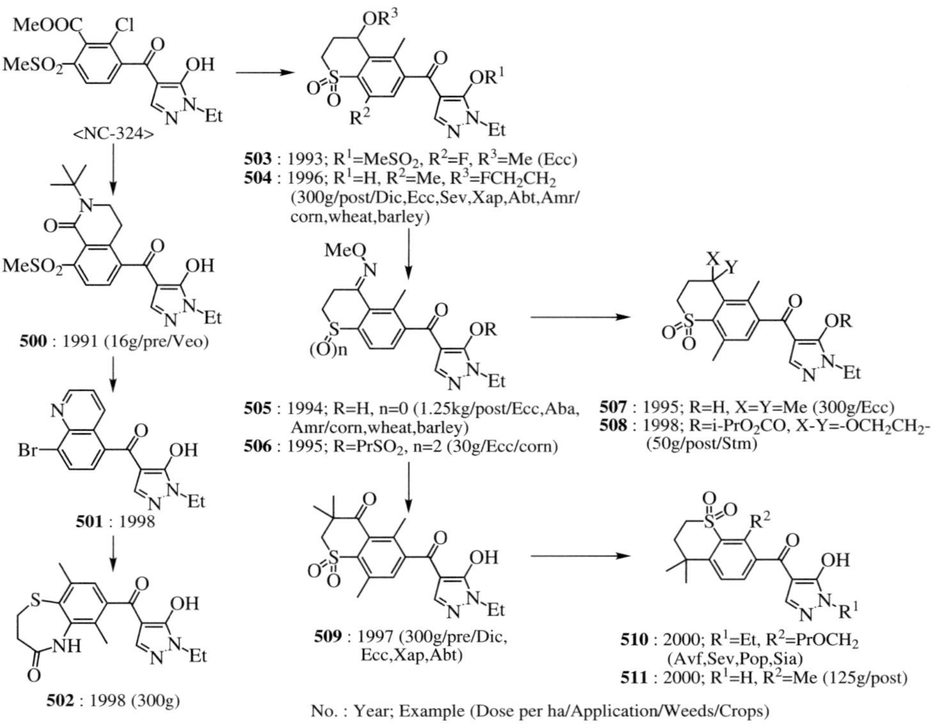

Fig. 20. Structural evolution of pyrazole 4-HPPD inhibitors with bicyclic benzologues since 1990

herbicidal efficacy. Quite recently, various substituents including heterocycles have been introduced at the 3-position of the benzoyl group. Regarding the substituent at the 2-position of the pyrazole ring, some groups easily metabolized to a hydroxy group are introduced because the hydroxy form is considered to be an active herbicidal species.

Figure 20 shows structural evolution of pyrazolone derivatives bearing bicyclic benzologues such as thiochroman and quinoline. A sulfonyl moiety in the bicyclic benzologues is thought to correspond with the methylsulfonyl group of aforementioned pyrazole derivatives.

Structural modifications of triketone 4-HPPD inhibitors are indicated in Figs. 21, 22 and 23. Until today, many companies have exhaustively investigated triketone chemistry and more than 100 patents on triketone 4-HPPD inhibitors have been continuously made available to the public. The substituents at the benzoyl groups are mainly modified in a similar manner to pyrazole 4-HPPD inhibitors. Substituents at the 2- and 4-positions are selected mostly from halogen atoms, methyl, nitro and methyl sulfonyl groups, and various groups and heterocycles have been proposed for substitution at the 3-position.

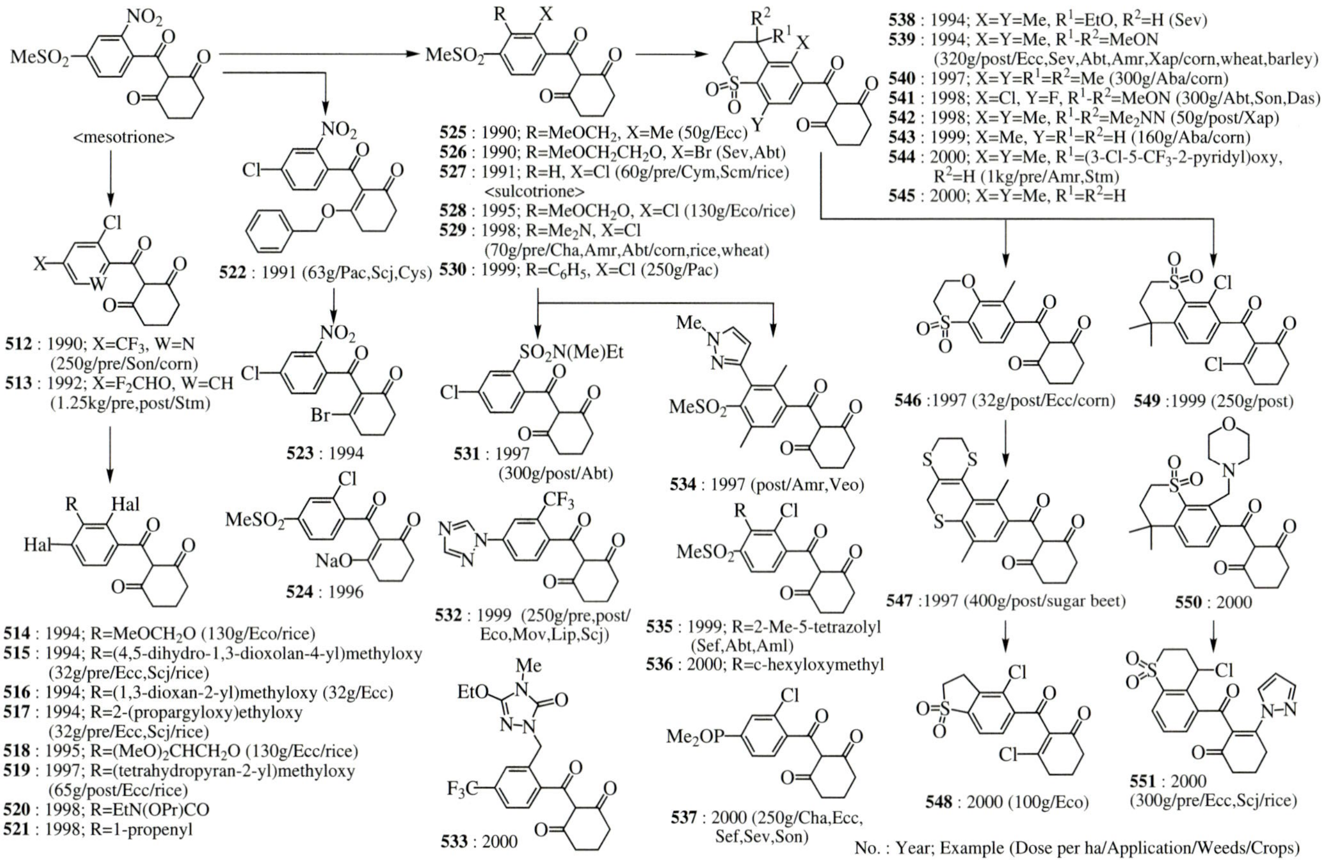

Fig. 21. Structural evolution of triketone 4-HPPD inhibitors since 1990 (1)

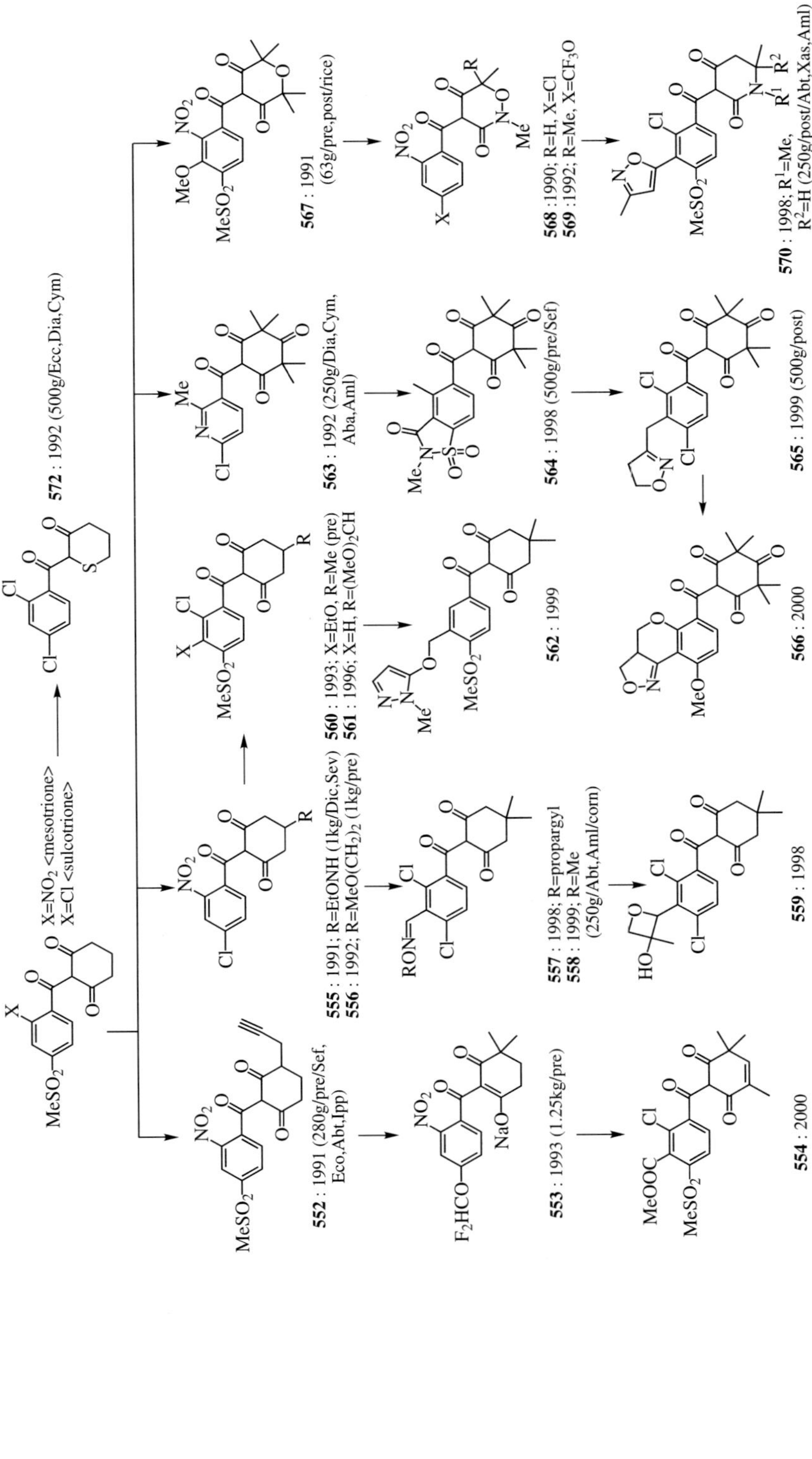

No. : Year; Example (Dose per ha/Application/Weeds/Crops)

Fig. 22. Structural evolution of triketone 4-HPPD inhibitors since 1990 (2)

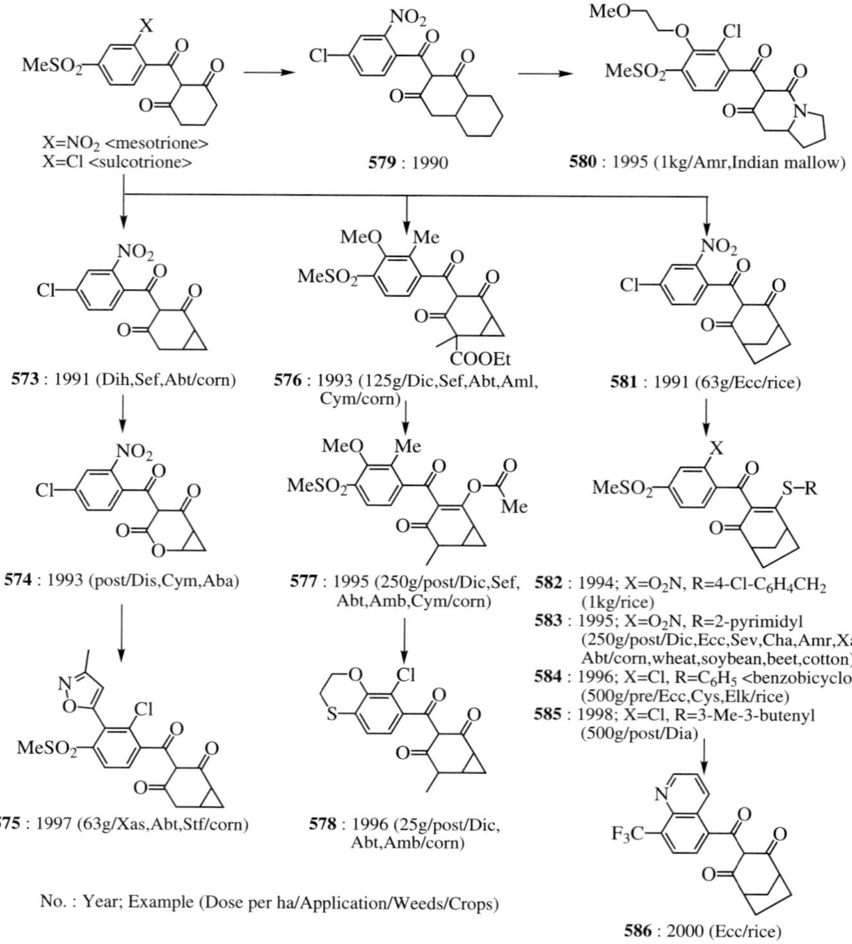

Fig. 23. Structural evolution of triketone 4-HPPD inhibitors since 1990 (3)

Substitutions by heteroatoms, introduction of another carbonyl moiety and/or plural methyl groups on the cyclohexanedione ring are arranged in Fig. 22.

Figure 23 focuses on structural modifications of the cyclohexanedione ring into bicycloalkanes such as bicycloheptane and bicyclooctane. The Nippon Soda group has intensively studied bicyclo[4.1.0]heptane-2,4-dione derivatives [**573–578**], which are readily synthesized by mesylation of 5-hydroxymethyl-cyclohexane-1,3-diones followed by cyclopropanation in the presence of a base. In the course of studies on bicyclo[3.2.1]octane-2,4-dione derivatives [**581–586**], a new triketone 4-HPPD inhibitor, benzobicyclon [**584**], in which the carbonyl group is masked as a phenylthio ether, was disclosed by SDS Biotech.

Isoxazole chemistry is an unrivaled area of Rhône-Poulenc and many compounds have been proposed since 1990, as shown in Fig. 24. While a cyclopropyl group is usually introduced at the 5-position of the isoxazole ring, the benzoyl groups are variously modified like other 4-HPPD inhibitors. Figure 24 includes several non-cyclic derivatives [**614–621**] due to a substitution pattern similar to isoxazole 4-HPPD inhibitors. Among them, RPA-203038, which is easily synthesized from the key intermediate of isoxaflutole (Scheme 8) is a novel herbicide developed since 1997.

10.3.2.3
Major Synthetic Routes for 4-Hydroxyphenylpyruvate Dioxygenase Inhibitors

Because pyrazolate, pyrazoxyfen and benzofenap exhibit little structural difference, synthetic methodologies of these compounds are quite similar. For example, Scheme 8 shows the major synthetic route for benzofenap. 2,4-Dichloro-3-methylbenzoyl chloride prepared from 2,6-dichlorotoluene reacts with 1,3-dimethylpyrazolone in the presence of calcium hydroxide to give 4-benzoylpyrazole (*80*). The compound (*80*) reacts with 4-methylphenacyl bromide (*81*) to yield benzofenap. Using *p*-tosyl chloride instead of *81* gives a methyl-analogue of pyrazolate.

Sulcotrione and mesotrione are synthesized by reaction of the corresponding benzoyl chlorides with 1,3-cyclohexanedione. In these reactions, *O*-benzoyl derivatives predominantly obtained are rearranged to the desired *C*-benzoyl forms using acetone cyanohydrin. In the synthesis of benzobicyclon, bicyclo[3.2.1]octane-2,4-dione (*82*) is used instead of 1,3-cyclohexanedione and *O*-benzoyl precursor (*83*) is transformed to triketone (*84*) in the presence of acetone cyanohydrin and triethylamine. Triketone (*84*) is chlorinated by sulfuryl chloride or oxalyl chloride followed by the reaction with thiophenol to give benzobicyclon as shown in Scheme 8.

Synthetic routes for isoxazole herbicides such as isoxaflutole and isoxachlortole are illustrated in Scheme 8. Benzoyl chloride (*85*) is treated with *tert*-butyl 3-cyclopropyl-3-oxopropionate in the presence of magnesium and carbon tetrachloride to give 2-benzoyl-3-cyclopropyl-3-oxopropionate (*86*), followed by decarboxylation, ethoxymethylidenation and cyclocondensation with hydroxylamine to give isoxaflutole ($X = CF_3$) and isoxachlortole ($X = Cl$). Intermediate (*87*), which is the same precursor for the synthesis RPA-203038 can be transferred into the ethoxyaminomethylidene derivative.

10.3.3
Other Carotenogenesis Inhibitors

This section deals with agrochemical characteristics of bleaching herbicides whose target enzymes are still unknown (Table 8).

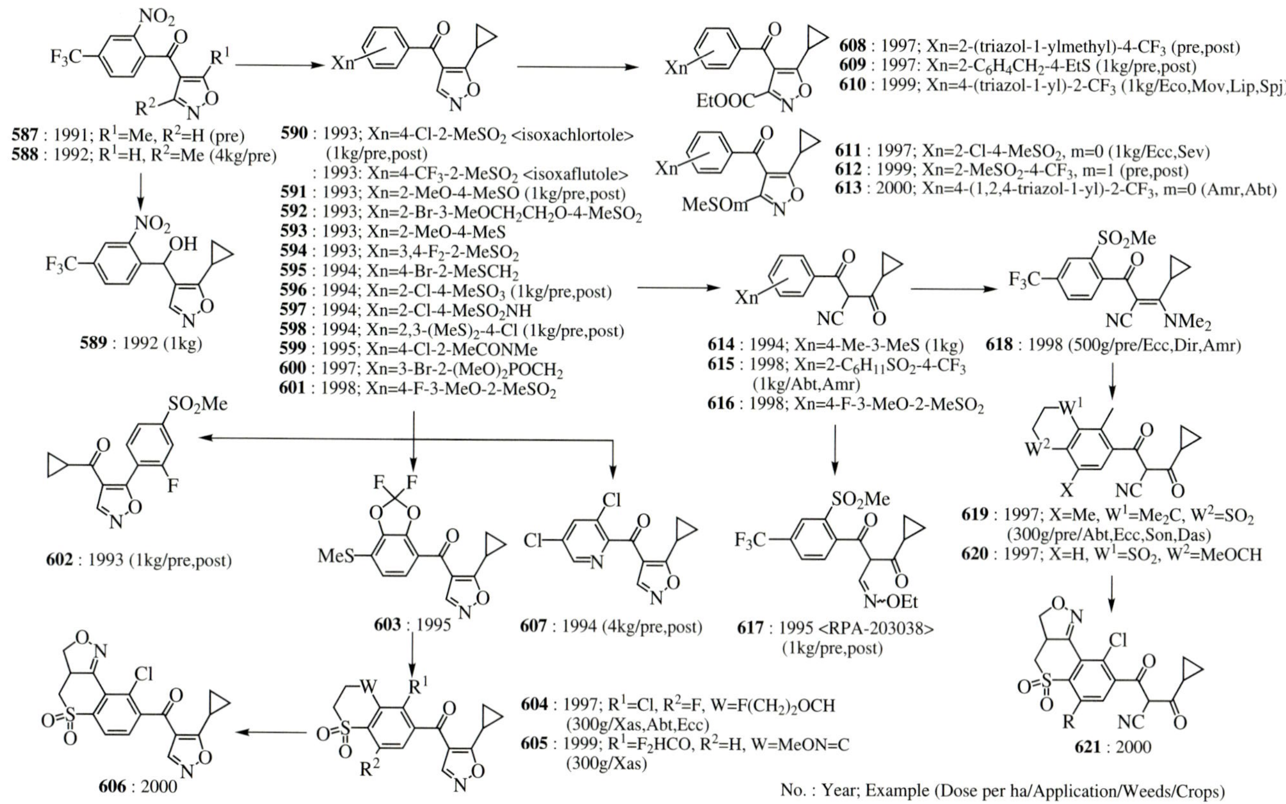

Fig. 24. Structural evolution of isoxazole 4-HPPD inhibitors and related compounds since 1990

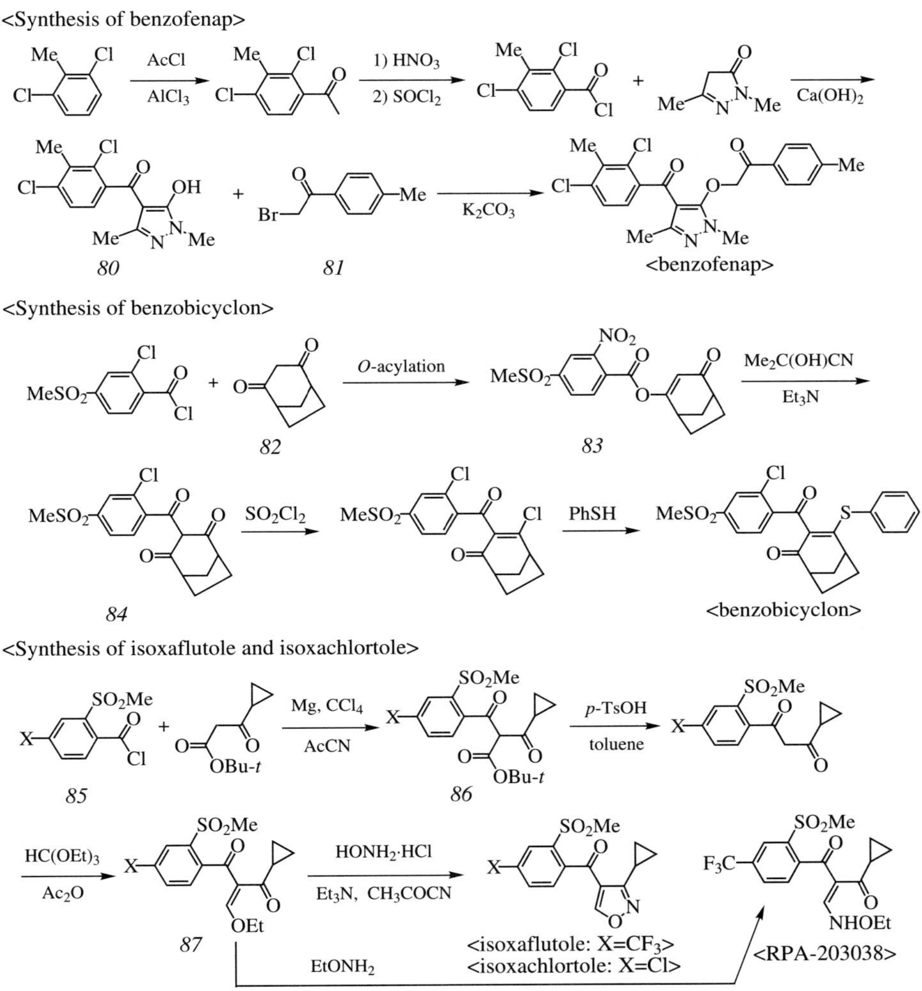

Scheme 8. Major synthetic routes for benzofenap, benzobicyclon, isoxaflutole and isoxachlortole

Amitrole is a pre- or post-emergence herbicide for fruit trees and bushes in orchards, and applied on fallow land and other non-crop areas, such as paths, railway tracks and industrial areas, at the use rate of 1–20 kg/ha. It is also used for control of aquatic weeds in marshes, drainage ditches, etc. Fluometuron can be applied pre-emergence for weed control before planting or post-emergence after target crops and weeds come up. It controls annual broadleaf and grass weeds in cotton and sugarcane at 1–1.5 kg/ha.

Clomazone is a broad-spectrum herbicide. It controls annual grasses and broadleaf weeds at 0.75–1.25 kg/ha in cotton, peas, soybeans, tobacco,

Table 8. Other carotenogenesis inhibitors (unknown target)

Chemical structure	ISO name Code No. Company	Dose Appl. method Target crops	Patent No.
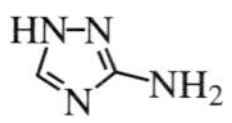	amitrole Rhône-Poulenc	1~20 kg/ha pre, post apple	US2670282 Brit.799709 US2875209
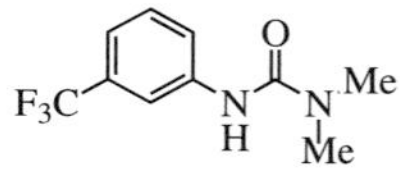	fluometuron C-2059 Ciba	1~1.5 kg/ha pre, post cotton, sugarcane	BE594227 GB914779
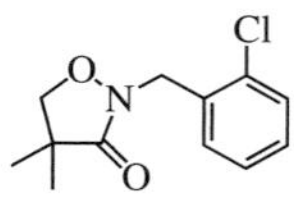	clomazone FMC-57020 FMC	0.75~1.25 kg/ha pre soybeans, corn sugarcane	US4405357 FR2483406
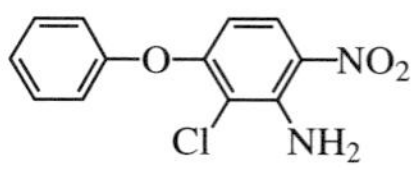	aclonifen MK-140, CME-127 Cela	2~3 kg/ha pre cereals, potatoes corn, sunflowers	US4394159 JP5519260

and wheat fields, etc. It can be used early pre-plant, pre-emergence or pre-plant-incorporated depending on the crops. Clomazone causes bleaching damage to crops, it must be carefully handled to avoid drift or vapors when applied. It is thought that a clomazone metabolize inhibits a prenyltransferase involved in the conversion of isopropenyl pyrophosphate to geranyl pyrophosphate in carotenoid biosynthesis. Aclonifen is a slow-acting diphenyl ether with a moderate inhibition of PPO and photosystem II. Accordingly, 2–3 kg/ha is required for a pre-emergence control of broadleaf and grass weeds in winter wheat, potatoes, sunflowers, peas, corn and other crops.

10.4 Aromatic Amino Acid Biosynthesis Inhibitors

Glyphosate is a competitive inhibitor of 5-enolpyruvylshikimate-3-phosphate (EPSP) synthase in shikimate biosynthesis (cf. Chap. 3). Only two compounds, glyphosate-sodium and sulfosate, are classified into group G by HRAC (Table 9). They are effective on deep-rooted perennial species, and annual and biennial species of grasses, sedges and broadleaf weeds. Glyphosate-sodium is used in fruit orchards, vineyards and many plantation crops. It is also used as a post-weed-emergence herbicide but as a pre-crop-emergence herbicide in a wide range of crops including vegetables, beets, soybeans, cotton, etc. It can be applied on non-crop areas. Sulfosate is used as a plant growth regulator on sugarcane.

Major synthetic routes for glyphosate-sodium are shown in Scheme 9. Reaction of phosphorus trichloride with formaldehyde in acetic acid forms a key intermediate *88*, which is treated with hydrochloric acid to give

Table 9. Practical EPSP and GS inhibitors

Chemical structure	ISO name Code No. Company	Patent No.
<EPSP inhibitors>		
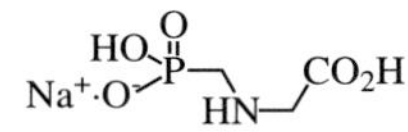	glyphosate-sodium MON-0459 Monsanto	DE2152826 DE2314134
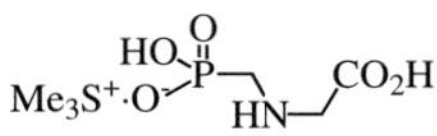	sulfosate SC0224, ICIA0224 ICI	US4315765 EP54382
<GS inhibitors>		
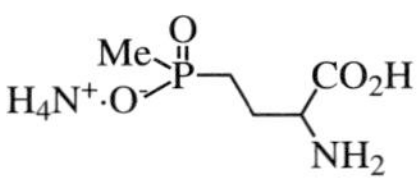	glufosinate-ammonium (pre) Hoe-39866 Hoechst	JP4885538 JP4891019 JP52139727 JP54101427
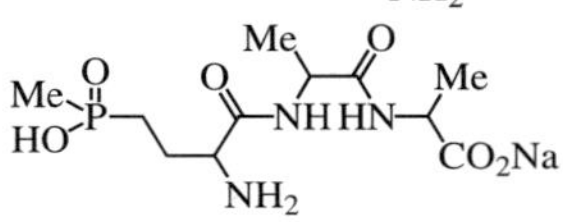	bialaphos-sodium (post) MW-801, SF-1293 Meiji Seika	JP62058998 JP62205789

<Synthesis of glyphosate-sodium (Method-1)>

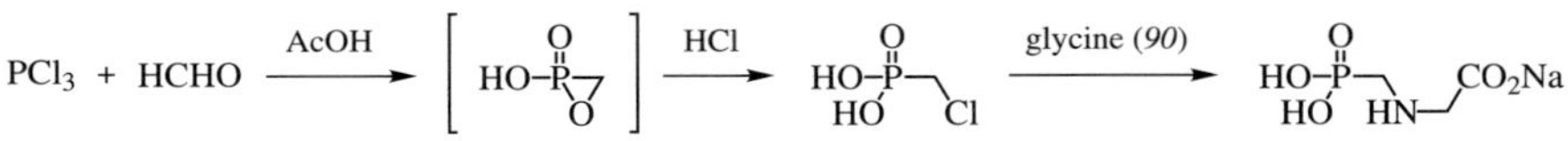

<Synthesis of glyphosate-sodium (Method-2)>

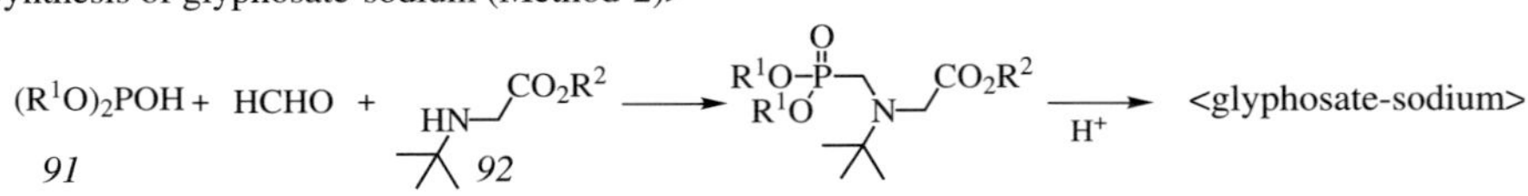

<Synthesis of glufosinate (Method-1)>

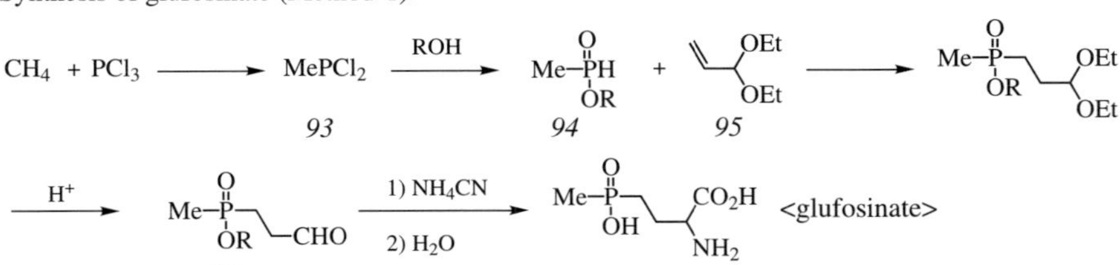

<Synthesis of glufosinate (Method-2)>

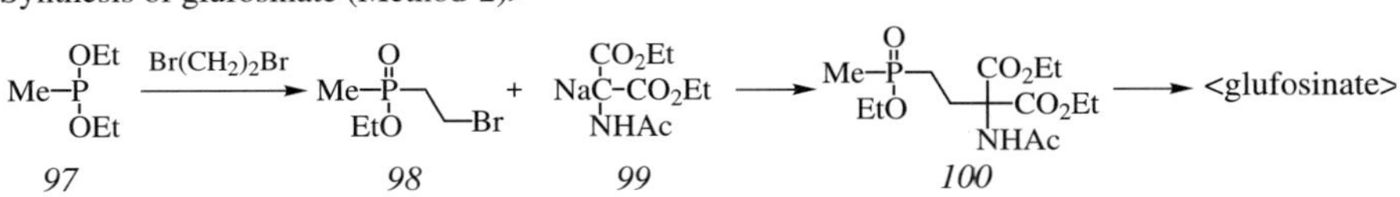

Scheme 9. Major synthetic routes for glyphosate-sodium and glufosinate

chloromethylphosphoric acid (*89*). It reacts with glycine (*90*) in aqueous sodium hydroxide affording glyphosate-sodium.

Another synthetic method uses *N*-phosphonomethylation of *N*-*tert*-butylglycine (*92*), which is performed with dialkylphosphite (*91*) and formaldehyde, and deprotection of the *tert*-butyl group by hydrobromic acid and subsequent hydrolysis give glyphosate-sodium.

10.5 Glutamine Synthetase Inhibitors

The glufosinate class acts by inhibition of glutamine synthetase (GS) in L-glutamine biosynthesis, causing ammonia accumulation and inhibition of photosynthesis (cf. Chap. 4). Only two practical herbicides, glufosinate and bialaphos-sodium, are noted in group H of HRAC classification (Table 9). Glufosinate is used to control a wide range of annual and perennial broad-leaf weeds and grasses in fruit orchards, vineyards and non-crop areas, and pre-emergence in vegetables. Bialaphos-sodium is used as a pre-emergence herbicide against annual weeds in vines or apples, and perennial weeds in non-cultivated land.

Among various synthetic methods for glufosinate, two major routes are shown in Scheme 9. Alcoholysis of dichloromethylphosphine (*93*), which is prepared by reacting phosphorous trichloride with methane at 600°C, gives methylphosphinate (*94*). The additional reaction of *94* to acrolein diethylacetal (*95*) and deprotection are accomplished to yield 2-formylethyl(methyl)phosphinate (*96*). The Strecker reaction of *96* gives glufosinate. The other method is as follows; diethyl diethoxy(methyl)phosphine (*97*) reacts with ethylene dibromide to give ethyl 2-bromoethyl(methyl)phosphinate (*98*), which is condensed with diethyl acetamidomalonate sodium salt (*99*), and produces ethyl 3,3-di(ethoxycarbonyl)-3-(acetamido)propyl(methyl)phosphinate (*100*). Hydrolysis and decarboxylation of *100* with hydrochloric acid give glufosinate.

10.6 Acetyl CoA Carboxylase (ACCase) Inhibitors

ACCase inhibitors including 4-aryloxyphenoxypropionates and cyclohexanediones act by interfering acetyl CoA carboxylase (ACCase) in the biosynthesis of fatty acids (cf. Chap. 5). Group A of HRAC classification has 16 kinds of ACCase inhibitors. These herbicides are divided into two groups, noncompetitive and competitive inhibitors. The former includes 4-aryloxyphenoxypropionate class herbicides and the latter the cyclohexanedione class.

Incipient 4-aryloxyphenoxypropionate ACCase inhibitors were racemic mixtures in spite of the existence of an asymmetric carbon at the 2-position of the propionate moiety. Since then, much more effective *R*-isomers have been

Table 10. Practical 4-aryloxyphenoxypropionate ACCase inhibitors

Chemical structure	ISO name Code No. Company	Dose Appl. method Target crops	Patent No.
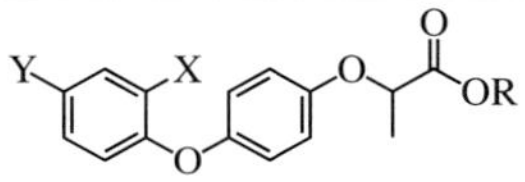	diclofop-methyl (RS) Hoe-23408 Hoechst (X=Y=Cl, R=Me)	0.8~1.6 kg/ha post wheat, soybeans, vegetables	DE213682 DE2223894
	cyhalofop-butyl (R) DEH-112, XDE-537 DowElanco (X=F, Y=CN, R=Bu)	50~100 g/ha post rice	EP302203 US4894085
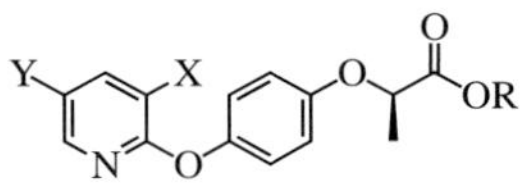	fluazifop-P-butyl ICI-A-0005 ICI (X=H, Y=CF_3, R=Bu)	250~500 g/ha post rape, potatoes, soybeans	JP-51106735
	clodinafop-propargyl CGA-184927 Ciba-Geigy (X=F, Y=Cl, R=propargyl)	40~80 g/ha post wheat, cereals	EP191736
	haloxyfop-R-methyl Dowco-453 DowElanco (X=Cl, Y=CF_3, R=Me)	62.5~125 g/ha pre, post soybeans, cotton	EP3890
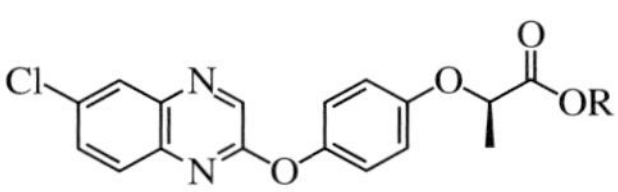	propaquizafop Ro-17-3664 Ciba-Geigy (R=Me_2C=$NO(CH_2)_2$)	60~280 g/ha post vegetables, soybeans, cotton, potatoes	GB1599121 US4687849 EP52798
	quizalofop-P-ethyl NC-302 Nissan (R=Et)	50~500 g/ha post cotton, soybeans, vegetables, orchards	JP56016475
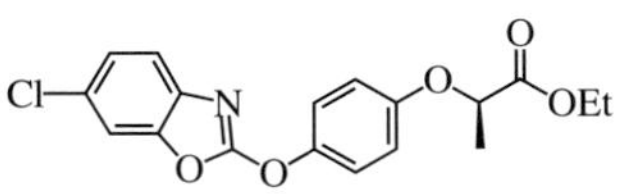	fenoxaprop-P-ethyl Hoe-046360 Hoechst	post soybeans, cereals vegetables	BE873844 DE2758002 EP2800

developed, and now, all practical 4-aryloxyphenoxypropionate inhibitors, except for diclofop-methyl, are *R*-isomers. The first practical herbicide belonging to the cyclohexanedione class is alloxydim-sodium. These class herbicides are used for the control of grass weeds. Seven compounds have already been launched and fenoxaprop-P-ethyl is under development.

10.6.1 Practical Acetyl CoA Carboxylase Inhibitors

Practical ACCase inhibitors are depicted in Table 10. Diclofop-methyl is only one racemic herbicide in 4-aryloxyphenoxypropionate classes. It is used for the post-emergence control of wild oat, wild millet and other annual grasses in broadleaf crops such as soybeans, peanuts, sugar beet and tomatoes at 0.8–1.6 kg/ha. Furthermore, diclofop-methyl can be also applied in wheat, barley and rye because it is rapidly metabolically detoxicated in grass crops. Cyhalofop-butyl substituted with a 4-cyano-2-fluorophenoxy group was launched by DowElanco in 1996. It is a post-emergence rice herbicide for the control of grass weeds, especially barnyardgrass at 50–100 g/ha.

The first optically active phenoxypropionate herbicide, fluazifop-P-butyl, was developed by ICI. It is a post-emergence herbicide in soybeans at 250–500 g/ha and controls wild oat, volunteer cereals, and annual and perennial grass weeds. Structural modifications of fluazifop-P-butyl succeeded in enhancing herbicidal activity and lead to clodinafop-propargyl and haloxyfop-R-methyl. Clodinafop-propargyl, commercialized as a post-emergence herbicide, is used in wheat at 40–80 g/ha. It controls annual grass weeds including *Avena*, *Lolium*, *Setaria*, *Phalaris* and *Alopecurus*. Also, haloxyfop-R-methyl is effective at 62.5–125 g/ha with pre- or post-emergence application. It controls annual and perennial grass weeds in sugar beet, oilseed rape, potatoes, vegetables, soybeans and other crops.

Propaquizafop and quizalofop-P-ethyl are distinguished by a 6-chloroquinoxaline ring. Propaquizafop having a unique ester moiety controls annual grass weeds at 60–120 g/ha and perennial weeds such as johnsongrass, quackgrass and bermudagrass at 140–280 g/ha in soybeans, cotton, sugar beet, potatoes, oilseed rape and vegetables. Quizalofop-P-ethyl is a post-emergence herbicide against perennial grass weeds, johnsongrass, quackgrass and bermudagrass in broadleaf crops such as sugar beet, oilseed rape, sunflowers, soybeans, cotton, peanuts and some vegetables. It controls weeds at 50–500 g/ha. Fenoxaprop-P-ethyl with a benzoxazole ring was introduced by Hoechst. It is a post-emergence control of annual and perennial grasses in potatoes, soybeans, beet, vegetables and cotton.

Since cyclohexanedione ACCase inhibitors are effective on grass weeds, only these classes are typical graminicides. Until now, seven compounds were launched (Table 11). Alloxydim-sodium, which is a pioneer compound in the cyclohexanedione class, was commercialized by Nippon Soda in 1980. It controls southern crabgrass, green foxtail, annual bluegrass, water foxtail, goosegrass, eulaliagrass, johnsongrass and barnyardgrass by post-emergence application.

Sethoxydim is more active than alloxydim-sodium. It is effective on perennial weeds including johnsongrass as well as annual weeds at 0.2–1.0 kg/ha in broadleaf crops such as cotton, oilseed rape, soybeans and vegetables. Since the discovery of sethoxydim, exhaustive structural modifications of cyclohexanedione and oxime moieties have been performed to find more efficient ACCase inhibitors. Nippon Soda also introduced a cotton herbicide, clethodim, with a 2-(ethylthio)propyl group at the 5-position of the cyclohexanedione ring. It shows good herbicidal activity against annual and perennial grass weeds in a wide range of broadleaf crops. On the other hand, BASF developed three kinds of cyclohexanedione ACCase inhibitors.

Table 11. Practical cyclohexanedione ACCase inhibitors

Chemical structure	ISO name Code No. Company	Dose Appl. method Target crops	Patent No
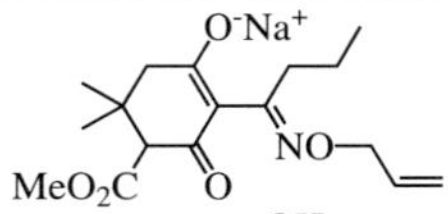	alloxydim-sodium NP-48-Na Nisso	0.5~1.5 kg/ha pre, post soybeans, rape, cotton, potatoes	JP5295636
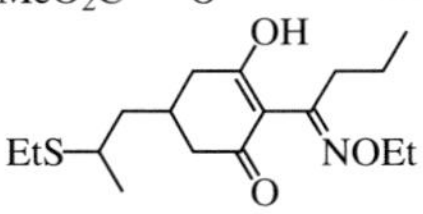	sethoxydim NP-55 Nisso	0.2~1.0 kg/ha post soybeans, rape, cotton, vegetables	JP52112945
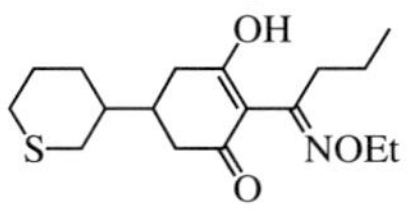	cycloxydim BAS-517-H BASF	post soybeans, rape, cotton	EP70370 EP71707 US4422864
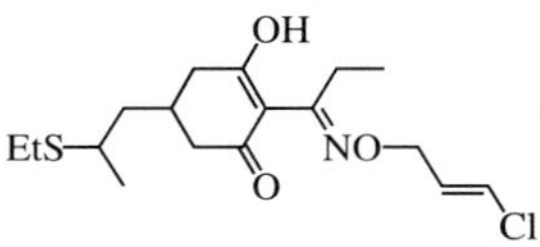	clethodim RE-45601 Nisso	post cotton, soybeans, potatoes, peanuts	GB2090246
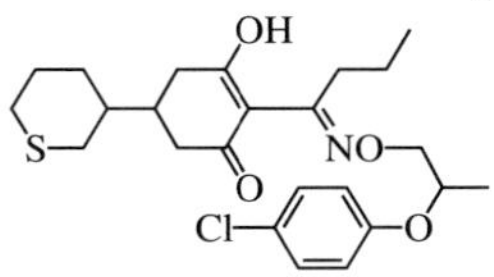	profoxydim BAS-625-H BASF	75~200 g/ha rice	DE4126999 WO97/20807
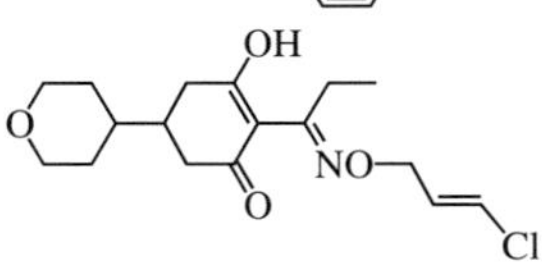	tepraloxydim BAS-620-H BASF	25~100 g/ha post soybeans, cotton, rape	DE4222261
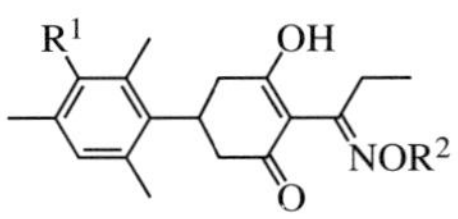	tralkoxydim PP-604 ICI (R^1=H, R^2=Et)	200~350 g/ha post cereals	EP85530
	butroxydim ICI-A0500 Zeneca (R^1=PrCO, R^2=Et)	40~50 g/ha post cereals, soybeans	WO9221649 EP444769

Cycloxydim and profoxydim are substituted with a tetrahydrothiopyran ring at the 5-position of the 1,3-cyclohexanedione ring and tepraloxydim with a tetrahydropyrane ring. Cycloxydim is used as a post-emergence herbicide in soybeans, cotton and oilseed rape. It is active on annual and perennial grass weeds with post-emergence application. Profoxydim, commercialized in 1998, controls grass weeds such as crab grass, barnyardgrass and yellow bristlegrass in seeded and transplanted rice. A dose of 75–200 g/ha controls a broad spectrum of grass weeds, and it can be applied to rice growth stages 13–25. Tepraloxydim is a post-emergence herbicide for use on soybeans, oilseed rape, cotton and sugar beet at rates of 25–100 g/ha.

Zeneca (ICI) developed novel cyclohexanedione ACCase inhibitors, tralkoxydim and butroxydim, with substituted phenyl groups at the 5-position of the cyclohexanedione rings. Tralkoxydim, which is substituted with a mesityl group, is a post-emergence herbicide. It shows good selectivity between wheat or barley and annual grass weeds including *Lolium* sp., green foxtail, *Phalaris* sp., *Alopecurus myosuroides* and *Apera spica-venti* at rates of 200–350 g/ha. Butroxydim, launched in 1996, is more herbicidally active than tralkoxydim due to the further introduction of a butyryl group at the mesityl ring. It is applied at 40–50 g/ha with post-emergence application.

10.6.2
Structural Evolution of Acetyl CoA Carboxylase Inhibitors

4-Aryloxyphenoxypropionate ACCase inhibitors have been actively investigated because of their good systemic activity with post-emergence application, and many kinds of compounds have been synthesized. Figure 25 shows structural evolution of 4-aryloxyphenoxypropionates disclosed since 1990. Judging from Fig. 25, electron-withdrawing groups at *para*-position and halogen atoms at *ortho*-position of the aryloxy group seem to be effective for highly herbicidal activity. Compounds [**635–640**], which do not belong to the 4-aryloxyphenoxypropionate class, are illustrated for reference. Other miscellaneous compounds [**641–648**] are also depicted in Fig. 25.

Cyclohexanedione ACCase inhibitors have been actively investigated as graminicides (Figs. 26 and 27). Substituents at the cyclohexanedione ring and the oxime-oxygen atom have been mainly modified. Cyclohexanedione synthesized in the 1980s has a lower alkyl or alkenyl group at the oxime-oxygen atom; however, subsequent investigations have made clear that bulky substituents such as aralkyl and arylalkenyl groups are more effective for activity. On the other hand, with regard to the substituents at the 5-position of the cyclohexanedione ring, for example, cyclopropyl [**649–651**], 3-thiopyranyl [**655–666**], 3- and 4-pyranyl [**667–672**] and substituted phenyl groups [**679–684**] were introduced.

10.6.3
Major Synthetic Routes for Acetyl CoA Carboxylase Inhibitors

Most 4-aryloxyphenoxypropionates are generally synthesized by reacting 2-chloropropionates with 4-aryloxyphenoxides which are prepared by a coupling reaction of 4-hydroxyphenoxides with halogenated-benzenes or -pyridines. Synthetic routes for fluazifop-P-butyl are shown in Scheme 10 as examples of general synthetic methods for 4-aryloxyphenoxypropionates.

2-Chloro-5-methylpyridine (*101*) is chlorinated with chlorine gas to give 2-chloro-5-trichloromethylpyridine (*102*), of which the trichloromethyl group is perfluorinated by antimony trifluoride to produce the key intermediate, 2-chloro-5-trifluoromethylpyridine (*103*). The phenoxide derived from

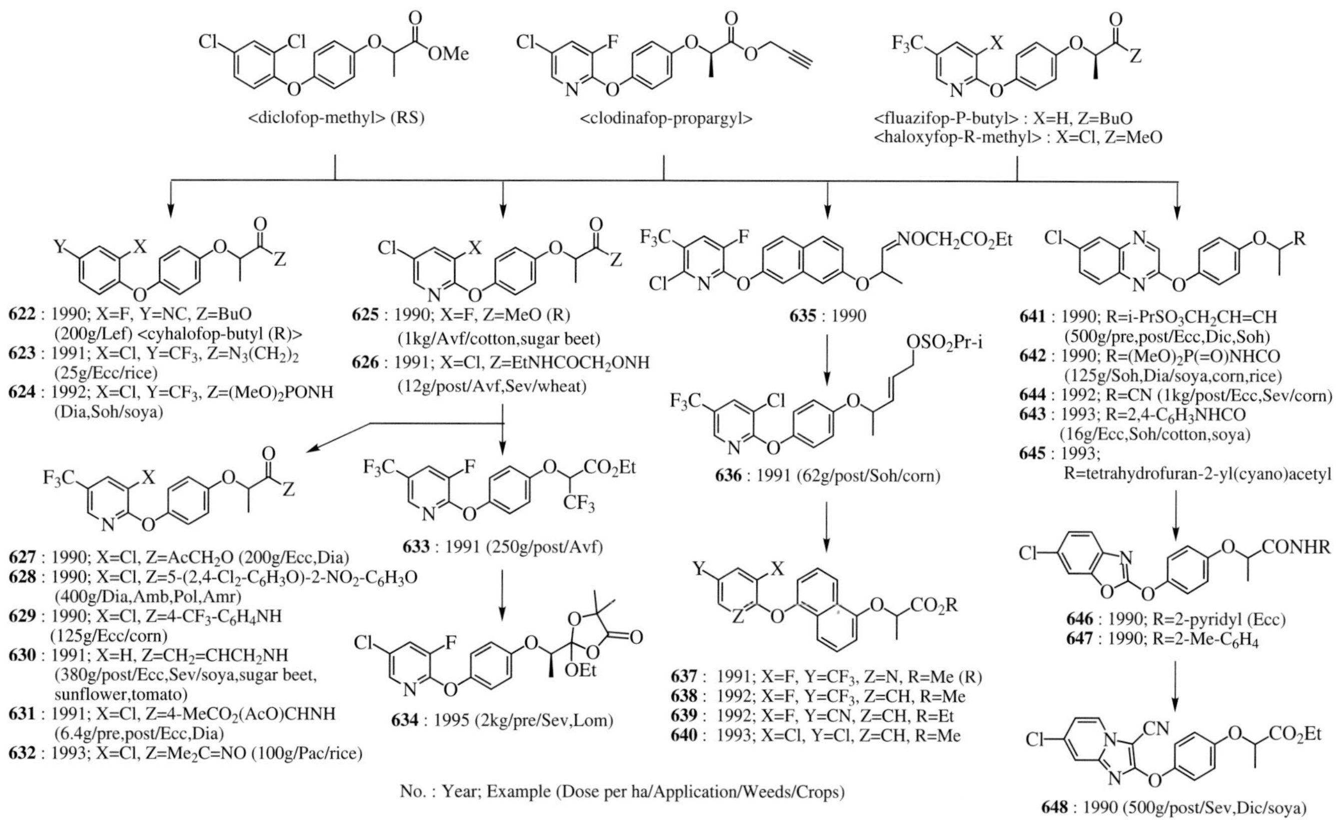

Fig. 25. Structural evolution of 4-aryloxyphenoxypropionate ACCase inhibitors and related compounds since 1990

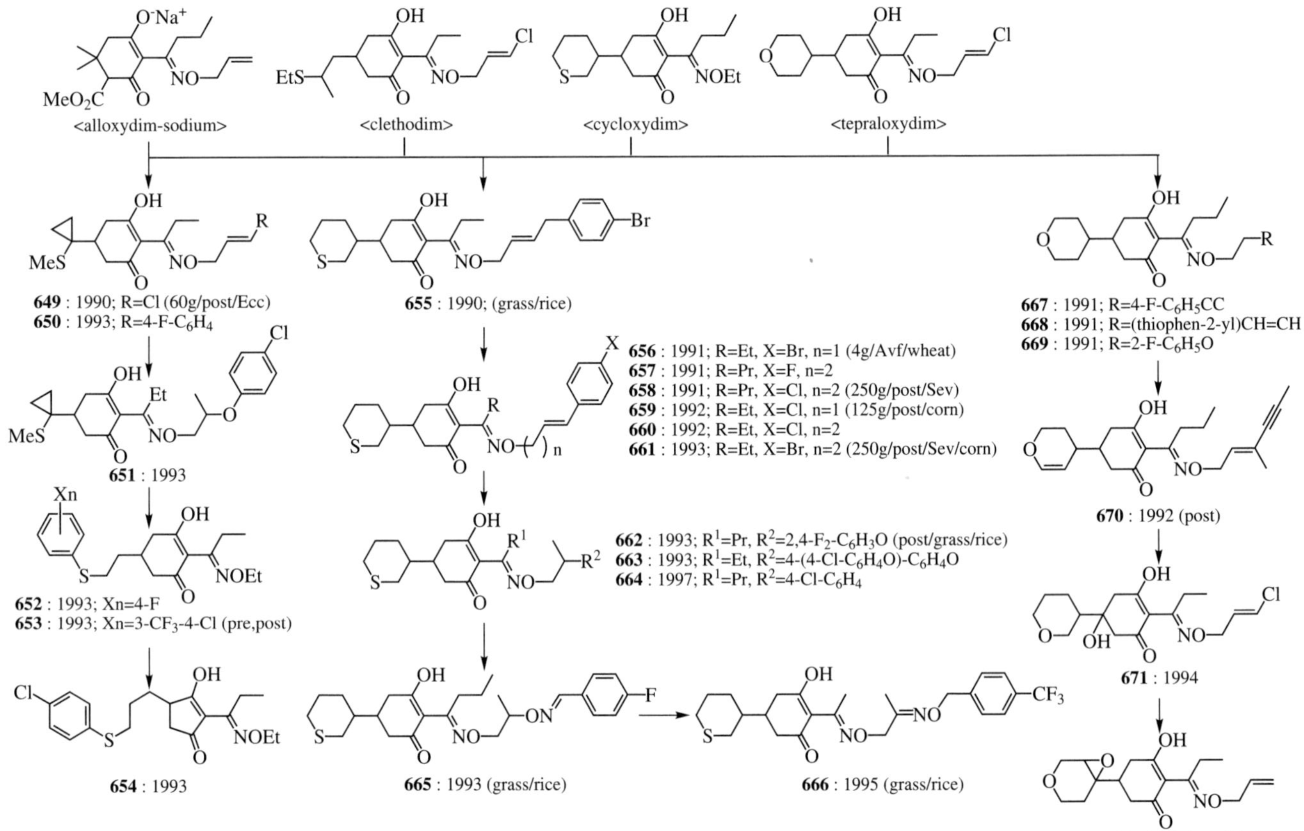

No. : Year; Example (Dose per ha/Application/Weeds/Crops)

Fig. 26. Structural evolution of cyclohexanedione ACCase inhibitors since 1990

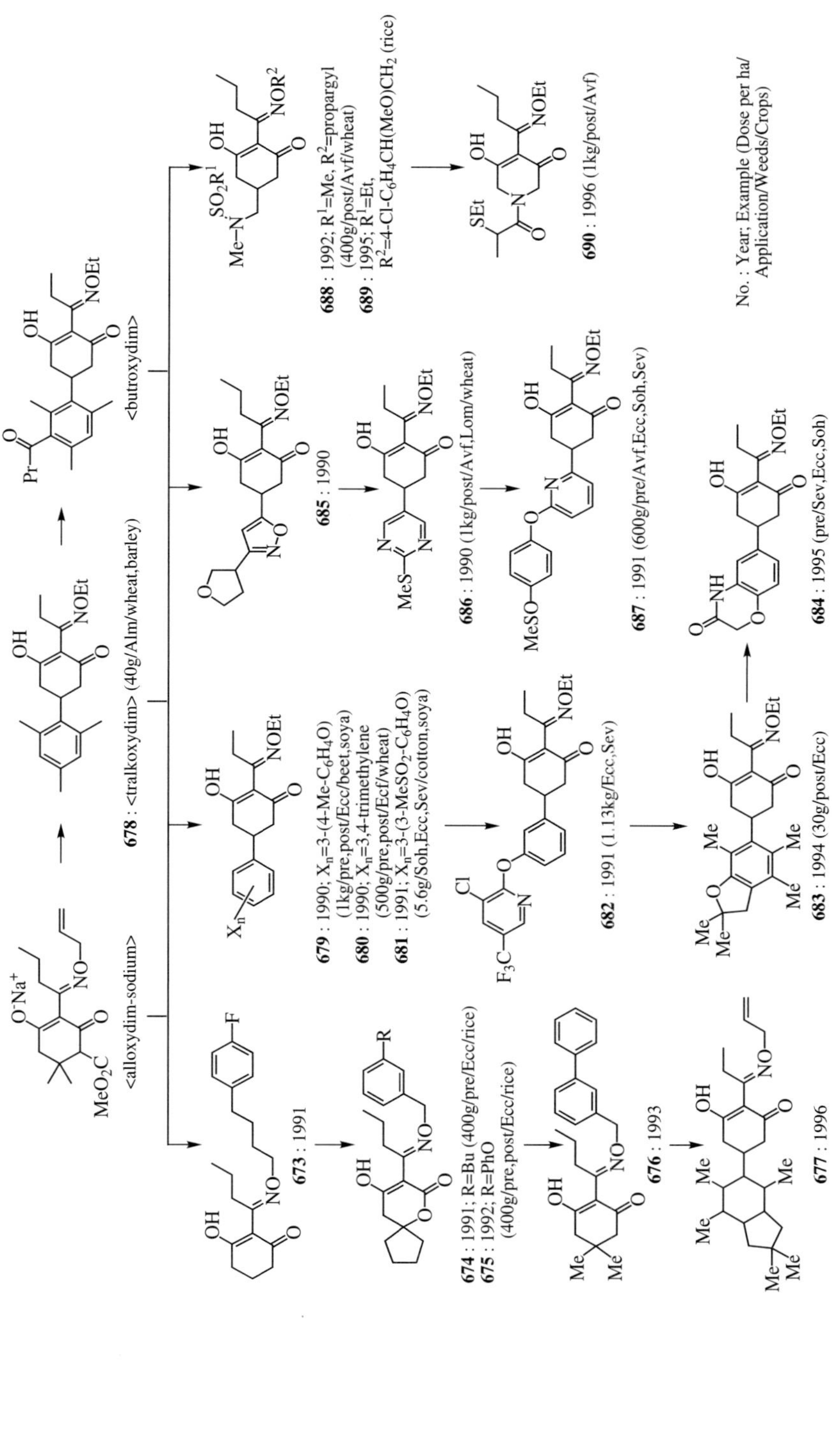

Fig. 27. Structural evolution of cyclohexanedione ACCase inhibitors and related compounds since 1990

<Method *via* 3-trifluoromethylpyridine>

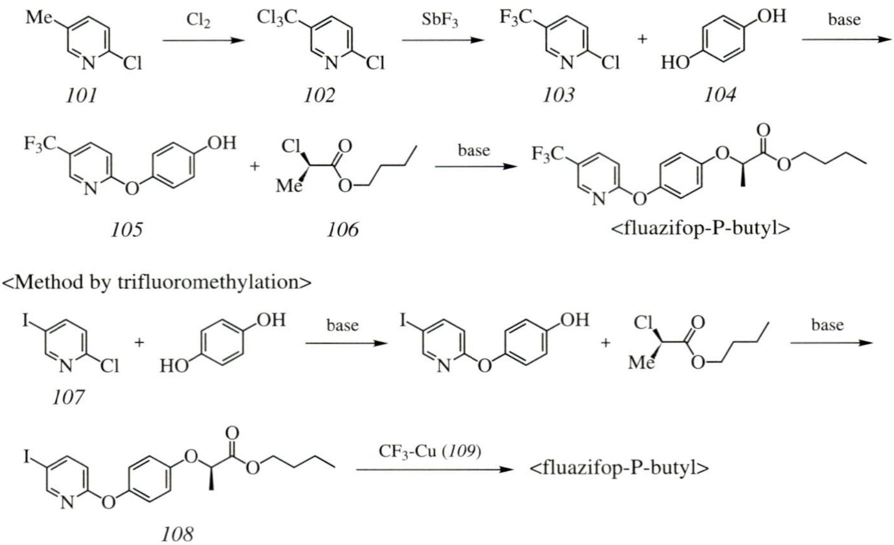

Scheme 10. Synthetic routes for fluazifop-P-butyl

hydroquinone (*104*) reacts with *103* to yield 4-(5-trifluoromethyl-2-pyridyloxy)phenol (*105*), which reacts with *R*-butyl 2-chloropropionate (*106*) to yield fluazifop-P-butyl. 4-(2-Chloro-5-iodo-2-pyridyloxy)phenoxypropionate (*108*), prepared with 2-chloro-5-iodopyridine (*107*) instead of 2-chloro-5-trifluoromethylpyridine, is trifluoromethylated with copper salt (*109*), affording fluazifop-P-butyl.

General synthetic routes for cyclohexanedione ACCase inhibitors are shown in Scheme 11. Desired cyclohexanediones (*114*) are derived from 5-substituted-1,3-cyclohexanedione (*112*) *via* the 2-acyl derivative (*113*). Aldol condensation of the aldehyde (*110*) with carbonyl compounds such as acetone and acetoacetic acid sodium salt yields α,β-unsaturated ketones (*111*), which are cyclized with ethyl malonate to give *112*. Using a variety of aldehydes (*110*) this method makes it possible to modify the 5-position of 1,3-cyclohexanediones. Modifications of the substituents led to the discovery of novel practical cyclohexanedione ACCase inhibitors.

A synthetic scheme for sethoxydim, is also depicted in Scheme 11 together with that of butroxydim. Crotonaldehyde (*115*) reacts with ethanethiol to obtain 3-ethylthiobutylaldehyde (*116*), which reacts with sodium acetoacetate and subsequently with diethyl malonate to produce 5-(2-ethylthiopropyl)-1,3-cyclohexanedione (*117*). Cyclohexanedione (*117*) reacts with butyryl chloride to provide an undesired *O*-acyl derivative (*118*). However, this compound (*118*)

<General synthetic route of cyclohexanedione ACCase inhibitors>

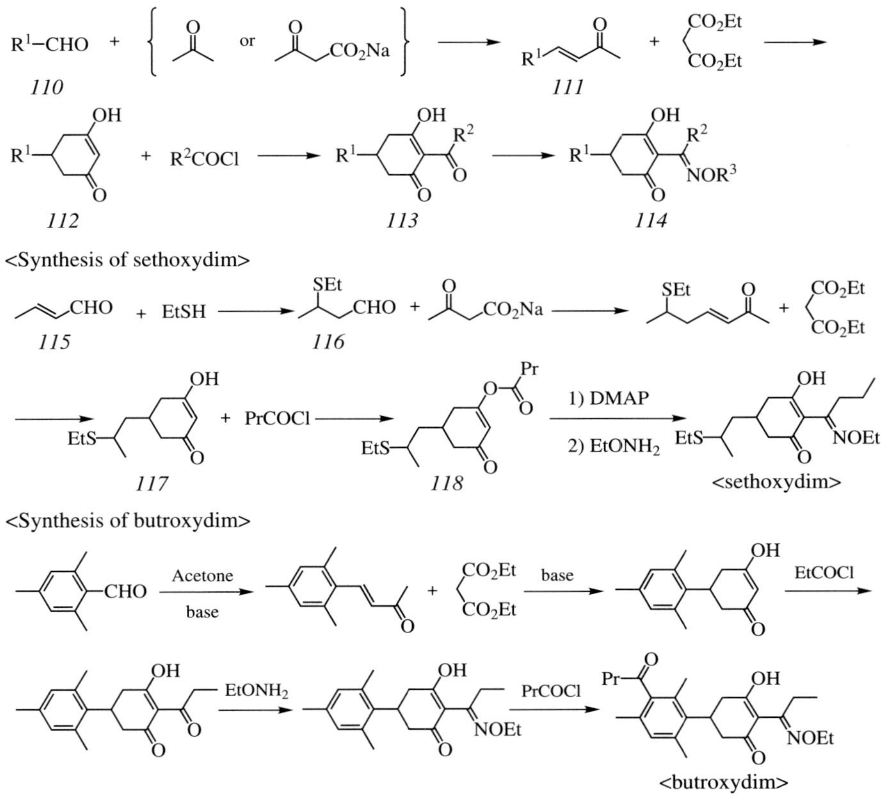

Scheme 11. General synthetic routes for cyclohexanedione ACCase inhibitors and major synthetic routes for sethoxydim and butroxydim

is readily isomerized in the presence of dimethylaminopyridine (DMAP) to yield the desired triketone, which is condensed with *O*-ethyl hydroxylamine to give sethoxydim. Butroxydim is synthesized from 2,4,6-trimethylbenzaldehyde in a similar manner.

10.7 Very Long-Chain Fatty Acids Biosynthesis Inhibitors

There are many herbicides that inhibit biosynthesis of very long-chain fatty acids (VLCFAs) with more than 20 carbon chains (cf. Chap. 6, 13). These herbicides belong to group K_3 in HRAC classification, and are divided into four groups, namely, the chloroacetamide, thiophosphate, amide and carbamoyl heterocycle classes. This section deals with agrochemical properties and major synthetic pathways for representative VLCFAs inhibitors together with the structural evolution of newer herbicides disclosed since 1990.

10.7.1 Practical Chloroacetamide Very Long-Chain Fatty Acids Biosynthesis Inhibitors

One of the pioneer chloroacetamide VLCFAs inhibitors is allidochlor (*N,N*-diallyl-2-chloroacetamide) released by Monsanto in 1958 (US2864638), but it was not used practically. The first successful herbicide was propachlor followed by alachlor and butachlor. The most remarkable characteristics of chloroacetamide VLCFAs inhibitors are their long residual activity and selective phytotoxicity against grass weeds. Until today, eleven chloroacetamide herbicides have been launched on the market (Table 12).

Propachlor applied at 3–5 kg/ha is used as a pre-emergence or post-emergence herbicide to control annual grass weeds and some broadleaf weeds in soybeans, cotton, peanuts, corn, onions or sugarcane. In chloroacetamide VLCFAs inhibitors, introduction of lower alkyl groups such as methyl and ethyl groups at both *ortho*-positions of the benzene ring tends to enhance herbicidal activity. Alachlor, which has a 2,6-diethylphenyl group, controls annual grass weeds and many broadleaf weeds including southern crabgrass, barnyardgrass, green foxtail, annual bluegrass, water foxtail and purple nutsedges in cotton, corn, oilseed rape, peanuts, radishes, soybeans and sugarcane. It is used as a pre-emergence herbicide at rates as low as 0.5–4 kg/ha than propachlor. Butachlor is used with pre-emergence application for control of annual grass weeds and some broadleaf weeds including barnyardgrass, umbrella plant, needle spikerush, Japanese bulrush, *Monochoria vaginalis*, *Indian toothcup*, common false pimpernel, *Alisma canaliculatum* in both seeded and transplanted rice. It shows good selectivity for barley, cotton, wheat, peanuts and sugar beet at rates of 1–4.5 kg/ha.

Dimethachlor gives a pre-emergence control of annual grass weeds such as *Alopecurus myosuroides*, *Apera spica-venti* and annual bluegrass, and broadleaf weeds in rape. Pretilachlor is effective against annual grass and broadleaf weeds such as early watergrass, smallflower, umbrella sedge, Japanese bulrush, needle spikerush, *Monochoria vaginalis*, *Alisma canaliculatum*, *Eriocaulon miquelianum* and Indian toothcup in rice with pre-emergence application at 600 g/ha. Propisochlor is a pre-emergence herbicide for the control of annual grass weeds and some broadleaf weeds in corn, sunflowers, soybeans and potatoes, and it can be used at the application rates of 1.15–2 kg/ha. Acetochlor introduced by Monsanto is used as a pre-emergence or pre-plant herbicide to control annual grass weeds, some broadleaf weeds and yellow nutsedges in cotton, soybeans, potatoes, peanuts, corn and sugarcane at rates of 0.75–1.5 kg/ha.

Metazachlor substituted with a pyrazolylmethyl group on the amide-nitrogen atom was developed by BASF. It gives pre-emergence and early

Table 12. Practical chloroacetamide VLCFAs biosynthesis

Chemical structure	ISO name Code No. Company	Dose Appl. method Target weeds	Patent No.
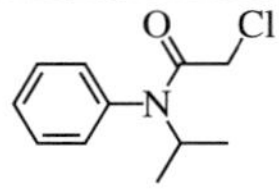	propachlor CP-3139 Monsanto	3~5 kg/ha pre, post corn, cotton, soybeans	US2863752 DE1014380
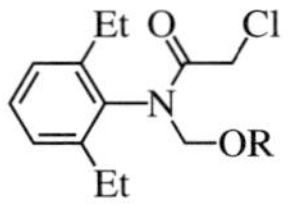	alachlor CP-50144 Monsanto (R=Me)	0.5~4 kg/ha pre corn, vegetables	US3442945 US3547620 NL6602564
	butachlor CP-53619 Monsanto (R=Bu)	1~4.5 kg/ha pre rice	US3442945 US3547620 BE677201 FR1481107
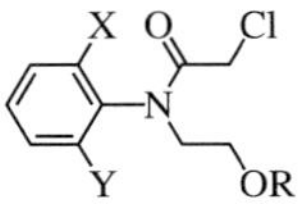	dimethachlor CGA-17020 Ciba-Geigy (X=Y=R=Me)	2~2.5 kg/ha pre rape	BE795021 GB1422473
	pretilachlor CG-113, CG-26423 Ciba-Geigy (X=Y=Et, R=Pr)	600 g/ha pre rice	BE800471 GB1438311 GB1438312
	propisochlor MG-87 Nitrokenia (X=Me, Y=Et, R=i-Pr)	1.15~2 kg/ha pre corn, soybeans, vegetables	HU208224 FR2514611
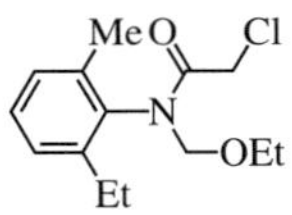	acetochlor NON-097, CP-55097 Monsanto	0.75~1.5 kg/ha pre corn, cotton, soybeans	US3941783
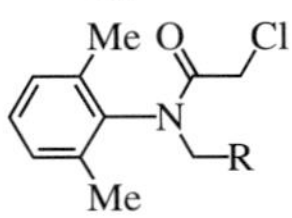	metazachlor BAS-47900-H, BAS-479-H BASF (R=pyrazol-1-yl)	0.5~1.5 kg/ha pre, post soybeans, rape	BR7800632 DE2704281 DE2742583
	thenylchlor NSK-850, DTH-105 Tokuyana (R=3-MeO-thiofen-2-yl)	180~270 g/ha pre rice, turf	US4802907 JP60-4181
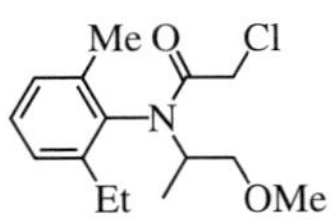	metolachlor CG-119, CGA-24705 Ciba-Geigy	0.75~2 kg/ha pre corn, cotton, turf, vegetables	BE800471 GB1438311 GB1438312 DE2328340
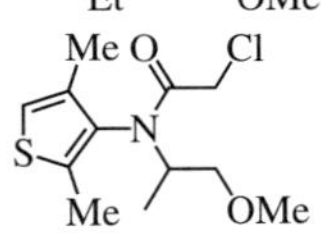	dimethenamid SAN-582H Sandoz	0.75~1.5 kg/ha pre corn, soybeans	GB2114566
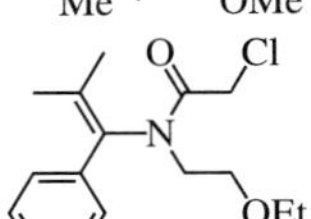	pethoxamid NSK-688 Tokuyama		EP206251 JP60134387 JP61293956

post-emergence control of annual grass weeds such as *Alopecurus myosuroides*, *Apera spica-venti*, wild oat, *Digitaria sanguinalis*, barnyardgrass, annual bluegrass and green foxtail, and broadleaf weeds such as *Amaranthus* sp., *Anthemis* sp., *Matricaria* sp., *Polygonum* sp., *Sinapis* sp., *Solanum* sp. and *Veronica* sp. in oilseed rape, soybeans, peanuts, maize, sugarcane, cotton and tobacco. Thenylchlor bearing a (3-methoxythiophene-2-yl)methyl group was developed by Tokuyama Soda. It controls annual grass weeds and annual broadleaf weeds in paddy fields with pre-emergence application at rates of 180–270 g/ha. Especially, thenylchlor is active on barnyardgrass up to the two-leaf stage.

Metolachlor shows good herbicidal activity on grass weeds, such as *Setaria* sp., *Digitaria* sp., *Echinochloa* sp., *Panicum* sp., *Sorghum halepense* and *Cyperus* sp. as well as broadleaf weeds, *Amaranthus* sp. and *Solanum* sp. in corn, cotton, sugar beet, sugarcane, potatoes, peanuts, soybeans, sunflowers, some vegetables and woody ornamentals. It is applied with pre-emergence at rates of 0.75–2 kg/ha. The *S*-isomer of metolachlor has more than twice the activity of the racemic form, and *S*-metolachlor will completely replace racemic metolachlor in the near future.

Dimethenamid was developed by Sandoz. It has a 2,4-dimethylthiophene ring instead of the benzene ring, which is applied for pre-emergence control of annual grass weeds and some broadleaf weeds in corn and soybeans at rates of 0.75–1.5 kg/ha. The *S*-isomer is now used as dimethenamid-P at rates of 400–820 g/ha because it is much more active than the racemic form. Pethoxamid, which has a 2-methyl-1-phenyl-1-propenyl group at the amide-nitrogen atom, is under development by Tokuyama Soda.

Generally, most chloroacetamides are easily synthesized by chloroacetylation and alkylation of anilines, irrespective of the order. In the case of *N*-alkoxymethyl derivatives, an alternative efficient method is adopted. The synthetic route for propisochlor is a good example of this method (Scheme 12). *N*-Methylidene-aniline (*119*) is acylated by chloroacetyl chloride, then chlorination of the imino moiety proceeds simultaneously to give the key precursor, *N*-chloroacetyl-*N*-chloromethylaniline (*120*), and subsequent treatment of *120* with sodium isopropoxide yields propisochlor. Scheme 12 demonstrates the synthetic method for pethoxamid. Isopropyl(phenyl)ketone reacts with 2-ethoxyethylamine to give 2-ethoxyethyl(2-methyl-1-phenylpropylidene)amine (*121*), which is chloroacylated with chloroacetic acid to obtain pethoxamid.

10.7.2
Other Very Long-Chain Fatty Acids Biosynthesis Inhibitors

Other VLCFAs biosynthesis inhibitors are depicted in Table 13. As there is no structural similarity among the compounds summarized, these herbicides are classified into thiophosphate-, acetamide- and carbamoyl-heterocycle classes based on their chemical structures.

In the thiophosphate class, piperophos and anilofos have been shown to be VLCFAs biosynthesis inhibitors. Piperophos is a selective herbicide active on

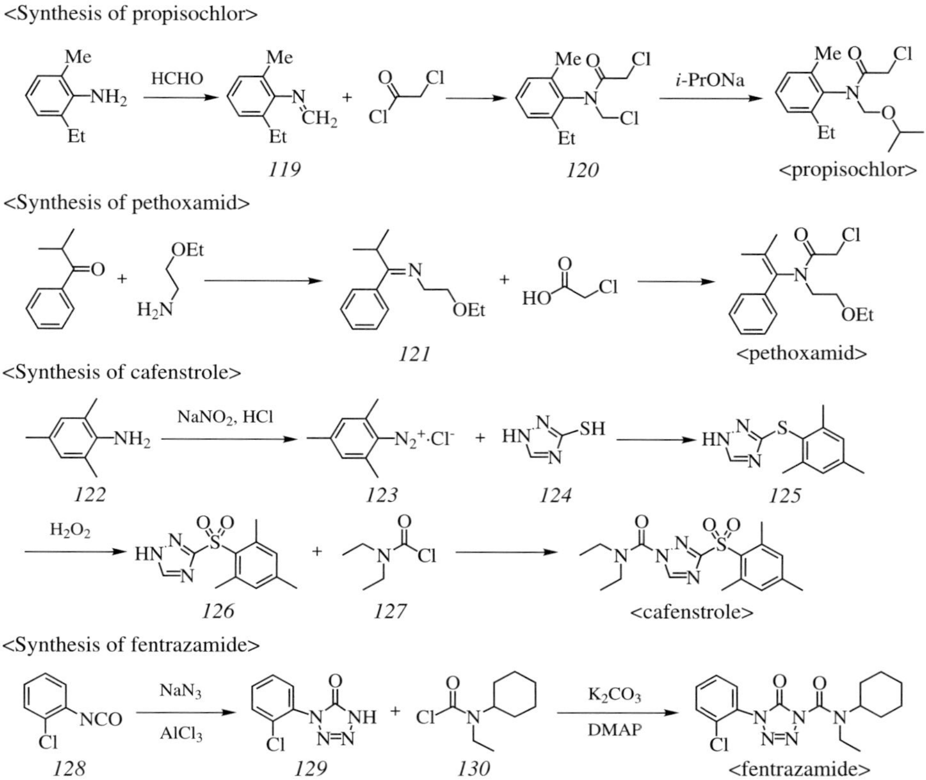

Scheme 12. Major synthetic routes for propisochlor, pethoxamid, cafenstrole and fentrazamide

annual grass weeds such as barnyardgrass, umbrella plant, *Monochoria vaginalis* and needle spikerush in seeded or flooded rice. Anilofos controls annual grass weeds including barnyardgrass and smallflower umbrella sedge in transplanted rice.

The acetamide class includes diphenamid, naproanilide, napropamide, mefenacet and flufenacet. Diphenamid is a pre-emergence herbicide in a wide range of crops, including cotton, potatoes, soybeans, vegetables, etc. It controls annual grass weeds and certain broadleaf weeds at rates of 3–6 kg/ha. A rice herbicide, naproanilide, was developed by Mitsui. It controls annual and some perennial grass weeds such as *Monochoria vaginalis*, Indian toothup, *Sagittaria pygmaea*, *Alisma canaliculatum* and needle spikerush. Naproanilde has long-lasting activity although its onset activity is rather slow.

By pre-emergence napropamide controls annual grass and broadleaf weeds, such as southern crabgrass, green foxtail, common lambsquarters, barnyardgrass, tufted knotweeds, common chickweed, livid amaranth, henbit, etc. in oilseed rape, tomatoes, potatoes, nuts, sunflowers, tobacco, turf and other crops

Table 13. Other VLCFAs biosynthesis inhibitors

Chemical structure	ISO name Code No. Company	Dose Appl. method Target crops	Patent No.
<Thiophosphate class>			
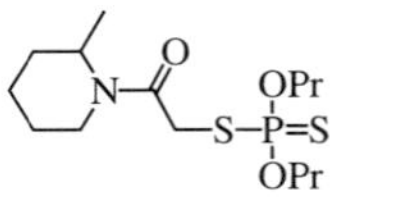	piperophos CG-7102, C-19490 Ciba-Geigy	2~3 kg/ha pre, post rice	BE725995 GB1255946
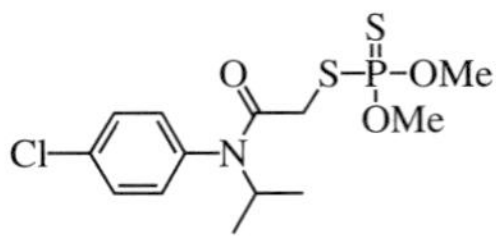	anilofos Hoe-574, HJ-9301 Hoechst	300~450 g/ha post rice	EP302334
<Acetamide class>			
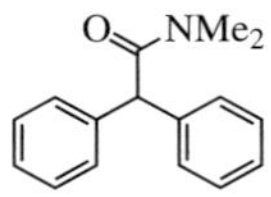	diphenamid L-34414 Eli-Lilly	3~6 kg/ha pre soybeans, cotton, vegetables	US3120434 US3141041
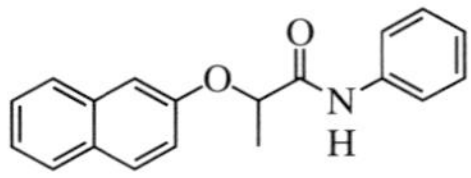	naproanilide NT-101 Mitsui	2~3 kg/ha pre rice, turf	JP49035533 DE2247076
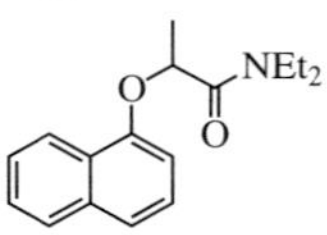	napropamide R-7465 Stauffer Chemical	2~4 kg/ha pre vegetables, potatoes, rape, sunflower, turf	US3480671 US3718455 FR1471683 GB1066606
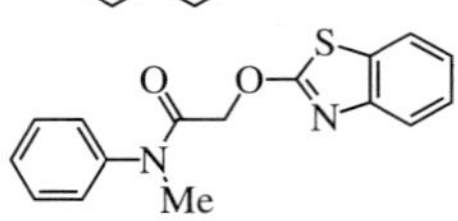	mefenacet NTN-801 Bayer	1~1.6 kg/ha pre, post rice	DE2822155 DE2903966 DE3038636 DE3323334
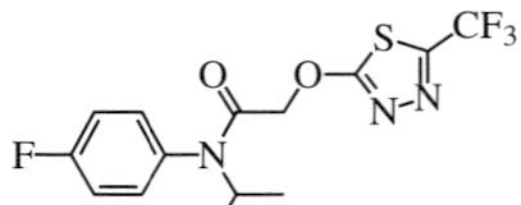	flufenacet BAY-FOE-5043 Bayer	75~100 g/ha pre, post corn, soybeans, cereals	DE4223465 WO9617519 JP0245475
<Carbamoyl-heterocycles>			
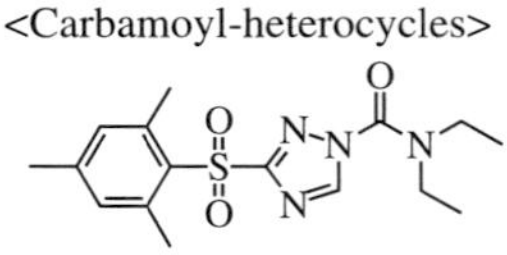	cafenstrole CH-900 Chugai	200~300 g/ha pre, post rice, turf	EP332133
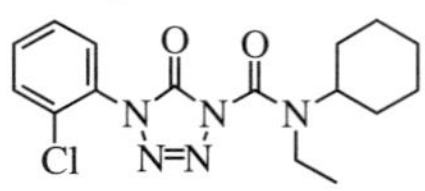	fentrazamide NBA-061, BAY-YRC-2388 Nihon Bayer Agrochem	200~300 g/ha rice	EP612735 EP146279 JP06306061 JP0782258

at rates of 2–4 kg/ha. The *R*-isomer of napropamide is eight times more active than the *S*-isomer. Mefenacet with a benzothiazole ring is a rice herbicide. It controls weeds up to the 2.5-leaf stage of early watergrass at rates of 1.2 kg/ha. Furthermore, mefenacet shows good herbicidal activity against harmful weeds such as needle spikerush and *Alisma canaliculatum*. Flufenacet controls grass weeds including barnyardgrass, green foxtail and southern crabgrass, and some broadleaf weeds in corn, soybeans, small-grain cereals, potatoes, rice and other crops with pre-emergence application rates of 75–100 g/ha.

In the carbamoyl-heterocycle class, cafenstrole and fentrazamide have been recognized as VLCFAs biosynthesis inhibitors. Cafenstrole was developed by Chugai and controls weeds up to the 2.5-leaf stage of *Monochoria vaginalis* in paddy rice. Moreover, it is effective against perennial broadleaf weeds and Japanese bulrush at rates of 200–300 g/ha.

Scheme 12 provides the major synthetic route for cafenstrole. 3-Mercapto-1,2,4-triazole (*124*) is mesitylated with diazonium salt (*123*) of 2,4,6-trimethylaniline (*122*) to give 3-mesitylthio-1,2,4-triazole (*125*). After *S*-oxidation by hydrogen peroxide, 3-mesitylsulfonyl-1,2,4-triazole (*126*) reacts with diethylcarbamoyl chloride (*127*) to yield cafenstrole.

The diethylcarbamoyl group of cafenstrole seems to be the most suitable substituent for herbicidal activity and the sulfonyl moiety is of herbicidal significance. Therefore, most of the novel triazole derivatives disclosed since 1990 possess such diethylcarbamoyl and sulfonyl moieties. Accordingly, structural modifications of the triazole class have been performed to optimize the substituents at the sulfonyl moiety as shown in Fig. 28.

Fentrazamide, which was launched by Nihon Bayer Agrochem in 1999, controls barnyardgrass and annual sedges in rice. Fentrazamide synthesis is shown in Scheme 12. 2-Chlorophenyl isocyanate (*128*) reacts with sodium azide in the presence of aluminum trichloride to give tetrazolinone (*129*). In the beginning, costly trimethylsilyl azide was used instead of sodium azide in this reaction. Compound *129* is converted to fentrazamide by reacting with *N*-cyclohexyl-*N*-ethylcarbamoyl chloride (*130*). In the second step, *O*-carbamoyl product is formed as a by-product; however, it was found that addition of a catalytic amount of DMAP suppresses this side reaction effectively.

In the tetrazolinone class, various substituents have been introduced at the carbamoyl group (Fig. 29), while the diethylcarbamoyl group is mainly substituted in the triazole class. Substituents at the other nitrogen atom are not always limited to the phenyl group like fentrazamide. For example, substituted alkyl groups and heterocycles such as pyrimidine and pyrazole rings have been proposed since 1990. Recent compounds [**747–750**], which are not tetrazolinone derivatives, are also illustrated in Fig. 29.

10.8 Cellulose Biosynthesis Inhibitors

10.8.1 Practical Cellulose Biosynthesis Inhibitors

Four practical herbicides, dichlobenil, chlorthiamid, isoxaben and flupoxam, are known as cellulose biosynthesis inhibitors (Table 14). These herbicides

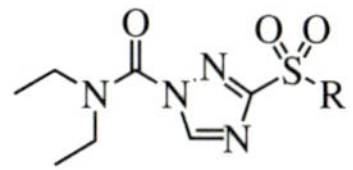

691 : 1991; R=4-Cl-$C_6H_4CH_2$
692 : 1991; R=2,4,6-Me_3-C_6H_2 (200g/Ecc/rice,turf) <carfenstrole>
693 : 1992; R=2-CF_3CH_2O-C_6H_4 (300g/Cha,Ecc,Dis/wheat,soya)
694 : 1994; R=cyclopentylmethyl (Ecc,Soh,Sev)
695 : 1999; R=i-Pr(i-PrOOC)CH (500g/Scj/rice)
696 : 2000; R=CF_2Br

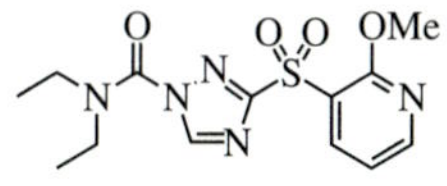

697 : 1993 (500g/Dia,Amb, Cha/soya,cotton)

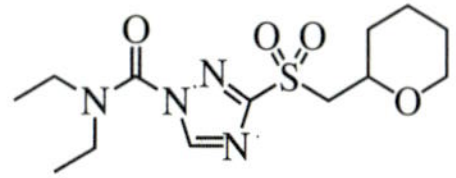

698 : 1998

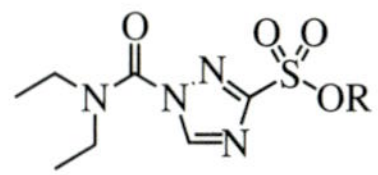

699 : 1993; R=4-Me-C_6H_4O (32g/Ecc,Scj/rice)
700 : 1993; R=3-pylidyloxy (130g/Ecc/rice)

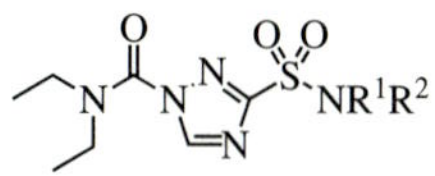

701 : 1992; R^1=Me, R^2=$C_6H_5CH_2$ (100g/Ecc,Mov)
702 : 1992; R^1=propargyl, R^2=$C_6H_5CH_2$ (300g/Ecc/rice)
703 : 1993; R^1=CF_3CH_2, R^2=$C_6H_5CH_2$ (1kg/pre/Sev,Cha,Pob/soya,corn,beet,rape,wheat)
704 : 1993; R^1=2-CF_3-C_6H_4CO, R^2=Me (500g/Scj,Mov,Pac/rice)
705 : 1993; R^1=propargyl, R^2=thienyl (100g/Pac,Scj,Mov/rice)
706 : 1993; R^1=propargyl, R^2=Et_2NCOCH_2 (50g/Pac,Scj,Mov/rice)

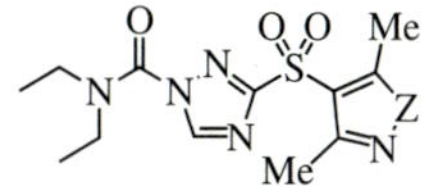

707 : 1991; Z=EtN (130g/post/Ecc/rice)
708 : 1994; Z=O

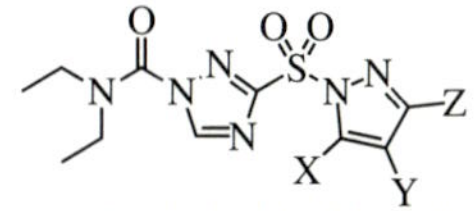

709 : 1993; X=Z=Me, Y=$MeNHCOCH_2$ (64g)
710 : 1993; X=BuNH, Y=H, Z=t-Bu (64g/rice)
711 : 1993; X=Z=Me, Y=H (130g/Ecc)
712 : 1993; X=Me, Y=H, Z=cyclohexyl (64g/rice)
713 : 1993; X=Z=Me, Y=MeS (64g/rice)

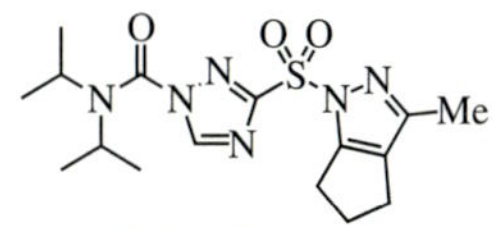

714 : 1993

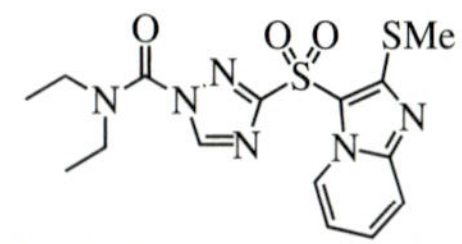

715 : 1997 (1kg/post/rice,orchard)

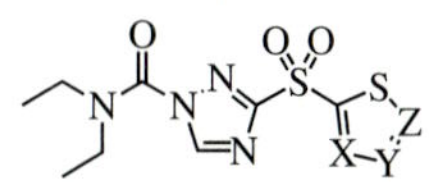

716 : 1997; X=t-BuC, Y=Z=N (250g/Ecc,Scj,Mov,Cyd,Lip/rice)
717 : 1998; X=N, Y=t-BuC, Z=MeC (125g/rice)

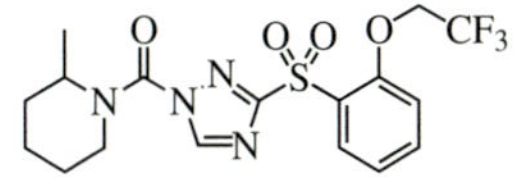

718 : 1991 (25g/Mov/rice)

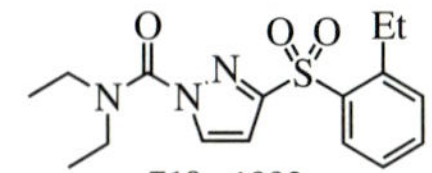

719 : 1992

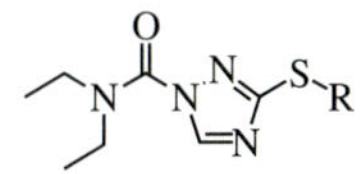

720 : 1991; R=$ClCH=CH_2$
721 : 1993; R=4-NC-2,5-F_2-C_6H_2

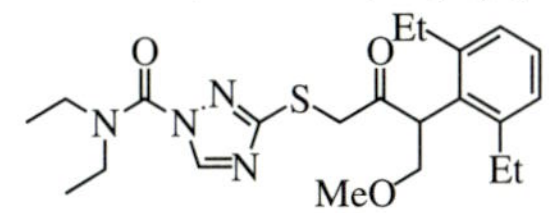

722 : 1993 (1kg/pre/Ecc,Scj/rice)

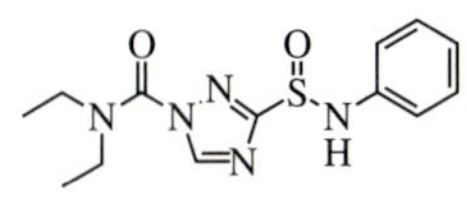

723 : 1993 (pre/Ecc)

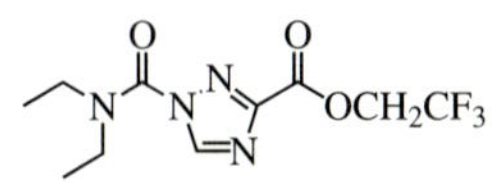

724 : 1993 (400g/pre/Ecc,Sev, Blackgrass,Stm,Avf/sugar beet)

No. : Year; Example (Dose per ha/Application/Weeds/Crops)

Fig. 28. Structural evolution of triazole VLCFAs biosynthesis inhibitors since 1990

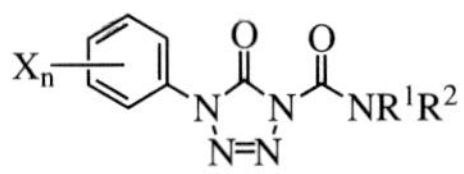

725 : 1993; $R^1=R^2$=allyl, X_n=3-Cl-4-CF_3O (500g/rice)
726 : 1993; $R^1=R^2$=Pr, X_n=3-Cl-4-CF_3 (500g/rice)
727 : 1993; $R^1=R^2$=allyl, X_n=3-Cl-4-i-Pr (250/Ecc,Scj/rice)
728 : 1994; $R^1=R^2$=Et, X_n=2-Cl (150g/rice)
729 : 1995; R^1=Et, R^2=cyclohexyl, X_n=2-Br (200g/Mov,Ecc/rice)
730 : 1998; R^1=Et, R^2=4-tetrahydropyranyl, X_n=2-Cl (Ecc, Dic,Sev/rice)
731 : 1998; $R^1=Me_2N$, R^2=1-cyclohexenyl, X_n=2-Cl (200g/Sev)
732 : 1999; $R^1=NCCH_2$, R^2=4-tetrahydropyranyl, X_n=2-Cl (1kg/post/Ecc)

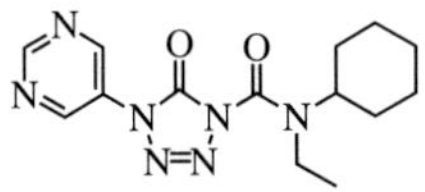

733 : 1997 (1kg/Ecc,Amr)

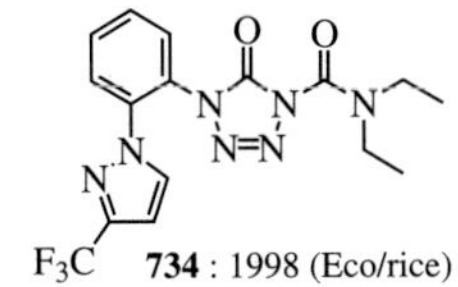

734 : 1998 (Eco/rice)

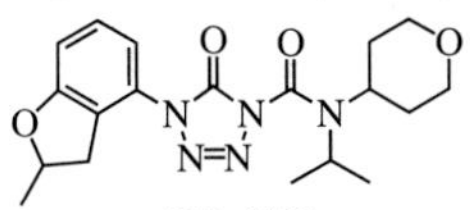

735 : 1999

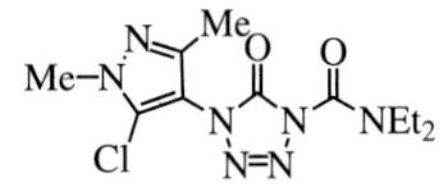

736 : 1996 (1kg/pre/Ecc,Amv)

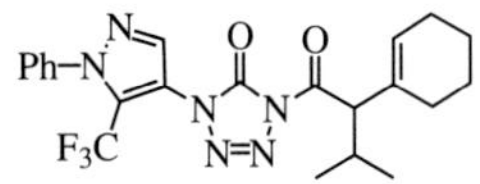

737 : 1999

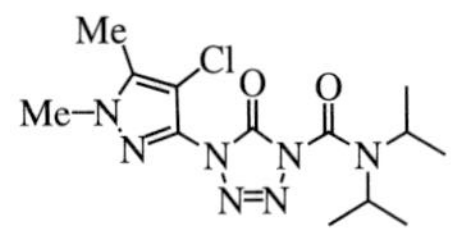

738 : 1997 (250g/Ecc/rice)

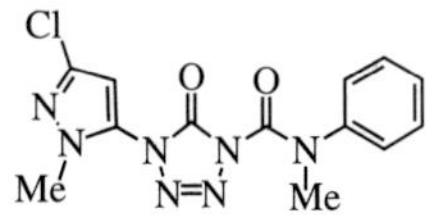

739 : 1997 (250g/Ecc/rice)

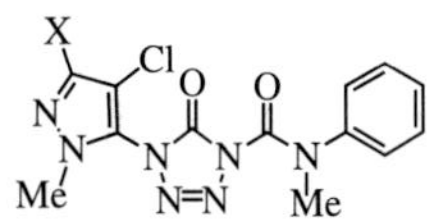

740 : 1997; X=Me (64g/Ecc/rice)
741 : 1997; X=Cl (60g/Scj/rice)

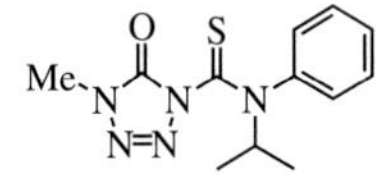

742 : 1997 (1kg/Ecc)

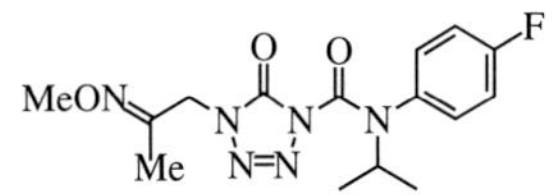

743 : 1998 (1kg/Ecc,Amr)

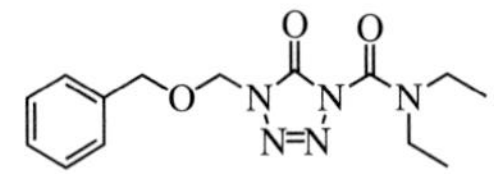

744 : 1998 (250g/pre/Ecc,Mov,Scj,Roi)

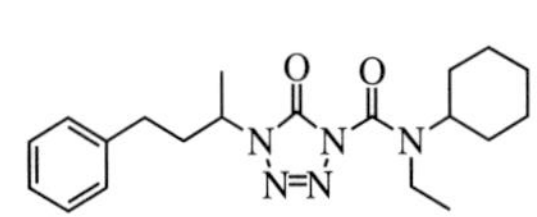

745 : 2000 (rice)

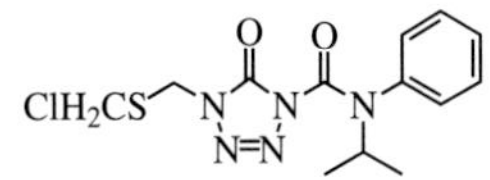

746 : 1999 (252g/rice)

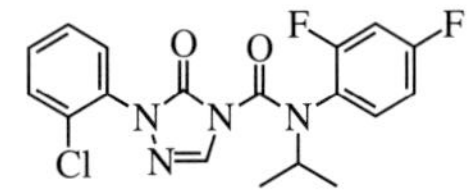

747 : 1998 (150g/Ecc)

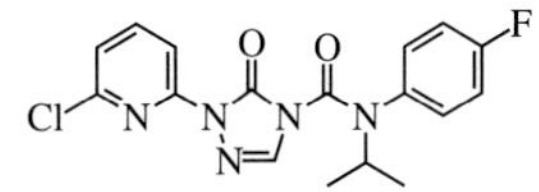

748 : 2000

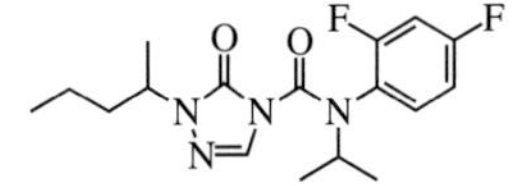

749 : 2000

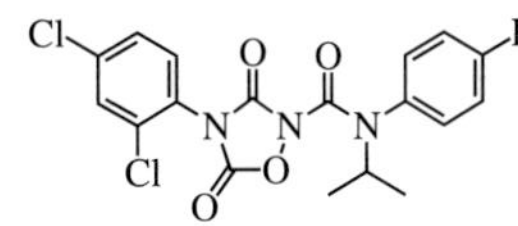

750 : 2000

No. : Year; Example (Dose per ha/Application/Weeds/Crops)

Fig. 29. Structural evolution of tetrazolinone VLCFAs biosynthesis inhibitors since 1990

Table 14. Practical cellulose biosynthesis inhibitors

Chemical structure	ISO name Code No. Company	Dose Appl. method Target crops	Patent No.
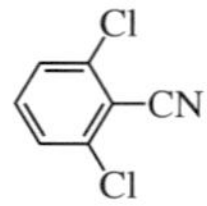	dichlobenil H 133 Philips Duphar	2.7~5.4 kg/ha pre, post rice, wheat	US3027248 BE587164 BE593212
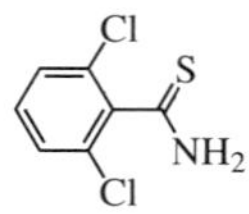	chlorthiamid WL 5792 Shell	pre, post rice, citrus	GB987253
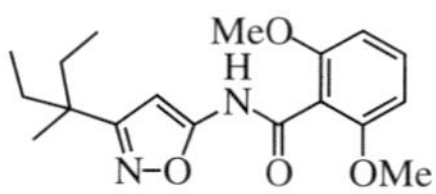	isoxaben EL-107 DowElanco	0.5~3 kg/ha pre cereals, corn, turf	FR2483406
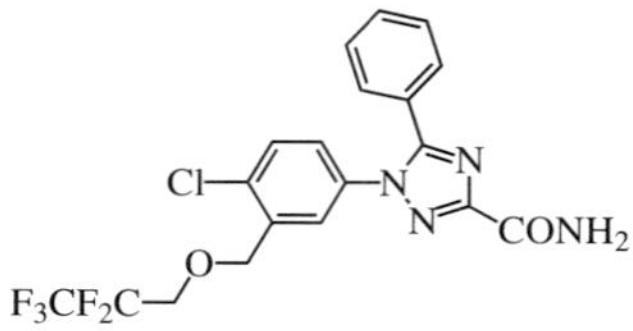	flupoxam KNW-739 MON-18500 Kureha	100~200 g/ha pre, post winter wheat, barley	EP612473

show a relatively low toxicity to mammals, fishes and shellfishes than other herbicides, because cellulose biosynthesis is peculiar to plants (cf. Chap. 7).

Dichlobenil is a pre- or post-emergence herbicide that gives good efficacy at 2.7–5.4 kg/ha. It controls annual and many perennial weeds in woody ornamentals, fruit orchards, vineyards, bush fruit and forest plantations. In the beginning, it was applied only in rice fields, especially to control slender spikerush, but the herbicide was not used widely because of damage to rice. Chlorthiamid is a pre- or post-emergence herbicide used in fruit orchards, e.g., pome fruit, currants, gooseberries, raspberries, citrus fruit, vines, olives etc. The herbicide is a so-called pro-drug, since it is converted to dichlobenil in soil and gradually undergoes microbial degradation to 2,6-dichlobenzamide and 2,6-dichlorobenzoic acid.

Isoxaben is a pre-emergence herbicide developed by DowElanco in 1984. It cannot be used in rice, because it is photodegraded in water. Isoxaben controls a wide range of broadleaf weeds including chamomile, chickweed, smartweed, ironweed and violet in winter cereals, turf, fruits, vines, ornamental trees and shrubs at 0.5–3 kg/ha without injuring winter barley, wheat, rye and oats. Flupoxam was discovered by Kureha and is under joint development with Monsanto for the European cereal market. It is a pre- or early post-emergence herbicide that controls annual broadleaf weeds in winter wheat and barley at 100–200 g/ha.

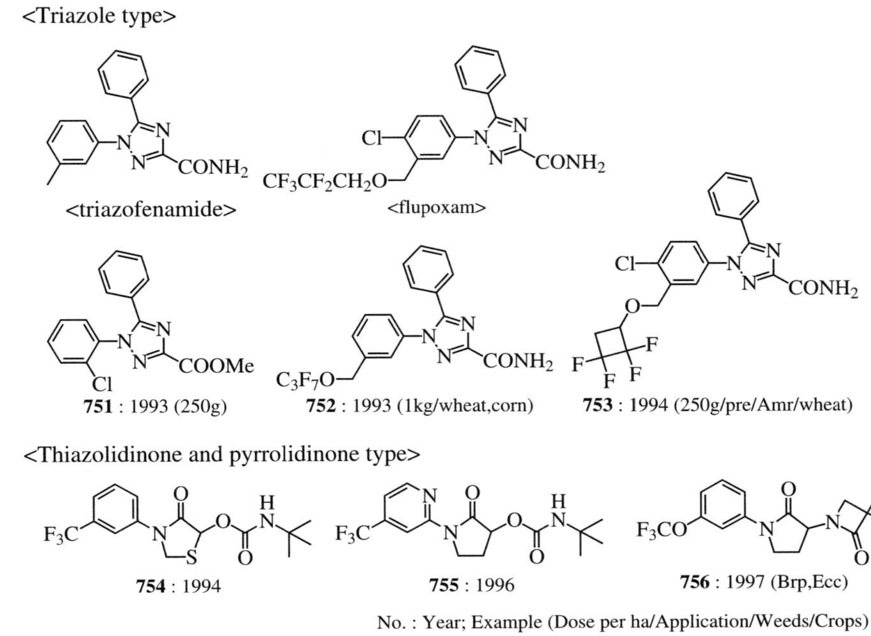

Fig. 30. Structural evolution of cellulose biosynthesis inhibitors since 1990

10.8.2
Structural Evolution of Cellulose Biosynthesis Inhibitors

Since 1990, several cellulose biosynthesis inhibitors such as triazoles, thiazolidinones and pyrrolidinones have been proposed (Fig. 30). Flupoxam was developed by Kureha through an ancient lead compound, triazofenamide, discovered in 1982. Since then, only three patents have been disclosed until now. Two pyrrolidinone derivatives [**755**, **756**] are thought to be derived by changing the hetero ring systems of thiazolidinone [**754**]. The mode of action of these pyrrolidinones [**755**, **756**] as cellulose biosynthesis inhibitors is not yet confirmed. The chemical structures of **755** and **756** are very similar to the PDS inhibitor, flurochloridone.

10.8.3
Major Synthetic Routes for Cellulose Biosynthesis Inhibitors

Isoxaben synthesis is illustrated in Scheme 13. The isoxazole moiety (*133*) is easily synthesized by cyclocondensation of hydroxylamine with 1-ethyl-1-methylpropyl(cyanomethyl)ketone (*132*) which is prepared by reaction of *131* with acetonitrile following methylation of methyl 2-ethylbutylate. The

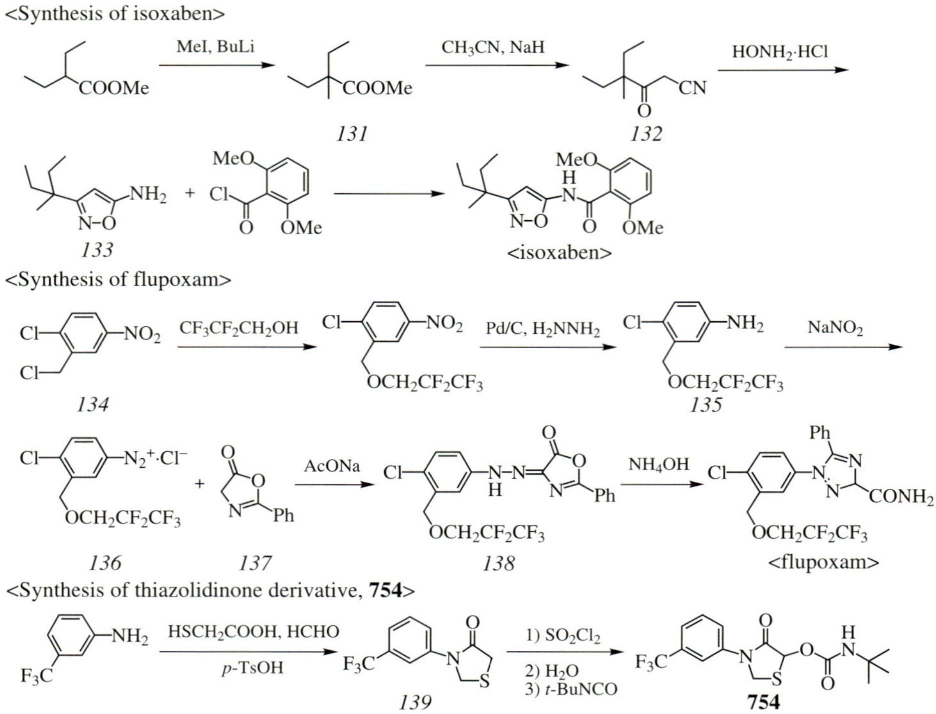

Scheme 13. Major synthetic routes for isoxaben, flupoxam and thiazolidinone derivative, **754**

compound (*133*) reacts with 2,6-dimethoxybenzoyl chloride to form isoxaben. Scheme 13 also shows the synthetic scheme for flupoxam. Benzyl chloride (*134*) reacts with pentafluoropropyl alcohol followed by reduction of the nitro group to give 3-pentafluoropropyloxymethylaniline (*135*). Diazonium salt (*136*), prepared by reacting *135* with sodium nitrite, reacts with 2-phenyl-2-oxazolin-5-one (*137*) to obtain 4-hydrazono-oxazolinone (*138*). Finally, the intermediate (*138*) is re-cyclized with ammonia affording flupoxam.

A new thiazolidinone-type herbicide opened to the public in 1994 by Zeneca is also classified as a cellulose biosynthesis inhibitor. The synthetic pathway for thiazolidinone [**754**] is outlined in Scheme 13. Cyclocondensation of 3-trifluoromethylaniline, thioglycolic acid and formaldehyde in the presence of *p*-toluenesulfonic acid gives the thiazolidinone derivative (*139*). Chlorination at 5-position of the thiazolidinone ring followed by hydrolysis and carbamoylation with *tert*-butyl isocyanate is accomplished to yield the target compound [**754**].

10.9 Protoporphyrinogen-IX Oxidase Inhibitors

This type of herbicide causes cell components to be destroyed by light-induced peroxidation which results in cell leakage, interruption of photosynthesis, and finally bleaching of chloroplast pigments. For this reason, protoporphyrinogen oxidase inhibitors are also called peroxidizers or photobleaching herbicides (cf. Chap. 8).

In contrast to other herbicides, PPO inhibitors have some general characteristics: (1) PPO inhibitors give long-lasting control for up to 30 days after the application. (2) PPO inhibitors are effective on currently difficult to control weeds, such as velvetleaf, cleavers, speedwells and deadnettle, ALS inhibitor-resistant waterhemp, *Kochia* and cleavers, and triazine-resistant pigweed and lambsquarters. (3) PPO inhibitors act more rapidly than many other herbicides, causing necrosis within 24h and death in 2–5 days. (4) PPO inhibitors in combination with other herbicides can provide one-shot weed control with a wide window of application.

PPO inhibitors can be grouped roughly into two classes, diphenyl ethers (DPE) and cyclic imide. HRAC and WSSA currently categorize eight DPEs and eighteen cyclic imides into this group. DPEs were the first widely used family of PPO inhibitor herbicides. Since the 1970s, the first-generation DPEs such as bifenox, chlomethoxyfen bearing a chlorine atom at the 4-position of the benzene ring have been changed into the second-generation forms by substitution with a trifluoromethyl group. This modification greatly improved the herbicidal efficacy. In the early 1980s, many DPEs have been proposed through continuous chemical modifications at the 3′-position of the other benzene ring; however, further evolution of DPEs has declined since 1990. Nowadays, the second-generation DPEs such as acifluorfen-sodium, fomesafen, halosafen, lactofen, oxyfluorfen are mainly used for weed control in soybeans and cotton. Details of DPEs have been published previously (see *Peroxidizing Herbicides* edited by P. Böger and K. Wakabayashi, Springer, Berlin Heidelberg New York, 1999). This chapter gives only a brief description of chemical structures of practical DPE class PPO inhibitors as shown in Fig. 31.

Active studies on cyclic imide class PPO inhibitors using the pioneer lead compounds of oxadiazon and chlorophthalim resulted in discovery of the next generation herbicide, flumiclorac-pentyl, which led to a major breakthrough in the later explosive development of cyclic imide herbicides. Study on new herbicides of this family is currently a very aggressive area of research in many chemical companies. More than ten kinds of new herbicides have been approved since 1990 and are now used worldwide. By the way, the general term of "*cyclic imide*" is thought to be misleading for modern PPO inhibitors because almost all compounds except for tetrahydrophthalimides do not strictly have an "*imide*" moiety at the hetero rings. Therefore, tentatively the

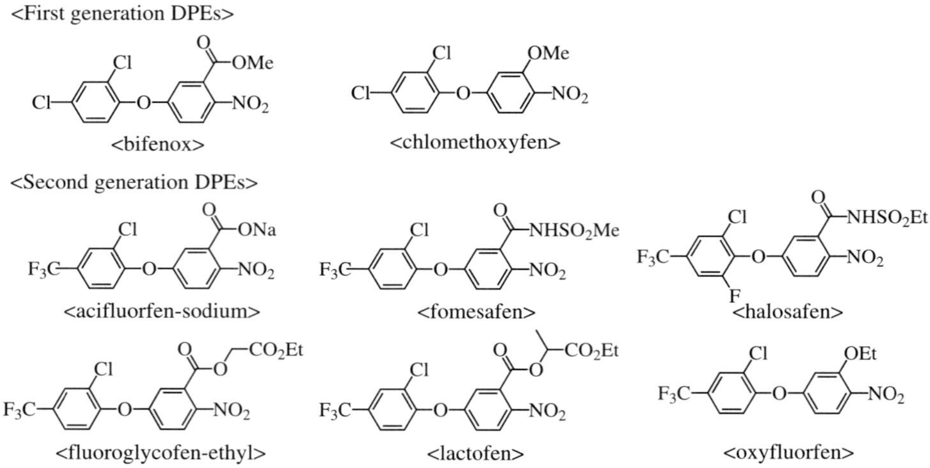

Fig. 31. Practical DPE class PPO inhibitors

term "*heterocycle PPO inhibitor*" is subsequently used in contrast to the DPE class PPO inhibitors.

10.9.1 Heterocycle Protoporphyrinogen-IX Oxidase Inhibitors

Common names and chemical structures of practical "*heterocycle PPO inhibitors*" are depicted in Tables 15 and 16. This section describes briefly agrochemical characteristics and major synthetic pathways for these heterocycle PPO inhibitors because most of them are fully summarized in the abovementioned volume, *Peroxidizing Herbicides* (see previous section). Exhaustive investigations looking for even more efficient PPO inhibitors have since been carried out and more than 200 patents on PPO inhibitors have been continuously released. Accordingly, the structural evolution of such newer PPO inhibitors released since 1995 is chronologically reviewed from the viewpoint of molecular design.

Most heterocycle PPO inhibitors developed in the 1980s have a cyclic methylene chain or terminal methyl groups such as *tert*-butyl and isopropylidene groups at their hetero ring parts, and such hydrophobic substituents have been considered to be essential for activity. However, in the latest heterocycle PPO inhibitors, those alkylene and alkyl moieties are replaced by fluorinated methyl groups such as a trifluoromethyl or difluoromethyl, which, of course, are also hydrophobic and may play an important role in improving the affinity to PPO. Based on these structural differences, heterocycle PPO inhibitors can be divided into two generations; the former including oxadiazon and

Table 15. First-generation heterocycle PPO inhibitors

Chemical structure	ISO name Code No. Company	Dose Appl. method Target crops	Patent No.
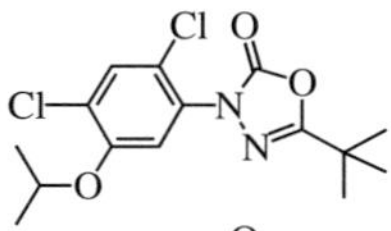	oxadiazon RP-17623 Rhône-Poulenc	1~4 kg/ha pre rice, turf	GB1110500 US3385862
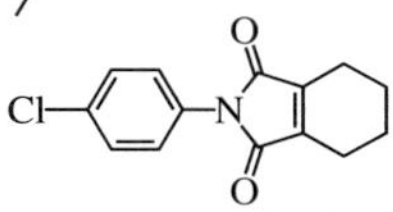	chlorphthalim MK-616 Mitsubishi	turf	JP4811940
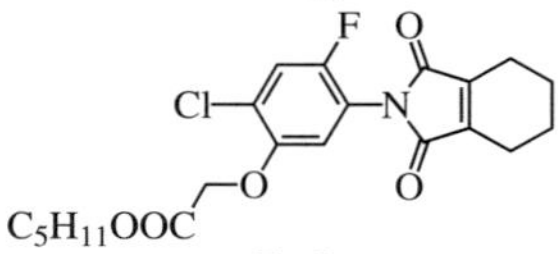	flumiclorac-pentyl S-23031,V-23031 Sumitomo/Valent	40~100 g/ha post soybeans	EP83055
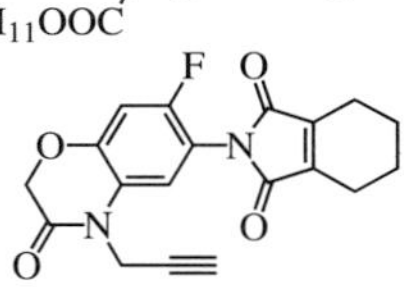	flumioxazin S-53482,V-53482 Sumitomo	50~100 g/ha pre soybeans, peanuts	EP170191
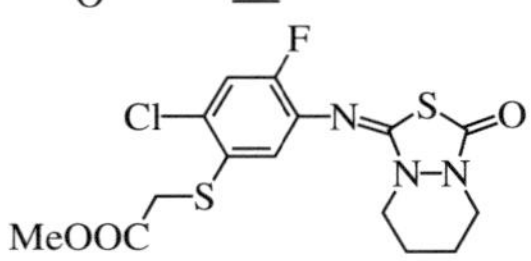	fluthiacet-methyl KIH-9201 Kumiai/Novartis	5~15 g/ha post corn, soybeans	EP273417
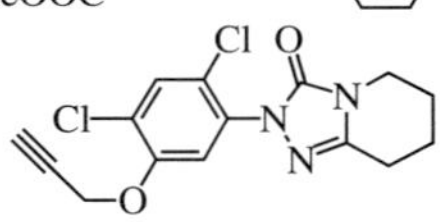	azafenidin DPX-R6447 DuPont	100~240 g/ha post fruit trees	US5332718 WO94/22828 EP784053
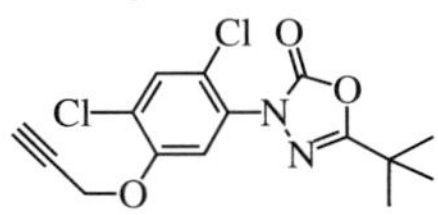	oxadiargyl RP-20630, RYH-118 Rhône-Poulenc	45~2000 g/ha pre, early post rice, sunflower, sugarcane	US3385862
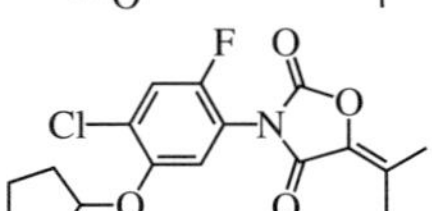	pentoxazone KPP-314 Kaken	150~450 g/ha pre, early post rice	WO87/02357
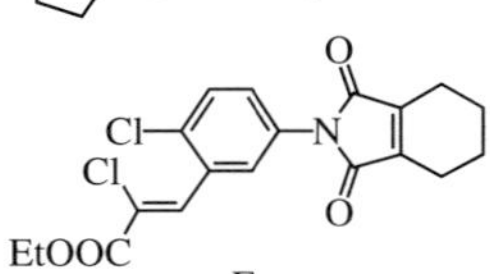	cinidon-ethyl BAS-615005 BASF	30~50 g/ha post winter wheat, winter barley	DE4209497 DE4042194
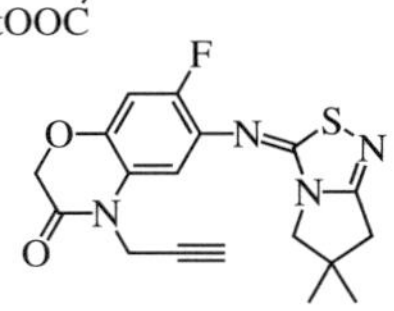	thidiazimin SN-124085 Schering	post winter cereals	EP311135

Table 16. Second-generation heterocycle PPO inhibitors

Chemical structure	ISO name Code No. Company	Dose Appl. method Target crops	Patent No.
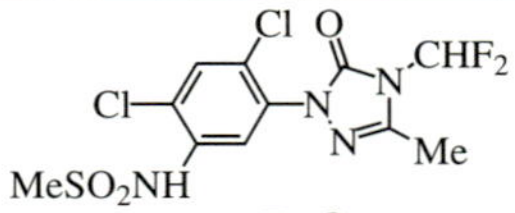	sulfentrazone F-6285 FMC	280~560 g/ha pre soybeans, sugercane, tabacco	WO87/03782 US4818275 GB2230261
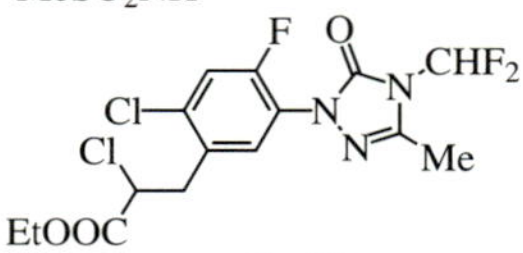	carfentrazone-ethyl F-8426 FMC	20~30 g/ha post corn, soybeans, cereals, rice	WO90/02120
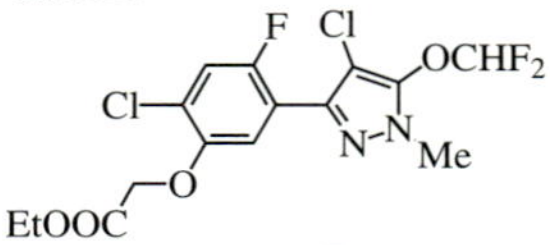	pyraflufen-ethyl ET-751 Nihon Nohyaku	6~12 g/ha post cereals	JP0372460 JP04225937
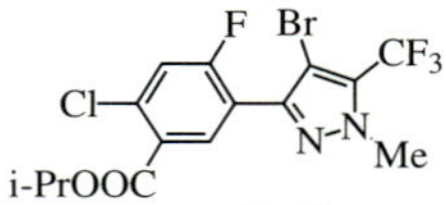	fluazolate JV-485, MON48500 Monsanto/Bayer	125~175 g/ha pre winter wheat	WO92/06962 US5489571 US5587485
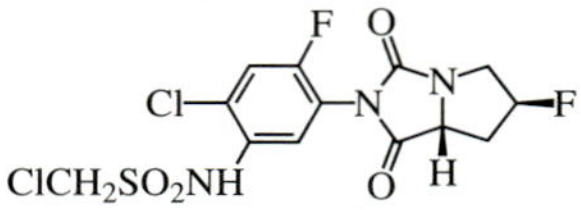	profluazol DuPont		WO97/15576
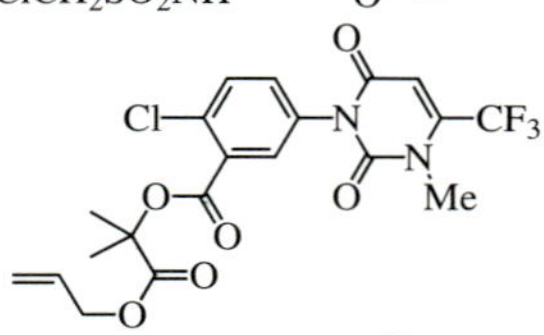	butafenacil-allyl CGA-276854 Novartis	herbicide-tolerant crops maize, wheat, rice, soybeans, etc	WO95/32952
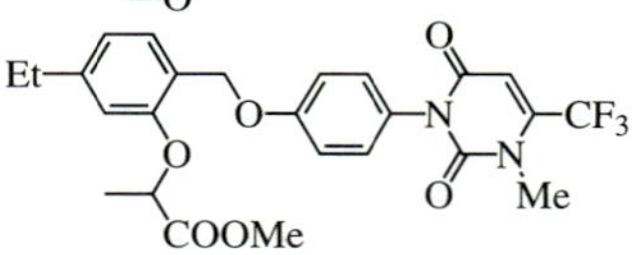	benzfendizone F-3686 FMC		WO95/17096
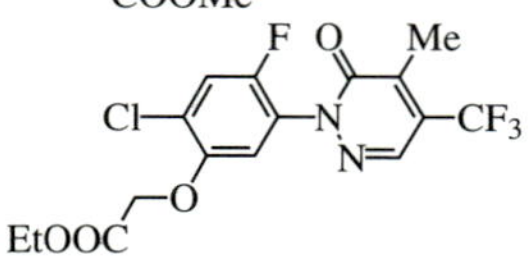	flufenpyr-ethyl S-3153 Sumitomo	corn, soybeans, sugarcane	WO97/07104
<Next-generation di-heterocycle PPO inhibitor>			
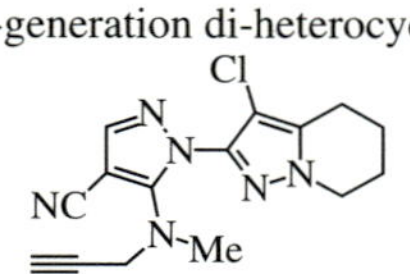	pyrazogyl HSA-961 Aventis	200 g/ha rice	WO94/08999

chlorophthalim as the first-generation, and the latter, including the latest heterocycle PPO inhibitors, as the second-generation.

10.9.1.1 First-Generation Heterocycle Protoporphyrinogen-IX Oxidase Inhibitors

Oxadiazon, disclosed by Rhône-Poulenc in the 1960s, is a pioneering compound in the development of PPO inhibitors. The herbicide contributed to an increased yield of rice for more than 20 years in Japan and has now been used mainly as a soybean herbicide in the USA. Oxadiargyl is a propargyl analogue of oxadiazon. Chlorophthalim is the origin of "*cyclic imide*" herbicides and has been continuously used as a turf herbicide.

Flumiclorac-pentyl is the first practical herbicide belonging to the next-generation PPO inhibitors. It gives post-emergence control of broadleaf weeds, e.g., velvetleaf, lambsquarters, cocklebur, ragweeds and morningglory at 30–60 g/ha. Particularly, the post-emergence application at 60 g/ha enables sufficient control of velvetleaf even at the six- to ten-leaf stages. Among tetrahydrophthalimide derivatives bearing a bicyclic benzologue moiety fused with various heterocycles at the 4- and 5-positions of the benzene ring, flumioxazin is herbicidally the most active analogue, which has a benzoxazine ring with a propargyl group at the nitrogen atom of the 3-position. Like flumiclorac-pentyl, flumioxazin is a leading herbicide for development of the next-generation PPO inhibitors. It provides effective control of annual grasses and broadleaf weeds such as velvetleaf, lambsquarters, amaranths, morningglory and prickly sida at 50–100 g/ha with contact activity and residual soil activity.

Kumiai developed fluthiacet-methyl in collaboration with Novartis (Syngenta) for post-emergence control of a wide range of broadleaf weeds in corn and soybeans. The herbicide was first approved for use on soybeans in 1999 in the USA. It controls problem broadleaf weeds such as velvetleaf and lambsquarters with an application rate of 4–5 g/ha; fully grown velvetleaf of 60 cm in height can be especially well controlled. Fluthiacet-methyl is thought to be a so-called pro-drug and its active species is the corresponding 2-thioxo-1,3,4-triazolidin-5-one that is biologically generated by isomerization of the 1,3,4-thiadiazolidin-5-one ring catalyzed by glutathion-*S*-transferase. Thidiazimin is a bioanalogue of fluthiacet-methyl although the substituents at the benzene and heterocyclic rings are different. That is, in theory, the thiocarbamate (NC(=O)—S) moiety in fluthiacet-methyl may transform bioanalogously to a thioimino (C=N—S) moiety. However, the thidiazimin structure does not change and the authentic thidiazimin is the herbicidally active species.

Bicyclic triazolinone, azafenidin, is a long-residue herbicide active on broadleaf weeds at 100–240 g/ha and is now under development as a broad spectrum post-emergence herbicide on fruit trees (citrus), cultivated sunflowers, cucurbits, basil and asparagus. Degree of residue and carry-over may determine the registration for sunflowers and other tolerant crops.

An oxadiazole herbicide, oxadiargyl was developed for pre- and early post-emergence application on annual grass and broadleaf weeds in several crops such as rice, sunflowers, sugarcane, potatoes and other vegetables. It was first launched in Latin America in 1996 and then in Asia in 1997 for pre-emergence use in rice. It exhibits potent activity against barnyardgrass at the 1.5-leaf stage and other annual lowland weeds at 35–80 g/ha under soil incorporation conditions.

Pentoxazone developed by Kaken is effective against small-seeded annual weeds such as barnyardgrass, *Monochoria vaginalis*, umbrellasedge and a wide-range of broadleaf weeds when applied pre- or early post-emergence at the rates of 150–450 g/ha. Many perennial sedge weeds except arrowhead and harmful weeds such as *Eleocharis kuroguwai* are also controlled or suppressed. Application timings of pre- or early post-emergence (–4 to +5) at less than the 1.0-leaf stage of weeds are most effective for full efficacy. Pentoxazone gives rapid burn down and residual control of barnyardgrass for up to 45 days. In combination with sulfonylureas such as imazosulfuron, pyrazosulfuron-ethyl and cyclosulfamuron, pentoxazone is further effective as a one-shot herbicide with a wide window of application timings for the control of annual and perennial weeds in transplanted rice. As a PPO inhibitor, it gives an alternative mode of action to control ALS inhibitor-resistant weeds such as *false pimpernel* and *Monochoria korsakowii*.

A fluorine atom at *ortho*-position of the benzene ring in PPO inhibitors plays an important role in enhancing the herbicidal efficacy. However, it was recently reported that non-fluorinated tetrahydrophthalimide derivatives inhibited the target enzyme stronger than the corresponding fluorinated ones *in vitro*. The fluorine atom seems to be not always indispensable for high herbicidal activity and is merely required to form a suitable dihedral angle between the two rings owing to its steric hindrance with the carbonyl group at the hetero ring. This is supported by the recent approval of cinidon-ethyl or butafenacil-allyl having non-fluorinated phenyl groups. Cinidon-ethyl developed by BASF is particularly active post-emergence on cleavers, among other broadleaf weeds, in small grains at 30–50 g/ha. It controls cleavers, speedwells, deadnettle and hempnettle in winter wheat and winter barley.

10.9.1.2
Second-Generation Heterocycle Protoporphyrinogen-IX Oxidase Inhibitors

Sulfentrazone and carfentrazone-ethyl developed by FMC are the first herbicides of the second-generation PPO inhibitors. Both herbicides as well as azafenidin belong to the aryl triazolinone family, but unlike azafenidin, they are monocyclic triazolinone heterocycles substituted by a difluoromethyl group at 4-position of the 1,2,4-triazoline ring. Sulfentrazone is a broad-spectrum, pre-emergence herbicide for use on soybeans, sugarcane and tobacco, and shows strong herbicidal activity on a wide range of broadleaf weeds such as morningglory, pigweeds, lambsquarters, velvetleaf and knotweeds as well as some grasses and nutsedges at 280–560 g/ha. It was launched in the US in combination with chlorimuron-ethyl in 1997 following the first registration in Brazil in 1996.

Carfentrazone-ethyl, unlike sulfentrazone, is used as a post-emergence non-residual herbicide in corn, cereals, rice and soybeans. Even at rates as low as 20–30 g/ha, it controls numerous harmful weeds, for example, cleavers, speedwells and deadnettle in cereals; velvetleaf, redroot pigweed, black nightshade, morningglories and lambsquarters in US maize crops; and other weeds including waterhemp and cocklebur. Carfentrazone-ethyl can be used alone or tank-mixed with most other herbicides, especially with existing cereal herbicides such as flupyrsulfuron-methyl-sodium for further enhancement of the activity and weed spectrum. The herbicide was first launched in Europe in 1997 followed by US approval for use on maize, soybeans and wheat the next year. It was also used as a herbicide for rice and cereals in Asia and sold as a potato haulm desiccant in Europe.

Generally, in heterocycle PPO inhibitors, the aromatic character and planarity of the hetero ring consisting of sp^2 carbons and heteroatoms are necessary for the potent herbicidal efficacy. However, most of the bicyclohydantoin derivatives are exceptionally active despite their non-aromatic character. Profluazol developed by DuPont is such a bicyclohydantoin PPO inhibitor with a fluorine atom at a unique position. The fluorine atom and its position are considered to be significant for the herbicidal performance and crop safety like the difluoromethyl group of sulfentrazone and carfentrazone-ethyl.

Pyraflufen-ethyl has an essentially different structure from the aforementioned PPO inhibitors, i.e., the pyrazole and benzene rings are connected by a C—C bond instead of the usual C—N bond, and a difluoromethyloxy group is substituted at the 5-position at the pyrazole ring. Its structural novelty together with a high PPO inhibitory activity was the driving force behind further explorations for new herbicides. Pyraflufen-ethyl is a post-emergence herbicide for use on cereals; it controls broadleaf weeds such as cleavers, henbit, chickweeds and wild chamomile at rates as low as 6–12 g/ha. In particular, it provides effective control of the five to six-leaf stages of cleavers. The herbicide has been developed by Nihon Nohyaku and was first launched in 1999 in eastern Europe and Japan, and subsequently registered in the EU in 1999 in combination with bifenox. The non-selective herbicide in combination with glyphosate-trimesium was also registered for use in orchards and non-agricultural areas in Japan in 1999. Fluazolate was discovered by Monsanto and is being developed through the Monsanto/Bayer joint venture. Like pyraflufen-ethyl, the benzene and pyrazole rings are connected by a C—C bond. Fluazolate provides effective control of cleavers and blackgrass in wheat and it is recommended for pre-emergence use with a 125–175 g/ha application rate.

Since the 1990s, research energies have been focused on 6-membered heterocycle PPO inhibitors particularly because of the unexpected efficacy of 6-trifluoromethyluracils. Although the mode of action of these 6-membered heterocycles has not been confirmed throughout, modern compounds such as butafenacil-allyl, benzfendizone and flufenpyr-ethyl are evidently PPO inhibitors based on their chemical structures. The spectacular success of the molecular design of most 6-membered heterocycle PPO inhibitors is that a trifluoromethyl group is introduced at the site opposite to the phenyl group. The trifluoromethyl group is the most important and favorable substituent, and enhances herbicidal activity extremely. From the viewpoint of structural distinction, these herbicides are also classified into the second-generation PPO inhibitors.

Among trifluoromethyluracil derivatives, butafenacil-allyl is the most promising herbicide that is developed by Novartis (Syngenta) for use on their herbicide-tolerant Acuron crops. It is an effective pre-emergence herbicide, and provides long-lasting weed control for 30–60 days and rapid weed kill in post-emergence use. Like other PPO inhibitors, it controls various broadleaf and grass weeds including ALS inhibitor-resistant weeds and triazine-resistant pigweed and lambsquarters. It will be introduced alone and in combination with other herbicides to give broad-spectrum weed control over a wide window of application timings. The Acuron crop is a transgenic crop that exhibits resistance to PPO inhibitors such as butafenacil-allyl and products of other companies already on the market. Transgenic crops to be developed are maize, wheat, rice, soybeans, canola, cotton, sorghum, sugar beet and other crops. Currently, Syngenta is targeting corn for its first introduction. The first approval of butafenacil-allyl in the USA is expected to coincide with the launch of the Acuron corn in 2004.

Flupropacil (UCC C-4243), which is illustrated in Fig. 36, is also a trifluoromethyluracil PPO inhibitor with a slightly different substituent from butafenacil-allyl. It is a pre-emergence herbicide effective on annual weeds in wheat and as a desiccant for potatoes; however, it seems that the development has been discontinued recently. Benzfendizone is the third PPO inhibitor of FMC in the trifluoromethyluracil family. A uniquely substituted benzyl group like *N*-{4-(4-chlorobenzyloxy)phenyl}-3,4,5,6-tetrahydrophthalimide (MK-129), the incipient tetrahydrophthalimide herbicide, modifies the *para*-position of the benzene ring.

Flufenpyr-ethyl is a novel 6-membered heterocycle PPO inhibitor, where the hetero ring is an unusual pyridazin-6-one substituted by a trifluoromethyl group at the opposite position to the benzene ring. Therefore, the herbicide should be classified as a second-generation PPO inhibitor. It gives excellent control of velvetleaf and morningglory and has been developed by Sumitomo for use on corn, soybeans or sugarcane.

10.9.2 Structural Evolution of Protoporphyrinogen-IX Oxidase Inhibitors Since 1995

All compounds released from 1980 up to the end of 1997 are described in detail in the volume mentioned (see Sect. 10.9, *Peroxidizing Herbicides*). Nevertheless compounds released since 1995 are cited in this section to facilitate the understanding of chronological changes of each class. Since 1995, more than 200 patents on PPO inhibitors have been disclosed. Among them, there are only about 30 patents on the first-generation PPO inhibitors, the rest relate to the second-generation PPO inhibitors.

10.9.2.1
Structural Evolution of First-Generation Heterocycle Protoporphyrinogen-IX Oxidase Inhibitors

The structural evolution of tetrahydrophthalimides and related compounds is indicated in Fig. 32. Drastic modifications are not recognized in tetrahydrophthalimides except for the tricyclic derivatives [**762–764**]. A series of phthalimide derivatives [**766–769**] are distinguished from the conventional ones by introducing a fluorine atom at the phthalimide ring. *N*-acylated tetrahydrophthalamic acid ester [**771**], in which recyclization to the corresponding phthalimide is blocked by the acyl group, has been developed for pre-emergence use in rice.

Figure 33 shows the structural evolution of thiadiazolidinones, thiadiazolinones and bicyclic pyrazoles since 1995. Based on the model fluthiacet-methyl, some thiadiazolidinones [**775–778**] and its ring-opened derivatives [**773, 774**] have been proposed. Unfortunately, at present, no compound is especially mentioned. Thiadiazolinones [**779, 780**] are sulfur-modified analogues of oxadiazon or oxadiargyl.

10.9.2.2
Structural Evolution of Second-Generation Heterocycle Protoporphyrinogen-IX Oxidase Inhibitors

Because of their structural novelties and herbicidal performance, active investigations on the 3-arylpyrazole family have proceeded since the consecutive discoveries of pyraflufen-ethyl and fluazolate. As shown in Fig. 34, more than ten derivatives have been proposed, even for the short period since 1995. However, low flexibility in structural modification of the pyrazole ring limits the variety of the substituents. In contrast, the benzene ring is decorated in many ways, particularly; compounds [**791, 792**] are characterized by their unique benzene ring systems. That is, each benzene ring is fused with a thiazole or oxazole ring at the 5- and 6-positions, unlike flumioxazin. These unconventional ring systems are often observed in the latest trifluoromethyluracil PPO inhibitors. 4-Cyanophenyl-substituted pyrazoles [**786, 795**] should be placed in the series of 4-cyanophenyl-heterocycle PPO inhibitors that have been intensively developed by Bayer.

The triazolinone family has turned out three practical herbicides, namely sulfentrazone, carfentrazone-ethyl and azafenidin. More than 50 patents have been opened to the public since the early 1980s but there has been no

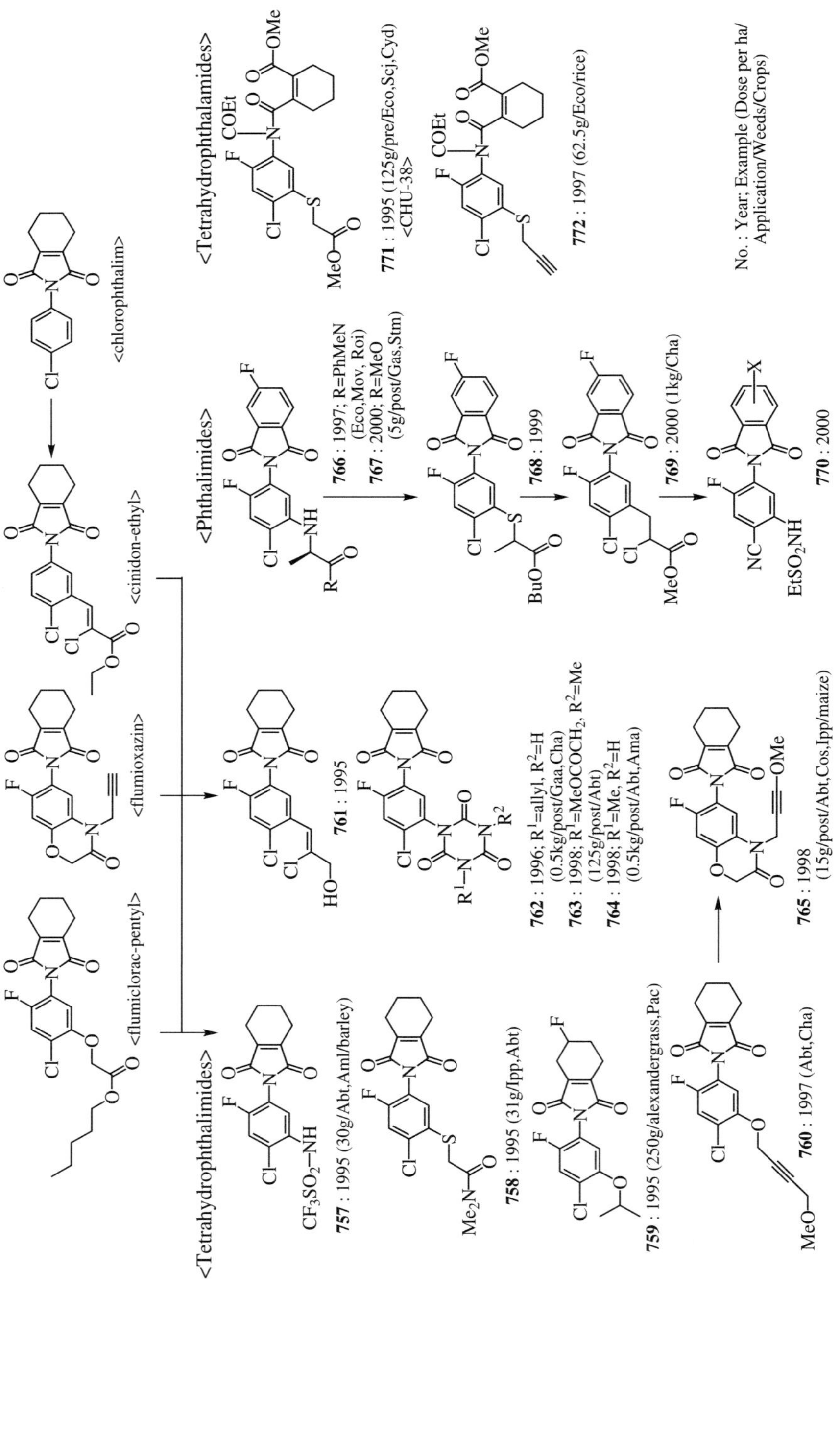

Fig. 32. Structural evolution of tetrahydrophthalimide PPO inhibitors and related compounds since 1995

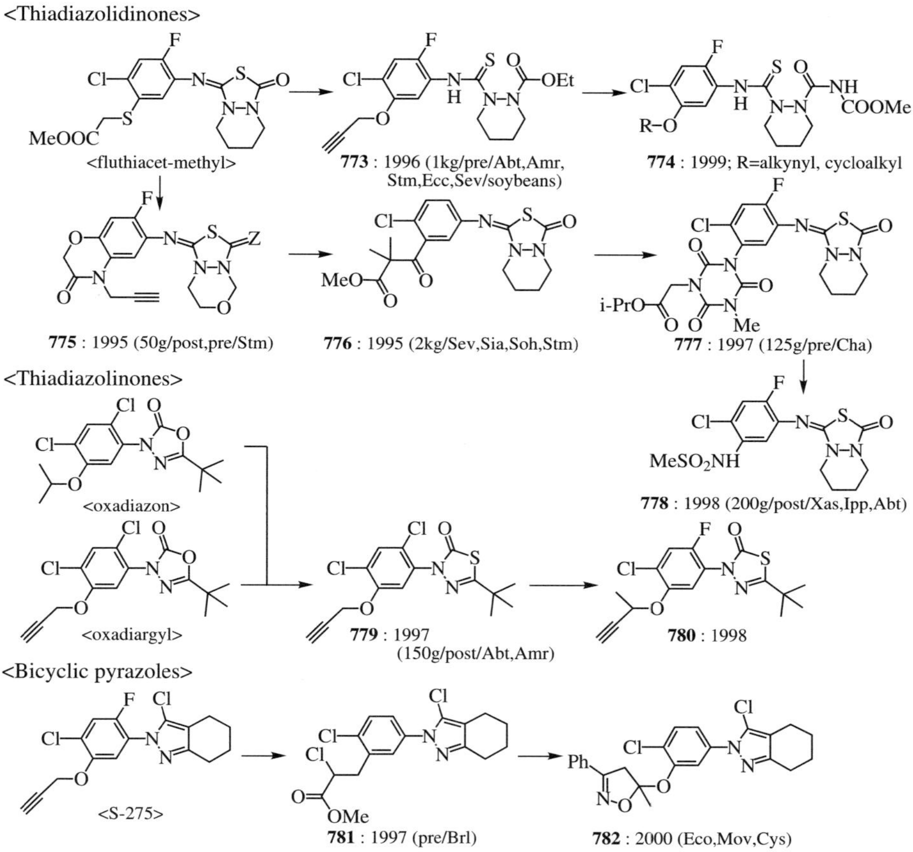

No. : Year; Example (Dose per ha/Application/Weeds/Crops)

Fig. 33. Structural evolution of thiadiazolidinone, thiadiazolinone and bicyclic pyrazole PPO inhibitors since 1995

disclosure in the last 3 years as shown in Fig. 35. Triazolinone derivatives [**806–810**] are a new type of PPO inhibitors with a benzofuran ring. Especially, 5-trifluoromethyltriazolinone (OK-9701) is being developed as a long-residual rice herbicide active on barnyardgrass, *Monochoria vaginalis*, *Lindernia pyxidaria* and *Cyperus difformis* at 80–120 g/ha.

Trifluoromethyluracils have been most exhaustively investigated as active PPO inhibitors, which lead to the high performance candidates, butafenacil-allyl, benzfendizone and flupropacil. In almost all derivatives synthesized, the substituents at the uracil ring are fixed as the trifluoromethyl group at 6-position and the methyl group at 1*N*-position except for a few 1*N*-amino derivatives, therefore, research energies have been devoted entirely to the structural modifications of the benzene ring systems. Among the new deriva-

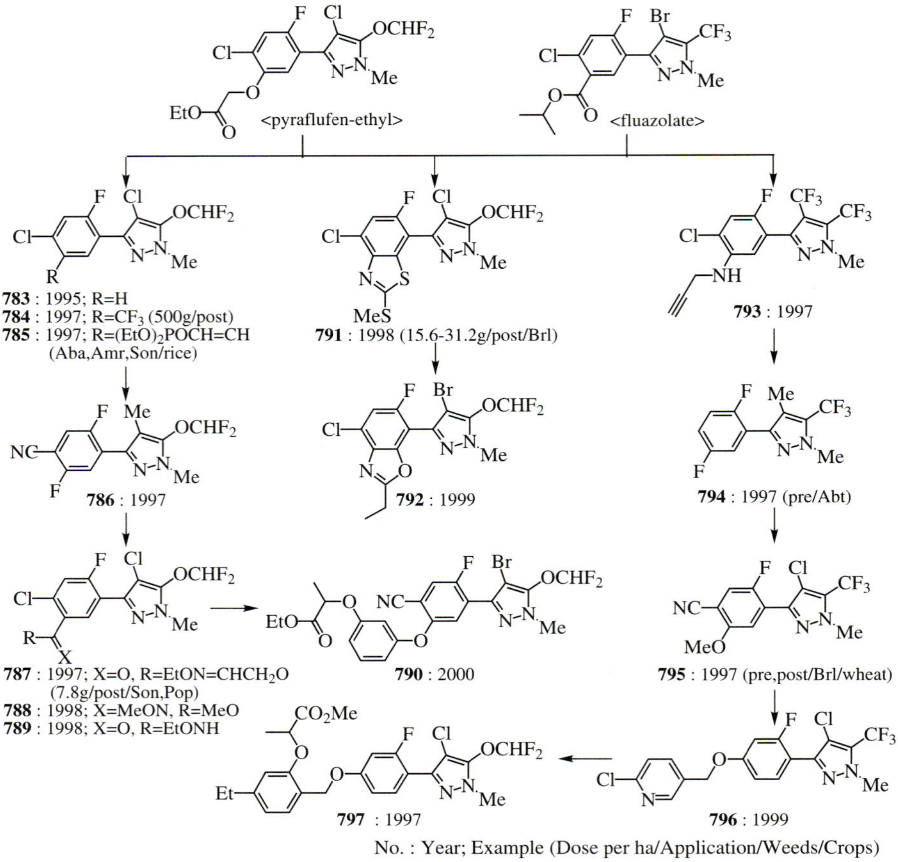

Fig. 34. Structural evolution of pyrazole PPO inhibitors since 1995

tives in Fig. 36, compounds [**826, 827**] have quite unconventional groups at the *ortho*-position.

It is noteworthy that various 3-(4-cyanophenyl)-6-trifluoromethyluracils have been developed by Bayer and other companies. There was no derivative before 1995 and all compounds until now are illustrated in Fig. 37. Currently, through the 2,5-difluoro-4-cyanophenyl derivative [**828**] as a key intermediate, about 20 kinds of new trifluoromethyluracils have been mainly synthesized by nucleophilic substitution of the 5-fluorine atom with nitrogen or oxygen nucleophiles and the successive reactions. It is thought that these compounds have a great potential for practical herbicides.

Figure 38 shows structural modifications of bicyclic benzologue-substituted trifluoromethyluracils. As the benzoxazine derivative [**851**] released in 1991 is the first example for this series, all compounds disclosed continuously since

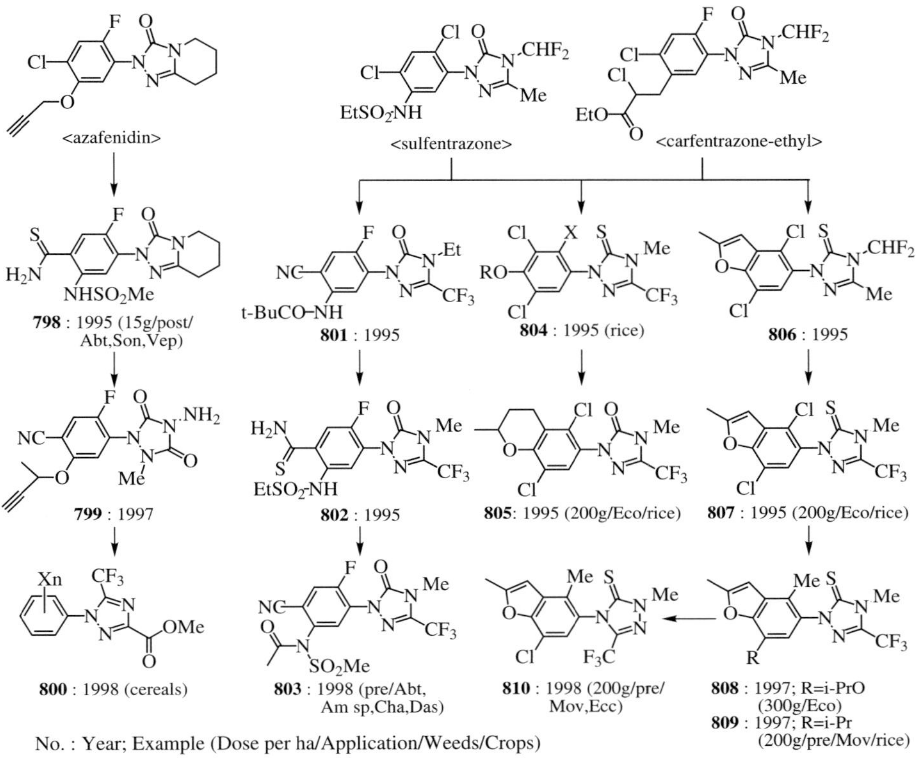

Fig. 35. Structural evolution of triazolinone PPO inhibitors and related compounds since 1995

1991 are cited in this figure. Benzologue systems of uracils [**851–867**] are as usual as other PPO inhibitors; however, special attention should be paid to a series of uracils [**868–885**] because of their unprecedented benzologue systems. That is, a variety of 5-membered heterocycles are fused at the 5- and 6-positions of the benzene rings. Further investigations and evaluations will reveal which kinds of benzologues are suitable for potent herbicides.

The structural evolution of other trifluoromethyluracils [**886–891**] is demonstrated in Fig. 39. The left column shows 4-(substituted benzyloxy)phenyluracils. The importance of the substituted benzyloxy group at the benzene rings is not yet clear, but the substitution system, like MK-129, is attractive due to its simple structure. The middle column shows benzyluracils [**892–898**] whose mode of action is not identified as PPO inhibition as well as those illustrated on the right column [**899–903**].

In addition to trifluoromethyluracils, several 6-membered heterocycle PPO inhibitors have been investigated since 1995; as shown in Fig. 40. Most compounds can be classified into the group of second-generation PPO inhibitors because electron-withdrawing substituents such as trifluoromethyl,

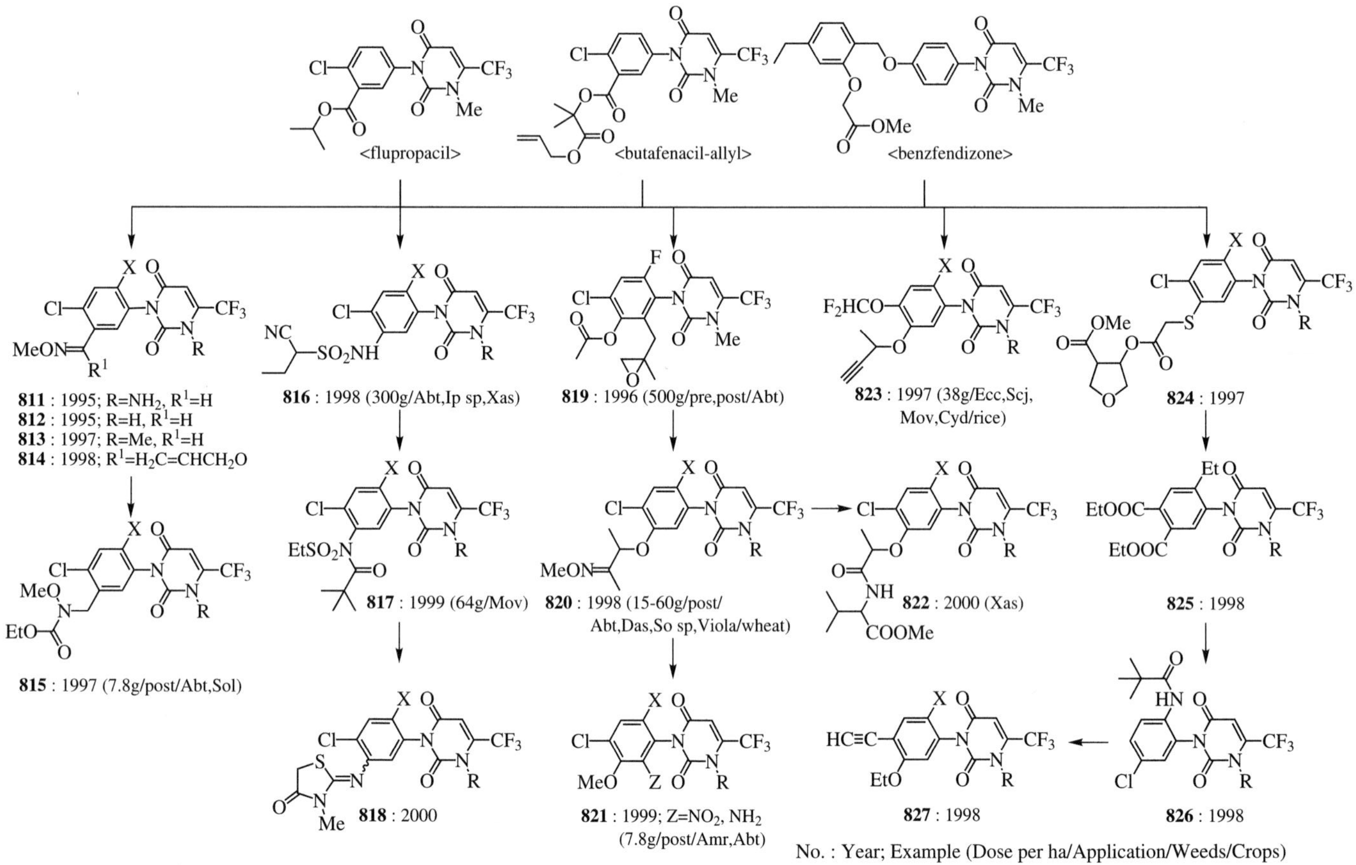

Fig. 36. Structural evolution of trifluoromethyluracil PPO inhibitors since 1995

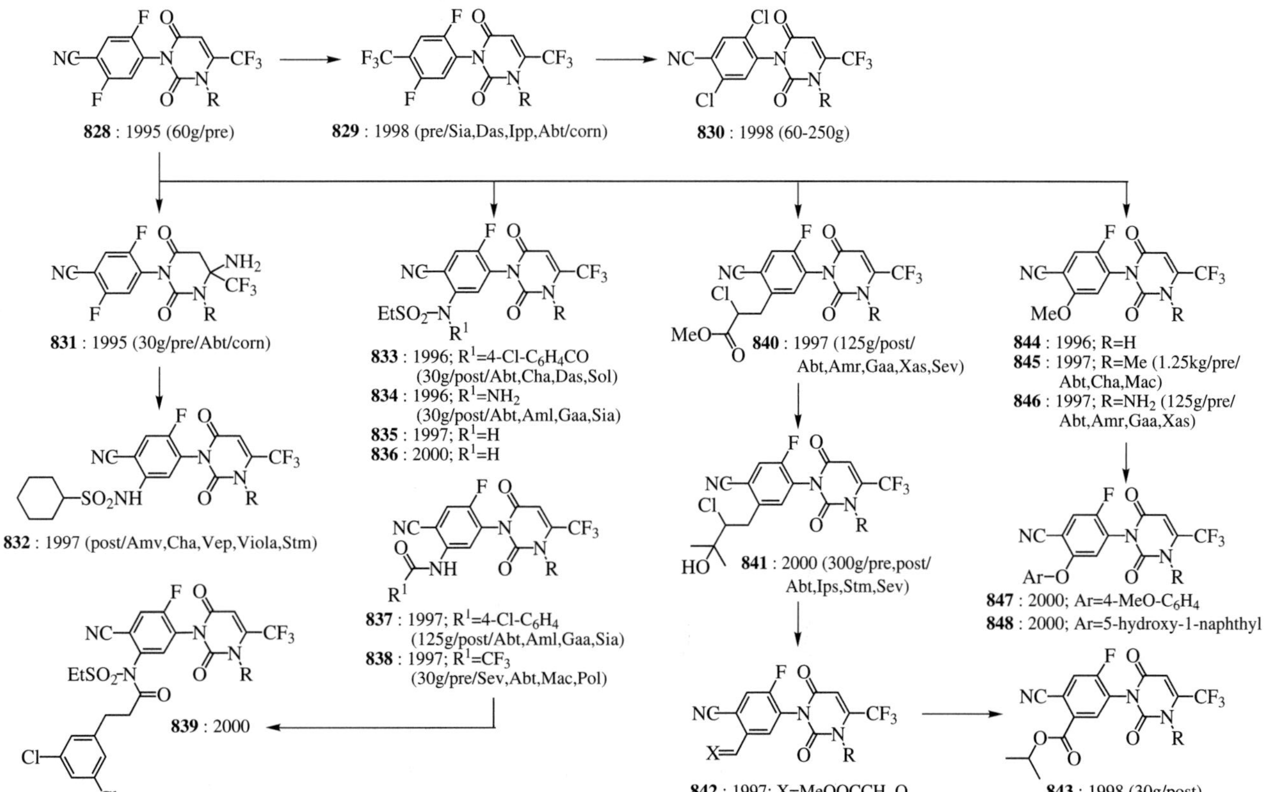

No. : Year; Example (Dose per ha/Application/Weeds/Crops)

Fig. 37. Structural evolution of 3-(4-cyanophenyl)-6-trifluoromethyluracil PPO inhibitors since 1995

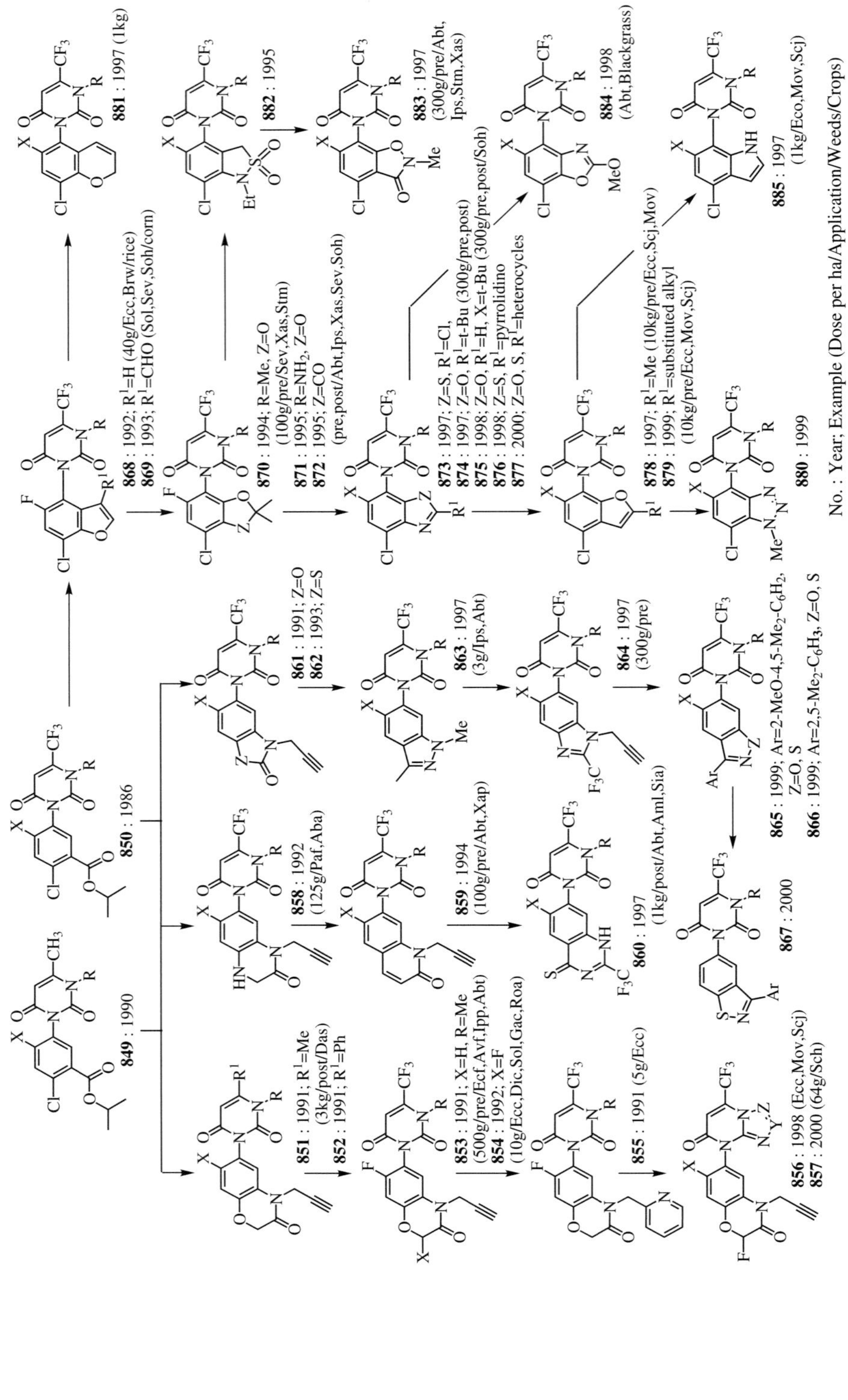

No. : Year; Example (Dose per ha/Application/Weeds/Crops)

Fig. 38. Structural evolution of bicyclic benzologue-substituted trifluoromethyluracil PPO inhibitors since 1990

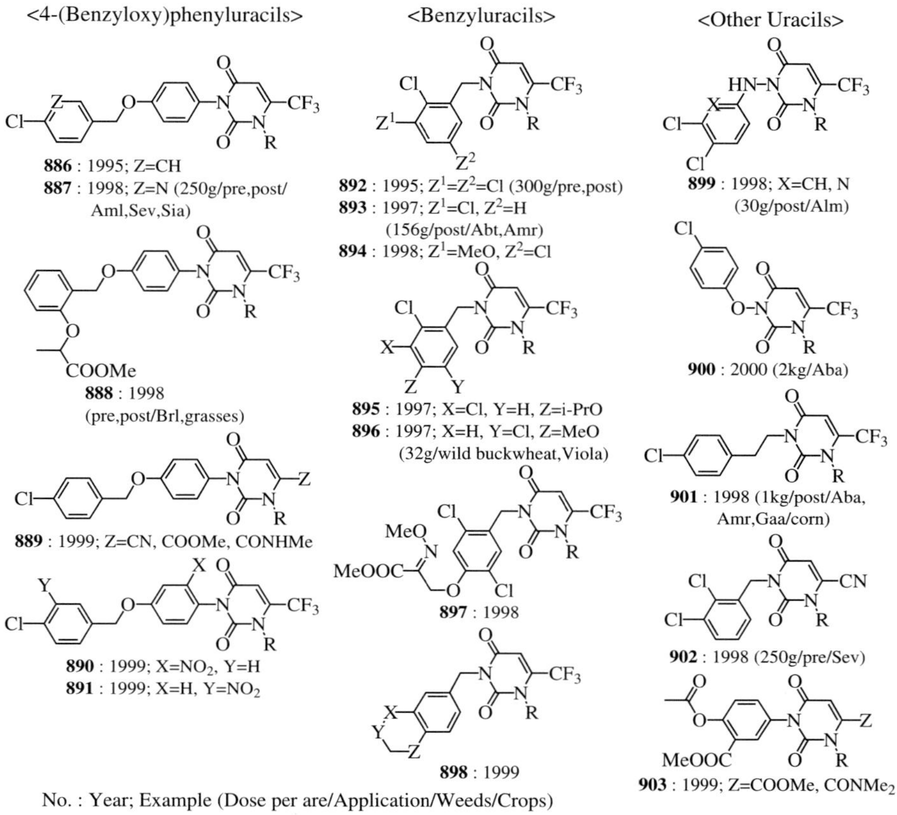

Fig. 39. Structural evolution of other trifluoromethyluracil PPO inhibitors since 1995

methylsulfonyl, carbonyl and thiocarbonyl groups are introduced at opposite sites to the benzene rings. Above all, pyridazine derivatives [**915–921**] have been actively studied, mostly by Sumitomo, and a promising candidate, flufenpyr-ethyl [**915**], is being developed for use on corn, soybeans or sugarcane.

10.9.2.3
Next-Generation Heterocycle Protoporphyrinogen-IX Oxidase Inhibitors

The aforementioned heterocycle PPO inhibitors consist of substituted benzene rings and a variety of heterocycles, and both rings are believed to be necessary for potent herbicidal activity. However, since 1995, unique heterocycle PPO inhibitors assembled with a pair of heterocycles have been disclosed, i.e., the benzene rings of usual heterocycle PPO inhibitors are replaced with nitrogen containing heterocycles such as pyrazole, thiazole, pyridine and pyridone rings (Fig. 41). Leading compounds of these “*di-heterocycle PPO inhibitors*” are

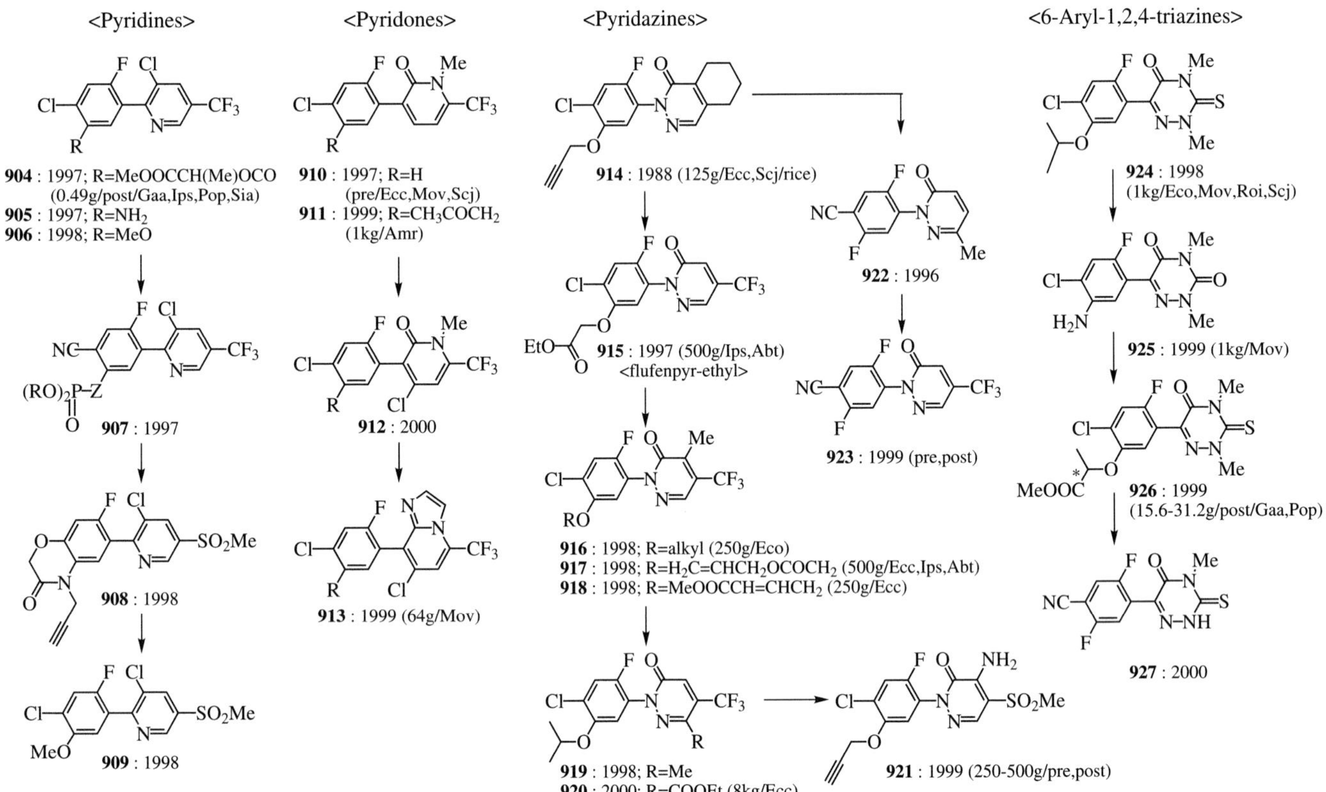

No. : Year; Example (Dose per ha/Application/Weeds/Crops)

Fig. 40. Structural evolution of pyridine, pyridone, pyridazine and 1,2,4-triazine PPO inhibitors since 1995

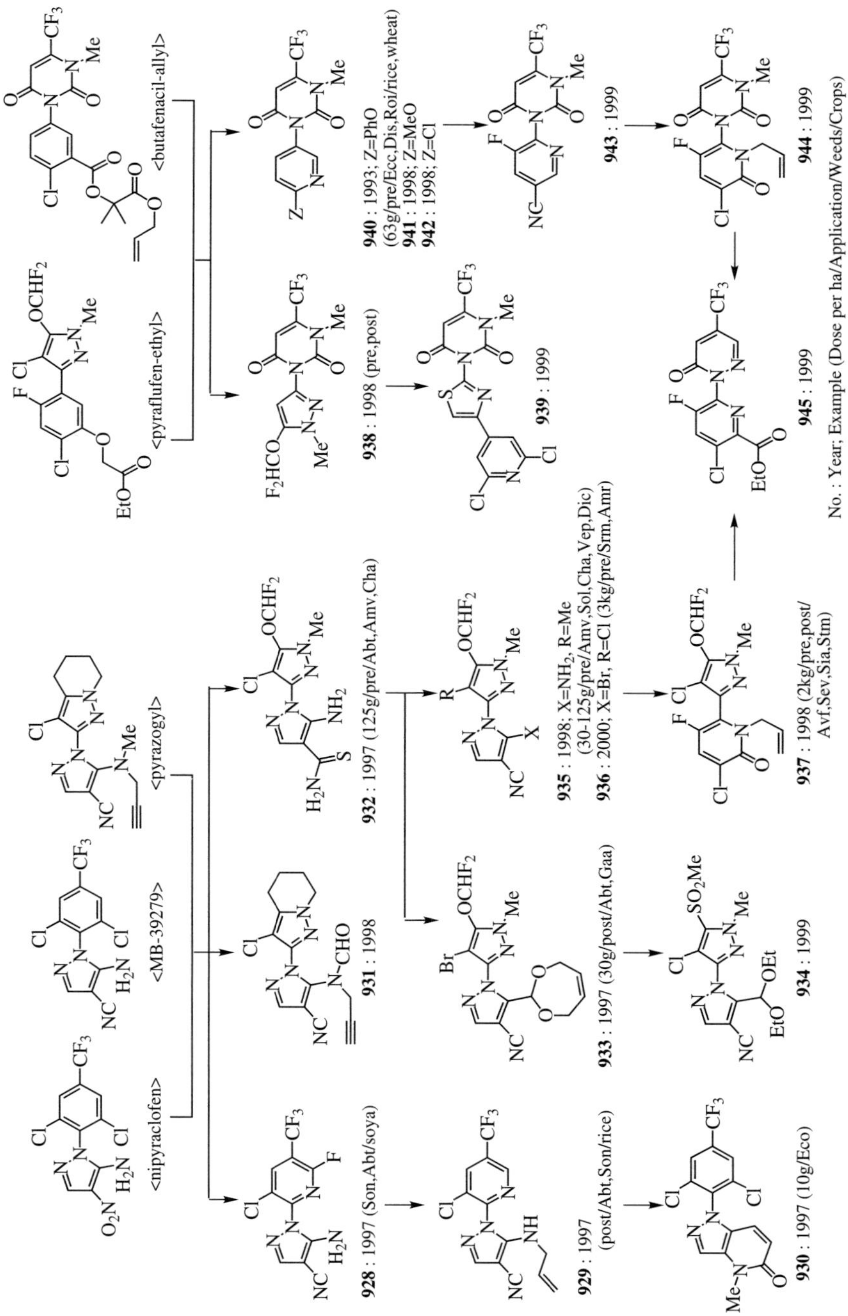

Fig. 41. Structural evolution of di-heterocycle PPO inhibitors since 1990

nipyraclofen and MB-39279, whose pyrazole rings probably correspond to the so-called benzene ring as well as their analogue [**930**] and pyridine derivatives [**928**, **929**]. Furthermore, it should be kept in mind that the 2,6-dichloro-4-trifluoromethylphenyl group in nipyraclofen, MB-39279 and **930** corresponds to the so-called hetero ring of usual heterocycle PPO inhibitors.

Further modifications produced a promising herbicide, pyrazogyl. It is effective on barnyardgrass in rice at a low rate, although the development has been discontinued recently. Subsequently, less than 20 kinds of di-heterocycle PPO inhibitors have been proposed from several companies. Currently, pyrazole-pyrazole [**931–936**] and pyridine-uracil [**940–944**] are representative combinations and their pyrazole and uracil rings, corresponding to the so-called hetero rings of heterocycle PPO inhibitors, are designed on the basis of pyraflufen-ethyl and butafenacil-allyl to elicit powerful herbicidal activity. Further evolutions of this type of compounds certainly should lead to the discovery of the next-generation PPO inhibitors in the near future.

10.9.3
Major Synthetic Routes for Protoporphyrinogen-IX Oxidase Inhibitors

Synthetic routes for many PPO inhibitors including oxadiazon, oxadiargyl, flumiclorac-pentyl, flumioxazin, fluthiacet-methyl, azafenidin, cinidon-ethyl, thidiazimin, sulfentrazone, carfentrazone-ethyl and pyraflufen-ethyl have been described in detail in the previous volume, *Peroxidizing Herbicides*. Therefore, this chapter focuses on several 5- and 6-membered heterocycle PPO inhibitors of which synthetic pathways are new or have been improved.

The 5-isopropylidene-1,3-oxazolidine-2,4-dione skeleton of pentoxazone can be constructed by the addition reaction of 2-hydroxy-3-methyl-3-butenoate (*142*) to 4-chloro-5-cyclopentyloxy-2-fluorophenyl isocyanate followed by intramolecular cyclization and olefin isomerization in basic conditions. In this pathway, however, preparation of the phenyl isocyanate required toxic phosgene or its dimer. Thermal cyclization of *N*-phenylcarbamate (*141*) with butenoate (*142*) provided an efficient new method for the synthesis of pentoxazone. The *N*-phenylcarbamate (*141*) was prepared by carbamation, hydrolysis and cyclopentylation of *O*-protected aniline (*140*) in a one-pot reaction as shown in Scheme 14.

Pyrazole PPO inhibitors represented by pyraflufen-ethyl and fluazolate have chemical structures entirely different from other C—N bridged cyclic imide classes, i.e., the pyrazole and phenyl rings bond together by a C—C bond. Therefore, a somewhat complicated synthetic strategy, via benzoylacetate or benzoylacetone as a key intermediate, has been employed for them. Synthesis of fluazolate requires cyclocondensation of trifluoroacetylacetophenone (*144*) with hydrazine followed by methylation on the nitrogen atom and bromination at 4-position of the pyrazole ring. The acetophenone (*144*) is prepared by Aldol-type condensation of 5-acetyl-2-chloro-4-fluorobenzoate (*143*) with ethyl trifluoroacetate as described in Scheme 14.

<Synthesis of pentoxazone>

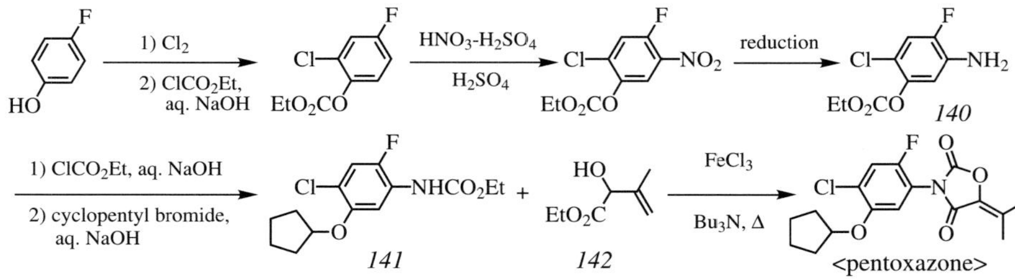

<Synthesis of fluazolate>

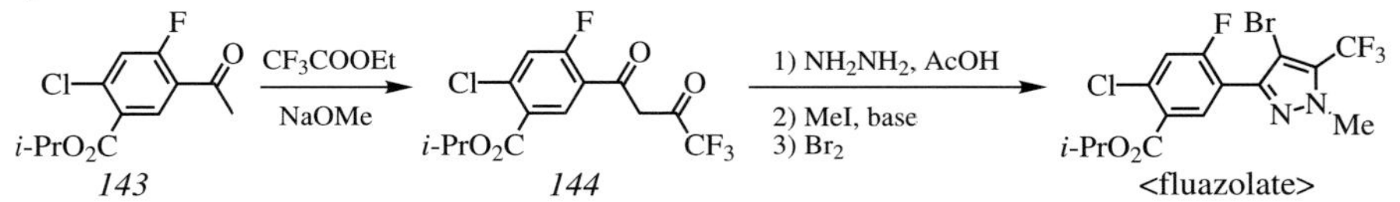

<Synthesis of profluazol>

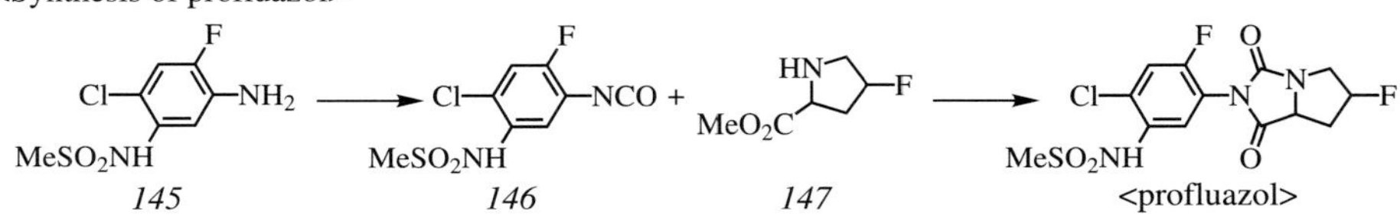

Scheme 14. Major synthetic routes for pentoxazone, fluazolate and profluazol

The bicyclic hydantoin derivative, profluazol, is simply synthesized by cycloaddition of aryl isocyanate (*146*) with fluorinated proline (*147*). Aryl isocyanate is prepared by phosgenation of the corresponding aniline (*145*) which is synthesized by chlorination, nitration and reduction of *N*-methylsulfonated 4-fluoroaniline (Scheme 14). Fluoroproline (*147*) is prepared from hydroxyproline by a fluorination reagent such as DAST. Using L-hydroxyproline gives an optically active fluoroproline that leads to optically active profluazol.

With regard to the synthetic method for the 6-trifluoromethyluracil classes, three representative synthetic pathways have been developed. Cycloaddition of aryl isocyanate with 3-amino-4,4,4-trifluorocrotonate (*149*) is the simplest method and, for example, flupropacil is readily synthesized by this method using isopropyl 2-chloro-5-isocyanatobenzoate (*148*) followed by methylation as shown in Scheme 15. *N*-arylcarbamate (*150*) instead of aryl isocyanate can be applied to the cyclocondensation with *149* yielding 3-aryl-6-trifluoromethyluracil (*151*). This method has the advantage of not using toxic phosgene for preparing aryl isocyanate. A third method has been developed for the synthesis of butafenacil-allyl. 5-Alkoxycarbonyl-4-chloro-2-fluoroaniline (*155*) reacts with trifluoroacetoacetate (*156*) affording 2-(trifluoroacetyl)acetanilide (*157*), which is subjected to amination with ammonium acetate and cyclization with phosgene, and then accomplished by methylation to give butafenacil-allyl. Aniline (*155*) is prepared by selective

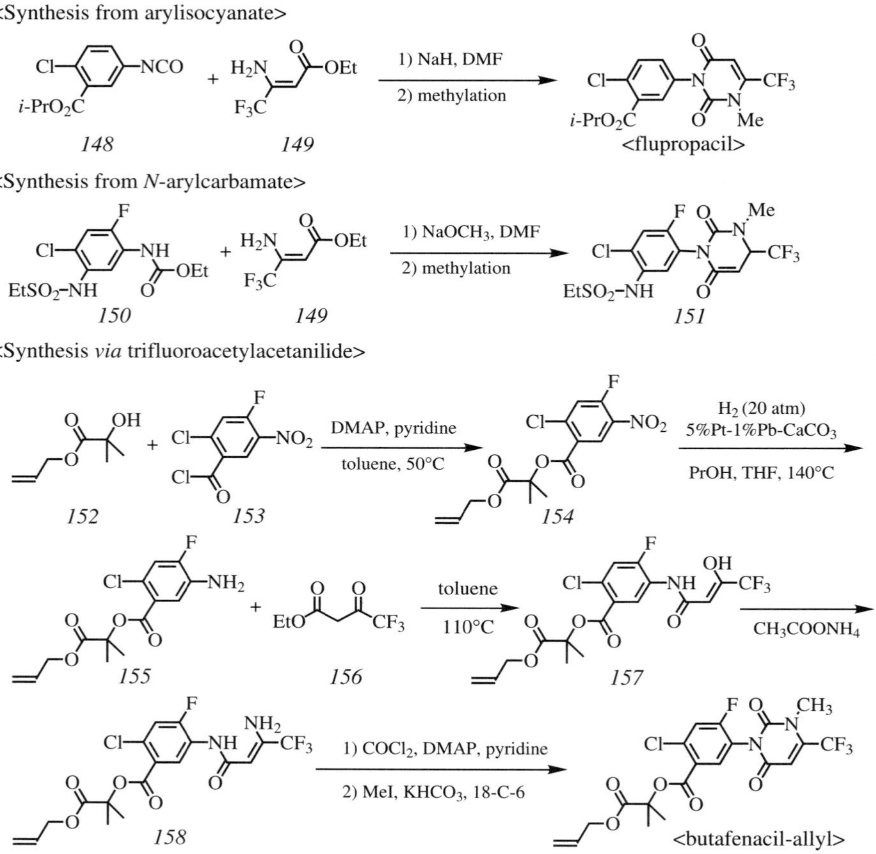

Scheme 15. Major synthetic routes for flupropacil, butafenacil-allyl and related compound

reduction of the nitro group of 5-nitrobenzoate (*154*), obtained by the reaction of alcohol (*152*) with benzoyl chloride (*153*). The reaction of 3-(4-hydroxyphenyl)-1-methyl-6-trifluoromethyluracil with substituted benzyl chloride in potassium carbonate containing DMF gives benzfendizone.

2-Aryl-5-trifluoromethyl-2*H*-pyridazin-3-one classes have been aggressively developed by Sumitomo. Among their compounds, flufenpyr-ethyl is considered to be the most promising candidate. Reaction of phenylhydrazine (*159*) with 1,1-dibromo-3,3,3-trifluoroacetone in the presence of sodium acetate in water to give hydrazone (*160*), which condenses with methylmalonic acid affording 2-(4-chloro-3-hydroxyphenyl)-4-methyl-5-trifluoromethyl-2*H*-pyridazin-3-one (*162*) *via* an adduct (*161*). Finally, *O*-alkylation with chloroacetate gives flufenpyr-ethyl (Scheme 16).

The Japp-Klingemann reaction is effective for the formation of hydrazones. Thus, coupling of aryldiazonium salt (*164*) with an active methylene com-

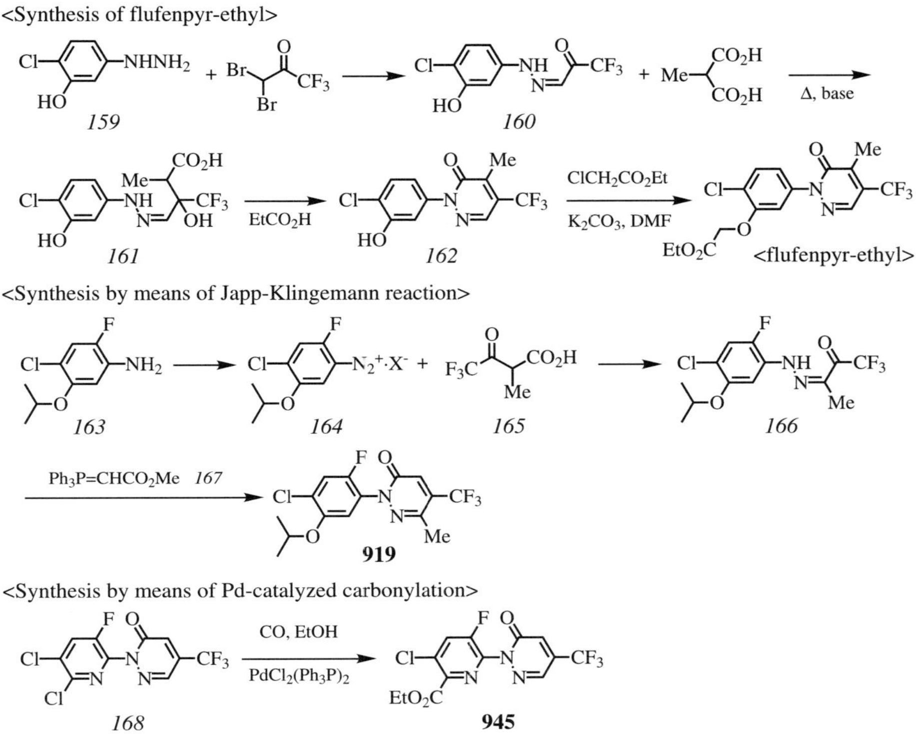

Scheme 16. Major synthetic routes for flufenpyr-ethyl and related compounds

pound such as *165*, in which at least one of the activating groups is an acyl or carboxy group, leads to the hydrazone (*166*). Cyclocondensation of *166* with Wittig reagent (*167*) gives the desired 5-trifluoromethyl-2*H*-pyridazin-3-one [**919**]. By this method, various substituents can be introduced at 6-position of the pyridazine ring.

The synthetic method for 2-pyridylpyridazine [**945**] is an unusual methodology, since an important ethoxycarbonyl group is introduced at the *meta*-position of the pyridine ring by Pd-catalyzed carbonylation of the chloropyridine precursor (*168*). Pyrazogyl is a pioneering herbicide belonging to the di-heterocycle PPO inhibitors. The fundamental bicyclic structure (*171*) of pyrazogyl is constructed by reacting ethoxymethylenemalononitrile with 2-hydrazino-4,5,6,7-tetorahydropyrazo[4,5-*a*]pyridine (*170*) which is prepared by double-cyclization using 1,1,7-trichloro-1-hepten-3-one (*169*) and hydrazine as shown in Scheme 17. After chlorinating, consecutive *N*-methylation and *N*-propargylation give pyrazogyl.

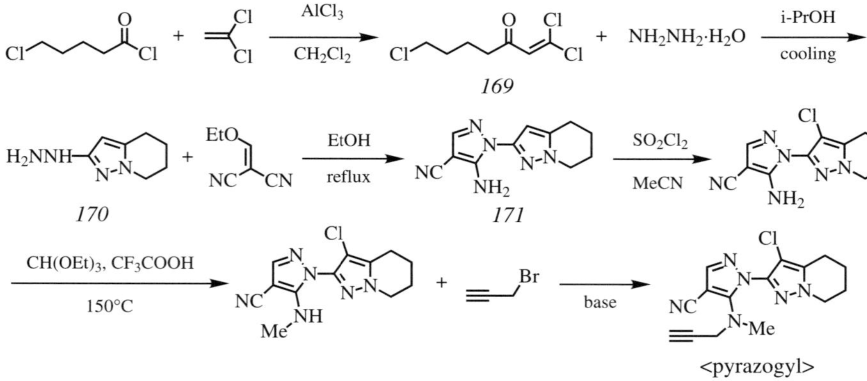

Scheme 17. Major synthetic route for pyrazogyl

10.10 Notes

The purpose of this chapter is to summarize the agrochemical characteristics and major synthetic routes for all practical herbicides cited in Chapters 1–8, and to list almost all related herbicides disclosed since 1990, except for the PPO inhibitors after 1995, and to provide a comprehensive guide to the patent literature. As an ambiguous presentation in the patent literature does not always reveal the most active compound, the compounds shown in the figures are selected as representative examples. However, the structural evolution of all classes of herbicides is clearly demonstrated. Hopefully, this review will encourage further progress of the underexplored classes and will stimulate new ideas for development of next-generation herbicides.

For easy access to references, the number assigned to each compound corresponds to its reference number. Company names used in this chapter are abbreviated as follows; ACC: American Cyanamid Co.; A-Kanesho: Agro-Kanesho Co.; Aventis: Aventis CropScience; BASF: BASF A.-G.; Bayer: Bayer A.-G.; Central Glass: Central Glass Co. Ltd.; Chugai: Chugai Pharm. Co. Ltd.; Ciba: Ciba Corp.; Ciba-Geigy: Ciba-Geigy A.-G.; Degussa: Degussa A.-G.; DIC: Dainippon Ink & Chemicals, Inc.; Dow: Dow Chemical Co.; DowAgro: Dow Agrosciences LLC; DowElanco: Dow Elanco; Dunlena: Dunlena Pty Ltd.; DuPont: E. I. DuPont de Nemours and Co.; FMC: FMC Corp.; Hoechst: Hoechst A.-G.; Hoechst-S-A: Hoechst Schering AgrEvo; H-La-Roche: Hoffman-La Roche, F.; Hokko: Hokko Chemical Ind. Co. Ltd.; ICI: Imperial Chemical Ind.; Idemitsu: Idemitsu Kosan Co. Ltd.; Ihara: Ihara Chemical Ind. Co. Ltd.; Isagro: Isagro Ricerca S.R.L.; Ishihara: Ishihara Sangyo K.K.; Kaken: Kaken Pharmaceutical Co. Ltd.; Korea Research: Korea Research Institute of Chemical Technology; Kumiai: Kumiai Chemical Ind. Co. Ltd.; LG Chemical: LG Chemical

Ltd.; Kuraray: Kuraray Co. Ltd.; Kureha: Kureha Chemical Ind. Co. Ltd.; Mitsubishi: Mitsubishi Chemical Co.; Mitsui: Mitsui Chemicals Inc. or Mitsui Toatsu Chemicals Inc.; Monsanto: Monsanto Co.; Nichino: Nihon Nohyaku Co. Ltd.; Nihon Bayer: Nihon Bayer Agrochem K.K.; Nissan: Nissan Chemical Ind. Ltd.; Nisso: Nippon Soda Co. Ltd.; Novartis: Novartis A.-G.; Otsuka: Otsuka Chemical Co. Ltd.; R & H: Rohm and Haas Co.; RP: Rhône-Poulenc Agrochimie; RP-Yuka: Rhône-Poulenc Yuka Agro; Sagami: Sagami Chemical Research Center; Sandoz: Sandoz A.-G.; Sankyo: Sankyo Co. Ltd.; Schering: Schering A.-G.; Schering-Ag: Schering Agrochemicals Ltd.; SDS: SDS Biotech K.K.; Shell: Shell International Research M.B.V.; Shionogi: Shionogi and Co.; Sumitomo: Sumitomo Chemical Co. Ltd.; Suntory: Suntory Ltd.; Takeda: Takeda Chemical Ind. Ltd.; Teijin: Teijin Ltd.; Tokuno: Nihon Tokushu Nohyaku Seizo K.K.; Tokuyama: Tokuyama Soda Co. Ltd.; Tosoh: Tosoh Corp.; Ube: Ube Ind. Ltd.; Uniroyal: Uniroyal Chemical Co.; Yatoron: Yatoron Laboratories, Inc.; Zeneca: Zeneca Agrochemicals.

Weed names in the parentheses in each figure are abbreviated as their acronyms of the botanical name as follows; Aba: *Abutilon avicennae*, Abt: *Abutilon theophrasti*, Agt: *Agrostis tenuis*, Ala: *Alopecurus aequalis*, Alm: *Alopecurus myosuroides*, Ama: *Ambrosia artemisiaefolia*, Amb: *Amaranthus blitum*, Aml: *Amaranthus lividus*, Amr: *Amaranthus retroflexus*, Aps: *Apera spica-venti*, Avf: *Avena fatua*, Avs: *Avena sativa*, Bip: *Bidens pilosa*, Brl: Broadleaf weeds, Brp: *Brachiaria platyphylla*, Brt: *Bromus tectorum*, Cab: *Capsella bursa-pastoris*, Cao: *Cassia obtusifolia*, Cha: *Chenopodium album*, Chs: *Chrysanthemum segetum*, Cos: *Convolvulus sepium*, Cyd: *Cyperus difformis*, Cye: *Cyperus esculentus*, Cyi: *Cyperus iria*, Cym: *Cyperus microiria*, Cys: *Cyperus serotinus*, Das: *Datura stramonium*, Dia: *Digitaria adscendens*, Dic: *Digitaria ciliaris*, Dih: *Digitaria henryi*, Dir: *Digitaria retroflexus*, Dis: *Digitaria sanguinalis*, Ecc: *Echinochloa crus-galli*, Ecf: *Echinochloa frumentacea*, Eco: *Echinochloa oryzicola*, Ela: *Eleocharis acicularis*, Elk: *Eleocharis kuroguwai*, Gaa: *Galium aparine*, Gac: *Galinsoga ciliata*, Gas: *Galium spurium*, Ipi: *Ipomoea indica*, Ipl: *Ipomoea lacumosa*, Ipp: *Ipomoea purpurea*, Ips: *Ipomoea subspecies*, Laa: *Lamium amplexicaule*, Lef: *Leptochloa filiformis*, Lip: *Lindernia procumbens*, Lom: *Lolium multiflorum*, Mac: *Matricaria chamomilla*, Mai: *Matricaria inodora*, Mob: *Monochoria bulrush*, Mov: *Monochoria vaginalis*, Nas: *Nasturtium officinale*, Pac: *Panicum crus-galli*, Paf: *Panicum frumentaceum*, Pam: *Panicum miliaceum*, Phn: *Pharbitis nil*, Php: *Pharbitis purpurea*, Pob: *Polygonum blumei*, Pol: *Polygonum lapathifolium*, Pon: *Polygonum nobosum*, Pop: *Polygonum persicaria*, Ras: *Raphanus sativus*, Roa: *Rorippa austriaca*, Roi: *Rorippa indica*, Sap: *Sagittaria pygmaea*, Sat: *Sagittaria trifolia*, Sch: *Scirpus hotarui*, Scj: *Scirpus juncoides*, Scm: *Scirpus maritimus*, See: *Sesbaniu exaltata*, Sef: *Setaria faberii*, Sev: *Setaria viridis*, Sia: *Sinapis arvensis*, Sis: *Sida spinosa*, Soh: *Sorghum halepense*, Sol: *Solanum lycopersicum*, Son: *Solanum nigrum*, Spj: *Sirpus juncoides*, Stf: *Setaria faberii*, Stm: *Stellaria media*, Stn: *Stellaria neglecta*, Tra: *Triticum aestivum*, Veo: *Veronica officinalis*, Vep: *Veronica persica*, Vim: *Viola mandshurica*, Xap: *Xanthium pennsylvanicum*, Xas: *Xanthium strumarium*.

Patent Literatures

[**1**] EP388873 (1990) BASF. [**2**] EP469460 (1992), DE4024755 (1992) BASF. [**3**] DE4029484 (1992) BASF. [**4**] EP496701 (1992) Ciba-Geigy. [**5**] WO92/13845 (1992) Hoechst. [**6**] EP562575 (1993) Kureha. [**7**] EP394059 (1990) DuPont. [**8**] DE4038430 (1992) BASF. [**9**] EP528212 (1993) Bayer. [**10**] US5084086 (1992) DuPont. [**11**] WO97/32861 (1997) Bayer & Hoechst-S-A. [**12**] DE4018349 (1991) Bayer. [**13**] WO90/06308 (1990) DuPont. [**14**] DE4102905 (1991) Nehezvegyipari Kutato Intezet. [**15**] Yingyong Huaxue, 9(4) 64–66 (1992). [**16**] EP413221 (1991) Bayer. [**17**] DE4110881 (1992) Bayer. [**18**] HU59567 (1992) MTA Novenyvedelmi Kutato Intezete. [**19**] DE4104154 (1992) Bayer. [**20**] EP507172 (1992) Bayer. [**21**] EP457581 (1991) ICI. [**22**] WO92/06965 (1992) ICI. [**23**] DE3842621 (1990) Bayer. [**24**] DE4023680 (1991) BASF. [**25**] DE4007683 (1991), DE4238175 (1994) BASF. [**26**] DE4304864 (1993) Ciba-Geigy. [**27**] DE4337847 (1995) Bayer. [**28**] US4892946 (1990) DuPont. [**29**] EP382473 (1990) Kureha. [**30**] EP382436 (1990) Kureha. [**31**] DE4230933 (1994) Hoechst. [**32**] DE4236902 (1994) Hoechst. [**33**] DE4335297 (1995) Hoechst-S-A. [**34**] WO95/32950 (1995) Hoechst-S-A. [**35**] DE19611355 (1997), DE19650955 (1998) Hoechst-S-A. [**36**] US5723409 (1998) Hoechst-S-A. [**37**] DE19748470 (1999) Hoechst-S-A. [**38**] DE19702200 (1998) Hoechst-S-A. [**39**] EP384602 (1990) Kureha. [**40**] DE4927453 (1990) DuPont. [**41**] DE3900472 (1990) BASF. [**42**] WO91/15478 (1991) DuPont. [**43**] DE4322067 (1995) Hoechst. [**44**] WO00/47566 (2000) Aventis. [**45**] DE4415049 (1995) Hoechst-S-A. [**46**] DE4302701, EP609733 (1993) Bayer. [**47**] EP528211 (1993), DE4126423 (1993) Bayer. [**48**] DE4128441 (1993) BASF. [**49**] DE4206145 (1993) BASF. [**50**] DE4341454 (1995) Bayer. [**51**] DE4439675 (1996) Hoechst-S-A. [**52**] DE19521668 (1996) Hoechst-S-A. [**53**] DE19616445 (1997) Hoechst-S-A. [**54**] WO92/15568 (1992) Korea Research. [**55**] DE4105518 (1992) BASF. [**56**] DE4206146 (1993) BASF. [**57**] DE4241303 (1994) Bayer. [**58**] DE4322726 (1995) BASF. [**59**] EP770603 (1997) Isagro. [**60**] DE19543648 (1997) Hoechst-S-A. [**61**] WO97/40021 (1997) Bayer. [**62**] JP02191275 (1990) ICI. [**63**] EP559044 (1993) Bayer. [**64**] EP645386 (1995) Bayer. [**65**] EP395251 (1990) ICI. [**66**] EP364141, JP02174777 (1990) ICI. [**67**] EP385775 (1990) ICI. [**68**] EP393999 (1990) ICI. [**69**] JP0495091 (1992) Sumitomo. [**70**] EP402316 (1990), DD299294 (1992) Ciba-Geigy. [**71**] WO91/10660 (1991) Hoechst. [**72**] JP04139170 (1992) Ishihara. [**73**] EP582021 (1994) Ciba-Geigy. [**74**] EP520740 (1992) DuPont. [**75**] EP558445 (1993) Ciba-Geigy. [**76**] EP451468 (1991) Ishihara. [**77**] EP459949 (1991) Ciba-Geigy. [**78**] EP508348 (1992), EP521500 (1993) Hoechst. [**79**] EP555770 (1993) Hoechst. [**80**] DE4311787 (1994), US5663118 (1997) Hoechst-S-A. [**81**] WO95/06049 (1995) Hoechst, DE4328397 (1995) Hoechst-S-A. [**82**] DE19520602 (1995) Ciba-Geigy. [**83**] JP04253974 (1992) Ishihara. [**84**] EP496608 (1992) Ishihara. [**85**] JP04257580 (1992) Ishihara. [**86**] EP562731 (1993) Ishihara. [**87**] US5348933 (1994) Ishihara. [**88**] DE4000503 (1991) Hoechst. [**89**] WO92/16522 (1992) Ciba-Geigy. [**90**] WO93/17015, WO93/17016 (1993) Ciba-Geigy. [**91**] DE4324060 (1995) Hoechst-S-A. [**92**] EP581738 (1994) Ciba-Geigy. [**93**] WO97/41112

(1997) Novartis. [**94**] WO92/14728 (1992) Korea Research. [**95**] WO97/31913 (1997) Korea Research. [**96**] EP357345 (1990) DuPont. [**97**] WO92/16525 (1992) Korea Research. [**98**] DE4304288 (1994) Hoechst. [**99**] WO97/32875 (1997) Bayer. [**100**] DE4232417 (1994) Bayer. [**101**] DE19651037 (1998) Bayer. [**102**] WO95/13276 (1995) Korea Research. [**103**] US4943312 (1990) DuPont. [**104**] JP07196654 (1995) Nissan. [**105**] WO95/32202 (1995) Nissan. [**106**] JP07252253, JP07278144 (1995) Nissan. [**107**] WO95/18806 (1995) Nissan. [**108**] JP08277289 (1996) Nissan. [**109**] JP09132574 (1997) Nissan. [**110**] JP1087660 (1998) Nissan. [**111**] JP04198182 (1992) Tosoh. [**112**] WO93/00336 (1993) Nissan. [**113**] EP414067 (1991) Bayer. [**114**] DE4017460 (1991) Bayer. [**115**] DE4129317, DE4129876, EP529292 (1993) Bayer. [**116**] JP1135580 (1999) Yatoron & Otsuka. [**117**] JP05194495 (1993) DuPont. [**118**] JP0609314 (1994) DuPont. [**119**] JP1029990 (1998) Kureha. [**120**] WO92/04319 (1992) DuPont. [**121**] EP648764 (1995) Kureha. [**122**] US5472933 (1995) DuPont. [**123**] WO91/10668 (1991) DuPont. [**124**] JP0372407 (1991) Takeda. [**125**] EP477808 (1992) Takeda. [**126**] WO95/00509 (1995) Kumiai & Ihara. [**127**] JP1001478 (1997) Kumiai & Ihara. [**128**] DE3927788 (1990) Bayer. [**129**] DE3935277 (1991) Hoechst. [**130**] DE4008627 (1991) Bayer. [**131**] WO91/06546 (1991) Nissan. [**132**] EP467252 (1992) Hoechst. [**133**] JP05310709 (1993) Nissan. [**134**] EP609733 (1993) Bayer. [**135**] DE4302700, DE4302702 (1994) Bayer. [**136**] WO98/40361 (1998) Isagro. [**137**] DE4414840 (1995) Bayer. [**138**] EP679646 (1995) ACC. [**139**] EP661275 (1995) ACC. [**140**] DE3909053 (1990) Hoechst. [**141**] EP507093 (1992) Hoechst. [**142**] JP06329642 (1994) RP-Yuka. [**143**] DE3826230 (1990) Hoechst. [**144**] EP425948 (1991) Bayer. [**145**] EP507171 (1992) Bayer. [**146**] DE4231801 (1994) Bayer. [**147**] DE19525162, DE19525973 (1997) Bayer. [**148**] WO97/32876 (1997) Bayer. [**149**] DE19621685 (1997) Bayer. [**150**] DE19632945 (1998) Bayer. [**151**] DE3936623 (1991) Bayer. [**152**] DE19508119 (1996) Bayer. [**153**] DE19525974 (1997) Bayer. [**154**] CN1169723 (1998) Bayer. [**155**] EP422469 (1991) Bayer. [**156**] DE4343595 (1995) Bayer. [**157**] DE19517505 (1996) Bayer. [**158**] DE19823131 (1999) Bayer. [**159**] DE4411913 (1995) Bayer. [**160**] DE19540737 (1997) Bayer. [**161**] DE19650196 (1998) Bayer. [**162**] DE19615355 (1997) Bayer. [**163**] WO00/17196 (2000) Bayer. [**164**] JP03173887 (1991) Nissan. [**165**] JP03173884 (1991) Nissan. [**166**] JP04253971 (1992) Nissan. [**167**] WO91/13884 (1991) Nissan. [**168**] EP459244 (1991) Bayer. [**169**] DD298729 (1992) Chemishe und Pharmazeutische Fabriken. [**170**] DE4029753 (1992) BASF. [**171**] DE4029753 (1992) BASF. [**172**] EP375076 (1990) Shell. [**173**] DE3843849 (1990) Bayer. [**174**] EP434624 (1991) Ciba-Geigy. [**175**] DE4403533 (1992), WO93/16079 (1993) Ciba-Geigy. [**176**] DE3825041 (1990) Schering. [**177**] EP378508 (1990) Ciba-Geigy. [**178**] US5163995 (1992) DowElanco. [**179**] US5201938 (1993) DowElanco. [**180**] US5447905 (1995) DowElanco. [**181**] US5461161 (1995) DowElanco. [**182**] US5614469 (1997) DowElanco. [**183**] WO98/30564 (1998) Korea Research. [**184**] US5763359 (1999), WO97/08172 (1997) DowAgro. [**185**] US5858924 (1999) DowAgro. [**186**] DE3825043 (1990) Schering. [**187**] US4992091 (1991) DowElanco. [**188**] DE3916469 (1990) BASF. [**189**] DE3921271 (1991) BASF. [**190**] DE4106100 (1991) Bayer. [**191**]

CN1156145 (1997) China Agricultural Univ. [**192**] EP355885 (1990) Bayer. [**193**] JP02160785 (1990) Nisso. [**194**] JP02160782 (1990) Nisso. [**195**] EP384244 (1990) Bayer. [**196**] US4940482 (1990) Ciba-Geigy. [**197**] EP384250 (1990) Bayer. [**198**] JP023040822 (1990) A-Kanesyo. [**199**] DE3928605 (1991) Bayer. [**200**] EP417875 (1991) Schering. [**201**] JP03200772 (1991) Kumiai. [**202**] EP444286 (1991) Bayer. [**203**] HU59795 (1992) MTA Novenyvedelmi Kutatointezet. [**204**] DE4035141 (1992) Bayer. [**205**] EP733629 (1996) Ishihara. [**206**] US4932999 (1990) Kumiai & Ihara. [**207**] EP402751 (1990) BASF. [**208**] DE3942476 (1991) BASF. [**209**] WO91/13065 (1991) FMC. [**210**] DE4030929 (1992) BASF. [**211**] DE19536809 (1992) BASF. [**212**] JP04108777 (1992) Kumiai. [**213**] DE4034045 (1992) BASF. [**214**] WO93/03017 (1993) Kumiai. [**215**] US5401711 (1995) Kumiai. [**216**] DE4337323 (1995) BASF. [**217**] WO96/36613 (1996) Nisso. [**218**] JP09110839 (1997) Kumiai & Ihara. [**219**] DE19539637 (1997) BASF. [**220**] DE19620404 (1997) BASF. [**221**] JP02108674 (1990) Schering. [**222**] EP549344 (1993) Sumitomo. [**223**] EP363040 (1990) Schering-Ag. [**224**] JP03240787 (1991) Sumitomo. [**225**] JP0673022 (1994) Kumiai & Ihara. [**226**] WO94/17059 (1994) Nisso. [**227**] GB2237570 (1991) ICI. [**228**] EP426476 (1991) Kumiai. [**229**] EP435170 (1991) Kumiai & Ihara. [**230**] JP03291271 (1991) Sumitomo. [**231**] EP457505 (1991) Sumitomo. [**232**] DE4021441 (1992) Bayer. [**233**] EP468690 (1992) Sumitomo. [**234**] EP469711 (1992) Sumitomo. [**235**] JP04112876 (1992) Kumiai. [**236**] GB2250985 (1992) Ciba-Geigy. [**237**] US5149357 (1992) FMC. [**238**] WO92/22538 (1992) Shell. [**239**] DE4126936, DE4126937 (1993) BASF. [**240**] JP0570440 (1993) Sumitomo. [**241**] JP05194503 (1993) Kumiai. [**242**] JP05213904 (1993) Kumiai & Ihara. [**243**] EP564920 (1993) Bayer. [**244**] EP593252 (1994) Sumitomo. [**245**] EP608862 (1994) LG Chemical. [**246**] WO95/06039 (1995) Hoechst-S-A. [**247**] DE4337322 (1995) BASF. [**248**] JP07215948 (1995) Hokko. [**249**] JP07215949 (1995) Hokko. [**250**] JP07233011 (1995) Kumiai. [**251**] WO96/37419 (1996) Nisso. [**252**] WO96/994 (1996) Nisso. [**253**] EP374839 (1990) Mitsui. [**254**] JP03193766 (1991) Mitsui. [**255**] JP03264567, EP374839 (1991) Mitsui. [**256**] EP459243 (1991) Bayer. [**257**] WO92/13846 (1992) Ciba-Geigy. [**258**] JP04342547 (1992) Mitsui. [**259**] JP04342574 (1992) Mitsui. [**260**] JP0352873 (1991) Hokko. [**261**] JP03200784 (1991) Ishihara. [**262**] JP04342586, JP04342586 (1992) Ishihara. [**263**] JP04342575 (1992) Mitsui. [**264**] JP05202038 (1993) Sumitomo. [**265**] JP06321911 (1994) RP-Yuka. [**266**] EP643048 (1995) Nihon Bayer. [**267**] JP07179439 (1995) Ihara. [**268**] JP06172324 (1995) Hokko. [**269**] EP490224 (1992) BASF. [**270**] JP0256469 (1990) Kumiai. [**271**] EP451653 (1991) Bayer. [**272**] WO94/07868 (1994) Sumitomo. [**273**] DE3927382 (1991) BASF. [**274**] JP0477487 (1992) RP-Yuka. [**275**] WO94/01415 (1994) Kumiai & Ihara. [**276**] JP1143409 (1999) Novartis. [**277**] DE3910635 (1990) BASF. [**278**] EP372329 (1990) BASF. [**279**] EP435186 (1991) Mitsui. [**280**] JP0429980 (1992) Hokko. [**281**] JP04221372 (1992) Mitsui. [**282**] JP04221372 (1992) Mitsui. [**283**] WO93/12109 (1993), WO92/07849, JP06316574 (1994), US5391537 (1995) Kumiai & Ihara. [**284**] JP06345758 (1994) RP-Yuka. [**285**] US5403816 (1995) Kumiai & Ihara. [**286**] DE4025338 (1992) Bayer. [**287**] WO92/17468 (1992)

Kumiai. [**288**] DE4022478 (1992) Bayer. [**289**] JP0641116 (1994) Kumiai & Ihara. [**290**] EP567133 (1993), JP0641118 (1994) Kumiai & Ihara. [**291**] JP06199840 (1994) Kumiai & Ihara. [**292**] JP06345759 (1994) RP-Yuka. [**293**] JP00226386 (2000) Nissan. [**294**] DE3832237 (1990) BASF. [**295**] JP03232884 (1991) Kumiai. [**296**] JP03284676 (1991) Kumiai. [**297**] JP04145081 (1992) Kumiai. [**298**] WO93/13078 (1993) Kumiai. [**299**] US4973354 (1990) Nissan. [**300**] JP04356469 (1992) Kumiai & Ihara. [**301**] EP468766 (1990) RP-Yuka. [**302**] JP05202001 (1993) RP-Yuka. [**303**] JP06116262 (1994) RP-Yuka. [**304**] JP06128110 (1994) RP-Yuka. [**305**] JP06135940 (1994) RP-Yuka. [**306**] JP06220024 (1994) RP-Yuka. [**307**] JP06305912 (1994) RP-Yuka. [**308**] JP07179430 (1995) Mitsubishi. [**309**] JP05279209 (1993) RP-Yuka. [**310**] JP06192252, JP06211808 (1994) RP-Yuka. [**311**] JP0761974 (1995) Mitsubishi. [**312**] EP439243 (1991) Schering. [**313**] JP03271279 (1991) RP-Yuka, EP431707 (1991) Shell. [**314**] JP03271279 (1991) RP-Yuka. [**315**] DE4034295 (1992) Schering. [**316**] JP06228110 (1994) RP-Yuka. [**317**] JP06172322 (1994) RP-Yuka. [**318**] US5426090 (1995) Shell. [**319**] JP0477475 (1992) RP-Yuka. [**320**] DE4129876 (1993) Bayer. [**321**] EP581132 (1994) Bayer. [**322**] WO98/13365 (1998) Hoechst-S-A. [**323**] JP0578331 (1993) RP-Yuka. [**324**] JP05194422 (1993) RP-Yuka. [**325**] JP05301862 (1993) RP-Yuka. [**326**] JP06228176 (1994) RP-Yuka. [**327**] JP06321912 (1994) RP-Yuka. [**328**] JP06172322 (1994) RP-Yuka. [**329**] JP0597818 (1993) RP-Yuka. [**330**] JP05213898 (1993) RP-Yuka. [**331**] JP05213899 (1993) RP-Yuka. [**332**] JP05301861 (1993) RP-Yuka. [**333**] JP0789942 (1995) Mitsubishi. [**334**] EP409368 (1991) Schering. [**335**] JP03240777 (1991) Ube. [**336**] DE4040118 (1992) Schering, US5151113 (1992) DowElanco, JP03193765 (1991) Kumiai & Ihara. [**337**] EP517215 (1992) Ube. [**338**] JP04356470 (1992) Ube. [**339**] EP541041 (1993) Hoechst. [**340**] EP562510 (1993) Hoechst. [**341**] EP573837 (1993) Nihon Bayer. [**342**] WO93/25540 (1993) Ciba-Geigy. [**343**] DE4313412, DE4313411, DE4313413 (1994) BASF. [**344**] JP06329690 (1994) RP-Yuka. [**345**] DE4335950 (1995) BASF. [**346**] DE4329911 (1995) BASF. [**347**] JP08325246 (1996) Ube. [**348**] DE4035758 (1992) Schering. [**349**] JP04334372 (1992) RP-Yuka. [**350**] WO93/12094 (1993) Hoechst. [**351**] DE4313411 (1994) BASF. [**352**] WO94/10156 (1994) Ciba-Geigy. [**353**] EP409369 (1991) Schering, DE3924260 (1991) Schering. [**354**] JP0352872 (1991) RP-Yuka. [**355**] EP571856 (1993) Nihon Bayer. [**356**] ZA9207782 (1993) Ciba-Geigy. [**357**] EP549079 (1993) Shell. [**358**] EP567014 (1993) Ube. [**359**] EP581184 (1994) Ube. [**360**] JP0331266 (1991) Nissan. [**361**] DE4201875 (1993) BASF. [**362**] EP593998 (1994) Nihon Bayer. [**363**] EP611759 (1994) Nihon Bayer. [**364**] EP630890 (1994) Nichino. [**365**] JP06329643 (1994) RP-Yuka. [**366**] WO91/10653 (1991) DuPont. [**367**] WO95/25730 (1995) Kumiai & Ihara. [**368**] DE4026177 (1992) Schering. [**369**] DE19521653 (1996) Hoechst-S-A. [**370**] JP1160562 (1999) Kumiai & Ihara. [**371**] WO00/06553 (2000) Kumiai & Ihara. [**372**] EP410590 (1991) Schering-Ag. [**373**] EP422751 (1991) Schering. [**374**] EP481512 (1992) Ube. [**375**] JP0454168 (1992) Kumiai & Ihara. [**376**] WO92/09584 (1992) Kumiai. [**377**] WO92/16511 (1992) Schering-Ag. [**378**] JP0641090 (1994) Kumiai & Ihara. [**379**] WO94/05644 (1994) Schering. [**380**]

US5418212 (1995) Kumiai & Ihara. [**381**] JP09179434 (1995) Mitsubishi. [**382**] JP04275201 (1992) Kumiai. [**383**] JP06316565 (1994) Kumiai & Ihara. [**384**] DE4030041 (1992) Bayer. [**385**] JP0532639 (1993) Kumiai & Ihara. [**386**] EP658549 (1995) Lucky Ltd, CA2194080 (1997) LG Chemical. [**387**] EP461079 (1991) Sandoz. [**388**] WO93/09099 (1993) Schering. [**389**] JP05213902 (1993) Kumiai & Ihara. [**390**] JP0592971 (1993) Kumiai. [**391**] WO92/19603 (1992) RP-Yuka. [**392**] JP07179433 (1995) Mitsubishi. [**393**] JP0559015 (1993) Otsuka. [**394**] JP07233012 (1995) Kuraray. [**395**] WO98/57957 (1998) Nissan. [**396**] JP0504973 (1993) RP-Yuka. [**397**] JP0253787 (1990) Mitsubishi Kasei Co. [**398**] US4925944 (1990) ACC. [**399**] JP02142774 (1990) Mitsui. [**400**] EP375910 (1990) ACC. [**401**] JP02202886 (1990) Mitsui. [**402**] ACS Symp Ser., 443, p133–143 (1991) ACC. [**403**] EP434965 (1991) ACC. [**404**] WO91/13069 (1991) Korea Research. [**405**] EP517052 (1992), DE4224163 (1994) Bayer. [**406**] EP539676 (1993) ACC. [**407**] US5256629 (1993) ACC. [**408**] DE4232852 (1994) Bayer. [**409**] DE4233028 (1994) Bayer. [**410**] EP580042 (1994) Bayer. [**411**] JP07206829 (1995) Nisso. [**412**] US4956006 (1990) ICI. [**413**] JP02237970 (1990) Mitsui. [**414**] JP02115161 (1990) Mitsui. [**415**] JP02164862 (1990) Mitsui. [**416**] JP0543543 (1993) Mitsui. [**417**] JP05125042 (1993) Mitsui. [**418**] WO93/24483 (1993) Ciba-Geigy. [**419**] EP397602 (1990) Ciba-Geigy. [**420**] EP417044 (1991) Ciba-Geigy. [**421**] US4960457 (1990) ICI. [**422**] EP387869 (1990) Mitsui. [**423**] EP477626 (1992) Mitsui. [**424**] JP04112869 (1992) Mitsui. [**425**] JP04117357 (1992) Mitsui. [**426**] JP04154760 (1992) Mitsui. [**427**] JP04154759 (1992) Mitsui. [**428**] JP04282360 (1992) Mitsui. [**429**] DE4141399 (1993) Bayer. [**430**] EP550024 (1993) Mitsui. [**431**] US5076832 (1991) ICI. [**432**] JP04117355 (1992) Mitsui. [**433**] JP04193877 (1992) Mitsui. [**434**] US5302726 (1994) ICI. [**435**] EP600507 (1994) Mitsui. [**436**] JP07291926 (1995) Mitsui. [**437**] JP007124596 (1997) Mitsui. [**438**] EP354766 (1990), EP425247 (1991) Sumitomo. [**439**] JP02275865 (1990) Sumitomo. [**440**] US5022915 (1991) ICI. [**441**] JP03148264 (1991) Sumitomo. [**442**] JP0466578 (1992) Sumitomo. [**443**] JP03148265 (1991) Sumitomo. [**444**] JP0436284 (1992) Sumitomo. [**445**] JP02149566 (1990) Sumitomo. [**446**] JP0449279 (1992) Nihon Bayer. [**447**] JP0733748 (1995) Nissan. [**448**] WO99/28301 (1999) DuPont. [**449**] DE4011361 (1991) Bayer. [**450**] EP463492 (1992) Bayer. [**451**] EP410238 (1991) Bayer. [**452**] JP0551369 (1993) Sumitomo. [**453**] DE4130833 (1993) Bayer. [**454**] JP02178266 (1990) Chugai. [**455**] EP422470 (1991) Bayer. [**456**] EP467473 (1992) Shell. [**457**] EP447004 (1991) Shell. [**458**] EP646566 (1995) Shell. [**459**] EP488474 (1992) Shell. [**460**] CA2277930 (1994) Shell. [**461**] DE19854081 (2000) Bayer. [**462**] WO94/27983 (1994) Shell. [**463**] WO94/27974 (1994) Shell. [**464**] WO97/24330 (1997) Kureha. [**465**] WO00/75112 (2000) Kureha. [**466**] CA2078026 (1993) Shell. [**467**] WO94/08991 (1994) Shell. [**468**] EP572093 (1993) Shell. [**469**] EP692474 (1996) Kureha. [**470**] GB2285045 (1995) Shell. [**471**] JP0338586 (1991) Nisso. [**472**] EP410552 (1991) Schering. [**473**] WO98/50366 (1998) BASF. [**474**] WO98/50379 (1998) BASF. [**475**] JP1112275 (1999) Nisso. [**476**] WO99/21852 (1999) Nisso. [**477**] DE19914140 (2000) Bayer. [**478**] EP352543 (1990) Nissan. [**479**] WO98/56766 (1998), US5807806 Nisso. [**480**] WO98/42678 (1998) Dow. [**481**]

WO98/45273 (1998) Nisso. [**482**] US5824802 (1998) Dow. [**483**] WO99/10328 (1999) BASF. [**484**] JP1143480 (1999) Nisso. [**485**] JP11193275 (1999) Nisso. [**486**] JP11236376 (1999) Nisso. [**487**] WO00/17194 (2000) Dow. [**488**] WO98/52926 (1998) Dow. [**489**] JP11292849 (1999) Nisso. [**490**] WO97/41116, WO97/41117, WO97/41118 (1997) Nisso. [**491**] WO97/41105 (1997) Nisso. [**492**] WO99/54328 (1999) Nisso. [**493**] WO98/31681 (1998) BASF. [**494**] JP11240872 (1999), WO99/23094 Nisso. [**495**] WO00/34272 (2000) BASF. [**496**] WO00/34273, WO00/34274 (2000) BASF. [**497**] WO99/26930 (1999) BASF. [**498**] WO97/41106 (1997) Ishihara. [**499**] WO00/03993 (2000) Ishihara. [**500**] WO95/04054 (1995) Idemitsu. [**501**] WO98/49159 (1998) DuPont. [**502**] JP10130267 (1998) Idemitsu. [**503**] WO93/18031 (1993) Idemitsu. [**504**] WO96/31507 (1996) Idemitsu. [**505**] WO94/01431 (1994) Idemitsu. [**506**] WO97/12885 (1997) Idemitsu. [**507**] WO97/08164 (1997) DuPont. [**508**] WO98/12192 (1998) BASF. [**509**] WO95/13275 (1995), WO96/19470 (1996), JP09291088 (1997), JP10265472 (1998) Idemitsu. [**510**] WO00/34270 (2000) BASF. [**511**] WO00/34271 (2000) BASF. [**512**] JP0278662, EP353187 (1990) Ciba-Geigy. [**513**] EP502492 (1992) Hoechst. [**514**] JP06247891 (1994) Hokko. [**515**] JP06321932 (1994) Hokko. [**516**] JP06271562 (1994) Hokko. [**517**] JP06271498 (1994) Hokko. [**518**] JP0769963 (1995) Hokko. [**519**] WO97/45404 (1997) Hokko. [**520**] WO98/29383 (1998) BASF. [**521**] WO98/50337 (1998) BASF. [**522**] JP0305408 (1991) SDS. [**523**] DE4241999 (1994) Hoechst. [**524**] WO96/22958 (1996) Zeneca. [**525**] JP0200222 (1990) Nissan. [**526**] WO90/05712 (1990) ICI. [**527**] WO91/05470 (1991) Hoechst. [**528**] JP07206808 (1995) Hokko. [**529**] WO98/42648 (1998) Dow. [**530**] JP11193259 (1999) Nisso. [**531**] WO97/03045 (1997) DuPont. [**532**] WO99/03845 (1999) RP. [**533**] DE19921732 (2000) Bayer. [**534**] WO97/46530 (1997) DuPont. [**535**] JP1121280 (1999) Nisso. [**536**] DE19846792 (2000) Hoechst-S-A. [**537**] WO00/64912 (2000) BASF. [**538**] WO94/04524 (1994) Idemitsu. [**539**] WO94/08988 (1994) Idemitsu. [**540**] JP0925279 (1997) Idemitsu. [**541**] WO98/29406 (1998) Idemitsu. [**542**] WO98/35954 (1998) DuPont. [**543**] WO99/33820 (1999) Idemitsu. [**544**] DE19840337 (2000) Hoechst-S-A. [**545**] WO00/78146 (2000) Idemitsu. [**546**] WO97/30986 (1997) BASF. [**547**] EP860441 (1998) DuPont. [**548**] WO0020408 (2000) Idemitsu. [**549**] WO99/57111 (1999) BASF. [**550**] WO00/35903 (2000) BASF. [**551**] WO00/69853 (2000) Idemitsu. [**552**] US4997473 (1991) ICI. [**553**] EP563817 (1993) Hoechst. [**554**] DE19961466 (2000) Novartis. [**555**] WO91/01289 (1991) Nisso. [**556**] US5110343 (1992) Nisso. [**557**] WO98/29384 (1998) BASF. [**558**] JP1121274 (1999) Nisso. [**559**] WO98/50377 (1998) BASF. [**560**] WO93/03009 (1993) ICI. [**561**] DE4434987 (1996) BASF. [**562**] WO99/10327 (1999) BASF. [**563**] JP0429973 (1992) Nisso. [**564**] WO98/40366 (1998) BASF. [**565**] WO99/07688 (1999) BASF. [**566**] WO00/73311 (2000) BASF. [**567**] JP03120202 (1991), US5089046 (1992) Sandoz. [**568**] EP394889 (1990) Sandoz. [**569**] WO92/07837 (1992) Sandoz. [**570**] WO98/29412 (1998) Nisso. [**571**] WO00/50397 (2000) Nisso. [**572**] US5132434 (1992) Nissan. [**573**] WO91/00260 (1991) Nisso. [**574**] WO93/01171 (1993) Nisso. [**575**] WO97/35851 (1997) Nisso. [**576**] JP0570426 (1993) Nisso. [**577**] JP07196585 (1995) Nisso. [**578**] JP08245618 (1996) Nisso.

[**579**] US4921526 (1990) Sandoz. [**580**] JP07206863 (1995) Nisso. [**581**] JP03120203 (1991) Sandoz. [**582**] JP0625144 (1994) SDS. [**583**] JP0782240 (1995) SDS. [**584**] AU672058 (1996) SDS. [**585**] JP10109972 (1998) Ishihara. [**586**] JP200016982 (2000) Kumiai & Ihara. [**587**] EP418175 (1991) RP. [**588**] EP487357 (1992) RP. [**589**] EP470856 (1992) RP. [**590**] EP527036 (1993) RP. [**591**] EP527037 (1993) RP. [**592**] EP560483 (1993) RP. [**593**] JP05202009 (1993) RP. [**594**] EP560482 (1993) RP. [**595**] EP609798 (1994) RP. [**596**] EP609797 (1994) RP. [**597**] WO94/14782 (1994) RP. [**598**] WO94/18179 (1994) RP. [**599**] WO95/16678 (1995) RP. [**600**] WO/97/23491 (1997) RP. [**601**] WO98/51153 (1998) RP. [**602**] EP524018 (1993) RP. [**603**] EP636622 (1995) RP. [**604**] WO97/44340 (1997) Idemitsu. [**605**] JP11140084 (1999) Idemitsu. [**606**] DE19914948 (2000) BASF. [**607**] EP588357 (1994) RP. [**608**] WO/97/27187 (1997) RP. [**609**] WO/97/28136 (1997) RP. [**610**] WO99/03856 (1999) RP. [**611**] WO97/43270 (1997) Novartis. [**612**] WO99/51583 (1999) Novartis. [**613**] WO00/35916 (2000) Aventis. [**614**] EP625508 (1994) RP. [**615**] US5804532 (1998) RP. [**616**] WO98/51153 (1998) RP. [**617**] WO95/04716 (1995) RP. [**618**] JP10175937 (1998) Ishihara. [**619**] WO/9713765 (1997) Idemitsu. [**620**] DE19543641 (1997) BASF. [**621**] DE19914949 (2000) BASF. [**622**] US4894085 (1990) Dow. [**623**] JP03251570 (1991) RP-Yuka. [**624**] JP04149186 (1992) Teijin. [**625**] US4935051 (1990) Ciba-Geigy. [**626**] EP427549 (1991) Shionogi. [**627**] JP0225466 (1990) Suntory. [**628**] JP0225454 (1990) Suntory. [**629**] US4968343 (1990) Dow. [**630**] EP426491 (1991) Alkaloida Vegyeszeti Gyar. [**631**] JP03215470 (1991) Suntory & Shionogi. [**632**] JP05148235 (1993) RP-Yuka. [**633**] US5032168 (1991) DowElanco. [**634**] DE19510454 (1995) Ciba-Geigy. [**635**] DE3829586 (1990) Bayer. [**636**] US5034050 (1991) DowElanco. [**637**] DE4202053 (1993), DE4012711 (1991) Bayer. [**638**] EP508263 (1992) Bayer. [**639**] DE4111618 (1992) Bayer. [**640**] DE4131585 (1993) Bayer. [**641**] US4900354 (1990) Dow. [**642**] JP02300105 (1990) Teijin. [**643**] WO92/05700 (1992) Uniroyal. [**644**] EP527016 (1993) DuPont. [**645**] JP05339224 (1993) Tokuyama. [**646**] JP0217187 (1990) Kumiai. [**647**] JP0217187 (1990) Kumiai. [**648**] EP383319 (1990) Takeda. [**649**] US4909835 (1990) Ciba, DE3941160 (1990) Ciba-Geigy. [**650**] DE4126479 (1993) BASF. [**651**] DE4204206 (1993) BASF. [**652**] EP524525 (1993) Nihon Bayer. [**653**] EP554744 (1993) Nihon Bayer. [**654**] JP05117109 (1993) Tokuno. [**655**] EP368227 (1990) BASF. [**656**] EP456069 (1991), DE4222261 (1993) BASF. [**657**] DE4018499 (1991) BASF. [**658**] EP456090 (1991) BASF. [**659**] EP464542 (1992) BASF. [**660**] DE4039281 (1992) BASF. [**661**] DE4126999 (1993) BASF. [**662**] DE4204203 (1993) BASF. [**663**] DE4204204 (1992) BASF. [**664**] DE19545212 (1997) BASF. [**665**] DE4204205 (1993) BASF. [**666**] EP680949 (1995) BASF. [**667**] EP456068, DE4018508 (1991) BASF. [**668**] EP456118 (1991) BASF. [**669**] EP456112 (1991) BASF. [**670**] EP481354 (1992) BASF. [**671**] DE4227896 (1994) BASF. [**672**] WO98/30565 (1994) BASF. [**673**] EP456069 (1991) BASF. [**674**] WO91/10658 (1991) Dunlena. [**675**] WO92/14736 (1992) Dunlena. [**676**] WO93/10081 (1993) Dunlena. [**677**] WO96/06071 (1996) Korea Research. [**678**] CA1280768 (1991) ICI. [**679**] EP381291 (1990) Shell. [**680**] US4952722 (1990) ICI. [**681**] US5006159 (1991) DowElanco. [**682**]

US4983211 (1991) DowElanco. [**683**] WO94/03443 (1994) Korea Research. [**684**] US5399542 (1995) DowElanco. [**685**] DE3833264 (1990) BASF. [**686**] US4923989 (1990) ICI. [**687**] US5013352 (1991) DowElanco. [**688**] WO92/08696 (1992). [**689**] DE19510183 (1995) BASF. [**690**] EP384736 (1996) Agrimont S.p.A. [**691**] DE3929673 (1991) Bayer. [**692**] JP0390069 (1991) Chugai. [**693**] WO92/02512 (1992) Chugai. [**694**] US5308830 (1993) R & H. [**695**] WO99/25700 (1999) Kaken & Sagami. [**696**] WO00/42028 (2000) Monsanto. [**697**] JP05345779 (1993) Chugai. [**698**] WO98/12193 (1998) BASF. [**699**] JP04297464 (1992), JP0559020 (1993), RP-Yuka, US5211739 (1993) R & H, JP05140124 (1993) Hokko. [**700**] JP05140150 (1993) Hokko. [**701**] JP04187678 (1992) RP-Yuka. [**702**] JP04321671 (1992) RP-Yuka. [**703**] JP05140123 (1993) Hokko. [**704**] JP05163254 (1993) Hokko. [**705**] JP05163252 (1993) Hokko. [**706**] JP05163253 (1993) Hokko. [**707**] JP06116263 (1994) Hokko, WO93/09100 (1991) DuPont. [**708**] JP06157515 (1994) Ihara & Kumiai. [**709**] JP05255318 (1993) Hokko. [**710**] JP05255317 (1993) Hokko. [**711**] JP05194494 (1993) Hokko. [**712**] JP05255316 (1993) Hokko. [**713**] JP05255315 (1993) Hokko. [**714**] JP05255314 (1993) Hokko. [**715**] WO97/11075 (1997) Takeda. [**716**] JP09249665 (1999) DIC & Sagami. [**717**] JP10175975 (1998) DIC & Sagami. [**718**] JP03279368 (1991) RP-Yuka. [**718**] JP03279368 (1991) RP-Yuka. [**719**] EP493925 (1992) Sankyo. [**720**] EP433804 (1991) Bayer. [**721**] EP654468 (1995) Bayer. [**722**] JP05140125 (1993) Hokko. [**723**] US5258361 (1993) R & H. [**724**] WO93/21164 (1993) Bayer. [**725**] EP571854 (1993) Nihon Bayer. [**726**] EP571855 (1993) Nihon Bayer. [**727**] EP572855 (1993) Nihon Bayer. [**728**] EP578090 (1994) Nihon Bayer. [**729**] EP672663 (1995) Nihon Bayer. [**730**] WO98/35961 (1998) DuPont. [**731**] WO98/51683 (1998) DuPont. [**732**] WO99/07702 (1999) BASF. [**733**] EP771797 (1997) Nihon Bayer. [**734**] WO98/25912 (1998) DuPont. [**735**] EP902028 (1999) DuPont. [**736**] EP695748 (1996) Nihon Bayer. [**737**] WO99/48890 (1999) DuPont. [**738**] JP09110863 (1997) Nissan. [**739**] JP11180965 (1999) Nissan. [**740**] JP09100277 (1997) Nissan. [**741**] JP0987281 (1997) Nissan. [**742**] JP09100272 (1997) Nihon Bayer. [**743**] EP820994 (1998) Nihon Bayer. [**744**] WO98/01431 (1999) Nissan. [**745**] WO00/40568 (2000) Nihon Bayer. [**746**] JP11100371 (1999) Nissan. [**747**] WO98/38176 (1998) Hokko. [**748**] JP200063379 (2000) Hokko. [**749**] WO00/10984 (2000) Hokko. [**750**] WO00/43377 (2000) DuPont. [**751**] EP558448 (1993) Ciba-Geigy. [**752**] JP05170697 (1993) Kureha. [**753**] DE4320076 (1994) Bayer. [**754**] WO94/13652 (1994) Zeneca. [**755**] WO96/37466 (1996) Zeneca. [**756**] WO97/20838 (1997) Zeneca. [**757**] DE4414568 (1995) Bayer. [**758**] CA2130419 (1995) Sumitomo. [**759**] WO95/27698 (1995) DuPont. [**760**] EP796845 (1997) Isagro. [**761**] DE19517597 (1995) BASF. [**762**] EP745599 (1996) ACC. [**763**] US5705644 (1998) ACC. [**764**] US5726126 (1998) ACC. [**765**] WO98/35964 (1998) Isagro. [**766**] EP786453 (1997) Mitsubishi. [**767**] JP200026414 (2000) Mitsubishi. [**768**] JP11222472 (1999) Mitsubishi. [**769**] WO00/03985 (2000) Mitsubishi. [**770**] DE19837672 (2000) Bayer. [**771**] WO95/19962 (1995) Central Glass. [**772**] JP09227504 (1997) Central Glass. [**773**] JP08253455 (1996) A-Kanesho. [**774**] US5858923 (1999) A-Kanesho. [**775**] WO95/06643 (1995) DuPont, Degussa.

[**776**] WO95/21174 (1995) Ciba-Geigy. [**777**] US5679791 (1997) ACC. [**778**] WO98/47904 (1998) DuPont. [**779**] EP780385 (1997) Isagro. [**780**] WO98/54967 (1998) Isagro. [**781**] DE19613548 (1997) BASF. [**782**] WO00/5040 (2000) KRICT7. [**783**] WO95/19967 (1995) WO94/26109 (1994) Zeneca. [**784**] DE19542520 (1997) BASF. [**785**] WO97/30060 (1997) BASF. [**786**] DE19622189, DE19622189 (1997) Bayer. [**787**] DE19524623 (1997) BASF. [**788**] WO98/05649 (1998) BASF. [**789**] WO98/38169 (1998) BASF. [**790**] DE19838706 (2000) Bayer. [**791**] WO98/27090 (1998) BASF. [**792**] WO99/55702 (1999) BASF. [**793**] WO97/15559 (1997) BASF. [**794**] US5672715 (1997) Monsanto. [**795**] DE19615259 (1997) Bayer. [**796**] WO99/38861 (1999) Bayer. [**797**] WO97/05115 (1997) FMC. [**798**] WO95/30661 (1995) Bayer. [**799**] WO97/26248 (1997) Bayer. [**800**] DE19739208 (1998) Novartis. [**801**] DE4405614 (1995) Bayer. [**802**] DE19500439 (1995) Bayer. [**803**] DE19702786 (1998) Bayer. [**804**] JP0789813, JP07187919 (1995) Otsuka. [**805**] JP07188221 (1995) Otsuka. [**806**] WO95/32621 (1995) DuPont, Degussa. [**807**] JP07188220 (1995) Otsuka. [**808**] WO97/09326 (1997) Otsuka. [**809**] JP0971583 (1997) Otsuka. [**810**] JP1045754, WO98/46561 (1998) Otsuka. [**811**] DE4329537 (1995) BASF. [**812**] WO95/25725 (1995) Uniroyal. [**813**] WO97/02253 (1997) BASF. [**814**] WO98/42681 (1998) BASF. [**815**] DE19524617 (1997) BASF. [**816**] WO98/28280 (1998) FMC. [**817**] JP11140055 (1999) Nissan. [**818**] EP985670 (2000) ACC. [**819**] EP705829 (1996) Sumitomo. [**820**] WO98/54155 (1998) Bayer. [**821**] WO99/21837 (1999) Ishihara Americas. [**822**] JP09241245 (1997) DIC. [**823**] JP2000302764 (2000) Sumitomo. [**824**] JP09188676 (1997) Mitsubishi. [**825**] DE19627901 (1998) Bayer. [**826**] DE19708928 (1998) Bayer. [**827**] EP869123 (1998) Nihon Bayer. [**828**] DE4424401, DE4412079 (1995) Bayer. [**829**] DE19632005 (1998) Bayer. [**830**] DE19649094 (1998) Bayer. [**831**] WO95/31440 (1995) Bayer. [**832**] DE19617532 (1997) Bayer. [**833**] DE19516785 (1996) Bayer. [**834**] DE19517732 (1996) Bayer. [**835**] JP0948761 (1997) Nissan. [**836**] JP2000344732 (2000) Nissan. [**837**] DE19523640 (1996) Bayer. [**838**] DE19532344 (1997) Bayer. [**839**] DE19830694 (2000) Bayer. [**840**] DE19528186 (1997) Bayer. [**841**] US1923 (2000) FMC. [**842**] WO97/45418 (1997) Bayer. [**843**] DE19651036 (1998) Bayer. [**844**] DE19504188 (1996) BASF. [**845**] DE19547475 (1997) Bayer. [**846**] DE19527570 (1997) Bayer. [**847**] DE19853864 (2000) Bayer. [**848**] DE19927612 (2000) Bayer. [**849**] US4927451 (1990) Uniroyal. [**850**] EP195346 (1986) H-La-Roche. [**851**] US4981508 (1991) Uniroyal. [**852**] JP03215476 (1991) Nissan. [**853**] EP420194 (1991) Sumitomo. [**854**] US5084084 (1992) Nissan. [**855**] EP408382 (1991) Nissan. [**856**] WO98/14452 (1998) Nissan. [**857**] JP2000247975 (2000) Nissan. [**858**] JP08259523 (1996) Central Glass. [**859**] JP08259522 (1996) Central Glass. [**860**] DE19530451 (1997) Bayer. [**861**] JP08259521 (1996) Central Glass. [**862**] JP08259524 (1996) Central Glass. [**863**] WO97/12884 (1997) FMC. [**864**] US5661108 (1997) FMC. [**865**] WO99/14216 (1999) ACC. [**866**] EP908457 (1999) ACC. [**867**] WO00/49004, WO00/49002, WO00/49016 (2000) ACC. [**868**] EP476697 (1992) Sumitomo. [**869**] EP561319 (1993) Sumitomo. [**870**] US5346881 (1994) FMC. [**871**] US5441925 (1995) FMC. [**872**] JP0959113 (1997) Nichino. [**873**] DE19532048 (1997) BASF. [**874**] WO97/08170 (1997) FMC. [**875**]

US5753595 (1998) FMC. [**876**] WO98/33796 (1998) BASF. [**877**] WO00/28822 (2000) BASF. [**878**] WO97/29105 (1997) Kumiai. [**879**] JP11140083 (1999) Kumiai. [**880**] WO99/31091 (1999) BASF. [**881**] JP09301973 (1997) Kumiai. [**882**] WO95/33746 (1995) DuPont. [**883**] WO97/12886 (1997) FMC. [**884**] WO98/38188 (1998) FMC. [**885**] WO97/42188 (1997) Kumiai. [**886**] WO95/17096 (1995) FMC. [**887**] DE19715017 (1998) Bayer. [**888**] US5798316 (1998), WO00/04774, US6022830 (2000) FMC. [**889**] DE19728125 (1999) Bayer. [**890**] DE19752748 (1999) Bayer. [**891**] WO99/38851 (1999) Bayer. [**892**] US5391541 (1995) FMC, DE19523372 (1997) BASF. [**893**] DE19523372 (1997) BASF. [**894**] US1764 (1998) FMC. [**895**] WO97/35845 (1997) BASF. [**896**] US5683966 (1997) FMC. [**897**] WO98/42682 (1998) BASF. [**898**] DE19738084 (1999) BASF. [**899**] DE19705012 (1998) Hoechst. [**900**] WO00/21936 (2000) Sumitomo. [**901**] WO98/41093 (1998) Bayer. [**902**] WO98/55462 (1998) Bayer. [**903**] DE19739638 (1999) Bayer. [**904**] DE19534466 (1997) BASF. [**905**] WO97/34872 (1997) BASF. [**906**] WO98/54137 (1998) BASF. [**907**] WO97/30059 (1997) BASF. [**908**] WO98/07720 (1998) BASF. [**909**] WO98/07700 (1998) BASF. [**910**] WO97/28127 (1997) Kumiai. [**911**] JP11140054 (1999) Kumiai. [**912**] JP2000119252 (2000) Nissan. [**913**] WO99/40090 (1999) Nissan. [**914**] JP63156779 (1988) Nisso. [**915**] WO97/07104 (1997) Sumitomo. [**916**] WO98/17632, WO98/17633 (1998), EP963978 (1999), JP200053652, JP2000226375, JP2000229928 (2000) Sumitomo. [**917**] EP860435 (1998) Sumitomo. [**918**] EP860434 (1998) Sumitomo. [**919**] DE19754348 (1998) Novartis. [**920**] JP200026431 (2000) Sumitomo. [**921**] WO99/14201 (1999) BASF. [**922**] DE19520613 (1996) Bayer. [**923**] WO99/52878 (1999) Bayer. [**924**] JP1053508 (1998) Mitsubishi. [**925**] JP11180964 (1999) Mitsubishi. [**926**] WO99/59983 (1999) BASF. [**927**] DE19835943 (2000) Bayer. [**928**] EP759429 (1997) BASF. [**929**] DE19530606 (1997) BASF. [**930**] JP0959276 (1997) Kumiai. [**931**] DE19630555 (1998) Hoechst. [**932**] DE19532347 (1997) Bayer. [**933**] DE19623892 (1997) Hoechst. [**934**] DE19751943 (1999) Hoechst. [**935**] DE19631865 (1998) Bayer. [**936**] DE19834110 (2000) Hoechst. [**937**] WO98/42698, WO98/21199 (1998) Novartis. [**938**] DE19652426 (1998) Bayer. [**939**] DE19854082 (1999) Bayer. [**940**] JP05202031 (1993) Nissan. [**941**] DE19652429 (1998) Bayer. [**942**] DE19652431 (1998) Bayer. [**943**] WO99/52892 (1999) Novartis. [**944**] WO99/52893 (1999) Novartis. [**945**] WO99/55693 (1999) Novartis.

11 Diverse Response of Plants Towards Chiral Phytotoxic Chemicals

Hiroyoshi Omokawa

11.1 Introduction

At the beginning of weed control, simple and primitive organic chemicals were used for herbicides. Nowadays, molecular structures and modes of action of the currently commercialized herbicides are diverse, their physiological properties are almost ideal and of a high performance relating to activity, selectivity between crop and weeds, and safety towards the environment. Some concerns, however, remain. The major ones are environmental contamination, toxicity (Lewis et al. 1999), and the appearance of weeds resistant to herbicides (Powles and Holtum 1994; Duke 1996), resulting from the heavy and/or repeated input of herbicides.

Chirality, the right- and left-handedness, is one of the inherent concepts in molecular and life science (Ariens et al. 1988; Wainer and Drayer 1988; Holmstedt et al. 1989; Kurihara and Miyamoto 1998). The significance of molecular chirality in life science has long been recognized since the separation of the optical isomers of tartrate by Pasteur and the establishment of the tetrahedral theory by vant Hoff and LeBel on the stereochemistry of carbon compounds. All scientists recognized it again when we were confronted with the tragedy of thalidomide (Fabro et al. 1967). In agricultural science, we realized its importance by the following triple events. First, triazole compounds such as uniconazole controlled the growth of both fungi and plants, and its qualitatively different response results from molecular chirality (Krämer et al. 1983, Köller 1987), which raised great interest among agrochemists. Second, the less active *S*-aryloxyphenoxypropionate (APP) herbicide is converted into its active (*R*)-antipode in soil and shows herbicidal activity (Dicks et al. 1985). This finding, being a chiral inversion, was noted with interest by weed scientists. Third, metolachlor has different types of chirality (Moser et al. 1982; Blaser and Spindler 1997); one is a central chirality and the other is an axial chirality. Recent advances have been reviewed in a monograph (Kurihara and Miyamoto 1998).

In relation to the biological response, the influence of chirality is a well-known phenomenon, and the enantioselective and/or enantio-specific responses on activity, metabolism by microorganisms, plants, animals and also humans are critical, not only for chiral chemicals, but also for achiral chemicals. Usually, optically pure single enantiomers including naturally occurring

P. Böger, K. Wakabayashi, K. Hirai (Eds.)
Herbicide Classes in Development

materials, pharmaceuticals and agrochemicals are the most effective, since bioactive agents exhibit enantiomeric and enantiotropic selective responses to target organisms. The target site in organisms constitutes a special chiral environment, this being the handedness. Optical isomers actually respond to organisms with diversity. Until now, diverse relationships between chirality and biological response have been reported, not only for pharmaceuticals, but also for agrochemicals including insect hormones, pheromones, fungicides, herbicides and plant growth regulators. Among them, insect pheromones respond diversely (Mori 1998). A basic classification has been proposed with regard to pharmaceuticals (Powell et al. 1988): (1) all the biological activity may reside in one isomer. The second isomer in this example might be regarded as an isomeric impurity. (2) The isomers may have nearly identical qualitative and quantitative biological activity. (3) The isomers have qualitatively similar biological activity, but have different quantitative potency. (4) The isomers have qualitatively different biological activities.

Molecular structures, including stereochemistry, of herbicides are diverse. Although there are many chiral herbicides, optically active herbicides are only a part of them (Fig. 1). Since Fredga and Aberg (1965) foresightedly reviewed stereoisomerism in plant growth regulators of the auxin type, almost all studies on physiological responses of plants toward optically active chemicals, including plant hormones, plant growth regulators and herbicides, have reported a qualitatively similar action with different quantitative potency. During the last 10 years, various efficient and stereo-specific syntheses and purifications have become available by the development of enantio-specific and -selective catalysts. Chiral supports of column chromatography are still growing rapidly, leading to the commercialization of optically active herbicides (mecoprop-P, *S*-metolachlor, fenoxaprop-P-ethyl, etc.) and the establishment of three-dimensional structure-activity (3-D QSAR) studies based on molecular chirality.

The minimized input of herbicides to the environment including air, soil, water, crop and all organisms is essential. In order to conserve and hand our precious green earth over to future generations and minimize toxicity, it is essential to switch agrochemicals from an optically impure isomer (racemate or diastereomer) to an optically active one (Williams 1996). Molecular chirality is an essential concept to analyze the mode of action, and to design a drug with increased activity and decreased side effects to humans and the environment. Molecular chirality should provide a concept on the fact that organisms diversely react against chiral chemicals.

11.2 Diverse Response of Optically Active Herbicides

Chiral phytotoxic compounds exhibit diverse physiological properties to plants. Triazole compounds such as uniconazole, paclobutrazol and related chemicals exhibit both fungicidal and dwarfing activities based on chirality of

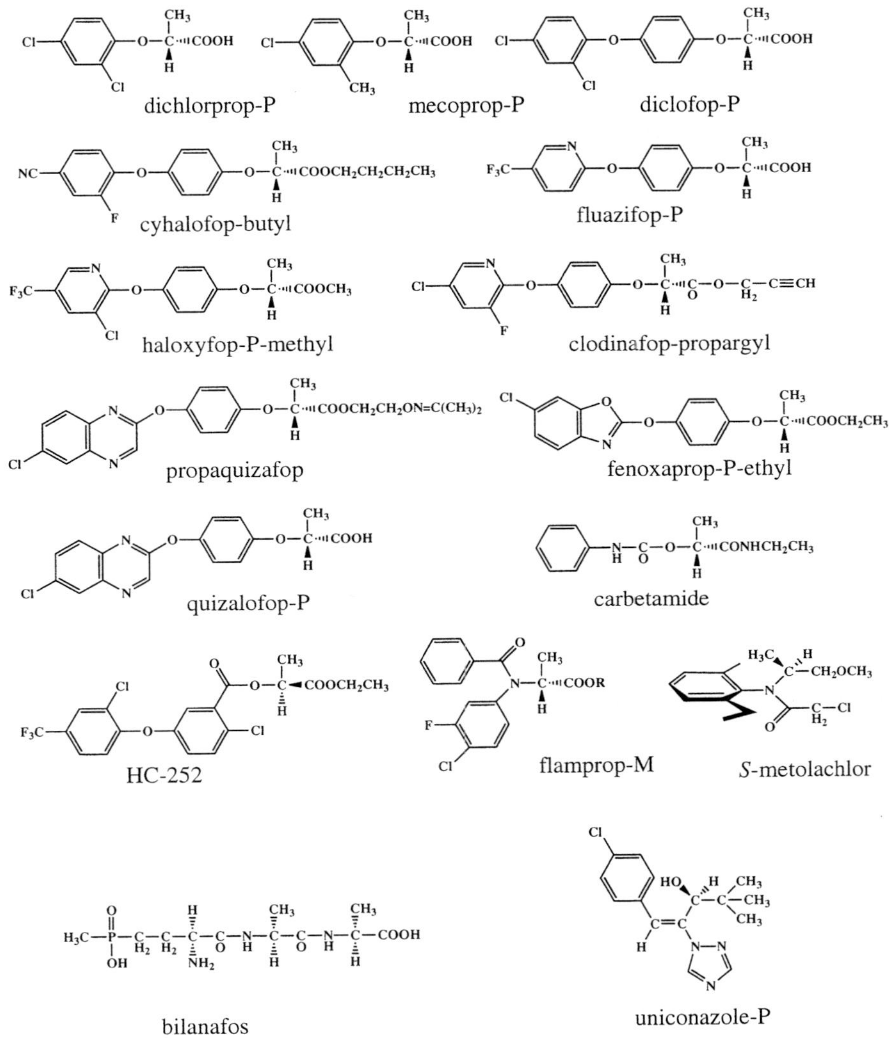

Fig. 1. Optically active herbicides and plant growth regulators. (Pesticide Manual, 11th edition, British Crop Protection Council, 1997)

their inherent molecular structures. Those biological responses are caused by the inhibition of differential biosynthetic pathways of sterol in fungi and gibberellin in plants by each different optically active enantiomer, respectively (Krämer et al. 1983; Takano et al. 1986). An acylalanine compound, CGA 29212, controlled the growth of fungi (by *R*-enantiomer) and plants (by *S*-enantiomer) enantio-selectively as similar as triazole compounds, and by further modification metalaxyl-M (*R*-enantiomer) and flamprop-M (*R*-

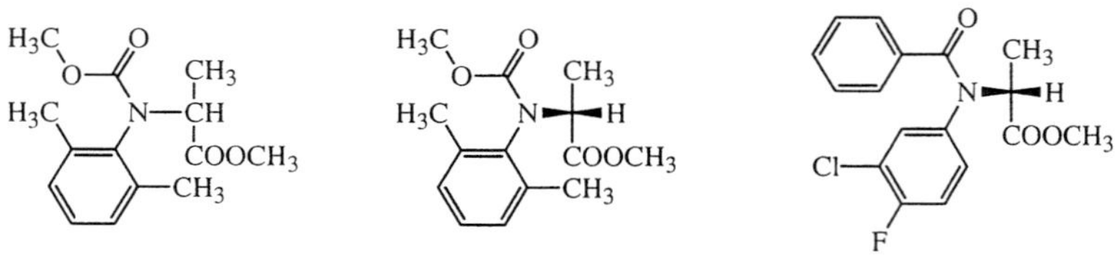

CGA 29212 Metalaxyl-M Flamprop-M

Fig. 2. Acylalanine fungicides and herbicide (flamprop-M) methyl

enantiomer) were developed as fungicide and herbicide, respectively (Fig. 2; Hubele et al. 1982). This qualitatively different response is between fungi and plants. In the field of herbicides, many interesting phenomena of chiral herbicides have been reported; they are not only qualitatively similar in enantioselective herbicidal activity, but also in (1) enantioselective metabolism (uptake, translation, and degradation), and (2) chiral inversion, and interesting studies on chirality and physiological responses are increasing.

11.2.1
Qualitatively Similar Enantioselective Action

Most of the optically active herbicides are synthesized with a chiral syntom of lactic acid and related compounds (Fig. 1). These are phenoxypropionate, aryloxyphenoxypropionate (APP) and acylalanine herbicides. Phenoxypropionate herbicides, of the auxin-type, exhibit enantioselectively (Chan et al. 1975, Barnwell and Cobb 1993), and optically pure enantiomers have been adapted as herbicides, dichlorprop-P and mecoprop-P. The APP herbicides enantioselectively control weeds through the inhibition of fatty acids syntheses and acetyl-CoA carboxylase. There is a valuable review of chiral APP herbicides by Haga et al. (1998). Optically pure *R*-enantiomers of the APP herbicides have been used as the herbicides clodinafop, cyhalofop, diclofop-P, fenoxaprop-P, fluazifop-P, haloxyfop-P and quizalofop-P. They inhibit the growth of plants enantioselectively (Sakata et al. 1985a,b; Uchiyama et al. 1986; Gerwick et al. 1988; Nakahira et al. 1988, 1990; Shimabukuro and Hoffer 1995). Optical isomers of DMPA enantioselectively regulated plant growth (Holmsen 1968). Chiral peroxidizing herbicides enantioselectively inhibit plant growth and protoporphyrinogen oxidase (Hallahan et al. 1992; Nandihalli et al. 1994; Hamper et al. 2000). The *S*-isomer of dimethenamid, a chloroacetamide herbicide, strongly inhibited algal growth and fatty acid elongation enantioselectively (Couderchet et al. 1997). Metolachor is of interest for its molecular structure; it has four optical isomers causing both central and axis chiralities (Fig. 3; Moser et al. 1982). An enantiomerically enriched form, DUAL MAGNUM, replaced the racemic mixture in 1997, since high performance enantioselective synthesis of S-metolachlor has been accomplished by the

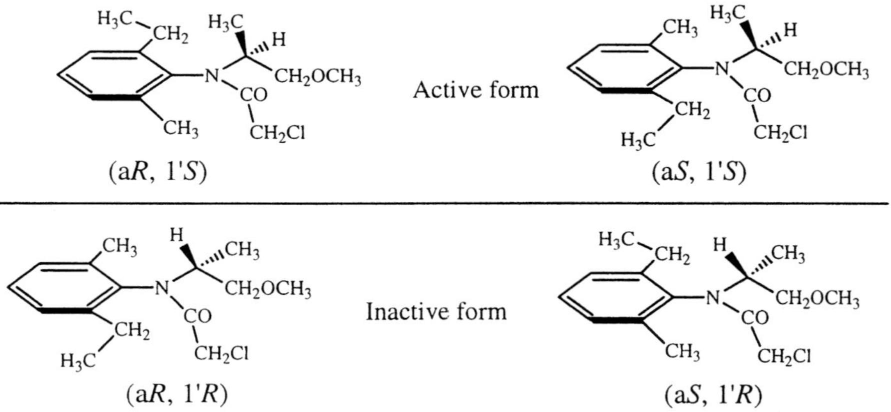

Fig. 3. Optical isomers of metolachlor

development of a hydrogenation catalyst (Blase and Spindler 1997; Spindler and Fruh 1998). A central chirality is an essential for generation of the activity, where S-configuration is a better stereo form.

11.2.2 Enantiomeric Metabolism

Haloxyfop-methyl, diclofop-methyl and fenoxaprop-ethyl were degraded enantioselectively in soil. Usually, the half-life of an *S*-APP herbicide is shorter than one of the *R*-antipodes. Fenoxaprop-ethyl was degraded with a half-life of 3 days in sand or in sandy loam soil (Wink and Luley 1988). Dichlorprop was degraded completely in soil, where the *S*-(−)-isomer degraded significantly faster than the *R*-(+)-isomer (Garrison et al. 1996; Zipper et al. 1998a). The difference in degradation between enantiomers results from enantioselective action of microorganisms in soil. Soil microbiotics (*Sphingomonas herbicidovorans* MH, *Comamonas acidovorans* MC1) degrade both racemic and optically pure APP herbicides with preferential degradation of the *S*-enantiomer vs. the *R*-antipode (Zipper et al. 1996; Muller et al. 1999; Kohler 1999). The case of mecoprop is similar (Nickel et al. 1997). Chiral phenoxypropionate herbicides (MCPP, DCPP) were also degraded enantioselectively, and the herbicidally active *R*-enantiomers were degraded significantly slower than the inactive *S*-antipodes (Zipper et al. 1996; Müller and Buser 1997). Enantioselective microbial degradation resulted in an increase in the enantiomeric ratio of *R*- to *S*-mecoprop during groundwater passage of the landfill leachate (Zipper et al. 1998b). Environmental changes in soils can alter the preferences of degradation of herbicides, and the preferences shift, due to different groups of related microbial genotypes. In Brazilian soils, almost all pasture samples preferentially transformed the nonherbicidal enantiomer of dichlorprop, while

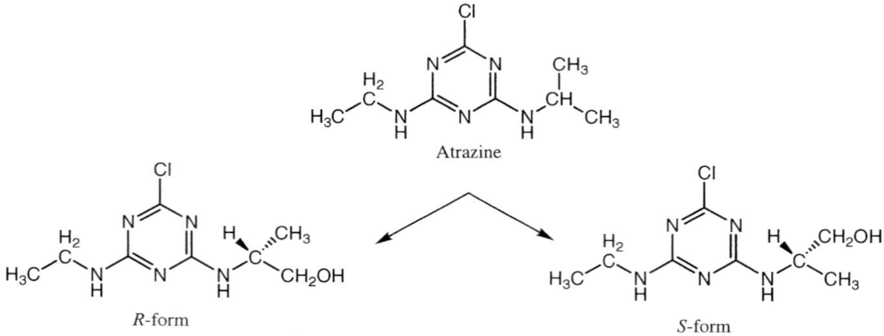

Fig. 4. Enantioselective metabolism of atrazine in rats, pigs and humans. (Lang et al. 1996)

most forest samples either transformed the herbicidal enantiomer more readily or as rapidly as the nonherbicidal enantiomer. Organic nutrient enrichments shifted enantioselectivity for methyl dichlorprop strongly towards a preferential removal of the nonherbicidal enantiomer in soil from Brazil and North America, potentially increasing phytotoxicity of its residues relative to that of the racemate (Lewis et al. 1999).

Enantioselective metabolism of chiral herbicides in animals is also observed. A single dose of *R*-flamprop-methyl to the rat was rapidly metabolized and 90% was eliminated in urine and feces within 48 h. The major metabolic routes were hydrolysis of the esters to the acids. The flamprop acid derived from the *R*-methyl ester was found to be partially converted to the *S*-form (Hutson et al. 1991). Achiral herbicides are also metabolized enantioselectively. The *s*-triazines herbicides (atrazine, terbuthylazine, ametryne, and terbutryne) were converted into their metabolites by *N*-monodealkylation, hydroxylation of the isoPr or *tert*Bu moiety, and sulfoxidation by liver microsomes from rats, pigs, and humans (Fig. 4; Lang et al. 1996). Species-specific stereoselective formation of a new chiral isopropyl-hydroxylated metabolite from atrazine was different in several species, with *S/R* ratios of 76:24 in rats, 49:51 in pigs, and 28:72 in humans.

11.2.3 Chiral Inversion

The chiral inversion reaction occurs in herbicides constituted with α-substituted propanoate chiral syntom. Differential chiral inversion of haloxyfop-methyl was observed in incubation in soil (Fig. 5; Dicks et al. 1985; Gerwick et al. 1988; Wink and Luley 1988). The *S*-APP showed a significantly higher inversion tendency than the acids with the *R*-form. Chiral inversion results in microbiotic action. The enantiomerization process of chiral phenoxy acid herbicides was studied; enantiomerization of MCPP and DCPA proceed in both directions, from *R* to *S* and from *S* to *R* favoring the herbicidally active

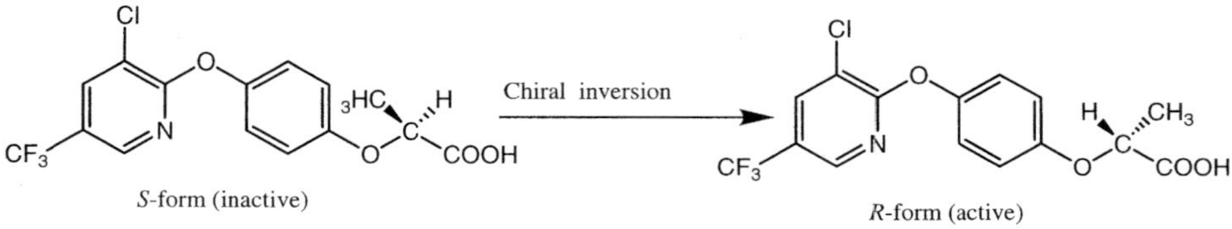

Fig. 5. Chiral inversion of haloxyfop in soil

R-enantiomers (Buser and Müller 1997). 2-Aryloxypropionic acid anti-inflammatory drugs such as ibuprofen change their configuration with unidirectional metabolic inversion via the stereoselective formation of the acyl-CoA thioesters and enzymatic or nonenzymatic inversion of the thioesters of the propionates. (Nakamura et al. 1981; Hutt and Caldwell 1983; Caldwell et al. 1988). A similar mechanism of the chiral inversion has been proposed for an α-substituted propanoate herbicide (Buser and Müller 1997).

11.3
Diverse Response of Plants Through Chirality

s-Triazine herbicides are one of the major compounds for weed control. Their phytotoxic key reaction site as a herbicide is inhibition of electron transfer in photosystem II (PS II). The typical phytotoxic symptoms of *s*-triazine herbicides are foliar chlorosis and subsequent necrosis in a later stage. In addition to the light-dependent inhibition, light-independent phytotoxic properties have been reported (Ashton and Crafts 1973). Also, light-independent stimulative properties of the *s*-triazine herbicides have been recorded; growth and yields, respiration, cytokinin-like activities, nitrogen metabolism, enzyme activity, chromatin-directed RNA synthesis, and protein synthesis. However, no inhibitory and stimulative properties, except PS II inhibition, have been adapted for a plant growth regulator. By chiral modification, light-independent physiological properties were additionally generated into *N*-α-methylbenzyl-2,4-diamio-6-chloro-*s*-triazine molecule by Omokawa et al. (1987).

Another major molecular herbicidal structure is a substituted urea. Traditional urea herbicides are classified as photosynthetic inhibitors. Thirty years ago, Moreland and Boots (1971) reported differential chiral responses of an optically active 1-(α-methylbenzyl)-3-(3,4-dichlorophenyl)urea, where each optical isomer enantioselectively, responded to respiration and/or photosynthetic reactions. Daimuron (α,α-dimethylbenzyl-*p*-tolylurea), was developed as a selective herbicide in Japan and has two kinds of physiological properties independently of PS II; a herbicidal activity against Cyperaceae weeds in rice paddy fields and a relieving (safener) action on rice injured with bensulfuron-methyl (Londax; Takematsu et al. 1975, Sakanishi et al. 1991, Honma and Teramura 1986). The different modes of action of daimuron were expressed by optically active desmethyl analogues of daimuron ((*R*)- or (*S*)-1-α-

methylbenzyl-*p*-tolylurea, MBTU; Omokawa et al. 1993). The former enantiomer inhibits the growth of Cyperaceae weeds significantly better than the *S*-antipode, whereas the latter effectively relieves the root growth reduction of rice seedlings damaged by Londax. Each methyl group (Pro*R*- and Pro*S*-methyl groups) on the benzylic position of daimuron may independently influence the affinity between the urea skeleton and the daimuron may independently influence the affinity between the urea skeleton and the receptor sites in both biological responses. Furthermore, both optical isomers show cross-intergenus selective phytotoxicity among Gramineae (Omokawa et al. 1993). One hundred and twenty derivatives (60 chiral pairs) of MBTU were synthesized and their physiological properties assessed towards plants (Omokawa et al. 1999). By chiral and substitutional modifications, the phytotoxic spectrum was spread, and some of them exhibited phytotoxic actions, not only to annual Cyperaceae paddy weeds, but also to *Echinochloa oryzicola* Vasing (barnyardgrass) and perennial *C. serotinus* Rottb (Ryoo et al. 1998).

11.3.1
Chiral *s*-Triazines

11.3.1.1
Light-Dependent and Light-Independent Growth Inhibition

Chiral and substitutional modification of the benzylamino-*s*-triazine provided two kinds of phytotoxic symptoms, leaf-burning and growth inhibition of germinating seedlings with concomitant greening and stunting; growth of barnyardgrass and *Cyperus serotinus* Rottb. was inhibited without any damage to transplanted rice seedlings under a greenhouse paddy pot test (Omokawa et al. 1987, 1988a,b). Modes of substituents at the benzylic position and alkylamino moiety of the benzylamino-*s*-triazine compounds are closely connected with a change of the symptoms (Table 1). The light-dependent inhibition is caused by the same action as that of the so-called *s*-triazine herbicides (Omokawa et al. 1989), while the light-independent one is a new type of action for the *s*-triazine herbicides. A high selectivity between rice and barnyardgrass has been expected among them, although typical PS II inhibiting *s*-triazine herbicides such as simetryn and atrazine are less selective between both plants. Both symptoms, light-dependent and -independent inhibitions, are strongly connected with molecular chirality of the α-methylbenzylamino moiety. The *S*-enantiomers showed leaf-burning and *R*-antipodes growth inhibition, but a relationship between chirality and modes of action is not necessarily evident (discussed later; Omokawa and Konnai 1992). Most of the optically active drugs are used for chiral switching avoiding undesirable side effects. However, the above results will provide a new concept on chirality and physiology in weed science, since the racemates were the most useful material for the herbicide with double symptoms. Further modification was successful to produce triaziflam as a turf herbicide. Only recently, a BASF research group

Table 1. Phytotoxic activity to rice and paddy weeds, and phytotoxic symptoms of *s*-triazines

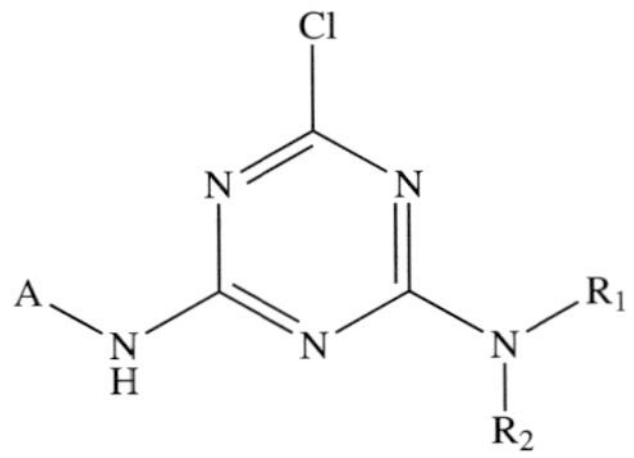

A[a]	Substituents R1, R2	Phytotoxic activity Rice[b]	ECHOR	CYPSE	Symptom[d]
Bz	Et, H	1	2 (1000)		LB
R-MBz	Et, H	2	4 (62)	–[c]	GI
S-MBz	Et, H	2	4 (31)	–	LB
RS-MBz	Et, H	2	4 (62)	5 (125)	GI/LB
RS-EBz	Et, H	2	5 (62)	–	GI/LB
RS-PrBz	Et, H	2	3 (62)	–	LB
RS-iPrBz	Et, H	0	2 (1000)		LB
R-MBz	Et, Et	1	4 (250)	–	GI
S-MBz	Et, Et	1	5 (32)	–	GI
RS-MBz	Et, Et	0	5 (32)	5 (62)	GI
R-MBz	isoPr, isoPr	0	1 (500)	–	GI
S-MBz	isoPr, isoPr	0	4 (16)	–	GI
RS-MBz	isoPr, isoPr	0	5 (32)	–	GI
DMBz	Et, H	0	4 (62)	–	GI
DMBz	*R*-*sec*Bu, H	0	4 (8)	–	GI
DMBz	*S*-*sec*Bu, H	0	4 (125)	–	LB
DMBz	RS-secBu, H	0	5 (16)	1 (1000)	GI
Simetryn		5	5 (32)	1 (250)	LB

The herbicidal and phytotoxic activities to rice seedlings and paddy weeds (*Echinochloa oryzicola* Vasing (ECHOR) and *Cyperus serotinus* Rottb. (CYPSE) were tested under paddy conditions, and their potential was visually evaluated by a 0 to 5 rating system (0 = no effect, 5 = complete killing) 3 WAT. ( ): the dose (mg a.i./m^2).

[a] Bz, benzyl; MBz, α-methylbenzyl; EBz, α-ethylbenzyl; PrBz, α-propylbenzyl; iPrBz, α-isopropylbenzyl; DMBz, α,α-dimethylbenzyl (Omokawa et al. 1987, 1988a, unpubl. data).

[b] Phytotoxicity to rice at 500 mg/m^2.

[c] –, Not tested.

[d] GI, Growth inhibition without leaf burning; LB, leaf burning.

has reported a reliable target site of triaziflam and its structurally related *s*-triazines (Grossmann et al. 2001). The *R*-enantiomer of triaziflam blocked cell division and declined cellulose deposition in the cell wall in maize root tips, suggesting that triaziflam and related *s*-triazines enantioselectively affect multiple sites of action which include PS II inhibitory activity, mitotic disruption by inhibiting microtubule formation and inhibition of cellulose synthesis.

From the similarity of the phytotoxic symptoms, (R)-α-methylbenzyl-*s*-triazines would cause disruption of mitosis of the Gramineae plants.

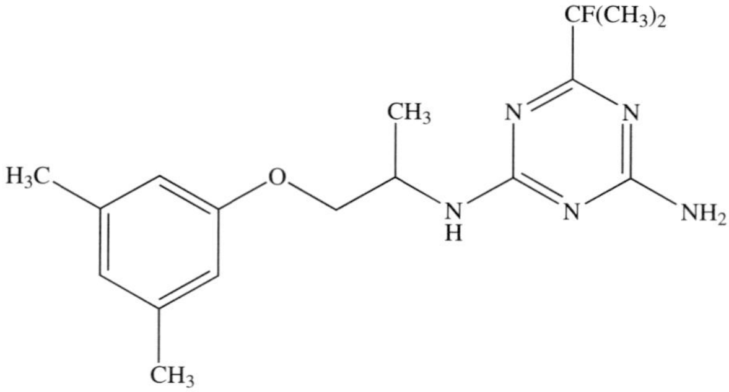

Triaziflam

11.3.1.2 Cytokinin-Like Activity

In potato plants cytokinins appear to be involved in the control of rhizome and stolon orientation (Palmer and Smith 1969; Sattelmacher and Marscher 1978). In nutsedge tubers, cytokinins break bud dormancy and apical dominance, stimulate sprouting and induce basal bulb formation in rhizomes (Teo et al. 1969; Bendixen 1975). The rhizome induction of *Cyperus serotinus* Rottb. was affected by the compounds showing cytokinin activity [benzyladenine (N^6-benzyl-1*H*-purine-6-amine, BA), kinetin (N^6-furfuryl-1*H*-purine-6-amine, forchlorfenuron (*N*-phenyl-*N'*–2-chloro-4-pyridylurea) and thidiazuron (*N*-phenyl-*N'*–1,2,3-thidiazol-5-ylurea)], and other plant hormones were not active (Omokawa et al. 1992). The *R*-enantiomers of α-methylbenzylamino-*s*-triazines strongly promoted secondary rhizome induction, and the response was highly enantioselective (Table 2; Omokawa et al. 1992, 1994). The activity is easily influenced by a slight structural change in the substituents of the amino moiety and at the benzylic position. Moreover, the activity of the methylthio-*s*-triazine compounds is higher than that of corresponding chloro-*s*-triazines. Cytokinin-like growth-promoting activities of *s*-triazine herbicides have already been reported for studies on seed germination, protein synthesis, callus growth and chlorophyll retention (Copping et al. 1972; Pillai and Davis 1973; Nader et al. 1975). However, these active *s*-triazine herbicides are inactive on a rhizome-inducing assay using *C. serotinus* (Table 2). Although the 2-(R)-α-methylbenzylamino-*s*-triazine compounds showed diverse physiological properties, namely cytokinin-like and growth inhibition towards *C. serotinus*, some differences in the structural activity relationships between both bioassay systems were found, suggesting that a light-independent herbicidal activity towards *C. serotinus* may not necessarily be correlated with cytokinin activity.

Table 2. Rhizome induction activity of cytokinins and *s*-triazines in *Cyperus serotinus* in the dark

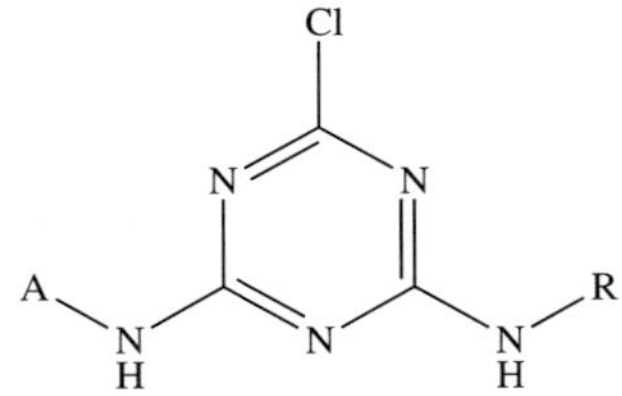

Compound		Concentration	Rhizome induction activity	
			No. of rhizomes	Length of rhizomes
A	R	(μM)	Per tuber	(mm)
Control			0.096	2.2
Bz	Et	30	1.2	6.5
RS-MBz	Me	3	2.3	14
R-MBz	Me	1	2.5	17
S-MBz	Me	30	0	
RS-MBz	Et	10	2.5	14
R-MBz	Et	1	1.2	4.8
S-MBz	Et	30	0	
RS-EBz	Me	1	1.8	19
RS-EBz	Et	1	1.4	8.2
RS-PBz	Me	1	1.3	10
RS-PBz	Et	1	0.9	4.3
RS-iPBz	Et	30	0.7	4.5
DMBz	Me	30	0	
DMBz	Et	30	0	
BA		3	2.3	12
Kinetin		10	2.3	8.7
Forchlorfenuron		1	2.6	7.5
Thidiazuron		0.1	2.4	7.2
Atrazine		30	0	
Ametryn		30	0	

(*R*)-α-Methylbenzyladenine (N^6–1-(1-phenyl)ethyl-1*H*-purine-6-amine, a BA analogue), enantioselectively exhibited cytokinin activity in shoot induction in tobacco callus tissue (Yamada et al. 1972). The above chiral requirement strongly suggests a similar mode of binding at the cytokinin receptor for the *s*-triazine and purine molecules. However, another chiral requirement for the cytokinin activity generation has been reported for a different assay system (Koshimizu et al. 1968; Matsubara et al. 1973). Only *S*-(−)-N^6–1-(1-naphthyl)ethyl-1*H*-purine-6-amine activated *Glycine max* L. callus growth and the protein kinase from barley leaf chromatin, whereas the *R*-isomer was less effective (Kulaeva et al. 1995).

11.3.2
Chiral Ureas

11.3.2.1
Enantioselective Phytotoxicity

Optically active 1-α-methylbenzyl-3-(substituted phenyl)ureas diversely affected the root growth of rice and *Echinochloa curs-galli* var. *frumentacea* (barnyard millet). A preferred configuration for the adequate inhibition was not necessarily the same for each plant, and activity and enantioselectivity were found dependent on the substituent at the phenyl ring (Table 3). The *R*-enantiomers with a lower alkyl group exhibited a significant activity towards both rice and barnyard millet. Among them, the *R*-2-F-5-CF_3 derivative severely inhibited the growth of barnyard millet. Most compounds with an *S*-configuration were inactive towards rice, whereas the *S*-2-F, *S*-2,3-F_2, *S*-2,5-F_2 and *S*-4-COOEt derivatives inhibited the root growth of barnyard millet.

Enantioselective high responses were observed to rice and to barnyard millet (Table 3). A good selective response between rice and *Echinochloa* is essential to develop a rice herbicide, since *Echinochloa* is a steady companion weed of rice. A distinguished difference in the response between rice and barnyard millet was observed (Fig. 6). For the *R*-enantiomers, a considerably high linear regression equation was obtained, and for the *S*-enantiomers each dot is concentrated at the top of the figure from left to right. The former relationship (*R*-enantiomers) suggests an increase in the activity to barnyard millet with an increase in the toxicity to rice indicating a similarity of responses between both plants. The latter relationship (*S*-enantiomers), by contrast, strongly indicates the possibility to develop a safety herbicide.

With most chiral pairs, the same configurational isomer affected the root growth of both rice and barnyard millet, where many derivatives were active as an *R*-configuration, but only the 2-F derivative was active in an *S*-configuration (Table 3). Interestingly, only two chiral pairs of the 3-Me and 4-Me derivatives exhibited a cross intergenus response to rice and barnyard millet, where their *S*-enantiomers affected the growth of barnyard millet more. The differential properties of rice and *Echinochloa* plants to the chiral response may provide a hint to develop excellent herbicides (Omokawa and Ryoo 2001).

The specific Cyperaceae weed killer daimuron successfully controlled the annual weeds, *C. difformis* (CYPDI) and *S. juncoides* (SCPJU), but it was not strong enough to control completely the perennial weed *C. serotinus* (CYPSE) even at the highest dose. Daimuron damaged the growth of the directly seeded rice, and only moderately controlled the growth of barnyardgrass. By chiral and substitutional modifications, the phytotoxic spectrum was spread (Ryoo et al. 1998). Optically active MBTUs exhibited inhibitory activities to annual Cyperaceae paddy weeds such as daimuron, and to perennial Cyperaceae paddy weeds and barnyardgrass additionally (Table 4). While the *R*–2-isoPr and *R*-2-*tert*Bu derivatives significantly controlled barnyardgrass and both

Table 3. Phytotoxic potential of optically active 1-α-methylbenzyl-3-phenylureas on root growth of rice and barnyard millet in the dark

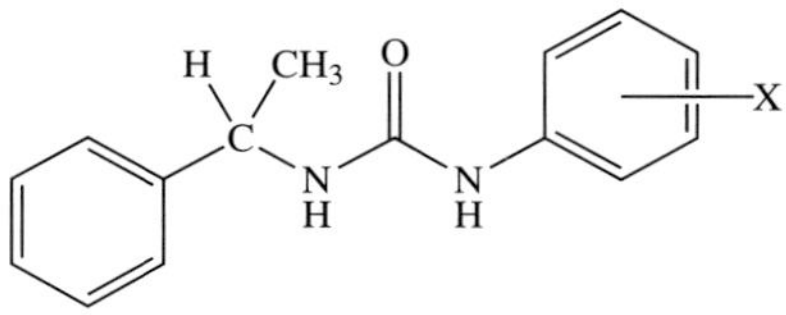

Compounds X	R/S	I_{50} (μM) Rice	Barnyard millet
H	R	(16.6)	30.4[a]
	S	(11.9)	11.8
2-Me	R	(15.1)	32.6
	S	(3.4)	(18.1)
3-Me	R	24.6	26.0
	S	(7.5)	20.0
4-Me	R	40.1[a]	22.7
	S	(5.1)	16.3
3-Et	R	2.9	3.6
	S	(13.9)	23.4
4-Et	R	2.6	4.6
	S	(13.6)	16.5
4-Pr	R	1.7	3.1
	S	(23)	(30.6)
2-isoPr	R	4.8	3.2
	S	(14.2)	47.2[a]
3-isoPr	R	5.9	4
	S	(20.3)	39.5[a]
4-isoPr	R	2.7	7.2
	S	(26)	21.9
4-Bu	R	1	3.3
	S	23.8	27.6
2-*tert*Bu	R	2.8	2.4
	S	(10.6)	(1.9)
2-CF_3	R	10.8	9.2
	S	(5.8)	(12.2)
4-COOEt	R	(7.8)	(21.7)
	S	(8.8)	6.4
2-F	R	(10.7)	(23.4)
	S	31.2	5.2
3,5-$(CF_3)_2$	R	4.7	12.1
	S	(10.5)	(25)
2,3-F_2	R	(6.7)	(13)
	S	(2.6)	6.9
2,5-F_2	R	(12.5)	(15.2)
	S	(5.9)	8
2-F-5-CF_3	R	44.7[a]	0.58
	S	(19.4)	20.8

[a] Extrapolated value; (%) rate of inhibition of root growth at 20 μM (Omokawa and Ryoo 2001).

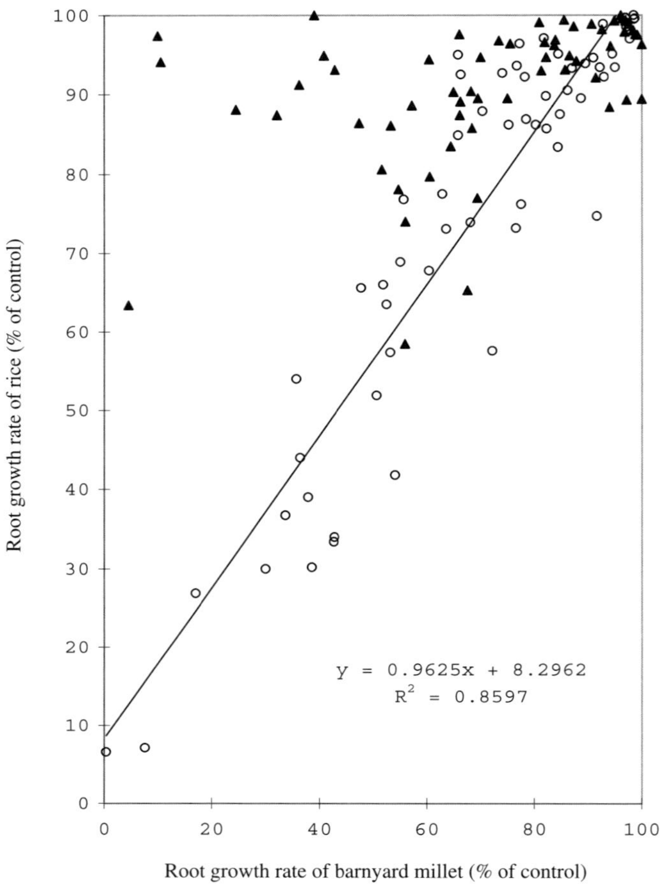

Fig. 6. ○: R-enantiomer; ▲: S-enantiomer Relationship between the root growth rates of rice and of barnyard millet treated with optically active α-methylbenzyl phenylureas at a 20 μM concentration. (Omokawa and Ryoo 2001)

annual and perennial Cyperaceae weeds with severe injury to the directly seeded rice plants, they scarcely affected growth of the transplanted rice plants. Enantioselectivity was significantly high. A configuration at the α-methylbenzyl chiral center was not necessarily the same for barnyardgrass, but alternatively for other test plants, suggesting different of mode(s) of action.

11.3.2.2
Stress-Relieving Activity

Crops are exposed to many kinds of stresses in their life cycles; temperature, water, salts, pests, agrochemicals and so on. One of them is herbicide stress. A

Table 4. Phytotoxic potential of optically active 1-(α-methylbenzyl)-3-phenylureas in a greenhouse pot test

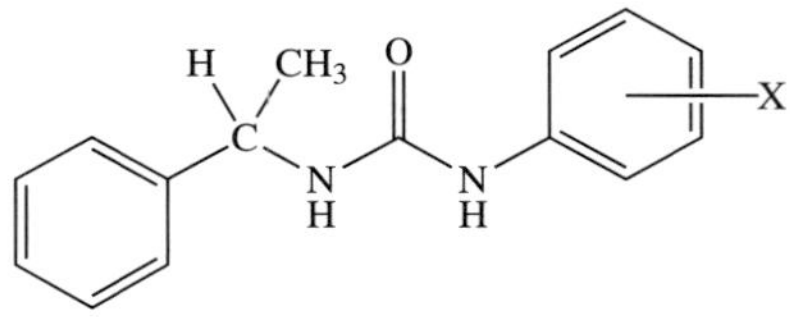

X	Chirality	ORYSA	ECHOR	CYPDI	SCPJU	CYPSE
H	R	0	2	2	1	3
	S	1	5	0	0	0
2-CF_3	R	1	2	6	6	8
	S	0	1	0	0	0
4-Me	R	0	1	5	4	2
	S	0	2	0	0	0
2-Et	R	0	0	5	1	8
	S	0	0	2	1	0
2-isoPr	R	6	7	7	7	8
	S	0	0	0	0	0
3-isoPr	R	0	0	5	2	6
	S	0	0	4	0	1
4-Pr	R	1	7	2	2	0
	S	0	1	2	0	0
2-*tert*Bu	R	7	7	7	7	8
	S	0	0	2	1	0
2-F	R	1	1	2	1	2
	S	2	7	1	0	0
2,3-F_2	R	0	0	1	0	–
	S	1	5	0	0	–
2,5-F_2	R	0	0	0	0	–
	S	1	5	0	0	–
Daimuron		1	4	7	8	2

The phytotoxic activity was visually evaluated by a 0 to 5 rating system (0 = no effect or growth as the untreated control, 5 = complete killing), and the phytotoxic potential of each test compound was ranked from 0 to 8. Each rank is defined by the phytotoxic activities and the dosage (g/a) in parentheses as follows; 0: 0 (100), 1: 0.5–1 (100), 2: 2–3 (100), 3: 4–5 (100), 4: 4–5 (50), 5: 4–5 (25), 6: 4–5 (12.2), 7: 4–5 (6.2) and 8: 4–5 (3.1). For example, rank 8 indicates a phytotoxic activity 4–5 at 3.1 g/ha. ORYSA *Oryza sativa* L.; ECHOR *Echinochloa oryzicola* Vasing; CYPDI: *Cyperus difformis* L.; SCPJU *Scirpus juncoides* Roxb. var. Hotarui Ohwi; CYPSE *Cyperus serotinus* Rottb. (Ryoo et al. 1998).

crop stress reliever, a safener, relieves the stress of herbicides which decrease growth of crops, and is becoming one of the useful agricultural materials nowadays (Hatzios and Hoagland 1989; Hatzios 1991; Wu et al. 1996). Daimuron exhibits a relieving action on rice injured by Londax, which is a typical sulfonylurea herbicide for the weed control of paddy fields (Sakanishi et al. 1991; Honma and Teramura 1986). *S*-MBTU, desmethyldaimuron, relieves the root

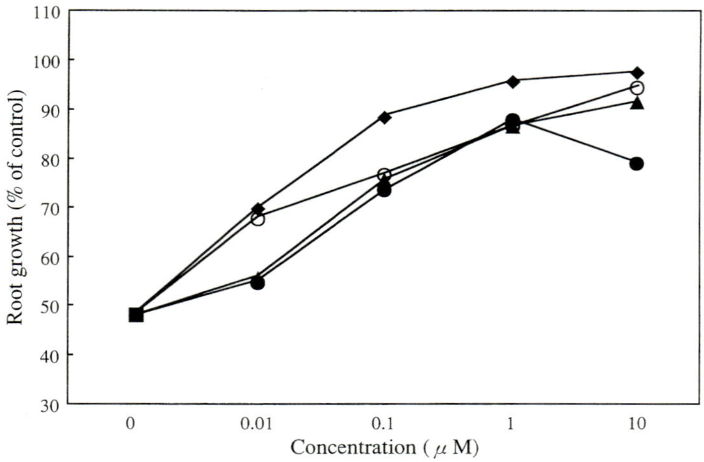

Fig. 7. Root growth rates of rice incubated in the dark in agar containing 24.4 nM of bensulfuron-methyl (■) and with *S*-4-COOEt (▲), *S*-2,3-Cl_2 (◆), *S*–2-F-4-CH_3 (○), or daimuron (●). (Omokawa et al. 1999)

growth reduction of rice seedlings damaged by Londax (Omokawa et al. 1993). The stress-relieving action of *S*-MBTU apparently results mainly from a drastic reduction in the uptake of Londax by stress-relieved rice seedlings. Daimuron and *R*-MBTU reduced the uptake of Londax by rice seedlings, but to a lesser extent than *S*-MBTU. This action is unique, since most of the commercially available stress relievers appear to protect monocot crops from herbicide injury by enhancing herbicide metabolism rather than reducing herbicide uptake and/or translocation (Omokawa et al. 1996). Among 60 chiral pairs assayed, 50 compounds relieved the root growth inhibition by bensulfuron-methyl more than daimuron did. Most of them were the *S*-isomers (Fig. 7; Omokawa et al. 1999). The *S*–2,3-Cl_2 compound was the most active being about ten times more active than daimuron.

Salinity-tolerant rice plants are expected in South Asia, where rice is cultivated near the seacoast. Some MBTUs relieve not only herbicide stress, but also salt (Omokawa and Aonuma 2002) and low-oxygen stresses.

11.3.2.3
Cross-Intergenus Selective Phytotoxicity Among Gramineae

There are many problems in the production of rice. One of them is selective control of barnyardgrass, especially at the early growth stage of rice both in planted and direct seeded cultivation. Cross-intergenus selective response has been found from the viewpoint of molecular chirality (Omokawa et al. 1993).

Gramineae plants may be classified into two groups based on chiral recognition of optically active MBTU exhibiting cross intergenus phytotoxic and

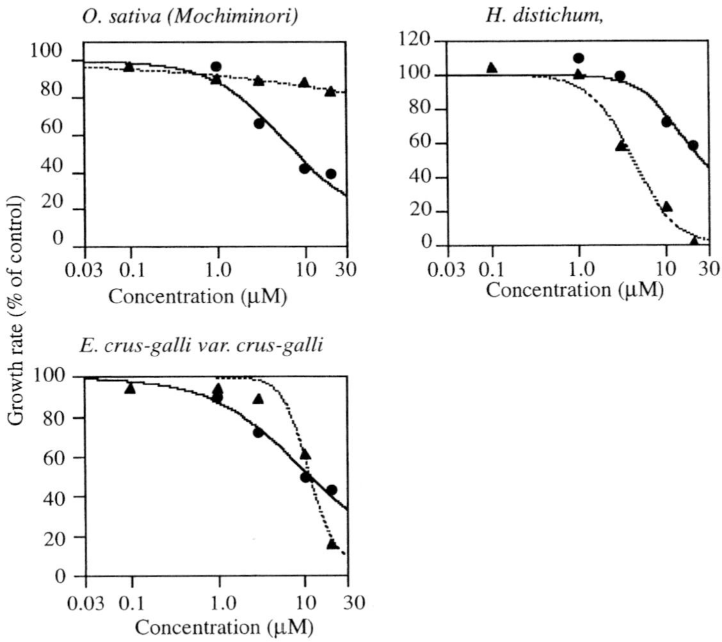

Fig. 8. Dose-response curve of *Gramineae* plants treated with the optically active urea compounds (● *R*-MBTU; ▲ *S*-MBTU). (Omokawa et al. 1993, 1997)

enantioselective actions. At the moment, the *R*-enantiomer inhibits root growth of the Oryzoideae (*Oryza, Zizania, Leersia* and *Chikusichloa*) more than the antipodal *S*-enantiomer. The *S*-enantiomer inhibits growth of the other Gramineae (*Triticum, Hordeum, Secale, Echinochloa, Zea, Lolium, Pennisetum* and *Phaenosperma*) more than the antipodal *R*-enantiomer (Fig. 8). In *Oryza* plants including the wild rice (*O. rufipogon, O. minuta, O. latifolia* etc.) and the cultivated rice plants (*O. sativa* and *O. glaberrima*), they respond similarly to the chiral recognition (Omokawa et al. 1997). As mentioned above, only chiral pairs of the 3-Me and 4-Me derivatives exhibited a cross-intergenus response between rice and barnyard millet. This interesting chiral response among Gramineae also strongly suggests enantiospecific and enantioselective response between *Oryza* and *Echinochloa* species. Further studies on the differential modes of action between rice and barnyardgrass may lead to the development of a high performance herbicide.

11.4 Chirality and Activity Relationship

Analysis of enantioselectivity is also a useful mathematical approach to focus the topological property of receptor site(s) and the influence of chirality on biological affinity. Eudismic analysis is one of them. Generally, a plot of activity difference between the chiral pairs (eudismic ratio or index) vs. potency of the activity of the more active chiral isomer (eutomer) gives straight lines with positive slopes, representing the increase in chiral discrimination per unit increase in affinity (Holmstedt et al. 1990). This relationship was first shown by Pfeiffer (1956), and known as Pfeiffer's rule. This analytical approach is not necessarily popular in structure-activity studies, and Testa (1990) already stated that Pfeiffer's rule is not always valid. Except for chiral *s*-triazine compounds analyzed by this author, no cases of chiral plant growth regulators were analyzed by the eudismic approach, although it was often done in pharmaceutical science.

11.4.1 Binding Direction of *s*-Triazines at the Photosystem II Reaction Center

The Quinone binding (QB) site is a well-established reaction site of triazine herbicides and their binding mechanism was investigated by many researchers (Trebst 1987; Lancaster and Michel 1999). Simple *s*-triazine herbicides have been constructed with a 2,4-diamino-*s*-triazine skeleton, and this common skeleton has a symmetrical plane across N3-C6 of the *s*-triazine molecule (Fig. 9). The binding direction of the *s*-triazine molecule at the photosystem II (PS) reaction center is critical to discussing activity. Topological analysis of the PS II binding site from the view of molecular chirality to optically active *s*-triazine compounds could provide an interesting aspect on a mode of binding direction of the *s*-triazines and the steric property of the binding site. Methyl substitution at the benzyl or alkyl group resulted in an increase or decrease in activity. There are two kinds of hydrogen atoms at the α-position of the benzyl or alkyl moiety of the *N*-benzyl- or *N*-alkyl-2,4-diamino-*s*-triazine structure (Bz-*s*-Tzs and alkyl-*s*-Tzs, respectively), pro*R*-hydrogen (H_R) and (pro*S*) hydrogen (H_S). In the case of the alkyl-*s*-Tzs, conversions of the Pr to a *sec*-Bu group (H_S–Me or H_R-Me) and the Et to Pr (α-methylation or β-methylation) resulted in an increase in activity (Table 5). In the case of the Bz-*s*-Tzs, both conversions resulted in a decrease and the decreased activity by H_R-CH_3 conversion (*R*-methylation) was greater than by H_S-CH_3 conversion (*S*-methylation). A similar activity change was shown by α-methyl introduction (Bz to MBz; Fig. 10). The H_R-CH_3 conversion resulted in a decrease of activity, and the H_S-CH_3 conversion resulted in an increase. Converting the MBz to dimethylbenzyl (DMBz), both conversions resulted in decreased activity, but the decrease by H_R-CH_3 conversion was greater than by H_S-CH_3 conversion. From these results on molecular chirality, a concept on structure-activity relationship and chiral discrimination of the binding site(s) on the adjacent space of the amino group of the 2,4-diamino-*s*-triazines was proposed for a series of triazine compounds by Shimizu et al. (1988); Omokawa and Konnai (1990); Omokawa and Takahashi (1994):

Plane of symmetry

Simazine

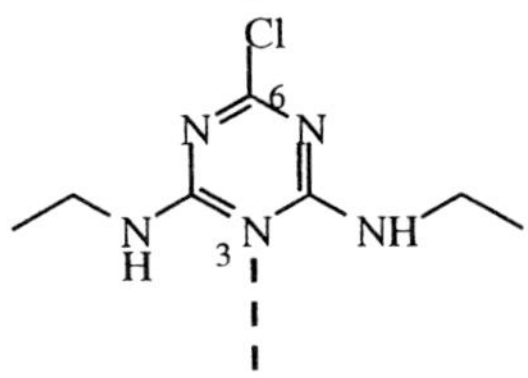

Symmetry against a plane passed N-3 and C-6 atoms

Unsymmetry

Atrazine

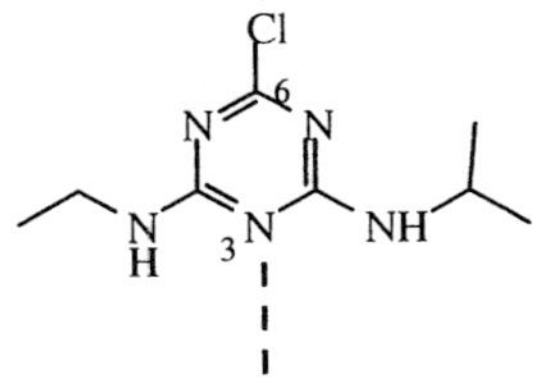

Slightly difference of the right part and the left part of the molecule separated by a plane passed N-3 and C-6 atoms

Unsymmetry

MBz-sTz

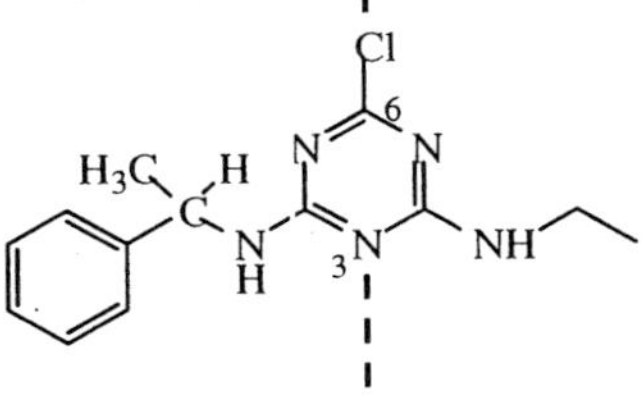

Big difference of the right part and the left part of the molecule separated by a plane passed N-3 and C-6 atoms

Fig. 9. Symmetrical or unsymmetrical molecules of *s*-triazines towards a plane crossing the N3-C6 axis

1. There are two different types of binding regions (the N and W regions) in the vicinity of Ser264 at the PS II receptor protein for an *s*-Tz inhibitor;
2. The steric requirement of the W region for the amino group is relatively large while that of the N region is small;
3. The chiral requirement of both regions is the *S*-configuration;
4. The binding direction of the inhibitor is determined by the size and/or steric hindrance of the substituent at each amino group, in particular at the carbon atoms adjacent to the N^2 or N^4-nitrogen atom; and
5. The level of optical discrimination in those regions is influenced by the steric suitability.

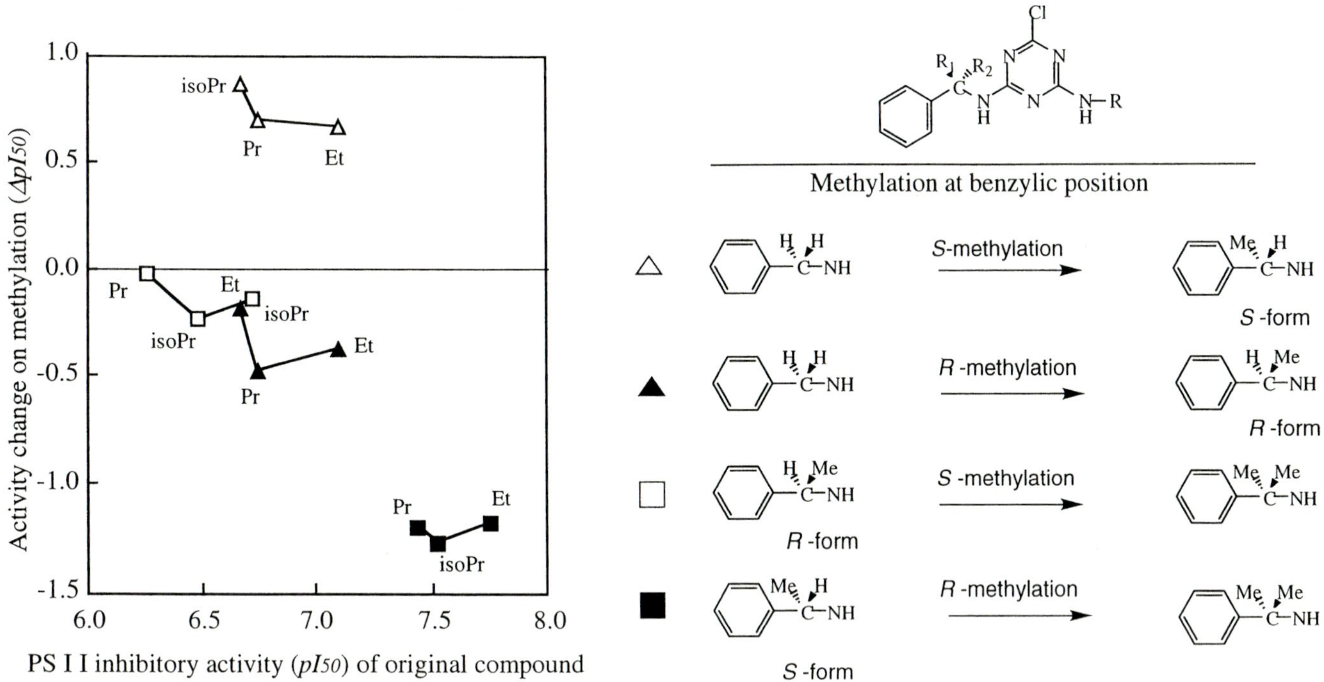

Fig. 10. Influence of methylation on the benzylic carbon of *s*-triazine compounds on PS II inhibitory activity change

Table 5. Influence of methyl introduction at the carbon of the *s*-triazines on the PS II inhibitory activity and rate of optical discrimination

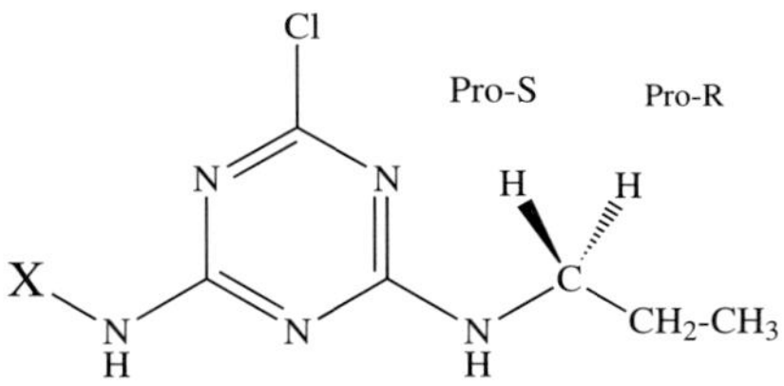

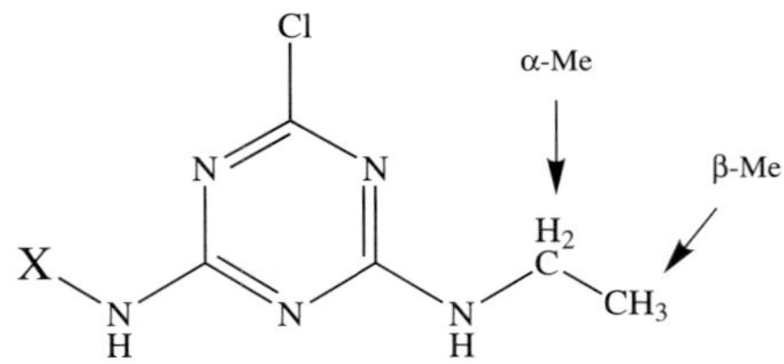

	Relative activity						
	Propyl	→	*sec*-Butyl		Ethyl	→	Propyls
X	$H_R:H_S$	H_R-Me	H_S-Me	S/R		isoPro (α-Me)	Pro (β-Me)
Et	1.0	2.2	2.3	1.0	1.0	1.6	1.7
Bz	1.0	0.1	0.44	4.3	1.0	0.37	0.44
R-MBz	1.0	0.16	0.51	3.3	1.0	0.48	0.36
S-MBx	1.0	0.41	0.52	1.3	1.0	0.53	0.57
DMBz	1.0	0.12	0.76	6.6	1.0	0.46	0.45

11.4.2 Eudismic Analysis

11.4.2.1 Photosystem II Inhibition

The relationship of PS II electron transfer inhibition of the chiral pairs of *s*-triazines afforded three types of linear regression equations on an α-methylbenzyl chiral syntom, and they are separated into two clusters (Fig. 11; Omokawa and Takahashi 1994). The first cluster (Eq-1) with a gentle slope contains derivatives including relatively small and less hindered alkyl substituents such as hydrogen, methyl, ethyl, propyl, isopropyl, isobutyl, and (*S*)-*sec*-butyl groups. The second cluster exhibiting a steeper slope included Eq-2 and Eq-3, where Eq-2 is composed of medium-sized and sterically hindered alkyl groups such as butyl, pentyl, 1-ethylpropyl, and (*R*)-*sec*-butyl, and Eq-3 of relatively long and lipophilic substituents such as tert-butyl, hexyl, heptyl, octyl, and decyl groups. The slopes of the regression line for Eq-2 and Eq-3 are almost the same and steeper than for Eq-1, and each eudismic index decreased gradually as the inhibitory activity of the eutomers (*S*-enantiomers) increased. These relationships, being the anti-Pfeiffer's rule, suggest a different mode of

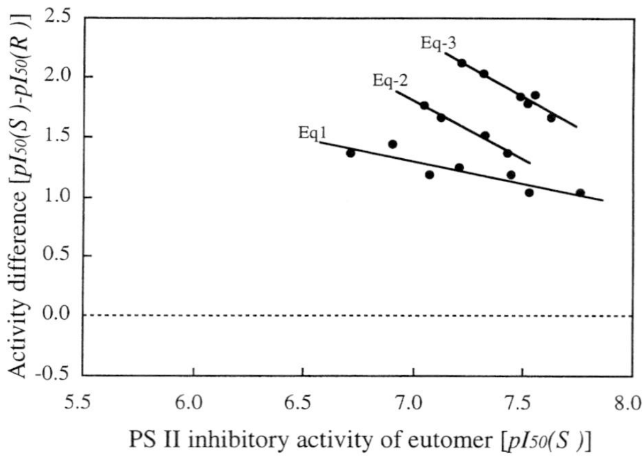

Fig. 11. Relationship between activity difference $[pI_{50}(S) - pI_{50}(R)]$ and inhibitory activity of the eutomer $[pI_{50}(S)]$ of chiral α-methylbenzyl-*s*-triazines on PS II electron-transfer system. (Omokawa and Takahashi 1994).

$$y = -0.378x + 3.94, \quad r^2 = 0.815, \quad n = 7$$

$$y = -0.933x + 8.76, \quad r^2 = 0.987, \quad n = 4$$

$$y = -1.02x + 9.51, \quad r^2 = 0.929, \quad n = 6$$

binding between both clusters and support a concept of binding direction at the PS II reaction site(s) described above. Analytical studies on three-dimensional QSAR of the optically active *s*-triazine compounds agreed with the above binding concept (Akamatsu et al. 1996).

Cyanoacrylate inhibitors with the α-methylbenzylamino moiety are potent PS II inhibitors and their *S*-isomers are active (Huppatz and Phillips 1987a,b). Eudismic analysis on atrazine susceptible and resistant *Brassica* chloroplasts based on the research by Phillips and Huppatz (1987) affords interesting relationships following Pfeiffer's rule (Fig. 12; Omokawa and Takahashi, 1994). The slopes for the wild types of both plants (*Brassica napus* and *Pisum sativum*) are almost the same, but different between the wild type and mutant of *Brassica*. This suggests that a different mode of binding resulted from different molecular structures of the PS II reaction center of wild type and mutant. There are many reports supporting a difference in structure of the reaction sites between susceptible and mutant (resistant) types (Trebst 1987, 1991). This relationship is contradictory to the chiral *s*-triazine compounds, despite the PS II inhibitors having the same chiral moiety and chiral requirement, suggesting a different binding mechanism at the PS II binding site(s) for the inhibitors, *s*-triazines and cyanoacrylates (Omokawa and Takahashi 1994).

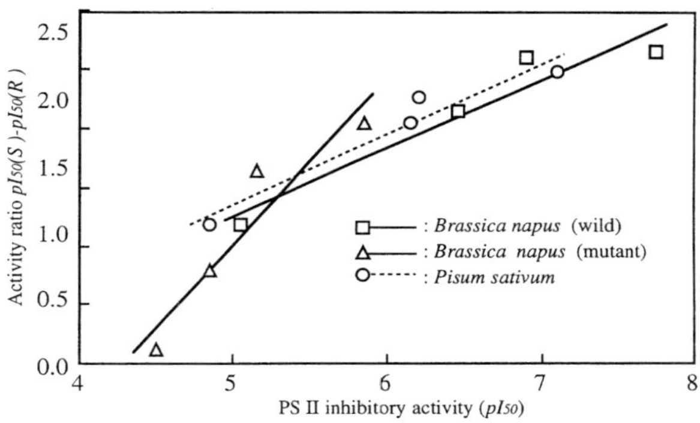

Fig. 12. Cyanoacrylate inhibitors of photosynthetic electron transport in atrazine-susceptible and atrazine-resistant *Brassica* chloroplasts. (Omokawa and Takahashi 1994). This figure is based on the data from Phillips and Huppatz (1987)

11.4.2.2
Light-Independent Inhibition

In the light-independent response of barnyard millet (*E. crus-galli* var. *frumentacea*) to the optically active α-methylbenzyl-*s*-triazines, a configuration of the active enantiomer (eutomer) is not an alternative, while for the dialkylamino derivatives their *S*-enantiomer was an eutomer, but in the monoalkylamino ones an eutomer changed. An analysis of the chirality-activity relationship for barnyard millet gave opposite results to the response in PS II;

$$(\mathrm{pI}_{50}(R)-\mathrm{pI}_{50}(S)) = 1.08\times \mathrm{pI}_{50}(R)-5.06, \quad r^2 = 0.814, \quad n = 8.$$

The activity difference $[\mathrm{pI}_{50}(R) - \mathrm{pI}_{50}(S)]$ increased with an increase in the inhibitory activity of the better-fitting *R*-enantiomers $[\mathrm{pI}_{50}(R)]$ (Omokawa and Takahashi 1994). Several chiral pairs had negative activity difference values. This would suggest that increasing the steric hindrance of the inhibitor at the binding site(s) causes a conformational change of the molecule to properly fit the surface of the binding site(s).

11.4.2.3
Stress Relief

On the stress-relieving activity of the chiral 1-α-methylbenzyl–3-phenylureas, the SIav (average safener index) values of the *S*-isomers were higher than those of the corresponding *R*-antipodes, except for a few chiral pairs, suggesting that the spatial property of the binding site(s) is suitable for the *S*-configuration. An eudismic analysis gave a good linear relationship following Pfeiffer's rule.

A regression analysis between the enantioselectivity [ES=SIav(*S*)-SIav(*R*)] and SIav(*S*) values yielded a linear equation (Omokawa et al. 1999).

$$ES = 0.582 \times SIav(S) - 8.97, \quad r^2 = 0.637, \quad n = 61$$

The enantioselectivity depends on the potency of the active enantiomer, increasing with increased inhibitory potency. The linear regression line also indicates the possibility of a change in the most suitable configuration as shown in the light-independent response of the *s*-triazines. In the case of 2-*tert*-Bu, 2-COOEt, 2,6-Et_2 and 2,5-*tert*-Bu_2 derivatives, the most suitable is the *R*-configuration.

References

Akamatsu M, Omokawa H, Shimizu R, Ueno T, Fujita T (1996) Three-dimensional QSAR of inhibitory activity on photosystem II of 1,3,5-triazines. 11th European Symposium on QSAR, Computer assisted lead finding and optimization, Lausanne, Switzerland

Ariens EJ, van Rensen JJS, Welling W (eds) (1988) Stereoselectivity of pesticides, biological and chemical problems. Elsevier, New York

Ashton FM, Crafts AS (1973) Mode of action of herbicides. Wiley, New York, pp 310–346

Barnwell P, Cobb AH (1993) An investigation of aryloxyphenoxypropionate antagonism of auxin-type herbicide action on proton efflux. Pestic Biochem Physiol 47:87–97

Bendixen LE (1975) Cytokinin effects induced in purple nutsedge by perfluidone. Weed Sci 23: 445–447

Blaser UH, Spindler F (1997) Enantioselective catalysis for agrochemicals: the case history of the DUAL MAGNUMR herbicide. Chimia 51:297–299

Buser HR, Müller MD (1997) Conversion reactions of various phenoxyalakanoic acid herbicides in soil. 2. Elucidation of the enantiomerization process of chiral phenoxy acids from incubation in a D2O/soil system. Environ Sci Technol 31:1960–1967

Caldwell J, Hutt AJ, Fournel-Gigleux S (1988) The metabolic chiral inversion and dispositional enantioselectivity of the 2-arylpropionic acids and their biological consequences. Biochem Pharmacol 37:105–114

Chan JHH, Walker F, Tseng CK, Baker DR, Arneklev DR (1975) Synthesis and herbicidal activity of *N,N'*-diethyl-2-(1-naphthyloxy)propionamide and its optical isomers. J Agric Food Chem 28:1008–1010

Copping LG, Davis DE, Pillai CGP (1972) Growth regulator-like activity of atrazine and ametryne. Weed Sci 20:274–277

Couderchet M, Bocion PF, Chollet R, Seckinger K, Böger P (1997) Biological activity of two stereoisomers of the *N*-thienyl chloroacetamide herbicide dimethenamid. Pestic Sci 50: 221–227

Dicks JW, Slater JW, Bewick DW (1985) PP005-the R-enantiomer of fluazifop butyl. Proceedings British Crop Protection Conference, Weeds, pp 171–79

Duke O (1996) Herbicide-resistant crops, agricultural, environmental, economical, regulatory, and technical aspects. Lewis, Boca Raton

Fabro S, Smith RL, Williams RT (1967) Toxicity and teratogenicity of optical isomers of thalidomide. Nature 215:296

Fredga A, Aberg B (1965) Stereoisomerism in plant growth regulators of the auxin type. Annu Rev Plant Physiol 16:53–72

Garrison AW, Schmitt P, Martens D, Kettrup A (1996) Enantiomeric selectivity in the environmental degradation of dichlorprop, as determined by high-performance capillary electrophoresis. Environ Sci Technol 30:2449–2455

Gerwick BC, Jackson LD, Handly J, Gray NR, Russell W (1988) Preemergence and postemergence activities of the (*R*) and (*S*) enantiomers of haloxyfop. Weed Sci 36:453–456

Grossmann K, Tresch S, Plath P (2001) Triaziflam and diaminotriazine derivatives affect enantioselectively multiple herbicide target sites. Z Naturforsch 56c:559–569

Haga T, Crosby KE, Schussler JR, Palmer CJ, Yoshii H, Kimura F (1998) Aryloxyphenoxypropanoate herbicides. In: Kurihara N, Miyamoto J (eds) Chirality in agrochemicals, Wiley, New York, pp 175–197

Hamper BC, Leschinsky KL, Mischke DA, Prosch SD (2000) Chiral 3-aryl-4-halo-5-(trifluoromethyl)pyrazoles. Synthesis and herbicidal activity of enantiomeric lactate derivatives of aryl-pyrazole herbicides. ACS Symp Ser 746, Washington, DC, pp 272–281

Hallahan BJ, Camilleri P, Smith A, Bowyer JR (1992) Mode of action studies on a chiral diphenyl ether peroxidizing herbicide. Correlation between differential inhibition of protoporphyrinogen IX oxidase activity and induction of tetrapyrrole accumulation by the enantiomers. Plant Physiol 100:1211–1216

Hatzios KK (1991) An overview of the mechanisms of action of herbicide safeners. Z Naturforsch 46c:819–827

Hatzios KK, Hoagland RE (1989) Crop safeners for herbicides: development, uses, and mechanisms of action. Academic Press, San Diego

Holmsen TW (1968) Plant growth regulator activity of optical isomers of DMPA. Weed Sci 16:187–188

Holmstedt B, Frank H, Testa B (1990) Chirality and biological activity, Alan R Liss, New York

Honma Y, Teramura M (1986) Herbicide composition. Japan Kokkai Tokkyo Koho, 61–112003

Hubele A, Kunz W, Eckhart W, Sturn E (1982) The fungicidal activity of acyl anilines. In: Miyamoto J, Kearney PC (eds) Pesticide chemistry: human welfare and the environment. Proc. 5th Int Congr Pestic Chem, Kyoto, Japan, 1982, vol 1. Pergamon Press, Oxford, pp 233–242

Huppatz JL, Phillips JN (1987a) Stereospecific inhibitor probes of the PS II herbicide binding site. Z Naturforsch 42c:674–678

Huppatz JL, Phillips JN (1987b) Cyanoacrylate inhibitors of the Hill reaction V. The effect of chirality on inhibitor binding. Z Naturforsch 42c:684–689

Hutson DH, Logan CJ, Taylor B (1991) The metabolism in the rat of the herbicidally active isomer (*R*)-flamprop-methyl in comparison with the racemic form. Pestic Sci 31:151–162

Hutt AJ, Caldwell J (1983) The metabolic chiral inversion of 2-aryloxypropionic acids. A novel route with pharmacological consequences. J Pharm Pharmacol 35:693–704

Köller W (1987) Isomers of sterol synthesis inhibitors: fungicidal effects and plant growth regulator activities. Pestic Sci 18:129–147

Kohler HPE (1999) *Spongomonas herbicidovorans* MH: a versatile phenoxyalkanoic acid herbicide degrader. I Ind Microbiol Biotechnol 23:336–340

Koshimizu K, Kobayashi A, Fujita T, Mitsui T (1968) Structure-activity relationships in optically active cytokinins. Phytochemistry 7:1989–1994

Krämer W, Büchel KH, Draber W (1983) Structure-activity correlation in the azoles. In: Miyamoto J, Kearney PC (eds) Proceedings of 5th Intern Congr Pestic Chem, Kyoto 1982, vol 1. Pergamon, Oxford, pp 223–232

Kulaeva ON, Corse J, Selivankina SY (1995) Effects of trans- and cis-zeatin and optical iosmers of synthetic cytokinins on protein kinase activity in vitro. J Plant Growth Regul 14:41–47

Kurihara N, Miyamoto J (1998) Chirality in agrochemicals. Wiley, New York

Lancaster CRD, Michel H (1999) Refined crystal structures of reaction centres from *Rhodopseudomonas viridis* in complexes with the herbicide atrazine and two chiral atrazine derivatives also lead to a new model of the bound carotenoid. J Mol Biol 286:883–898

Lang D, Criegee D, Grothusen A, Saalfrank RW, Boecker R (1996) In vivo metabolism of atrazine, terbuthylazine, ametryne, and terbutryne in rats, pigs, and humans. Drug Metab Dispos 24:859–865

Lewis DL, Garrison AW, Wommack KE, Whittemore A, Steudler P, Melillo J (1999) Influence of environmental changes on degradation of chiral pollutants in soil. Nature 401:898–901

Matsubara S, Koshimizu K, Fujita T (1973) Growth-promoting activity of optically active cytokinins in tobacco tissue culture. In: Tamura S (ed) Plant growth substances. Hirosawa Publishing Co, Tokyo, pp 450–461

Moreland DE, Boots MR (1971) Effects of optically active 1-(α-methylbenzyl)-3-(3,4-dichlorophenyl)urea on reactions of mitochondria and chloroplasts. Plant Physiol 47:53–58

Mori K (1998) Chirality in insect juvenile hormones and pheromones. In: Kurihara N, Miyamoto J (eds) Chirality in agrochemicals. Wiley, New York, pp 199–257

Moser H, Rihs G, Sauter H (1982) Der Einfluss von Atropisomerie und chiralem Zentrum auf die biologische Aktivität des Metolachlor. Z Naturforsch 37b:451–462

Müller MD, Buser HR (1997) Conversion reactions of various phenoxyalkanoic acid herbicides in soil. 1. Enantiomerization and enantioselective degradation of the chiral 2-phenoxypropionic acid herbicides. Environ Sci Technol 31:1953–1959

Muller RH, Jorks S, Kleinsteuber S, Babel W (1999) *Comamonas acidovorans* strain MC1. A new isolate capable of degrading the chiral herbicides dichlorprop and mecoprop and the herbicides 2,4-D and MCPA. Microbiol Res 154:241–246

Nader HM, Clegg MD, JW Maranville (1975) Promotion of sorghum callus growth by the *s*-triazine herbicides. Plant Physiol 56:747–751

Nakahira K, Uchiyama M, Ikai T, Igarashi H, Suzuki K (1988) Effect of (*R*)-(+)- and (*S*)-(−)-quizalofop-ethyl on lipid metabolism in excised corn stem-base meristems. J Pestic Sci 13:269–276

Nakahira K, Haga M, Uchiyama M, Suzuki K (1990) Comparative effects of quizalofop and its esters on acetyl-CoA carboxylase and fatty acid biosynthesis in corn leaf chloroplasts. J Pestic Sci 15:189–197

Nakamura Y, Yamaguchi T, Yakahashi S, Hashimoto S, Iwatani K, Nakagawa Y (1981) Optical isomerization mechanism of R(−)-hydratropic acid derivatives. J Pharmacobiol Dyn 4, S–1

Nandihalli UB, Duke MV, Ashmore JW, Musco VA, Clark RD, Duke SO (1994) Enantioselectivity of protoporphyrinogen oxidase-inhibiting herbicides. Pestic Sci 40:265–277

Nickel K, Suter MJF, Kohler HSE (1997) Involvement of two α-ketoglutarate-dependent dioxygenases in enantioselective degradation of (*R*)- and (*S*)-mecoprop. J Bacteriol 179:6647–6679

Omokawa H, Aonuma S (2002) Amelioration of the salt-stressed root growth of rice and normalization of the Na^+ distribution between the shoot and root by (*S*)-α-Methylbenzyl-2-fluoro-4-methylphenylurea. Biosci Biotechnol Biochem 66:336–343

Omokawa H, Konnai M (1990) PSII inhibitory activity of 2,4-diamino-6-chloro-*s*-triazines with a chiral *sec*-butyl and/or α-methylbenzyl group. Agric Biol Chem 54:2373–2378

Omokawa H, Konnai M (1992) Inhibition of *Echinochloa crus-galli* var. *frumentacea* seedling root elongation by chiral 1,3,5-triazines in the dark. Pestic Sci 35:83–86

Omokawa H, Takahashi M (1994) Reverse chiral discrimination relationships between the inhibitory activity of 1,3,5-triazines on photosystem II and light-independent root growth. Pestic Biochem Physiol 50:129–137

Omokawa H, Ryoo JH (2001) Enantioselective response of rice and barnyard millet on root growth inhibition by optically active α-methylbenzyl phenylureas. Pestic Biochem Physiol 70:1–6

Omokawa H, Ichizen N, Takematsu Y (1987) Phytotoxic properties of α-substituted benzylamino-*s*-triazines. Agric Biol Chem 51:2563–2568

Omokawa H, Ichizen N, Takematsu T (1988a) Phytotoxic activity of substituted α-methylbenzylamino derivatives of 2-chloro (or methylthio)-*s*-triazines. Agric Biol Chem 52:1047–1048

Omokawa H, Ichizen N, Konnai M, Takematsu T (1988b) Herbicidal activity and phytotoxic properties of *N*-alkyl-*N'*-(α,α-dimethylbenzyl)-2,4-diamino-6-chloro-*s*-triazines. Agric Biol Chem 52:1515–1519

Omokawa H, Kobayashi I, Konnai M (1989) Substituent effect on photosynthesis of N^2-α-substituted benzyl-N^4-alkyl-2,4-diamino-6-chloro-*s*-triazines and their phytotoxic property. Agric Biol Chem 53:2723–2729

Omokawa H, Takeuchi M, Konnai M (1992) Rhizome-induction activity of chiral 2-α-methylbenzylamino-4-alkylamino-6-chloro-1,3,5-triazines in *Cyperus serotinus* Rottb. Pestic Sci 35:87–90

Omokawa H, Kawata Y, Konnai M (1993) Interaction between optical isomerism and plant pharmacological action, change of modes of action and chirality of 1-(α-methylbenzyl)-3-*p*-tolylurea. Pestic Sci 37:107–112

Omokawa H, Kuramochi H, Kawata Y, Konnai M (1994) Structural characteristics of *s*-triazine compounds inducing rhizome in *Cyperus serotinus* Rottb. J Pestic Sci 19:25–32

Omokawa H, Wu J, Hatzios KK (1996) Mechanism of safening action of dymuron and its two monomethyl analogues against bensulfuron-methyl injury of rice (*Oryza sativa*). Pestic Biochem Physiol 55:54–63

Omokawa H, Murata H, Kobayashi S, Kuramochi H (1997) Gramineae plants may be classified into two groups on a chiral recognition of optical active urea compounds. In: Rajan A (ed) Proceedings 16th Asian-Pacific Weed Science Society Conf. Malaysian Plant Protection Society, ISBN 967-9942-19-8, pp 173–176

Omokawa H, Ryoo JH, Kashiwabara S (1999) Enantioselective relieving activity of α-methylbenzylphenylureas toward bensulfuron-methyl injury to rice (*Oryza sativa*). Biosci Biotechnol Biochem 63:349–355

Palmer CE, Smith OE (1969) Cytokinins and tuber initiation in the potato *Solanum tuberosum* L. Nature 221:279–280

Pfeiffer CC (1956) Optical isomerism and pharmacological action, a generalization. Science 124: 29–30

Phillips JN, Huppatz JL (1987) Cyanoacrylate inhibitors of photosynthetic electron transport in atrazine susceptible and atrazine resistant *Brassica* chloroplasts. Z Naturforsch 42c:670–673

Pillai CGP, Davis DE (1973) *s*-Triazine effects on seed germination and hypocotyl hook opening. Weed Sci 21:461–464

Powell JR, Ambre JJ, Ruo TI (1988) The efficacy and toxicity of drug stereoisomers. In: Wainer IW, Drayer DE (eds) Drug stereochemistry, analytical methods and pharmacology. Marcel Dekker, New York, pp 2456–470

Powles SB, Holtum JAM (1994) Herbicide resistance in plants, biology and biochemistry. Lewis, Boca Raton

Ryoo JH, Kuramochi H, Omokawa H (1998) Enantioselective herbicidal activity of chiral α-methylbenzylphenylureas against *Cyperaceae* and *Echinochloa* paddy weeds. Biosci Biotechnol Biochem 62:2189–2193

Sakanishi T, Nakahara S, Koyanagi H (1991) Effect of daimuron on safening action to rice injured with bensulfuron-methyl. Abstr Annu Meet Weed Sci Soc Jpn, pp 84–85 (in Japanese)

Sakata G, Makino K, Morimoto K, Ikai T, Hasebe S (1985a) Synthesis and herbicidal activity of optically active ethyl 2-[4-(6-chloro-2-quinoxalinyloxy)phenoxy-propanoate. J Pestic Sci 10:69–73

Sakata G, Makino K, Kusano K, Satow J, Ikai T, Suzuki K (1985b) Preparation of optically pure ethyl (*R*)-(+) and (*S*)-(−)-2-[4-(6-chloro-2-quinoxalinyloxy)phenoxypropanoate by resolution method and their herbicidal activities. J Pestic Sci 10:75–79

Sattelmacher B, Marscher H (1978) Cytokinin activity in stolons and tubers of *Solanum tuberosum* during the period of tuberization. Physiol Plant 44:69–72

Shimabukuro R, Hoffer BL (1995) Enantiomers of diclofop-methyl and their role in herbicide mechanism of action. Pestic Biochem Physiol 51:68–82

Shimizu R, Iwamura H, Fujita T (1988) Quantitative structure-activity relationships of photosystem II inhibitory anilides and triazines: topological aspects of their binding to the active site. J Agric Food Chem 36:1276–1283

Spindler F, Fruh T (1998) Chiral acylanilide and chiral triazole-related fungicides. In: Kurihara N, Miyamoto J (eds) Chirality in agrochemicals. Wiley, New York, pp 141–173

Takano H, Oguri Y, Kato T (1986) Antifungal and plant growth regulating activities of enantiomers of (*E*)-1-(2,4-dichlorophenyl)-4,4-dimethyl-2-(1,2,4-triazol-1-yl)-1-penten-3-ol (S-3308L). J Pestic Sci 11:373–378

Takematsu T, Konnai M, Akashiba T, Seki N (1975) Selective control of cyperaceous weeds with K-223. Weed Sci 23:15–19

Teo CKH, Bendixen LE, Nishimoto RK (1969) Bud sprouting and growth of purple nutsedge altered by benzyladenine. Weed Sci 21:19–23

Testa B (1990) Definitions and concepts in biochirality. In: Holmstedt B, Frank H, Testa B (eds) Chirality and biological activity. Alan R. Liss, New York, pp 15–32

Trebst A (1987) The three-dimensional structure of the herbicide binding niche on the reaction center polypeptides of photosystem II. Z Naturforsch 42c:742–750

Trebst A (1991) The molecular basis of resistance of photosystem II herbicides. In: Caseley JC, Cussans GW, Atkins RK (eds) Herbicide resistance in weeds and crops. Butterworth-Heinemann, Oxford, pp 145–164

Uchiyama M, Washio N, Ikai T, Igarashi H, Suzuki K (1986) Stereospecific responses to (*R*)-(+)- and (*S*)-(–)-quizalofop-ethyl in tissues of several plants. J Pestic Sci 11:459–467

Wainer IW, Drayer DE (1988) Drug stereochemistry, analytical methods and pharmacology. Marcel Dekker, New York

Williams A (1996) Opportunities for chiral agrochemicals. Pestic Sci 46:3–9

Wink O, Luley U (1988) Enantioselective transformation of the herbicides diclofop-methyl and fenoxaprop-ethyl in soil. Pestic Sci 22:31–40

Wu J, Omokawa H, Hatzios KK (1996) Glutathione *S*-transferase activity in unsafened and fenclorim-safened rice (*Oryza sativa*). Pestic Biochem Physiol 54:220–229

Yamada Y, Sekita J, Koshimizu K (1972) Cytokinin-induced shoot formation. Phytochemistry 11:1019–1021

Zipper C, Nickel K, Angst W, Kohler HSEJ (1996) Complete microbial degradation of both enantiomers of the chiral herbicide mecoprop [(*RS*)-2-(4-chloro-2-methylphenoxy)propanoic acid] in an enantioselective manner by *Sphingomonas herbicidovorans* sp. nov. Appl Environ Microbiol 62:4318–4322

Zipper C, Bunk M, Zehnder AJB, Kohler HSEJ (1998a) Enantioselective uptake and degradation of the chiral herbicide dichlorprop [(*RS*)-2-(2,4-dichlorophenoxy)propanoic acid] by *Spingomonas herbicidovorans* MH. J Bacteriol 180:3368–3374

Zipper C, Sulter MJF, Haderlein SB, Gruhl M, Kohler HPE (1998b) Changes in the enantiomeric ratio of (*R*)- to (*S*)-mecoprop indicate in situ biodegradation of this chiral herbicide in a polluted aquifer. Environ Sci Technol 32:2070–2076

12 Transcuticular Penetration of Foliar-Applied Pesticides – Its Analysis by a Logistic-Kinetic Penetration Model

Tadakazu Watanabe

12.1 Introduction

The surface of terrestrial parts of higher plants is generally covered by a thin extracellular structure called the plant cuticle. The plant cuticle (from the atmospheric side) consists of three layers, which are epicuticular waxes, a cuticularized layer and a cutinized layer. Epicuticular waxes are a thin deposition on the outer layer made up of mainly hydrohobic, very long-chain fatty acid derivatives, the cuticularized layer is a matrix of cutin polymer structure in which long-chain-fatty acids are embedded, and the cutinized layer is the innermost part of a cutin structure which contacts the subcuticular space composed of hemicellulosic and pectinous substances as an apoplastic space to the epidermal cells (Cutler et al. 1982; Kerstiens 1996).

The cuticle is a major barrier against the loss of inorganic and organic solutes from inner tissues and against invasion of foreign substances and microbes from the atmosphere. Therefore, when solutions containing pesticides and fertilizers are applied onto plants, foliar penetration of the active ingredients is essentially determined by the cuticles, and its penetration kinetics greatly influence the final biological efficacy.

The pathway for penetration of a foliar-applied pesticide is considered to be through the cuticle and stomata (Holloway and Stock 1990). Since stomatal infiltration is limited to the case of reduction in surface tension of the sprayed solution down to about 30 mN/m or less (Schönherr and Bukovac 1972; Stevens et al. 1991; Stevens 1993), cuticular penetration is thought to be the major route for most plants (Holloway and Stock 1990; Schönherr et al. 1990; Riederer and Schreiber 1995). A systemic pesticide applied first invades and then passes through the cuticular structure (hereinafter, CM; cuticular membrane) and migrates into the subcuticular (apoplastic) space, then moves into tissue cells and finally reaches its action sites. Among these processes, the transcuticular penetration process is considered to be the rate-limiting one (Schönherr and Riederer 1988; Schönherr et al. 1990; Schönherr and Baur 1994, 1996).

Since a transcuticular penetration can be markedly influenced by a combination of CM and pesticide and also by additives (adjuvants), formulations and ambient conditions (Chow et al. 1987; Foy 1992; Gaskin 1995; MacMullan 1998), its mechanistic quantification is mainly governed by a complex interaction among these factors.

P. Böger, K. Wakabayashi, K. Hirai (Eds.)
Herbicide Classes in Development

Thus, this review will (1) quantitatively analyze transcuticular penetration kinetics of foliar-applied pesticides, (2) investigate major parameter(s) governing the penetration as influenced by adjuvants, and (3) elucidate the mode of action of adjuvancy, based on a newly developed, non-steady state model, namely, "the logistic-kinetic penetration model".

12.2 Overview

The transcuticular penetration of a foliar-applied pesticide can be considered as a mass transfer based on a steady-state theory, in which the transfer system is composed of CM between the donor and receiver solutions (Hartley and Graham-Bryce 1980a; Reed and Tukey 1982; Kerler et al. 1984; Price and Anderson 1985; Schönherr and Riederer 1989; Schönherr et al. 1990; Salgado 1995). The mass transfer through CM has been generally quantified with the partition and diffusion coefficients as follows:

$$J = M/t = S \cdot P \cdot (C_{\mathrm{don}} - C_{\mathrm{rec}}) \tag{1}$$

where J is the rate of mass transfer per unit time and unit area, M the amount which diffuses through the contact area S of CM for time t, C_{don} the concentration of the pesticide in the donor solution and C_{rec} that in the receiver solution and P is the permeance. The term ($C_{\mathrm{don}} - C_{\mathrm{rec}}$) is a driving force for diffusion through CM. Since P is defined as $[(D \cdot K)/\Delta x)]$, where Δx is the thickness of CM and K the partition coefficient between the solution and CM, Eq. (1) is described as (2).

$$J = (S \cdot D \cdot K) \cdot (C_{\mathrm{don}} - C_{\mathrm{rec}})/\Delta x \tag{2}$$

where D is the diffusion coefficient in CM. As C_{rec} is similar to the concentration in a subcuticular space and can be considered to be actually almost zero, the Eq. (2) can be transformed to (3).

$$J = [(S \cdot D)/\Delta l] \cdot (K_{\mathrm{cw}} \cdot C_{\mathrm{don}}) \tag{3}$$

where K_{cw} is the partition coefficient at an equilibrium state between water and CM (K_{cw} at the outer surface of CM can differ, in a strict sense, from that at the inner surface because of a difference in its chemical constitution of CM, but both K_{cw} are assumed here to be the same) and Δl is the real diffusion path length in the CM.

Equation (4) is obtained using the Stokes-Einstein equation:

$$J = [(k \cdot S \cdot T)/(6\pi \cdot \mathrm{r} \cdot \eta \cdot \Delta l \cdot t)] \cdot (K_{\mathrm{cw}} \cdot C_{\mathrm{don}}) \tag{4}$$

where k is the Bolzmann constant, r the radius of the pesticide molecule, η the viscosity of CM and T the absolute temperature. This equation shows that the mass flow is proportional to S, T, K_{cw} and C_{don} and is inversely proportional to r, η and Δl.

For the penetration kinetics of a pesticide from the donor droplet to a large reservoir through a membrane, Hartley and Graham-Bryce (1980a,b) proposed Eq. (5):

$$(-P \cdot A \cdot t)/V_{\text{don}} = \ln(C_{\text{don}}/C_{\text{o}}) \tag{5}$$

where P is the permeance, A the contact area of a droplet with CM, t penetration time, C_{don} the remaining donor concentration at time t, V_{don} the volume of the donor, and C_o the initial concentration in the donor. This formula indicates that the penetration through a membrane is proportional to P, A and t and inversely proportional to V_{don}. However, P cannot be defined by a non-steady state mass flow.

Those theoretical considerations cannot realistically occur because of a non-equilibrium, dynamic property of the actual transcuticular penetration from droplets on the plant surface, although some fundamental implications shown above can be still available. In an actual application of pesticidal solutions, the droplets retained on plant surfaces have a certain volume and size and spread out to be in a state of equilibrium. These parameters of the droplet are greatly influenced by formulations and adjuvants. Just after retention, the droplet will begin to evaporate over several minutes to a couple of hours to form a dense thin liquid film on the contact area, and finally, dry up after a considerably long period. These processes occur sequentially and the pesticide in the droplet continues partitioning to and diffusing through the CM along a concentration gradient, as schematized in Fig. 1 (Schönherr and Riederer 1988; Schönherr and Baur 1994, 1996; Riederer and Schreiber 1995).

Thus, an actual transcuticular penetration should be treated as a non-equilibrium, dynamic state. To quantitatively describe it, some models have been proposed so far.

Price and Anderson (1985), Bucholz and Hess (1987), McCall (1988) and Breeze et al. (1992) adopted a rate-constant model, in which the whole process was divided into 3 to 7 compartments and the rate constants among the compartments were calculated. They recognized the importance of partition between the droplet and the CM and diffusion through the CM.

Stevens et al. (1991) proposed a regression equation as follows:

$$M_{\text{up}} = M - (P \cdot S^{t}) \tag{6}$$

where M_{up} is the amount penetrated through the CM for time t, M is the total amount penetrated, P the penetration-curve factor, S the nonlinear penetration factor, and t the time of penetration.

Buick et al. (1992) adopted the following regression equation derived from the Michaelis-Menten equation of an enzymatic reaction:

$$M_{\text{up}} = M/[1 + (D/t)] \tag{7}$$

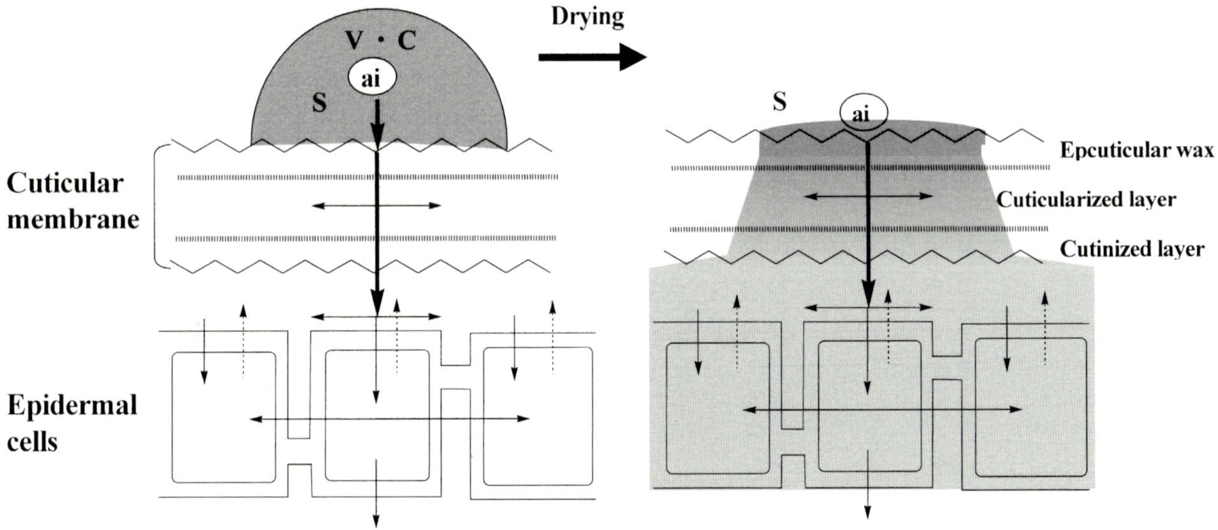

Fig. 1. Schematic transcuticular penetration of foliar-applied pesticides. *V* Volume of droplet, *C* concentration of ai, and *S* contact area of droplet

where D is the penetration-rate constant.

Lewis (1980), Price and Anderson (1985) and Breeze et al. (1992) proposed the first-order process models as expressed by Eqs. (6), (7) and (8), respectively.

$$C = C_O \cdot (1 - e^{-kt}) \quad (8)$$

$$C(\%) = 100 \cdot (1 - e^{-kt}) \quad (9)$$

$$M_{up} = M_i \cdot (K_{up}/K) \cdot (1 - e^{-kt}) \quad (10)$$

where C and M_{up} are the total amounts penetrated, C_o and M_i the amounts applied, k and K_{up} the penetration-rate constants, $K = M_{up} + K_{evap}$, and K_{evap} the evaporation-rate constants, respectively.

These models describe the total amount penetrated and its penetration rate together and show that they are independent of each other.

Bauer and Schönherr (1992) and Schönherr and Bauer (1992) proposed the UDOS (unilateral desorption from the outer surface of CM) method using an isolated CM to quantify the initial stage in transcuticular penetration kinetics as follows:

$$-\ln[1 - (M_t/M_o)]/t = k \quad (11)$$

where M_t is the amount penetrated, M_o the total amount applied and k the slope of the initial penetration (penetration-rate constant). They pointed out that the initial rate is linear and the kinetics is first-ordered. The molecular size of pesticide relates to its mobility and adjuvants can work as an accelerator by reducing an inner diffusion resistance of the CM.

Those models and treatments dealing with nonequilibrium transcuticular penetration kinetics, however, do not fully quantify all the kinetic parameters involved in penetration from a droplet. Therefore, a logistic-kinetic model is presented here to describe nonequilibrium, transcuticular penetration kinetics of foliar-applied pesticides with all the parameters involved to analyze their penetration kinetics and to characterize the mode of action of adjuvants.

12.3 Logistic-Kinetic Transcuticular Penetration Model of Foliar-Applied Pesticides

12.3.1 Scenario

The solution containing a pesticide, when sprayed, remains on plant surfaces and forms a droplet. Just after retention, the droplet spreads to form its final shape that is generally supposed to be a hemisphere with a circular area contacting the leaf surface. The pesticide contained in the droplet starts to parti-

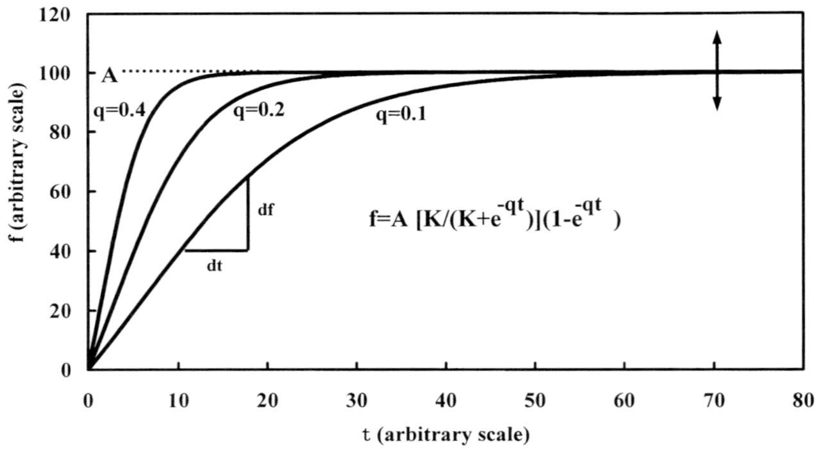

Fig. 2. Logistic-kinetic transcuticular penetration model. *A* Total amount of penetration, *q* penetration rate factor, *f* amount penetrated, and *K* constant

tion onto the outer surface layer of the CM and continues to do so until drying-up. At the same time, the droplet also begins to dry and the concentration of the pesticide increases. It then forms a homogeneous thin film in which the pesticide, water and adjuvants become condensed and will completely dry up later on. This drying process can be influenced by ambient conditions (temperature and humidity), formulation and/or adjuvants added. Partitioning of the pesticide onto the CM will continue until a complete dried state or an equilibrium with atmospheric humidity is attained. Immobilization will be reached when the pesticide does not partition any more after a longer period. The pesticide partitioned onto the CM starts to diffuse toward the inner surface of the CM along a chemical potential (concentration) gradient. In this sense, partitioning or sorption of a pesticide to the CM definitely plays the most important role in transcuticular penetration kinetics. The pesticide which reaches the inner surface of the CM will repartition into a subcuticular space. This repartitioning is not rate-limiting.

The transcuticular penetration kinetics consisting of the above processes can be modeled as in Fig. 2, referring to some preceding kinetic studies (Holloway and Stock 1990; Noveroske et al. 1992; Stock and Holloway 1993; Laerke and Streibig 1995). In Fig. 2, penetration actually starts with a lag time (usually 1–2 h) after application and increases linearly for an initial period (usually several to about 10 h), but afterwards the penetration rate gradually decreases and finally approaches asymptotically the total amount of penetration (A). Penetration will continue as long as partitioning lasts and the gradient through CM exists, but finally ends after a considerably longer period.

The above transcuticular penetration kinetics can be quantified as follows:

If a total quantity m exists that penetrates through the CM and the small quantity df that penetrates for a next short period dt with a rate-factor q is assumed to depend on both the terms of $(m - f)$ and f/m:

$$df = (m - f) \cdot (f/m) \cdot q \cdot dt \tag{12}$$

where t is the time after penetration. Equation (12) can be integrated to (13) under the condition of $t = 0$ and $f = 0$ (Watanabe and Yamaguchi 1993a).

$$f = A \cdot [K/(K + e^{-qt})] \cdot (1 - e^{-qt}) \tag{13}$$

where K is the integral constant ($K = 0.6$ or 0.7 is postulated for linearity or slight convexity in the initial period of penetration, respectively). A and q are supposed to be independent of each other. Other penetration parameters can be logically expressed by Eq. (14) (Watanabe and Yamaguchi 1994a):

$$A = V \cdot C \cdot S \cdot P_u \tag{14}$$

where V is the volume of the droplet applied, C the concentration of pesticide, S the contact area of the droplet and P_u the unit partition ratio of pesticide, i.e., the ratio of the pesticide partitioned from the droplet onto the CM to the total amount applied $(V \cdot C)$ per unit contact area, as defined by $P_u = A/(V \cdot C \cdot S)$. Thus, Eq. (14) can be transformed to (15) (Watanabe and Yamaguchi 1994b):

$$f = V \cdot C \cdot S \cdot P_u \cdot [K/(K + e^{-qt})] \cdot (1 - e^{-qt}) \tag{15}$$

Equation (15) describes quantitatively the complete transcuticular penetration kinetics of pesticides from a droplet through the CM. V and C can be artificially controlled, but other parameters can be regulated only by an interaction among the CM, pesticide and adjuvant and the ambient conditions (temperature and atmospheric humidity). Although the regular unit of each parameter should be the MSK system, the practically convenient units like V (μl), C (mol/l), S (cm), P_u (cm^{-2}), q (h^{-1}) and t (h) can be employed for an actual measurement and calculation.

Equation (15) shows that $(V \cdot C)$ is the total amount applied as the droplet, $(V \cdot C \cdot S \cdot P_u)$ the total amount partitioned to the CM to penetrate, $[(V \cdot C) \cdot (1 - S \cdot P_u)]$ is the amount not penetrated and remaining on or in the CM (if it does not evaporate), $[K/(K + e^{-qt})]$ is the partition rate from the droplet onto the CM, and $(1 - e^{-qt})$ the rate to diffuse through the CM. The equation also reveals that A and q cannot be decided only by the rate (df/dt) during an initial period of penetration. The rate q is a factor determining the penetration-rate distribution over the whole period of penetration. Ambient conditions and adjuvants added can influence the penetration affecting S, P_u and q (Watanabe and Yamaguchi 2000).

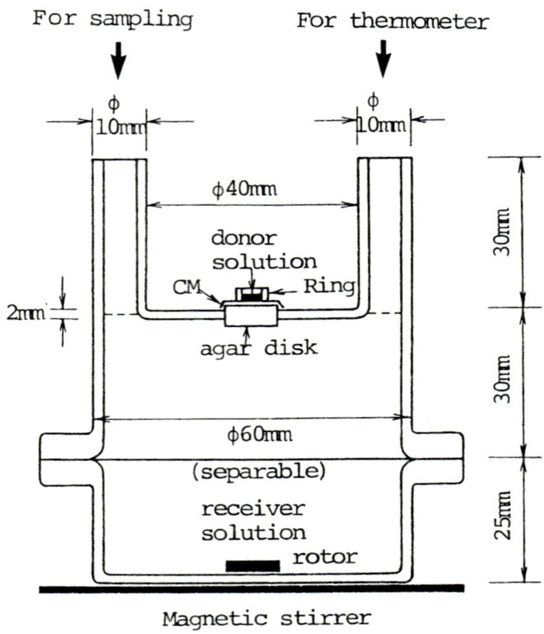

Fig. 3. Transcuticular penetration-measuring cell (not to scale)

12.3.2
Transcuticular Penetration-Measuring Cell

To measure penetration kinetics and analyze them by the above equations, a special transcuticular penetration-measuring cell was devised, as shown in Fig. 3 (Watanabe and Yamaguchi 1994c).

The cell is made of Pyrex glass, which can be taken apart, and the volume of the receiver solution is about 140 ml. A hole with a diameter of 9.0 mm in the upper part is filled by a disk of 1% agar gel 2.0 mm thick. The CM to be used is prepared with enzymes (pectinase and cellulase). It is taken from the upper surfaces, punched off the mature leaves of a test plant and carefully put onto the upper surface of the agar gel without any wrinkles and folds. All CMs used are microscopically checked to assure there are no pinholes, crevices and cracks. In the case of cracks or stomata, the kinetics, when extrapolated to $t = 0$, intercept with the ordinate. Then, the cell is filled with the receiver solution (buffer or distilled water) and is acclimatized for more than 12 h under measurement conditions (usually 25°C and 70% relative humidity). The receiver solution is stirred slowly with a magnetic stirrer. Usually, 5 or 10 μl of the test solution containing a ^{14}C-labeled pesticide at the desired concentration is placed carefully onto the center of the CM with a microsyringe and then 1.0 ml of the receiver solution is periodically sampled (1.0 ml of the fresh receiver solution is reversely supplemented).

The test solution which is usually employed consists of 1000 ppm of the ^{14}C-labeled pesticide, 5 ppm safranine, distilled water containing 25% 1,4-dioxane and, if necessary, adjuvant. 1,4-dioxane was chosen as a common solvent for pesticides throughout this series of measurement in order to avoid azeotropicity and it has a bp close to water. The diameter of the contact area of the droplet traced by safranine is measured vertically and horizontally. The sampled solution is mixed with a liquid scintillation cocktail and submitted to a liquid scintillation counter. For the inclusion of adjuvants in the test solution, two kinds of Pyrex glass rings (M: outer diameter 5.0 mm, inner diameter 3.4 mm and height 2.0 mm and L: 8.0, 5.9 and 2.0 mm, respectively) can be used to ensure its circularity and constancy in the contact area of the droplet if necessary. Repetition is usually ten for each measurement.

12.4 Parameters and Factors Governing Transcuticular Penetration Kinetics of Foliar-Applied Pesticides

12.4.1 Adaptability of the Logistic-Kinetic Penetration Model

To investigate the adaptability of Eq. (15) to an actual transcuticular penetration, ^{14}C-labeled amitrole (MW 84.1, mp 158°C, $\log P_{ow}$ −0.87), pyroquilon (MW 173.2, mp 112°C, $\log P_{ow}$ 1.31) and tricyclazole (MW 189.2, mp 185°C, $\log P_{ow}$ 1.60) at 1000 ppm were dissolved in distilled water including 25% 1,4-dioxane. CMs used were enzymatically isolated from mature leaves (adaxial surface) of citrus (*Citrus natsudaidai* Hayata) grown in a greenhouse. Ten μl of the test solution was applied into the glass ring on the CM to keep the contact area constant at 19.6 mm^2 and penetration was detected periodically (data not shown). Among the penetration parameters, S, A and P_u [$= A/(V \cdot C \cdot S)$], not artificially controlled, are easily derived from penetration data, but q is calculated from penetration kinetics by the following procedure; in the case that linearity in the initial period of penetration holds, q is calculated using the values of f and t in this linearity with $K = 0.6$ (or with $K = 0.7$ in the case of slightly upward curvilinearity) and A. Several values of q calculated are averaged.

The result showed that Eq. (13) well describes the actual transcuticular penetration as mentioned below and the parameters (A and q) changed with a combination of the CM and pesticide. A and q obtained with Natsudaidai's CM were 41% and 0.237 for amitrole, 53% and 0.177 for pyroquilon and 21% and 0.183 for tricyclazole, respectively. Although the measurement was performed using CM from pimento fruit (*Capsicum annuum* L. cv. Zuihoh), the penetration kinetics largely differed from the above result. Those parameters in the cases of no agar gel disk were almost the same as in A, but ca. 0.02 less in q than the values described above for all the pesticides with two kinds of CMs used. This means the agar gel disk slightly prohibited repartitioning to the receiver solution with pesticides having reached the inner surface of CM. However, prohibition is not substantial (Watanabe and Yamaguchi 1993a, 1994a).

12.4.2 Factors Influencing Transcuticular Penetration Kinetics

To detect the effect of other factors (concentration, contact area, temperature and humidity) on penetration kinetics, 10 μl of the test solutions of ^{14}C-labeled amitrole, pyroquilon and tricyclazole at 500 ppm, 1000 ppm and 2000 ppm, the glass rings M and L (contact area; 19.6 mm^2 for M and 50.2 mm^2 for L), temperatures 15, 25 and 35°C and relative humidity 50, 70 and 90% were combined and penetration measurement was performed with Natsudaidai's leaf CM. The effect of each combination of those factors on A and q were determined. The results analyzed by Eq. (13) for A and q are documented in Table 1. A change of pesticide and CM, of course, gave different penetration kinetics resulting in a change in A and q, however, an increase in S or C showed an increased A with an unchanged q. A higher temperature remarkably enhanced both A and q together, and a higher relative humidity led to increased A and decreased q. This result indicates that those factors exert their characteristic effects on partitioning and diffusion through the CM in different ways (Hartley and Graham-Bryce 1980a,b; Watanabe and Yamaguchi 1993b, 1994b).

12.4.3 Effect of Molecular Parameters of Pesticide on Transcuticular Penetration Kinetics

To examine the effect of molecular parameters of pesticides on penetration kinetics, ^{14}C-labeled linuron, carbaryl, tricyclazole, simazine, methomyl and oxamyl with MW ca. 200 (actually 162 to 249), but with different log P_{ow} (−0.47 to 2.76) and different mp (78 to 226°C) were selected and 10 μl of each test solution (a.i. 1000 ppm) was applied onto Natsudaidai's leaf CM without any glass ring at 25°C and 70% relative humidity. The penetration kinetics and parameters obtained are shown in Fig. 4 and Table 2. A has a positive relationship with P_u ($r^2 = 0.921$) and inversely with the logarithm of the mp ($r^2 = 0.804$), while q has a rough relationship with the magnitude of mp. A has no relationship with log P_{ow}, df/dt, MW, MV (Abraham and McGowan 1987) and q. This result clearly shows that A relates to the total amount partitioned to diffuse as a function of time and q seems to relate to fastness (speed) in the completion of partitioning to the CM. These are the characteristics of the non-equilibrium, non-steady state kinetics of transcuticular penetration (Watanabe and Yamaguchi 1997, 1999).

Table 1. Factors influencing transcuticular penetration kinetics (A and q) of foliar-applied pesticides

	Total amount of penetration (A)	Penetration rate factor (q)
(1) Change in pesticide	Change	Change
(2) Change in CM	Change	Change
(3) Increase in a.i. concentration	Increase	No change
(4) Increase in contact area	Increase	No change
(5) Ascendancy of temperature	Increase	Increase
(6) Increase in humidity	Increase	Decrease

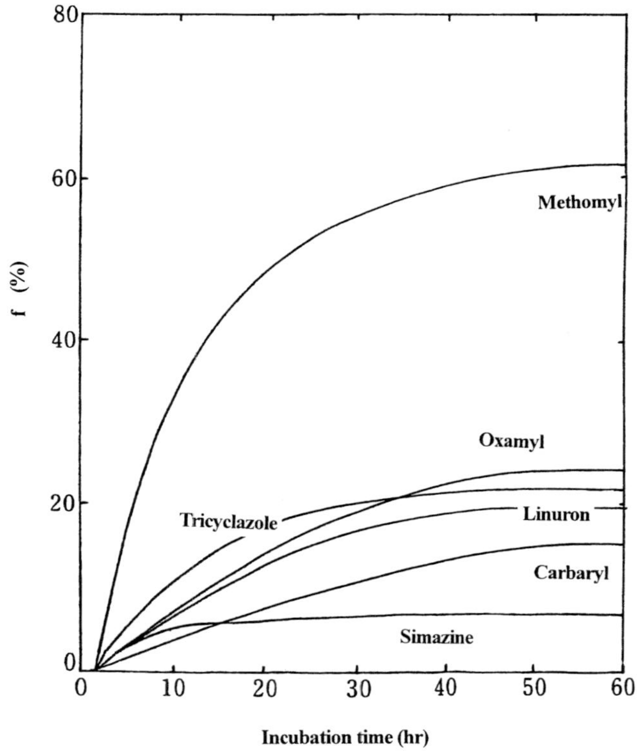

Fig. 4. Transcuticular penetration kinetics of some foliar-applied pesticides (through Natsudaidai's leaf CM at 25°C and 70% RH)

Table 2. Transcuticular penetration parameters of foliar-applied pesticides (through Natsudaidai's leaf CM at 25°C and 70% RH)

Pesticide	V (μl)	VC (μmol)	S (cm^2)	A (μmol)	q (h^{-1})	P_u (cm^{-2})
Linuron	10	0.040	0.338	0.0076 (19%)	0.095	0.56
Carbaryl	10	0.050	0.332	0.0065 (14%)	0.085	0.39
Tricyclazole	10	0.053	0.358	0.0106 (20%)	0.201	0.56
Simazine	10	0.050	0.344	0.0030 (6%)	0.405	0.17
Methomyl	10	0.046	0.294	0.0281 (61%)	0.198	2.08
Oxamyl	10	0.062	0.283	0.0149 (24%)	0.078	0.85

V, Volume of droplet applied; *C*, concentration of a.i. applied; *S*, contact area of droplet; *A*, total amount of penetration; *q*, penetration rate factor; and P_u, unit partition ratio.

12.5 Effect of Adjuvants on Transcuticular Penetration Kinetics of Foliar-Applied Pesticides

12.5.1 Analysis of Adjuvant Action (Adjuvancy) (Watanabe and Yamaguchi 1994d; Watanabe 2000a,b)

The penetration parameters of S, P_u, A and q not artificially controlled in Eq. (15) are very important to analyze the kinetics as influenced by adjuvants, because q can be ignored for a longer time of penetration, i.e., 30–50 h due to $e^{-qt} = 0$. Thus, A, S, P_u and q are the parameters available for the analysis of adjuvancy. Conventionally, Eq. (15) can be modified to Eqs. (16) and (17) as follows to obtain the penetration in %:

$$[(100 \cdot f)/(V \cdot C)](\%) = S \cdot P_u \cdot [K/(K + e^{-qt})] \cdot (1 - e^{-qt}) \tag{16}$$

$$[(100 \cdot A)/(V \cdot C)](\%) = S \cdot P_u \tag{17}$$

Equation (17) implies that the total amount of penetration (%) only depends on S and P_u. To analyze the effect of changing elements of adjuvant, pesticide and CM on penetration kinetics, Eq. (15) can be further modified to (18), using suffixes (control plot, 1 and test plot, 2):

$$f_2/f_1 = (V_2/V_1) \cdot (C_2/C_1) \cdot (S_2/S_1) \cdot (P_{u2}/P_{u1}) \cdot [(K + e^{-q1t})/(K + e^{-q2t})] \cdot [(1 - e^{-q2t})/(1 - e^{-q1t})] \tag{18}$$

Equation (18) indicates the contribution of a change in each parameter to a change in f. If V and C are artificially controlled for an identical figure in measurements, respectively, and t is a longer period for penetration, the equation can be simplified as in (19):

$$A_2/A_1 = (S_2/S_1) \cdot (P_{u2}/P_{u1}) \tag{19}$$

Equation (19) means that the total amount of penetration (A) influenced by adjuvants is finally determined only by changes in S and P_u, not by a change in q, as shown in Eq. (17). However, not only A, but also q can sometimes play an important role in terms of an expression rate in field efficacy of pesticides applied. This equation also means that if an increase in S causes a decrease in P_u, the resulting A cannot be changed and can even be reduced (antagonism). Those implications are useful for analysis of the transcuticular penetration kinetics as influenced by adjuvants.

12.5.2 Effect of Triton Surfactants

^{14}C-linuron as a lipophilic pesticide and ^{14}C-oxamyl as a hydrophilic one were chosen, and the effect of addition of Triton surfactants (polyoxyethylene octylphenyl ethers; Tritons X-45, X-114, X-100, X-165 and X-405) on their transcuticular penetration kinetics were investigated with Natsudaidai's leaf CM (25°C and 70% relative humidity). The concentration of a.i. (1000 ppm), composition of the test solution and its volume applied were the same as in Section 12.4.3 and the concentration of Tritons added was 0.1%. The penetration parameters obtained are shown in Table 3 for linuron. The total amount penetrated (A) was enhanced two to three times by all Tritons. Those cases enhanced by Tritons with lower HLB (hydrophilic-lipophilic balance) values were largely caused by enlargement in the contact area (S) and those by Tritons with a higher HLB, mainly by an increase in the unit partition ratio (P_u). Q tended to be slightly increased at a lower HLB. For oxamyl (data not shown) A was markedly enhanced two to seven times by all Tritons, however, that with Triton

Table 3. Transcuticular penetration parameters of foliar-applied linuron influenced by Triton surfactants (Natsudaidai's leaf CM, 25°C and 70% RH)

	Control	+Triton X-45 0.1%	+Triton X-114 0.1%	+Triton X-100 0.1%	+Triton X-165 0.1%	+Triton X-405 0.1%
A (%)	18	55	48	48	55	35
P_u (cm^{-2})	1.07	1.13	1.06	1.16	1.82	1.73
q (h^{-1})	0.054	0.081	0.060	0.046	0.042	0.055
A_2/A_1		2.75	2.20	2.30	2.80	1.85
S_2/S_1		2.21	2.13	2.13	1.71	1.16
P_{u2}/P_{u1}		1.24	1.13	1.08	1.70	1.60
q_2/q_1		1.50	1.11	0.85	0.78	1.00

Triton surfactants, polyoxyethylene octylphenyl ethers; A, total amount of penetration; P_u, unit partition ratio; q, penetration-rate factor, S, contact area of droplet; and *1* and *2*, control plot and test plot, respectively

X-100 reached the maximum (seven times), with a large increase in P_u involved, and those of other Tritons mainly depended on the increase in both S and P_u. Q tended to slightly increase at a lower HLB and to decrease at a higher HLB (Tritons X-165 and X-405), which implies a humectancy effect. This result shows that an intrinsic, molecular interaction between pesticide and surfactant (specified by its structure and HLB) determines penetration kinetics by changing P_u, S and q (Watanabe and Yamaguchi 2000).

12.5.3 Effect of Emulsifiable Oils

The effect of 0.5% of emulsifiable soybean oil, machine oil and Solvesso 150 (solvent naphtha, Exxon Chemical Co. Ltd.) containing 3 to 7% of the emulsifiers on the penetration kinetics of ^{14}C-linuron and ^{14}C-oxamyl through Natsudaidai's leaf CM was investigated in the same way as in the preceding section.

Among the penetration parameters for linuron (data not shown), A markedly increased, up to 1.3 to 2.7 times, in the order machine oil > soybean oil > Solvesso 150. The main reason for this was primarily caused by an enlargement in S and secondarily by an increase in P_u. A slight increase in q was seen only with Solvesso 150. For oxamyl (data not shown), A was enhanced in the order soybean oil ≥ machine oil > Solvesso 150 which seems to originate from a cooperation of both increases in S and P_u. Q increased only with Solvesso 150 as in linuron. This finding shows that the emulsified oils tended to increase A with an increase in S and P_u together, but did not significantly increase q. This means these oils mainly improve partitioning only (Watanabe and Yamaguchi 1998a).

12.5.4 Effect of Humectants

The effect of humectants like glycerol at 1.0% and POE-POP-POE block polymer with MW 3,000 [polyoxyethylene(MW: 600)-polyoxypropylene (MW: 1800)-polyoxyethylene (MW: 600) block polymer] and sodium polyacrylate with MW 400,000 at 0.5% on the penetration kinetics of ^{14}C-amitrole and ^{14}C-pyroquilon was investigated. The glass ring (M) was used to maintain a constant contact area of the droplet ($S = 19.6\,\text{mm}^2$). The concentration of pesticide was 1000 ppm and 10 μl of the test solution was applied. The other measuring conditions were the same as in Section 12.4.3. The result is shown in Table 4.

For both pesticides, A substantially increased and q significantly decreased with glycerol and POE-POP-POE but A remarkably decreased and q somewhat increased with polyacrylate. Humectancy shown by a decrease in q with glycerol and POE-POP-POE seems to greatly increase A with a prolonged partition originating from a delayed drying time. Polyacrylate, however, seems to notice-

Table 4. Effect of humectants on transcuticular penetration of foliar-applied amitrole and pyroquilon (through Natsudaidai's leaf CM with the glass ring M at 25°C and 70% RH)

		Control	+Glycerol 1.0%	+POE-POP-POE 0.5%	+Polyacrylate 0.5%
Amitrole	VC (μmol × 10^2)	11.9	11.9	11.9	11.9
	A (μmol × 10^2)	2.3	3.1	4.9	0.6
	q (h^{-1})	0.119	0.089	0.084	0.086
Pyroquilon	VC (μmol × 10^2)	5.8	5.8	5.8	5.8
	A (μmol × 10^2)	3.4	4.1	3.8	1.5
	q (h^{-1})	0.371	0.318	0.219	0.330

POE-POP-POE: polyoxyethylene (MW: 600)-polyoxypropylene (MW: 1800)-polyoxyethylene (MW: 600) block polymer

ably delay the drying time of the droplet with a decreased A, which means incorporation of the pesticides to its inner gel structure occurred, therefore, they could hardly partition onto the CM and so q was relatively large. Apparently, the pesticides being dissolved outside or in the peripheral part of the gel completed their partitioning so rapidly that immobilization quickly occurred, causing an increased q (Watanabe and Yamaguchi 1995; Baur 1999).

12.5.5 Effect of Amine Surfactants on Glyphosate Penetration

Amine-type surfactants characteristically enhance the herbicidal efficacy of foliar-applied glyphosate, which is usually formulated with them as a liquid. The effect of polyoxyethylene ($n = 15$) tallow amine (abbreviated as T-25), polyoxyethylene ($n = 10$) dimethyl coco-ammonium chloride (C-20) and polyoxyethylene ($n = 8$) lauryl ether (L-8) on the penetration kinetics of glyphosate isopropylamine (IPA) salt through Natsudaidai's leaf CM (25°C and 70% relative humidity) was investigated under the conditions of Section 12.4.3. The above surfactants at 2000 ppm were added to ^{14}C-glyphosate IPA amine salt (4000 ppm, aqueous solution) and 10 μl of it was applied. The result is shown in Table 5.

A increased markedly with the cationic T-25 and C-20, whose main cause was primarily a large increase in P_u and secondarily that in S. However, the fact that q for these surfactants was smaller as compared with the control means that humectancy operated to increase P_u. For the nonionic L-8, A slightly increased, but less than for the cationics and the main cause was an extreme increase in S and not in P_u, however, q became relatively smaller. This can be a kind of antagonism as predicted by Eq. (19). Thus, the adjuvancy of these amine surfactants seems to be predominantly caused by an extreme increase in P_u and a fair increase in S, and these two parameters worked cooperatively with q becoming smaller. This implies that the cationic surfactants associate

Table 5. Effect of amine surfactants on transcuticular penetration kinetics of foliar-applied glyphosate (through Natsudaidai's leaf CM at 25°C and 70% RH)

	Control	+T-25 0.2%	+C-20 0.2%	+L-8 0.2%
S	11.9 mm^2	20.1	19.6	42.1
A	0.002 μmol (0.8%)	0.007 (2.8%)	0.005 (2.0%)	0.004 (1.6%)
P_u	0.067/cm^2	0.139	0.102	0.038
q	0.258/h	0.062	0.115	0.113
A_2/A_1	–	3.5	2.5	2.0
S_2/S_1	–	1.69	1.65	3.54
P_{u2}/P_{u1}	–	2.07	1.52	0.57
q_2/q_1	–	0.24	0.45	0.44

T-25, polyoxyethylene (n = 15) tallow amine; C-20, polyoxyethylene (n = 10) coco-ammonium chloride; L-8, polyoxyethylene (n = 8) lauryl ether

or combine with glyphosate to increase partitioning by reducing its hydrophilicity and also operate to enlarge its partitioning area. A humectancy is also involved (Watanabe and Yamaguchi 1996, 1998b).

12.6 Discussion and Conclusions

The studies presented above on transcuticular penetration kinetics of some foliar-applied pesticides based on a non-steady state, non-equilibrium model ("the logistic-kinetic transcuticular penetration model") are considerably different from preceding ones which were based on several steady, equilibrium penetration models.

The results show that the total amount of penetration (A) is *not* correlated with log P_{ow}, the initial penetration rate (df/dt), penetration-rate factor (q) and molecular weight (MW) or molecular volume (MV), *but* with the unit partition ratio (P_u) and contact area of droplet (S) of the pesticides. They also reveal that the penetration-rate factor (q) *cannot* be directly related to the molecular weight (MW) or molecular volume (MV), *but* to the *time* up to complete partitioning of the pesticides. These facts imply that the whole transcuticular penetration kinetics are an integration in which the partition process onto and diffusion process through the CM correlate sequentially and mutually, and diffusion through the CM continues as long as partitioning to the CM lasts. Therefore, the penetration rate is *not only* related to the diffusion rate, *but also* depends on the partitioning rate.

Increases in both the concentration of a pesticide and the volume of the droplet can naturally increase A, but not q. Higher temperatures can increase both *A* and *q*, improve *S* of the droplet and make diffusion through the CM faster, instead of a possible decrease in partition. Higher humidity clearly shows an increase in *A* and a decrease in *q*, and this leads to a humectancy of adjuvants which causes prolonged drying of the droplet and therefore decreases *q* by a longer period of partitioning of the pesticide (Schönherr and Baur 1996; Baur 1999).

It is known that adjuvants added to the droplet can act in various ways to change penetration kinetics (Valkenburg (no year); Hartley and Graham-Bryce 1980a,b; Holloway et al. 1994; Cronfeld et al. 2001).

Surfactants often increase the volume of a droplet retaining (*V*) to plants and spread the contact area (*S*) by absorption to the CM to provide a larger partitioning site. Prolongation of the drying time of droplets, due to their hydrophilicity (specifically, non-ionics with a higher HLB), increases P_u and decreases *q*. Moreover, since they can solubilize pesticides into micelles and/or associate with them, the droplets can form a viscous thin layer film on the CM during evaporation to promote P_u and to reduce *q*. Surfactants also can penetrate into the CM to loosen, swell or hydrate its texture and increase *q* by reducing its diffusion resistance (activation).

Humectants sometimes play an interesting role by prolonging the drying time of a droplet to decrease *q* and increase P_u. Water soluble, hygroscopic substances like glycols and glycerol, polysaccharides, synthetic polymers and polymeric surfactants, except for polyacrylates that can build up a gel structure, act as both surfactant and humectant to increase P_u and even *S* and to decrease *q*.

Some emulsifiable oils counteract crystallization of the pesticide inside the oil emulsion by prolonging drying to decrease *q*, but the emulsifiers included also act as a surfactant to increase both P_u and *S*.

Cationic amine surfactants improve penetration of glyphosate by increasing primarily P_u and secondarily *S*, and both operate cooperatively. This means these surfactants associated with glyphosate reduced its hydrophilicity and increased P_u and also spread the droplet, but did not enhance *q* due to humectancy (Santier and Chamel 1992; De Ruiter et al. 1995; Garbow et al. 1995).

As to adjuvancy of surfactants in transcuticular penetration kinetics, three major categories of spreader, modifier and activator have been proposed (Hartley and Graham-Bryce 1980a,b; Schönherr et al. 1990; Holloway et al. 1994; Schönherr and Baur 1994; Riederer and Schreiber 1995; Cronfeld et al. 2001).

Their quantification or quantitative characterization can be proposed here from the present studies based on the logistic-kinetic penetration model (Table 6; Watanabe 2000a,b; Watanabe and Yamaguchi 2000).

A spreader can increase *S* where partitioning takes place to contribute to increasing *A* without changing P_u and *q*.

A modifier can change the physicochemical properties inside the droplet to increase *A* by humectancy, solubilization and association. Humectancy increases P_u, thereby decreasing *q* by a prolonged partitioning time. Solubi-

Table 6. Modes of action of adjuvants to transcuticular penetration kinetics of foliar-applied pesticides

	Classification in adjuvancy	Action site	Penetration parameter			
			A	P_u	S	q
1	Spreading (spreader)	Increase in contact area of droplet	Increase	No change	Increase	No change
2	Modification (modifier)	Modification of physical properties inside droplet (humectancy, accumulation, anticrystallization)	Increase	Increase	No change	No change, increase or decrease
3	Activation (activator)	Cuticular membrane	Increase	Increase	No change	Increase
4	Composite action	Above sites	Increase	Increase	Increase	No change, decrease or increase

lization by micelles and association with pesticides can perform improvement of partitioning itself to increase P_u with increasing, decreasing or unchanging q, because they can cause a prolonged drying time by micelles and trigger the amount of the pesticides partitioned to the CM to be increased as a function of time. Association can specifically bring an enhancement in q, as partly shown in glyphosate, introducing a larger driving force by making C_{don} and/or K_{cw} larger on the CM, but not reducing diffusion resistance inside the CM. This can be newly defined as "accumulation" or "pseudo-activation" because of the production of an increased driving force by making C_{don} and/or K_{cw} larger, with diffusion resistance itself unchanged inside the CM (Bukovac et al. 1990; Schönherr et al. 1991; Shafer and Bukovac 1991; Sharma 1995; Baur 1999). It still remains in the category of a modifier.

An activator penetrates into the CM to loosen, swell or hydrate its texture for reduction of diffusion resistance inside the CM and to fundamentally increase only q without increasing the driving force (activation) although it often promotes partitioning because of its surface activity (Holloway and Stock 1990; Schönherr et al. 1991; Stock and Holloway 1993; Coret and Chamel 1995). However, this activation should theoretically be distinguished as a part of acceleration from "accumulation" or "pseudo-activation" in terms of the mode of action of surfactants. Although it is not so easy to distinguish these activations, it can be suggested that an increase in q approximately paralleled or largely exceeded that in P_u by the adjuvants used and may indicate an "accumulation" or a "pseudo-activation". The reverse cases may mean an activation, but an increase in P_u accompanying a decrease in q can show an effect of humectancy. However, more detailed studies are needed to clarify these activation mechanisms.

Actually, most surfactants can operate by a combined action of spreader, modifier and activator because of their surface activity. However, the decisive action mechanisms of an adjuvant used can be characterized by the penetration parameters, as analyzed in this study.

This quantification seems to be useful for a better understanding of the transcuticular penetration kinetics of foliar-applied pesticides and the action mechanisms of adjuvants on them. It is also advantageous for formulational research for an intentional control of transcuticular penetration by adjuvants, although more studies remain to be performed (Chamel 1985; Schönherr and Riederer 1989; Stock and Holloway 1993; Holloway 1998; Kirkwood 1999).

References

Abraham MH, McGowan JC (1987) The use of characteristic volume to measure cavity terms in reversed phase liquid chromatography. Chromatographia 23:243–246

Bauer H, Schönherr J (1992) Determination of mobility of organic compounds in plant cuticles and correlation with molar volumes. Pestic Sci 35:1–11

Baur P (1999) Surfactant effects on cuticular penetration of neutral polar compounds: dependence on humidity and temperature. J Agric Food Chem 47:753–761

Breeze VC, Simmons JC, Roberts MO (1992) Evaporation and uptake of herbicide 2,4-D butyl applied to barley leaves. Pestic Sci 36:101–107

Bucholz DL, Hess FD (1987) A kinetic model for the absorption of 2,4-D acid and butoxyethanol ester into cabbage cotyledons. Pestic Biochem Physiol 28:1–8

Buick RD, Robson B, Field RJ (1992) A mechanistic model to describe organosilicone surfactant promotion of triclopyr uptake. Pestic Sci 36:127–133

Bukovac MJ, Petracek PD, Fader RG, Norse RD (1990) Sorption of organic compounds by plant cuticles. Weed Sci 38:289–298

Chamel A (1985) Foliar absorption of herbicides: study of the cuticular penetration using isolated cuticles. Physiol Vég 24:491–507

Chow PNP, Hinshal AN, Simundsson E (1987) (eds) Adjuvants and agrochemicals, vols 1 and 2. CRC Press, Boca Raton

Coret J, Chamel A (1995) Effects and possible mode of action of some nonionic surfactants on diffusion of ^{14}C-glyphosate and ^{14}C-chlorotoluron across isolated plant cuticles. Pestic Sci 43: 163–166

Cronfeld P, Larder K, Baur P (2001) Classification of adjuvants and adjuvant blends by effects on cuticular penetration. In: Viets AK, Tann RS, Mueninghoff JC (eds) Pesticide formulation and application system, vol 20. ASTM, Washington, DC, pp 81–94

Cutler DF, Alvin KL, Price CE (1982) (eds) The plant cuticle. Academic Press, London

De Ruiter H, Straatmenn K, Meinen W (1995) The influence of two fatty amine surfactants on foliar-absorption, translocation and efficacy of 2,4-D. J Agric Food Chem 43:3093–3097

Foy CL (1992) (ed) Adjuvants for agrochemicals. CRC Press, Boca Raton

Garbow JR, Wright DR, Hutton WC, Likos JJ (1995) Solution-state NMR studies of glyphosate/surfactant interaction. In: Gaskin RE (ed) Adjuvants for agrochemicals. New Zealand Forest Research Institute, Rotorua, pp 48–53

Gaskin RE (1995) (ed) Adjuvants for agrochemicals. New Zealand Forest Research Institute, Rotorua

Hartley GS, Graham-Bryce IJ (1980a) (eds) Physical principle of pesticide behaviour, vol 1. Academic Press, London, pp 127–137

Hartley GS, Graham-Bryce IJ (1980b) (eds) Physical principle of pesticide behavior, vol 2. Academic Press, London, pp 544–657

Holloway PJ (1998) Improving agrochemical performance: possible mechanisms for adjuvancy. In: Knowles DA (ed) Chemistry and technology of agrochemical formulations. Kluwer, Dordrecht, pp 232–263

Holloway PJ, Stock D (1990) Factors affecting the activation of foliar uptake of agrochemicals by surfactants. In: Korsa DR (ed) Industrial applications of surfactants II. Royal Soc Chem, Cambridge, UK, pp 303–337

Holloway PJ, Rees RT, Stock D (1994) Interactions between adjuvants, agrochemicals and target organisms. Ernst Schering Research Foundation. Springer, Berlin Heidelberg New York

Kerler F, Riederer M, Schönherr J (1984) Non-electrolyte permeability of plant cuticles; A critical evaluation of experimental methods. Physiol Plant 62:599–606

Kerstiens G (1996) (ed) Plant cuticle . Bios Scientific Publ, Oxford

Kirkwood RC (1999) Recent developments in our understanding of the plant cuticle as a barrier to the foliar-uptake of pesticides. Pestic Sci 55:69–77

Laerke PE, Streibig JC (1995) Foliar absorption of some glyphosate formulations and their efficacy on plants. Pestic Sci 44:107–116

Lewis CT (1980) (ed) Cuticle techniques in arthropods. Springer, Berlin Heidelberg New York, pp 367–388

McCall PJ (1988) Effect of chemical structure, temperature, crop oil concentrate and bentazon on the behavior of haloxyfop in yellow foxtail. Weed Sci 36:424–435

McMullan RM (1998) (ed) Proc 5th Int Symp, Adjuvants for Agrochemicals, Memphis, Tennessee

Noveroske RL, Keeney PN, Brown JG (1992) The characterization of uptake and transport with a radio-labeled aryloxyphenoxypropionate herbicide as influenced by adjuvants. In: Foy CL (ed) Adjuvants for agrochemicals. CRC Press, Boca Raton, pp 149–167

Price CE, Anderson NH (1985) Uptake of chemicals from foliar deposit: effects of plant species and molecular structure. Pestic Sci 16:369–377

Reed DW, Tukey HB (1982) Permeability of Brussels sprouts and carnation cuticles from leaves developed in different temperature and light intensities. In: Cutler DF, Alvin KL, Price CE (eds) The plant cuticle. Academic Press, London, pp 267–278

Riederer M, Schreiber L (1995) The transport barriers of plant cuticles. In: Hamilton RJ (ed) Waxes: chemistry, molecular biology and functions. The Oily Press, Dundee, pp 131–156

Salgado VL (1995) Steady-state and transient analysis of integument penetration by insects. Pestic Sci 44:59–67

Santier S, Chamel A (1992) Influence of an ethopropoxylated amine on the penetration of glyphosate across isolated tomato fruit cuticles. In: Foy CL (ed) Adjuvants for agrochemicals. CRC Press, Boca Raton, pp 101–108

Schönherr J, Bauer H (1992) Analysis of effects of surfactants on permeability of plant cuticles. In: Foy CL (ed) Adjuvants for agrochemicals. CRC Press, Boca Raton, pp 17–35

Schönherr J, Baur P (1994) Modeling of plant cuticles by crop protection agents and effects of adjuvants on their rates of penetration. Pestic Sci 42:185–208

Schönherr J, Baur P (1996) Effects of temperature, surfactants and other adjuvants on rates of uptake of organic compounds. In: Kerstiens G (ed) Plant cuticle. Bios Scientific Publ, Oxford, pp 135–155

Schönherr J, Bukovac MJ (1972) Penetration of stomata by liquids. Plant Physiol 49:813–819

Schönherr J, Riederer M (1988) Desorption of chemicals from plant cuticles: Evidence for asymmetry. Arch Environ Contam Toxicol 17:13–19

Schönherr J, Riederer M (1989) Foliar-penetration and accumulation of organic chemicals in plant cuticles. Rev Environ Contam Toxicol 108:1–70

Schönherr J, Baur P, Bucholz A (1990) Modeling foliar penetration: Its role in optimizing pesticide delivery. In: Blooks GT, Roberts TR (eds) Pesticide chemistry and bioscience. Royal Soc Chem, Cambridge, UK, pp 134–151

Schönherr J, Riederer M, Schreiber M, Bauer H (1991) Foliar uptake of pesticides and its activation by adjuvants: theories and methods for optimization. Pestic Sci 27:237–253

Shafer WE, Bukovac MJ (1991) Studies on octylphenoxy surfactants; effects of ethylene oxide chain length on sorption of NAA by isolated tomato fruit cuticles. J Agric Food Chem 37: 486–492

Sharma R (1995) (ed) Surfactant adsorption and surface solubilization. ASC Symposium Series 615, Washington, DC, pp 1–76

Stevens PJG (1993) Organosilicone surfactants as adjuvants for agrochemicals. Pestic Sci 38: 103–122

Stevens PJG, Gaskin RE, Hong SO, Zabkiewicz AJ (1991) Contributions of stomatal infiltration and cuticular penetration to enhancement of foliar-uptake by surfactants. Pestic Sci 33: 371–382

Stock D, Holloway PJ (1993) Possible mechanisms for surfactant-induced foliar uptake of agrochemicals. Pestic Sci 38:165–177

Valkenburg JWV (no year) Terminology, classification and chemistry. In: Adjuvants for herbicides. The Weed Science Society of America (ed), Champaign, Illinois, pp 1–8

Watanabe T (2000a) Factors governing retention on and penetration into crops and weeds of agrochemicals. J Pestic Sci 25:285–291

Watanabe T (2000b) Modes of action of adjuvants in transcuticular penetration of foliar-applied pesticides. Shokubutsu-Boueki 54:497–502

Watanabe T, Yamaguchi I (1993a) Studies on surfactant action to transcuticular penetration of pesticides, part 3. Logistic-kinetic model. 18th Annu Meeting Pestic Sci Soc Japan, p 147

Watanabe T, Yamaguchi I (1993b) Studies on surfactant action to transcuticular penetration of pesticides, part 4. A relationship between A (total amount penetrated) and q (penetration-rate factor) in the logistic-kinetic model. 13th Japan Agricultural Formulation and Application Symp, p 102

Watanabe T, Yamaguchi I (1994a) A kinetics of transcuticular movement of pesticides and surfactant action to it. 8th Int Congr Pestic Chem, Abstract vol 2, Washington, DC, p 722

Watanabe T, Yamaguchi I (1994b) Studies on surfactant action to transcuticular penetration of pesticides, part 5. Logistic-kinetic model and influencing factors. 19th Annu Meeting Pestic Sci Soc Japan, p 104

Watanabe T, Yamaguchi I (1994c) Japanese utility model. Jitsu kai hei:6-72635

Watanabe T, Yamaguchi I (1994d) Studies on surfactant action to transcuticular penetration of pesticides, part 6. Analysis of surfactant effects on transcuticular penetration of pesticides by the logistic-kinetic model. 14th Japan Agricultural Formulation and Application Symp, pp 43–44

Watanabe T, Yamaguchi I (1995) Studies on surfactant action to transcuticular penetration of pesticides, part 7. Analysis of effects of humectants on transcuticular penetration of foliar-applied pesticides. 20th Annu Meeting Pestic Sci Soc Japan, p 157

Watanabe T, Yamaguchi I (1996) Studies on surfactant action to transcuticular penetration of pesticides, part 8. Analysis of transcuticular uptake of glyphosate enhanced by amine surfactants. 21st Annu Meeting Pestic Sci Soc Japan, p 114

Watanabe T, Yamaguchi I (1997) Studies on surfactant action to transcuticular penetration of pesticides, part 9. Analysis of effects of log P_{ow} of pesticides on uptake behavior. 22nd Annu Meeting Pestic Sci Soc Japan, p 84

Watanabe T, Yamaguchi I (1998a) Studies on surfactant action to transcuticular penetration of pesticides, part 10. Analysis of effects of emulsifiable oils on transcuticular penetration of pesticides by the logistic-kinetic model. 18th Japan Agricultural Formulation and Application Symp, p 62

Watanabe Y, Yamaguchi I (1998b) Analysis of transcuticular uptake of glyphosate enhanced by amine surfactants. 37th Annu Meeting Weed Sci Soc Japan, p 212

Watanabe T, Yamaguchi I (1999) Studies on surfactant action to transcuticular penetration of pesticides, part 11. Analysis of transcuticular uptake kinetics through pimento fruit cuticle of foliar-applied pesticides. 24th Annu Meeting Pestic Sci Soc Japan, p 122

Watanabe T, Yamaguchi I (2000) Studies on surfactant action to transcuticular penetration of pesticides, part 12. Analysis of effect of Triton surfactants on transcuticular uptake kinetics of foliar-applied pesticides. 25th Annu Meeting Pestic Sci Soc Japan, p 66

13 Structure-Activity Correlation of Very Long-Chain Fatty Acid Biosynthesis Inhibitors

Ko Wakabayashi and Peter Böger

13.1 Introduction

Plant lipid biosynthesis up to fatty acid (C18:0) occurs in plastids. Interference with this fatty acid biosynthesis by herbicidal phenoxy-phenoxy propionic acids or cyclohexanediones is due to the inhibition of acetyl-CoA carboxylase, which explains their herbicidal mechanism of action (Devine et al. 1993). It has been recently concluded that herbicidal chloroacetamides (e.g., alachlor), oxyacetamides (like mefenacet), carbamoylated five-membered nitrogen heterocycles (fentrazamide or cafenstrole), oxiranes (e.g., indanofan) and some miscellaneous compounds (ethofumesate or prosulfocarb) act as specific inhibitors of elongase(s) involved in the extraplastidic biosynthesis of very long-chain fatty acids (VLCFAs) with alkyl chains longer than C_{18}, although the precise mechanism of interaction of the compounds at the molecular level has not been clarified yet (Couderchet et al. 1998; Matthes et al. 1998; Böger et al. 2000; Takahashi et al. 2001a). The relationship between herbicidal action and elongase inhibition by the compounds mentioned above has been discussed in Chapter 6 (Böger and Matthes, Chap. 6, this Vol.). A lack of VLCFAs in the plant cell membranes will lead to death.

Quite recently, a convenient assay system using a leek microsomal preparation has been developed for VLCFA biosynthesis inhibition (Böger et al. 2000; Schmalfuß et al. 2000). Although only limited numbers of inhibitory data by herbicidal compounds have been reported, some aspects concerning structure-activity considerations of VLCFA biosynthesis inhibitors are discussed in this chapter, facilitating the design of more active inhibitors and determine a more precise mechanism of action of these inhibitors.

13.2 Very Long-Chain Fatty Acid Biosynthesis Inhibition by Herbicides

The VLCFA biosynthesis inhibition has been quantitatively measured with cucumber cotyledons and unicellular microalgae (e.g., *Scenedesmus acutus*) using intact cells, as well as a cell-free leek microsomal preparation, to determine inhibition of VLCFA elongation in the presence of herbicides

P. Böger, K. Wakabayashi, K. Hirai (Eds.)
Herbicide Classes in Development

(Couderchet et al. 1998; Matthes et al. 1998; Schmalfuß et al. 1998, 2000). Thus, the VLCFA biosynthesis inhibition has been determined and compared in these three assays.

Herbicidal chloroacetamides (alachlor, metolachlor, metazachlor and thenylchlor), oxyacetamides (mefenacet andd flufenacet), carbamoylated five-membered nitrogen heterocycles (fentrazamide and cafenstrole), oxiranes (tridiphane and indanofan) and ethofumesate bearing a sulfonate moiety strongly inhibited C18:1 or C18:0 elongation at almost the same level, exhibiting approx. I_{50} values of 10^{-7} to 10^{-6} M. However, inhibitory activity by the old type of thiolcarbamates, such as EPTC or prosulfocarb, was found rather weak, indicating an I_{50} value higher than 10^{-5} M (see Table 1 and Fig. 1: Couderchet et al. 1998; Matthes et al. 1998; Takahashi et al. 2001a,b, 2002).

Reduction of fatty acid elongation is specific for herbicides with a chloroacetamide-like mode of action. The herbicides affecting acetolactate synthase, protoporphyrinogen oxidase, or photosynthetic electron transport do not inhibit oleic acid incorporation into VLCFAs present in a sporopollenin fraction of *Scenedesmus acutus* cells. Two correlations, phytotoxicity indicated by growth inhibition or chlorophyll decrease vs. inhibition of oleate incor-

Table 1. VLCFA biosynthesis inhibition by herbicides in cell and cell-free assays

	Inhibition of VLCFA elongation (% inhibition)						
	C18:0	C18:1			C18:0		
	Cucumber cotyledons	*Scenedesmus acutus* cells			Leek (*Allium porrum*) cell-free		
Herbicides	1 μM	0.01 μM	0.1 μM	1 μM	0.1 μM	1 μM	10 μM
Alachlor	–	29 (±7)	49 (±7)	62 (±5)	48 (±6)	62 (±7)	82 (±5)
Metolachlor	60 (±2)	28 (±1)	41 (±4)	50 (±5)	–	–	–
S-form	64 (±3)	–	–	–	–	–	–
R-form	1 (±6)	–	–	–	–	–	–
Metazachlor	89 (±1)	17 (±13)	51 (±12)	64 (±3)	25 (±5)	45 (±5)	61 (±10)
Dimethenamid	–	17 (±11)	41 (±10)	53 (±3)	–	–	–
Thenylchlor	–	2 (±1)	38 (±6)	68 (±6)	28 (±9)	71 (±8)	85 (±4)
Mefenacet	–	3 (±2)	24 (±10)	41 (±3)	–	–	–
Flufenacet	95 (±3)	32 (±5)	46 (±15)	65 (±5)	–	–	–
Cafenstrole	84 (±4)	30 (±5)	53 (±2)	73 (±3)	10 (±9)	33 (±8)	80 (±4)
Fentrazamide	83 (±3)	2 (±2)	32 (±8)	64 (±2)	–	–	–
Tridiphane	–	–	–	–	16 (±9)	28 (±18)	53 (±1)
Indanofan	–	0 (±2)	36 (±7)	69 (±5)	31 (±10)	44 (±6)	58 (±1)
S-form		–	–	–	47 (±2)	57 (±5)	63 (±6)
EPTC	–	0 (±1)	0 (±2)	8 (±3)	–	–	–
Prosulfocarb	–	17 (±9)	24 (±1)	26 (±8)	–	–	–
Ethofumesate	63 (±4)	–	–	–	–	–	–

Cell-free elongation inhibition assay; C18:0 substrate was activated with –CoA.

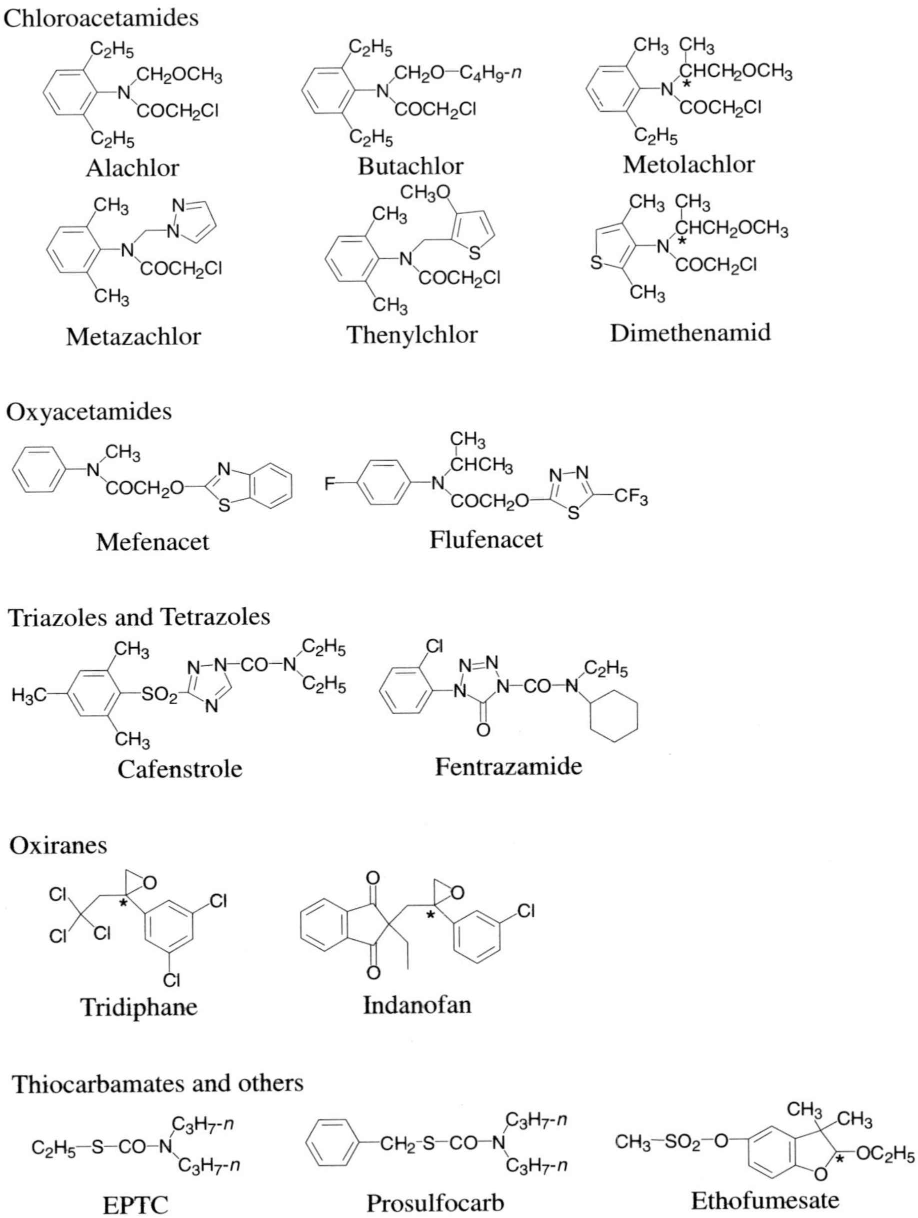

Fig. 1. Structures of herbicides acting as very long-chain fatty acid biosynthesis inhibitors; * indicates chiral center

Table 2. Phytotoxicity and inhibition of oleate incorporation into *Scenedesmus* cells by VLCFA biosynthesis inhibitors

No.	Herbicides	Phytotoxicity: Growth inhibition (%)	Phytotoxicity: Chlorophyll decrease (%)	Inhibition of C18:1 incorporation (% inhibition)
1	Alachlor	38 (±8)	73 (±8)	62 (±5)
2	Metolachlor	15 (±15)	51 (±19)	50 (±5)
3	Metazachlor	43 (±3)	73 (±7)	64 (±3)
4	Dimethenamid	23 (±9)	60 (±18)	53 (±3)
5	Mefenacet	11 (±12)	35 (±14)	41 (±3)
6	Flufenacet	47 (±1)	78 (±6)	65 (±5)
7	Cafenstrole	54 (±10)	84 (±6)	73 (±3)
8	Indanofan	49 (±10)	82 (±6)	69 (±5)
9	Butachlor	27 (±4)	48 (±3)	70 (±6)

Phytotoxic parameters were obtained from *Scenedesmus* cell cultures. Growth inhibition (packed cell volume) and chlorophyll decrease were used as phytotoxic parameters. Herbicide concentration was 1 μM in the oleate incorporation experiment and 5 μM for phytotoxic assays.

poration into sporopollenin of *Scenedesmus* cells, can be significantly calculated (data for calculation, see Table 2: Couderchet et al. 1998; Takahashi et al. 2002).

$$[\text{Growth inhibition}] = \underset{(\pm 0.294)}{1.506}[\text{Inhibition of C(18:1) elongation}] - 54.822 \quad [n = 8,\quad r = 0.981 \quad s = 3.408] \tag{1}$$

$$[\text{Chlorophyll decrease}] = \underset{(\pm 0.245)}{1.572}[\text{Inhibition of C(18:1) elongation}] - 26.753 \quad [n = 8,\quad r = 0.988 \quad s = 2.483] \tag{2}$$

Since butachlor herbicide was an outlier in the original regression line (Couderchet et al. 1998; Böger et al. 2000), indanofan, a new type of grass herbicide, has been included in the equations for butachlor. Both growth inhibition and chlorophyll decrease are positively correlated with the inhibition of oleate incorporation, as shown by Eqs. (1) and (2). This fact is attributed to a marked inhibition of oleic acid elongation to monounsaturated VLCFAs. The significant correlations of the equations indicate that the reduced formation of VLCFAs is responsible for the phytotoxicity of the herbicides.

It should be noted that I_{50} values of inhibition by metazachlor for each step of acyl-CoA elongation (the first step C16:0-CoA to C18:0-CoA, the second C18:0-CoA to C20:0-CoA, the third C20:0-CoA to C22:0-CoA and the fourth C22:0-CoA to C24:0-CoA) were reported to be 0.5 μM (for elongation C16:0-CoA), 1.7 μM (C18:0-CoA), 0.7 μM (C20:0-CoA) and 0.5 μM (for elongation of

C22:0-CoA), respectively, in the cell-free microsomal leek assay (Schmalfuß et al. 2000). The somewhat different I_{50} values obtained for the above-mentioned elongation steps may be due to two microsomal elongases present. Furthermore, a competition of acyl-CoA substrates with the inhibitor is assumed, which may be different for the elongation steps (see Schmalfuß et al. 2000). A future isolation and purification of the elongase enzyme(s) should clarify these questions (see Chap. 6).

Metolachlor and dimethenamid include two stereoisomers, the S-enantiomers being the active herbicides as well as VLCFA biosynthesis inhibitors, while R-isomers are not. The stereospecificity of inhibition of acyl-CoA elongation was investigated by the standard elongase assay with C18:0-CoA as primer substrate and metolachlor as a chiral chloroacetamide herbicide (Schmalfuß et al. 2000). *S*-Metolachlor (1 mM) inhibited the elongation almost completely and ca. 5 μM was sufficient for a 50% inhibition in the cell-free assay using the leek microsomal preparation, although the *R*-enantiomer had no influence on C18:0-CoA elongation, even at 0.1–1.0 mM. This evidence was also demonstrated with cucumber cotyledons (Matthes et al. 1998). Furthermore, a 50% inhibition of incorporation of labeled oleate into a sporopollenin-containing fraction of *Scenedesmus acutus* cells was attained with less than 10^{-7} M of the *S*-enantiomer of dimethenamid, while 10^{-5} M of the *R*-form achieved only a 40% inhibition (Couderchet et al. 1997). Possibly, this small inhibition was caused by some S-isomer present as an impurity in the *R*-dimethenamid sample.

In the following sections (Sects. 13.3–13.5), relationships between structures and acyl-CoA elongation inhibitory activity of three types of herbicidal VLCFA biosynthesis inhibitors, chloroacetamides (thenylchlor and its analogs), carbamoylated five-membered nitrogen heterocycles (cafenstrole and its analogs) and oxiranes (indanofan and its analogs), will be discussed.

13.3 Very Long-Chain Fatty Acid Biosynthesis Inhibition by Thenylchlor and Its Analogs

Recently, the structure and acyl-CoA elongation inhibitory activity of a new herbicidal chloroacetamide, thenylchlor, have been discussed (Takahashi et al. 2001a). Introduction of a thenyl group (R, see Table 3) to the nitrogen atom of 2-chloro-2′,6′-dimethylacetanilide (compound **1**) gave a remarkable increase in both herbicidal activity assayed with barnyardgrass (*Echinochloa oryzicola*) and stearic acid elongation inhibition in the leek cell-free assay, irrespective of whether R is 2-thenyl or 3-thenyl group (compounds **2** and **3**). A thenyl group instead of hydrogen yielded an approx. 16-fold stronger herbicidal activity against barnyardgrass than the nonsubstituted compound **1**, in accordance with the cell-free inhibition of stearic acid elongation. Apparently, the *N,N*-disubstituted amide moiety is important for both the herbicidal activity and

Table 3. Introduction of a thenyl group into 2-chloro-2′,6′-dimethylacetanilide and VLCFA biosynthesis inhibition

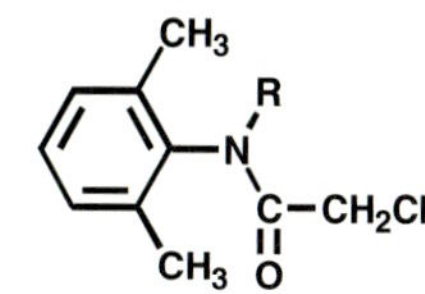

Compounds No.	R	Barnyardgrass[a] (ED_{90}, g/ha)	Inhibition of C18:0 elongation (% inhibition) Leek (*Allium porrum*) 0.1 μM	1 μM	10 μM
			10 (±4)	12 (±5)	16 (±10)
2	2-thenyl	62.5	0 (±0)	15 (±3)	51 (±1)
3	3-thenyl	62.5	11 (±3)	27 (±3)	54 (±6)

[a] Barnyardgrass (*Echinochloa oryzicola*). The compounds were applied in water 3 days after transplanting of rice. Herbicidal activity was estimated 21 days after application. The degree of inhibition was determined by fresh weight of the shoot. These data were used to calculate the ED_{90}.

Table 4. Inhibition of VLCFA biosynthesis by 2-chloro-*N*-(3-thenyl)acetanilides

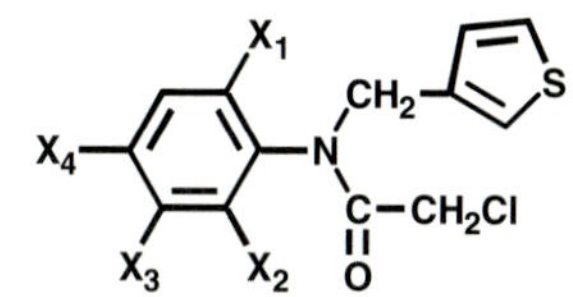

Compounds No.	X_1	X_2	X_3	X_4	Barnyardgrass[a] (ED_{90}, g/ha)	Inhibition of C18:0 elongation (% inhibition) Leek (Allium porrum) 0.1 μM	1 μM	10 μM
3	CH_3	CH_3	H	H	62.5	11 (±3)	27 (±8)	54 (±4)
4	H	H	F	H	500	12 (±1)	25 (±2)	62 (±1)
5	H	CH_3	H	H	500	16 (±0)	41 (±5)	44 (±10)
6	H	H	H	OC_2H_5	800	20 (±9)	32 (±8)	53 (±4)
7	H	Cl	Cl	H	1000	10 (±3)	30 (±3)	49 (±6)

[a] See footnote in Table 3.

elongation inhibition by chloroacetamide herbicides (Böger et al. 2000). In this case, one substituent is the aryl, the other one, the thenyl. However, 2-chloro-*N*-(3-thenyl)acetanilides substituted with alkyl, alkoxyl or halogen at the benzene ring, analogs of the active 2-chloro-*N*-(3-thenyl)-2′,6′-dimethyl-acetanilide (compound **3**), were herbicidally less active due to a lack of their metabolic stability in the 14-day herbicidal greenhouse test (Kato et al. 1998), despite their inhibition of C18:0-CoA elongation indicating I_{50} values of 10 μM (Table 4).

Inhibition of VLCFA biosynthesis by 2-chloro-*N*-(2-thenyl)acetanilides has also been considered (see Table 5). The influence of substituents at the thiophene ring of 2-chloro-*N*-(2-thenyl)-2′,6′-dimethylacetanilides on C18:0-CoA elongation inhibitory activity was examined together with herbicidal activity. Although the approx. I_{50} values of stearic acid elongation for compounds **2, 8, 9, 10, 15** and **16** were 10 μM, the I_{50} of compound **13** (thenylchlor) was below 1 μM, reflecting the stronger herbicidal activity against barnyardgrass (ED_{90}, below 25 g/ha). Although a relation between the inhibitory activity and electronic nature of the substituents at the thiophene ring cannot be discussed thoroughly, due to the limited variety of analogs available, substitution with an electron-donating group, such as methoxyl or methyl especially at 3-position of the thiophene ring, seems to produce a strong herbicidal activity and VLCFA biosynthesis inhibition.

Furthermore, substituents at the benzene moiety of 2-chloro-*N*-(2-thenyl)acetanilides and the inhibitory activities were investigated. A 2,6-dimethyl substitution caused severe inhibition of stearic acid elongation and strong herbicidal activity, e.g., compounds **2** and **13**. Compound **13** (thenylchlor) gave better inhibition than compound **2**. Dimethylation at the 2- and 6-positions of the benzene ring is essential to obtain a potent VLCFA biosynthesis inhibition, since monoalkylation or diethylation produces less active compounds (**17, 18, 19** and **20** in Table 5). Conventional effective chloroacetamide herbicides inhibiting VLCFA elongation are generally dialkylated at the 2- and 6-positions of their benzene moiety. and dimethenamid, also a strong inhibitor, is dimethylated at the 2- and 4-positions of the 3-thienyl ring which is due to bioisosterism with the benzene ring (see Table 1 and Fig. 1).

Activity–activity correlation between herbicidal activity and VLCFA biosynthesis inhibition by thenylchlor analogs can be expressed by Eq. (3), calculated with parameters from Table 5.

$$\begin{aligned}&[ED_{90}(\text{Barnyardgrass})]\\&\quad = \underset{(\pm 2.330)}{-5.521}[\text{Inhibition of C(18:0) elongation }(10\,\mu\text{M})] + 319.014\\&\qquad\qquad [n = 10, \quad r = 0.888 \quad s = 41.798]\end{aligned} \tag{3}$$

This significant equation indicates that inhibition of VLCFA biosynthesis closely relates to phytotoxic action of thenylchlor and its analogs. Thus phytotoxicity of newly designed analogs can be estimated in advance.

Table 5. Thenylchlor and its analogs: Inhibition of VLCFA biosynthesis by 2-chloro-*N*-(2-thenyl)-acetanilides

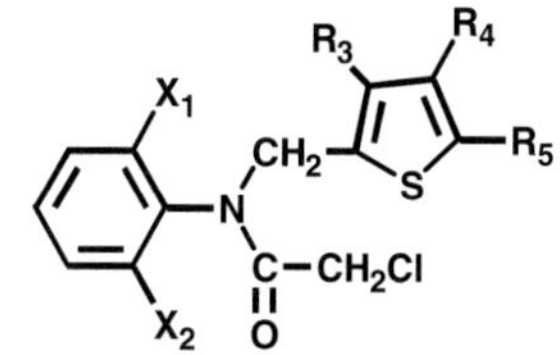

Compounds No.	X_1	X_2	R_3	R_4	R_5	Barnyardgrass[a] (ED_{90}, g/ha)	Inhibition of C18:0 elongation (% inhibition) Leek (*Allium porrum*) 0.1 μM	1 μM	10 μM
2	CH_3	CH_3	H	H	H	62.5	0 (±0)	15 (±0)	51 (±1)
8	CH_3	CH_3	CH_3	H	H	25	3 (±1)	9 (±1)	42 (±1)
9	CH_3	CH_3	H	CH_3	H	50	4 (±3)	15 (±3)	54 (±6)
10	CH_3	CH_3	H	H	CH_3	30	3 (±0)	13 (±3)	49 (±1)
11	CH_3	CH_3	Cl	H	H	25	–	–	–
12	CH_3	CH_3	H	H	Cl	62.5	–	–	–
13	CH_3	CH_3	OCH_3	H	H	<25	28 (±9)	71 (±8)	85 (±4)
(Thenylchlor)									
14	CH_3	CH_3	H	H	OC_2H_5	250	3 (±5)	7 (±7)	14 (±15)
15	CH_3	CH_3	CH_2OCH_3	H	H	25	2 (±3)	14 (±2)	52 (±1)
16	CH_3	CH_3	SCH_2CH_3	H	H	100	7 (±8)	12 (±10)	44 (±4)
17	CH_3	H	H	H	H	250	10 (±3)	15 (±3)	25 (±2)
18	C_2H_5	H	H	H	H	250	–	–	–
19	C_2H_5	C_2H_5	H	H	H	125	4 (±6)	9 (±2)	32 (±1)
20	CH_3	Cl	H	H	H	125	0 (±0)	1 (±4)	26 (±4)

[a] See footnote in Table 3.

13.4 Action of Cafenstrole and Its Analogs

The two herbicides, cafenstrole (Couderchet et al. 1998; Matthes et al. 1998; Takahashi et al. 2001b) and fentrazamide (Matthes et al. 1998), strongly inhibit VLCFA elongation in cucumber cotyledons, unicellular green microalgae (*Scenedesmus*) cells and leek microsomal preparation (Table 1). The two herbicides structurally belong to the carbamoylated five-membered nitrogen-heterocycles, bearing a dialkylcarbamoyl group as a common feature. Recently, a relationship between the structure of cafenstrole and its analogs and inhibition of VLCFA biosynthesis has been discussed using data obtained with *Scenedesmus* and leek microsomal assays (Takahashi et al. 2001b).

Cafenstrole, *N,N*-diethyl-3-mesitylsulfonyl-1*H*-1,2,4-triazole-1-carboxamide, is a new herbicide for rice cultivation which especially inhibits germination of monocot weeds, e.g., *Echinochloa oryzicola* and *Cyperus difformis* (Kanzaki et al. 1999, 2000). Its phytotoxic symptoms, namely impaired seedling emergence and stunting, have been reported as quite similar to those of chloroacetamide herbicides like alachlor or butachlor (Fukazawa et al. 1995). A strong inhibition of VLCFA biosynthesis by cafenstrole has been found in unicellular *Scenedesmus* cells (Couderchet et al. 1998) and in cucumber cotyledons (Matthes et al. 1998), and furthermore, in a cell-free preparation from leek seedlings (Takahashi et al. 2001b). Incorporation of exogenously applied [^{14}C]-oleate into a sporopollenin fraction of *Scenedesmus acutus* cells was drastically decreased by the herbicide. This is attributed to a marked inhibition of oleic acid elongation to monounsaturated VLCFAs. The inhibition of oleic acid elongation correlated with growth inhibition of the algal cells, indicating that the reduced formation of VLCFAs was responsible for the phytotoxicity of the herbicide [see Eqs. (1) and (2)].

Structure-activity correlation of cafenstrole and its analogs has been considered and reported on three items (Takahashi et al. 2001b): (1) sulfur linkage between aryl and triazole ring and VLCFA elongation inhibition, (2) benzene substituents and VLCFA elongation inhibition, and (3) dialkylcarbamoyl group and elongation inhibition.

1. Sulfur linkage: modification of the sulfur linkage of cafenstrole, *N,N*-diethyl-3-mesitylsulfonyl-1*H*-1,2,4-triazole-1-carboxamide, and C18:1, C18:0 elongation inhibition activity together with herbicidal activity (greenhouse test) was examined (Table 6). Compound **1** (cafenstrole with -SO_2- linkage) was the strongest inhibitor. In the *Scenedesmus* assay it inhibited elongation of C18:1 with an I_{50} value of approx. 10^{-7} M; for the leek cell-free elongation of C18:0 it was found to be about 10^{-6} M. For compound **1** an ED_{90} of 5 g a.i./ha was determined for *E. oryzicola*. Compounds **2** and **3** having an -SO-, or -S- linkage, respectively, were less active inhibitors both in the *Scenedesmus* and leek cell-free assays, and showed ED_{90} values of 60 and 300 g a.i./ha, respectively. Accordingly, the aryl and

Table 6. Cafenstrole and its analogs: Sulfur linkage and elongation inhibition assayed with *Scenedesmus* and microsomes

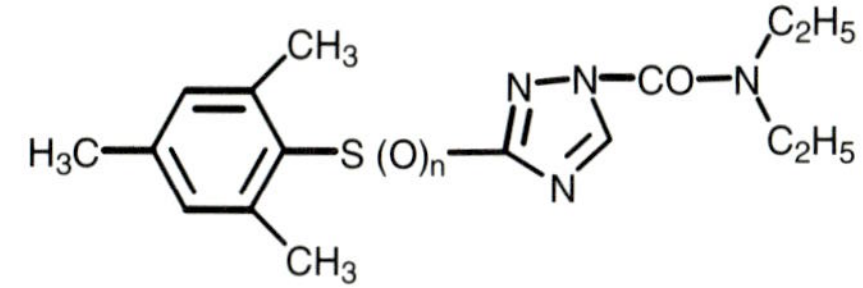

Compounds No.	n	Barnyardgrass[a] (ED_{90}, g/ha)	Inhibition of VLCFA elongation (% inhibition) C18:1 *Scenedesmus acutus* cells 0.01 μM	0.1 μM	1 μM	C18:0 Leek (*Allium porrum*) cell-free 0.1 μM	1 μM	10 μM
			30 (±14)	53 (±2)	73 (±4)	10 (±6)	33 (±14)	80 (±1)
2	1	60	12 (±6)	35 (±4)	67 (±3)	5 (±2)	6 (±3)	32 (±4)
3	0	300	–	–	–	2 (±5)	9 (±5)	11 (±6)

[a] See footnote in Table 3.

triazole rings should be connected by a sulfoxide linkage to achieve maximum inhibition.

2. Benzene substituents: Table 7A shows the benzene substituents of compound **1** (cafenstrole) and phytotoxicities (see compounds **1, 4–7**). The approx. I_{50} values for oleic acid elongation inhibition in *Scenedesmus* for compounds **4, 5** and **6** were about 10^{-6} M, reflecting the strong herbicidal activity against *Echinochloa oryzicola*. An electron-donating group such as methyl or 2,2,2-trifluoroethoxy at the benzene ring (compounds **1, 5** and **6**) apparently results in strong herbicidal activity and VLCFA biosynthesis inhibition. Introduction of an electron-withdrawing nitro group at the benzene ring, however, produced a herbicidally less active compound **7**. The approx. I_{50} values for the elongation inhibition of compounds **6** (benzene type in Table 7A) and **8** (pyridine type in Table 7B) were 10^{-6} M, while the herbicidal activity against *E. oryzicola* exhibited an ED_{90} of 10 g a.i./ha for compound **6** and ED_{90} of 200 g a.i./ha for compound **8**.
3. Dialkylcarbamoyl group: synthesis of 3-mesitylsulfonyl-1*H*-1,2,4-triazole-1-carboxamide failed, but mono- or dialkyl substitution of the carbamoyl nitrogen was possible (see Table 7C). The approx. I_{50} value for elongation inhibition for compounds **1** (cafenstrole), **10** and **11**, all substituted by a dialkyl group, was about 10^{-7} M. Compounds **9** and **12** exhibited a remarkable decrease in elongation inhibition (16 and 13% with 10^{-6} M inhibitor applied), since these were monoalkyl substituted at the carbamoyl nitrogen. Furthermore, compounds **9** and **12** were also found as poor inhibitors against *E. oryzicola*.In conclusion, a dialkyl-substituted carbamoyl nitrogen is essential to both herbicidal activity and elongation inhibition. This finding corroborates the conclusion that generally an *N,N*-disubstituted amide moiety is important for both herbicidal activity and VLCFA elongation inhibition by chloroacetamide herbicides (Böger et al. 2000). The dialkylcarbamoyl group of cafenstrole and *N,N*-disubstituted amide moiety of chloroacetamides seemingly play the same role in binding the inhibitor to elongases.

13.5 Action of Indanofan and Its Analogs

Tridiphane, bearing an oxirane structure, conjugates with reduced glutathione (GSH) in the presence of glutathione *S*-transferase (GST), and the conjugate strongly inhibits the enzyme. A synergistic effect to atrazine has been found, since this herbicide cannot be detoxified anymore by inhibited GST due to a lack of GSH (Lamoureux and Rusness 1986; Zorner et al. 1986). The new herbicide indanofan, 2-[2-(3-chlorophenyl) oxiran-2-ylmethyl]-2-ethylindan-1,3-dione, has been introduced by Hosokawa et al. (2001) which, in contrast to tridiphane, includes only one chlorine in its molecule.

Table 7. Oleate elongation inhibition by 3-arylsulfonyl-*N*,*N*-alkylated 1*H*-1,2,4-triazole-1-carboxamide

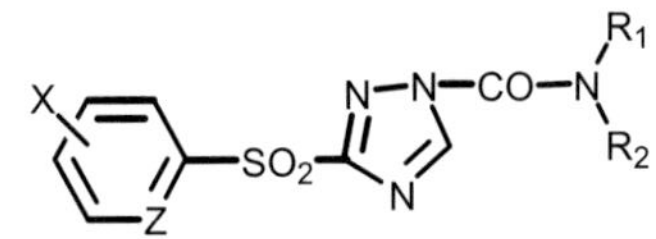

Structure	Compound No.	X	Z	R_1	R_2	Barnyardgrass[a] (ED_{90}, g/ha)	Inhibition of C18:1 elongation (% inhibition) *Scenedesmus acutus* 0.01 μM	0.1 μM	1 μM
A 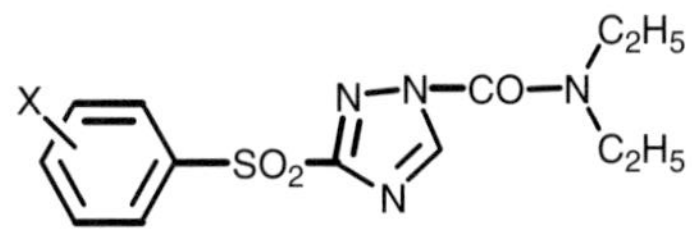	1	2,4,6-$(CH_3)_3$	CH	C_2H_5	C_2H_5	5	30 (±14)	53 (±2)	73 (±4)
	4	H	CH	C_2H_5	C_2H_5	100	4 (±7)	25 (±2)	73 (±4)
	5	3-CH_3	CH	C_2H_5	C_2H_5	40	0 (±14)	37 (±12)	64 (±4)
	6	2-OCH_2CF_3	CH	C_2H_5	C_2H_5	10	26 (±11)	37 (±8)	68 (±4)
	7	2-NO_2	CH	C_2H_5	C_2H_5	300	-	-	-
B 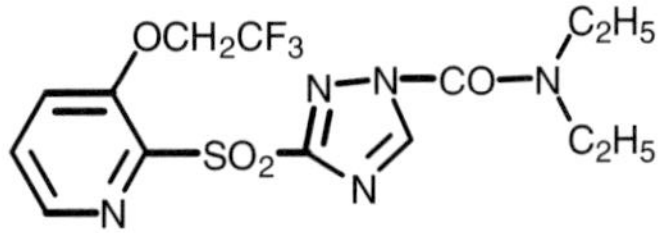	8	3-OCH_2CF_3	N	C_2H_5	C_2H_5	200	0 (±11)	30 (±3)	51 (±5)
C 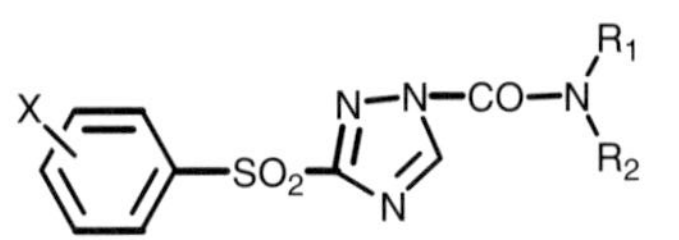	9	2,4,6-$(CH_3)_3$	CH	H	C_2H_5	>10,000	8 (±4)	9 (±6)	16 (±10)
	10	2-NO_2, 4-Cl	CH	CH_3	C_4H_9-*n*	750	18 (±8)	49 (±6)	69 (±3)
	11	2-NO_2, 4-Cl	CH	CH_3	$CH_2CH{=}CH_2$	300	13 (±10)	53 (±4)	72 (±4)
	12	2-NO_2, 4-Cl	CH	H	$CH_2CH{=}CH_2$	>10,000	2 (±9)	9 (±2)	13 (±3)

[a] See footnote in Table 3.

Very recently, arachidoyl-CoA elongation inhibition by indanofan and its four analogs has been reported including the inhibition by tridiphane. These compounds inhibit incorporation of [2-^{14}C]-malonyl-CoA into arachidoyl-CoA (Table 8; Takahashi et al. 2002). Both indanofan **2** and tridiphane **7** exhibit a ca. ten times stronger inhibition for C20:0 elongation than for C18:0 (stearoyl) elongation, possibly indicating inhibition of a microsomal elongase II by these two herbicides (see Schmalfuß et al. 2000). Compound **1**, 2-[2-phenyloxiran-2-ylmethyl]-2-ethylindan-1,3-dione, inhibited C20:0-elongation with an approx. I_{50} value of 10^{-6} M. The I_{50} values of compound **2** (indanofan), **4** and **5** can be estimated to be around 10^{-7} M. Chlorine substituent(s) at the benzene ring (A) appear(s) to be important for strong inhibition as well as the oxirane group. Replacement of the latter by a methylene group (compound **6**) does not result in an active inhibitor. An additional 4,5-dichlorination at the benzene ring (B) does not improve activity further (see compound **5**). The I_{50} value is about 10^{-7} M, the same as for indanofan. Changing the trichloromethyl group of tridiphane to an indan-1,3-dione moiety increases inhibitory activity about tenfold (compare compounds **7** and **4**). Compound **1** without any chlorine has about the same activity as compound **7** (tridiphane). The *S*-enantiomer **3** of indanofan, a herbicidally active isomer, exhibits an approx. ten times stronger inhibition for C18:0 elongation than the racemic indanofan **2**.

The greenhouse test exhibited strong herbicidal activity for compounds **1–5**, the activity scores coincided with cell-free elongation inhibition. This is particularly evident with *Echinochloa oryzicola*, while the herbicidal effect of compound **5** has been found somewhat less active. Compound **6** is considered as an outlier. Although the cell-free assay failed to indicate any activity, both *Digitaria* and *Echinochloa* are completely killed with 1 kg a.i./ha. This finding is difficult to compare with results from the leek assay, because the use rate of 1 kg a.i./ha is too high to allow for a score differentiation. Other unknown inhibitory effects may be involved, besides fatty acid elongation inhibition, when applied at high concentrations. Up-to-date available data allow us to conclude that the essential herbicidal effect of the oxirane compounds dealt with here is caused by inhibition of VLCFA formation.

13.6 Outlook

It is well known that the peroxidizing herbicides inhibiting protoporphyrinogen-IX oxidase, such as diphenyl ethers or cyclic imides, and the herbicides inhibiting acetolactate synthase, like sulfonylureas or pyrimidylthiobenzoates (Stetter 1994; Böger and Wakabayashi 1999) have quite different chemical structures. Accordingly, various chemical types have also been confirmed in the herbicide class inhibiting VLCFA biosynthesis. The old conventional herbicides (e.g., many chloroacetamides, oxyacetamides and thiolcarbamates) and

Table 8. Inhibition of VLCFA elongation and herbicidal activity by indanofan and its relatives

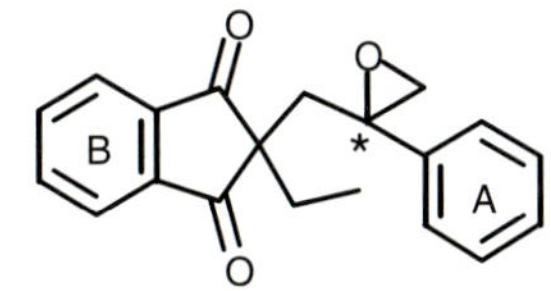

Compounds No	Structure	Herbicidal activity			Inhibition of C20:0 elongation[a] (% inhibition) Leek (*Allium porrum*)		
		Digitaria ciliaris	*Echinochloa oryzicola*				
		1.0 kg/ha	1.0 kg/ha	0.063 kg/ha	0.1 μM	1 μM	10 μM
1	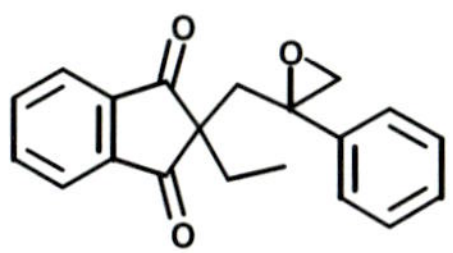	8	9	8	9 (±11)	43 (±0)	81 (±1)
2 (Indanofan)	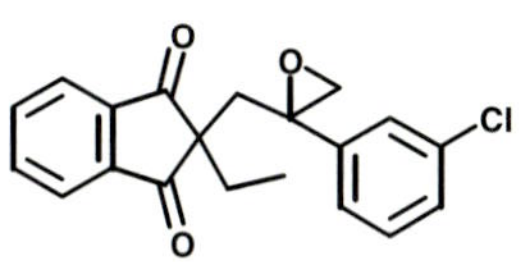	10	10	10	49 (±1) [31 (±10)	83 (±1) 44 (±6)	95 (±1) 58 (±1)]
3	Indanofan *S*-enantiomer	10	10	10	[47 (±2)	57 (±5)	63 (±6)]

4	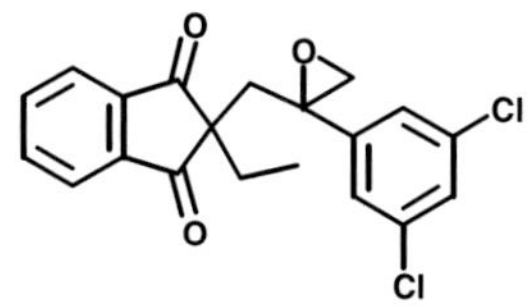	9	10	8	44 (±1)	83 (±2)	94 (±1)
5	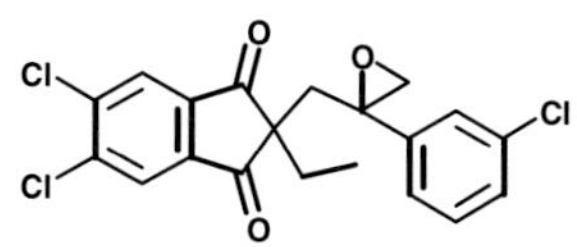	5	10	8	46 (±5)	86 (±5)	93 (±3)
6	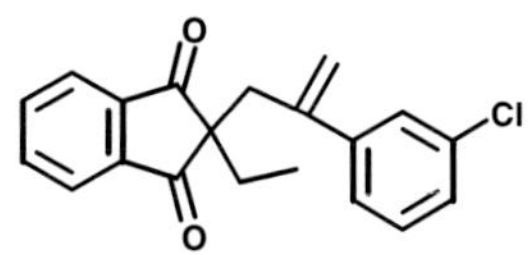	5	10	8	0 (±1)	2 (±1)	7 (±6)
7 (Tridiphane)	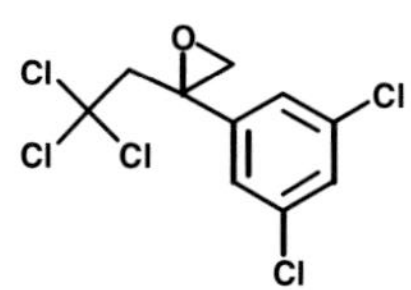	10	10	7	12 (±9) [16 (±9)	44 (±19) 28 (±18)	72 (±3) 53 (±1)]

* Chiral carbon

[a] Figures in brackets indicate inhibition of C18:0 elongation.

new herbicides with different structures (cafenstrole, fentrazamide, indanofan etc.) are all VLCFA-biosynthesis inhibitors. The general characteristics of these herbicides are attractive features to develop a number of novel target products by herbicide design. A variety of compounds with different core structures may be good VLCFA-elongase inhibitors, offering the synthetic chemist a fair chance to find out more. There are good prospects that this chance will be improved by more knowledge on enzymology of the VLCFA elongation.

References

Böger P, Wakabayashi K (eds) (1999) Peroxidizing herbicides. Springer, Berlin Heidelberg New York

Böger P, Matthes B, Schmalfuß J (2000) Towards the primary target of chloroacetamides – new findings pave the way. Pestic Manage Sci 56:497–508

Couderchet M, Bocion PF, Chollet R, Seckinger K, Böger P (1997) Biological activity of two stereoisomers of the *N*-thienyl chloroacetamide herbicide dimethenamid. Pestic Sci 50: 221–227

Couderchet M, Schmalfuß J, Böger P (1998) A specific and sensitive assay to quantify the herbicidal activity of chloroacetamides. Pestic Sci 52:381–387

Devine MD, Duke SO, Fedtke C (1993) Physiology of herbicide action. Prentice-Hall International, London

Fukazawa M, Toriu K, Kanzaki M, Ohishi H, Shirakawa N (1995) Herbicidal activity of new triazole carboxamide derivatives VII. J Weed Sci Technol 40:36–37 (in Japanese)

Hosokawa A, Shike T, Katsurada M, Ikeda O, Minami N, Jikihara T (2001) Synthesis and herbicidal activity of indanofan and related 2-(2-phenyloxiran-2-ylmethyl)indane dione derivatives. J Pestic Sci 26:41–47

Kanzaki M, Kojima S, Takeuchi M, Shirakawa N (1999) The development of cafenstrole, a new herbicide and its application technology. J Weed Sci Technol 45:118–123 (in Japanese)

Kanzaki M, Takeuchi M, Shirakawa N (2000) Chemical structure and herbicidal activity of 1,2,4-triazole-carboxamides. J Weed Sci Technol 44:139–143 (in Japanese)

Kato S, Suyama T, Takematsu T (1998) Development of paddy field rice herbicide, thenylchlor. J Syn Org Chem 56:221–227 (in Japanese)

Lamoureux GL, Rusness DG (1986) Tridiphane, [2-(3,5-dichlorophenyl)-2-(2,2,2-trichloromethyl)]oxirane, an atrazine synergist: enzymatic conversion to potent glutathione *S*-transferase inhibitor. Pestic Biochem Physiol 26:323–342

Matthes B, Schmalfuß J, Böger P (1998) Chloroacetamide mode of action II: inhibition of very-long-chain fatty acid synthesis in higher plants. Z Naturforsch 53c:1004–1011

Schmalfuß J, Matthes B, Mayer P, Böger P (1998) Chloroacetamide mode of action I: inhibition of very-long-chain fatty acid synthesis in *Scenedesmus acutus*. Z Naturforsch 53c:995–1003

Schmalfuß J, Matthes B, Knuth K, Böger P (2000) Inhibition of acyl-CoA elongation by chloroacetamide herbicides in microsomes from leek seedlings. Pestic Biochem Physiol 67:25–35

Stetter J (1994) Herbicides inhibiting branched-chain amino acid biosynthesis – recent developments. In: Stetter J (ed) Chemistry of plant protection, vol 10. Springer, Berlin Heidelberg New York

Takahashi H, Ohki A, Kato S, Tanaka A, Sato Y, Matthes B, Böger P, Wakabayashi K (2001a) Inhibition of very-long-chain fatty acid elongation by 2-chloro-*N*-(3-methoxy-2-thenyl)-2′,6′-dimethylacetanilide, thenylchlor, and its analogs. Pestic Biochem Physiol 71:140–146

Takahashi H, Ohki A, Kanzaki M, Tanaka A, Sato Y, Matthes B, Böger P, Wakabayashi K (2001b) Inhibition of very-long-chain fatty acid elongation by cafenstrole, *N*,*N*-diethyl-3-mesitylsulfonyl-1H-1,2,4-triazole-1-carboxamide and its analogs. Z Naturforsch 56c:781–786

Takahashi H, Schmalfuß J, Ohki A, Hosokawa A, Tanaka A, Sato Y, Matthes B, Böger P, Wakabayashi K (2002) Inhibition of very-long-chain fatty acid elongation by indanofan, 2-[2-(3-chlorophenyl)oxiran-2-ylmethyl]-2-ethylindan-1,3-dione, and its analogs. Z Naturforsch 57C:72–74

Zorner PS, Markley LD, Staffoed LE, Ray PG, Renga JM (1986) Physiological mechanism responsible for triazine synergism with pro-epoxides of tridiphane herbicide. Abstract 7D-09, 6th Int Congr Pestic Chem, Ottawa

Index

Printing: Druckhaus Berlin-Mitte
Binding: Buchbinderei Stein & Lehmann, Berlin